GEOLOGICAL STUDIES OF THE MINERAL DEPOSITS IN JAPAN AND EAST ASIA

GEOLOGICAL STUDIES OF THE MINERAL DEPOSITS IN JAPAN AND EAST ASIA

Hideki Imai

UNIVERSITY OF TOKYO PRESS

Exclusive Distributor:
ISBS, Inc.
P. O. Box 555
Forest Grove, OR 97116

Exclusive Distributor:
ISBS, Inc.
P. O. Box 555
Forest Grove, OR 97116

© UNIVERSITY OF TOKYO PRESS, 1978
UTP 3044–68759–5149

Printed in Japan

All rights reserved. No part of this publication may be reproduced or transmitted in any form or by any means, electronic or mechanical, including photocopy, recording, or any information storage and retrieval system, without permission in writing from the publisher.

ISBN 0–86008–208–3

PREFACE

In 1938, when I was a student in the Department of Geology, University of Tokyo, I began to study the geology of ore deposits under the tutelage of Professor Takeo Kato. After graduating in 1939, I became a research fellow under Professor Shigeru Nishio in the Department of Mining. In 1944 I became an associate professor, and in 1959 a full professor in that Department (now reorganized into the Department of Mineral Development Engineering) in charge of economic geology, succeeding Professor Nishio. In 1975, I retired under the age limit of 60 from the University. Since that time, I have been a lecturer at Waseda University and an advisor to the Central Research Laboratory of Mitsui Mining and Smelting Company Ltd.

During these 40 years, I studied the ore deposits in Japan and East Asia from the geological and mineralogical standpoints. In commemoration of my retirement, I wished to compile my own previously published papers and to publish them with revisions and supplements. This book is the result.

Parts I–VI of this book include my own papers and those written in collaboration with graduate students and research associates. Part VII consists of the contributions of some of my former students.

It may be said that a renaissance has occurred during the past three decades, both in structural geology (including tectonophysics) and in geochemistry. In my laboratory at the University, connected with the above field surveys of ore deposits and the related laboratory works, we studied the experimental geology of hydrothermal mineral syntheses, fluid inclusion studies and model analyses of the fissure formation. As the geology of ore deposits is an interdisciplinary science and technology, I wished to study ore deposits from three sides, i.e. mineralogy, structural geology and geochemistry in a wide sense (including petrology), trying to coordinate, harmonize and amalgamate these three with one another.

I believe that in geological sciences and technologies areal study or regional geology is fundamental and essential. The geology of ore deposits, especially mining geology, begins outside in the field and must finally return to the field. From this standpoint, I describe and discuss in this book the geological problems of the individual ore deposits in Japan and some in East Asia, laying emphasis on the geological structure, paragenesis and mineralization of each deposit.

Many of the mines which are studied in this book have recently shut down because of lack of ore in sight, financial change in the world economy and environmental problems. So, this book is an elegy for these mines. On the other side, it becomes consequently more necessary than ever, I think, to describe the present state of views on the geology and ore deposit of these mines. I believe that the case histories of the individual ore deposits are indispensable for the future development of economic geology as well as mining geology from both academic and practical viewpoints.

I am grateful to Professor Takayasu Uchio, Professor Sukune Takenouchi, Professor Tetsuya Shoji and Mr. Hiroaki Kaneda, my colleagues at the University, for their assistance in preparing this book. The publication of this book was supported by financial aid from many of my former students in the Departments of Mining and of Mineral Development Engineering, University of Tokyo, and by many of my friends in the geology field, who contributed to a fund at the time of my retirement from the University.

I am also indebted to Mr. Kazuo Ishii and Mr. Wataru Izumi of the University of Tokyo Press for their help in publishing the book.

Feb. 11, 1978

Hideki Imai

LIST OF CONTRIBUTORS

Author

Hideki Imai: Emeritus Professor, Department of Mineral Development Engineering, University of Tokyo, Tokyo, Japan

Coauthors

Sukune Takenouchi: Professor, Department of Mineral Development Engineering, University of Tokyo, Tokyo, Japan

Tetsuya Shoji: Associate Professor, Department of Mineral Development Engineering, University of Tokyo, Tokyo, Japan

Hiroaki Kaneda: Research Fellow, Department of Mineral Development Engineering, University of Tokyo, Tokyo, Japan

Collaborators and Contributors

Michiaki Bunno: University Museum, University of Tokyo, Tokyo, Japan

Hamed M. El Shatoury: Geology Department, University of Sana'a, Sana'a, Yemen, Republic of Arab

Yoshinori Fujiki: National Institute for Research in Inorganic Materials, Tsukuba, Japan

Sho'ichiro Hayashi: Power-Reactor and Nuclear Fuel Development Corporation, Tokyo, Japan

Kohei Iida: Nippon Mining Co., Ltd., Tokyo, Japan

Keisuke Ito: Department of Earth Sciences, Kobe University, Kobe, Japan

Yong Won John: Department of Mineral and Petroleum Engineering, Seoul National University, Seoul, Korea

Taka'aki Kashiwagi: Mitsubishi Metal Corporation, Tokyo, Japan

Takeshi Kihara: Toto Ltd., Kitakyushu City, Japan

Moon Yung Kim: Department of Geosciences, Osaka City University, Osaka, Japan

Hitoshi Koide: Geological Survey of Japan, Kawasaki, Japan

Min Sung Lee: Department of Earth Sciences, College of Education, Seoul National University, Seoul, Korea

Ken'ichi Nagano: Nippon Steel Corporation, Tokyo, Japan

Takehiro Sakimoto: Nittetsu Mining Co., Ltd., Tokyo, Japan

Edward Schreiber: Department of Earth and Environmental Science, Queens College, City University of New York, New York, U.S.A.

Taro Takahashi: Department of Geological Science, Columbia University, New York, U.S.A.

Shigeaki Tsukagoshi: Mitsui Metal Mining and Smelting Co., Ltd., Tokyo, Japan

Takayasu Uchio: Department of Mineral Development Engineering, University of Tokyo, Tokyo, Japan

Hideo Yamadera: Department of Chemistry, Nagoya University, Nagoya, Japan

CONTENTS

Preface v
List of Contributors vii

PART I
Introduction

Igneous Activity, Metamorphism and Metallogenesis in Japan
.... (H. Imai and T. Uchio).... 1

PART II
Vein-Type Deposits

A. GENERAL PROBLEMS 13
   1. Formation of Fissures and Their Mineralization ....(H. Imai).... 13
B. PEGMATITIC, HYPOTHERMAL AND MESOTHERMAL VEINS 20
   2. Ohya Mine, Miyagi Prefecture ....(H. Imai).... 20
   3. Takatori Mine, Ibaragi Prefecture
       .... (H. Imai, S. Hayashi and T. Kashiwagi).... 24
   4. Geologic Structure and Fluid Inclusion Study at the Ohtani and
      Kaneuchi Mines, Kyoto Prefecture
       .... (H. Imai, M. Y. Kim, Y. Fujiki and S. Takenouchi).... 27
   5. Komori Mine, Kyoto Prefecture ....(H. Imai and Y. Fujiki).... 40
   6. Geologic Structure and Fluid Inclusion Studies at the Taishu Mine,
      Nagasaki Prefecture ....(H. Imai, S. Takenouchi and T. Kihara).... 43
C. EPITHERMAL (SUBVOLCANIC) VEINS 54
   7. Sado Mine, Niigata Prefecture .... (H. Imai and M. Bunno).... 54
   8. Mikawa Mine, Niigata Prefecture ....(H. Imai).... 56
   9. Osarizawa Mine, Akita Prefecture ....(H. Imai).... 58
  10. Ani Mine, Akita Prefecture ....(H. Imai).... 59
  11. Tsuchihata Mine, Iwate Prefecture ....(H. Imai).... 62
  12. Yaso Mine, Fukushima Prefecture ...... (S. Takenouchi and H. Imai).... 63
  13. Geology and Vein System of the Hosokura Mine, Miyagi Prefecture
      ....(H. Imai).... 65
  14. Geologic Structure and Fluid Inclusion Study at the Toyoha Mine,
      Hokkaido .... (H. M. El Shatoury, S. Takenouchi and H. Imai).... 75
D. XENOTHERMAL VEINS 86
  15. Geologic Structure and Mineralization of Polymetallic Xenothermal
      Vein-type Deposits in Japan ....(H. Imai, M.
      S. Lee, S. Takenouchi, Y. Fujiki, K. Iida, T. Sakimoto and S. Tsukagoshi).... 86
  16. Obira Mine, Ohita Prefecture ....(H. Imai).... 123

E. GENETICAL PROBLEMS OF SULFIDE MINERALS 124
  17. Geology and Ore Deposit of the Ilkwang (Nikko) Mine,
      Korea, with Special Reference to the Genesis of Gudmundite
      . . . . (H. Imai and M. S. Lee) 124
  18. Paragenesis of Cu-Fe-S Minerals in the Komori Mine
      . . . . (Y. Fujiki, S. Takenouchi and H. Imai) 130
  19. "Wurtzite" from the Hosokura Mine . . . . (H. Imai) 135
  20. Problems of Enargite, Luzonite and Famatinite . . . . (H. Imai) 143
  21. Syntheses of the Cu-Fe-Sn-S Minerals
      . . . . (M. S. Lee, S. Takenouchi and H. Imai) 161

PART III
Pyrometasomatic Deposits

A. PYROMETASOMATIC DEPOSITS IN JAPAN AND KOREA 171
  1. Introduction . . . . (H. Imai) 171
  2. Kamioka Mine, Gifu Prefecture . . . . (H. Imai) 171
  3. Nakatatsu Mine, Fukui Prefecture . . . . (H. Imai) 176
  4. Kuga, Fujigatani and Kiwada Mines, Yamaguchi Prefecture
     . . . . (H. Imai, S. Takenouchi and K. Ito) 179
  5. Kamaishi Mine, Iwate Prefecture . . . . (H. Kaneda, T. Shoji and H. Imai) 183
  6. Chichibu Mine, Saitama Prefecture . . . . (T. Shoji and H. Imai) 190
  7. Sangdong Mine, Korea . . . . (Y. W. John) 196
B. GENETICAL PROBLEMS 201
  8. Skarn Formation . . . . (T. Shoji) 201
  9. Malayaite . . . . (S. Takenouchi) 212
  10. Cu-Fe-S Mineral Syntheses
      . . . . (H. Kaneda, T. Shoji, S. Takenouchi and H. Imai) 216
  11. Transition of Pyrite into Marcasite, Magnetite and Hematite by Chalcopyrite
      Mineralization at the Sasagatani Mine, Shimane Prefecture . . . . (H. Imai) 227

PART IV
Strata-Bound Deposits

  1. Introduction . . . . (H. Imai) 233
  2. Geology and Genesis of the Okuki Mine, Ehime Prefecture, and Other Related
     Cupriferous Pyrite Deposits in Southwest Japan . . . . (H. Imai) 234
  3. Besshi Mine, Ehime Prefecture . . . . (H. Imai) 256
  4. Yanahara Mine, Okayama Prefecture . . . . (H. Imai) 258
  5. Hitachi Mine, Ibaragi Prefecture . . . . (H. Imai) 260

PART V
Porphyry Copper Deposits

Porphyry Copper Deposits in the Southeast Asia, with Special Reference to Fluid
  Inclusion Study . . . . (H. Imai, S. Takenouchi, T. Shoji and K. Nagano) 265

PART VI
Sedimentary Deposits

The Peculiar Phosphate and Copper Deposits in Noto Peninsula
.... (H. Imai and H. Yamadera).... 281

PART VII
Contributions

1. Magma Genesis in a Dynamic Mantle .... (K. Ito).... 293
2. Tectonophysics as Related to the Structural Features of Ore Deposits: The Role of the Fluid Intrusion in the Formation of Vein Fractures .... (H. Koide).... 302
3. Fluid Inculsion Studies............................. (S. Takenouchi).... 312
4. A Thermodynamic Study of Manganese Minerals and its Applications to the Formation of Some Manganese Mineral Deposits in Japan
    .... (T. Takahashi and E. Schreiber).... 335

References 351
Index of Subjects 383
Index of Mines and Localities 390

PART I
INTRODUCTION

Igneous Activity, Metamorphism and Metallogenesis in Japan

H. Imai and T. Uchio

(A) Pre-Tertiary Period   (Imai, 1963b)

Japan is geologically divided into Southwest and Northeast Japan by Fossa Magna (fault and depression belt) which traverses the central part of Honshu from the Japan Sea coast to the Pacific coast (Fig. I–1). In Southwest Japan a conspicuous tectonic line called the Median Tectonic Line runs from central Honshu to Kyushu through Shikoku, nearly parallel to the island-arc. This line subdivides Southwest Japan into the Inner Zone (the Japan Sea side) and the Outer Zone (the Pacific Ocean side) (Fig. I–1).

Inner Zone of Southwest Japan

The geological basement of the Inner Zone is characterized by zonal arrangements of older formations and their metamorphic equivalents. A belt of gneisses and granitic rocks called the Hida Metamorphic Rocks lies on the Japan Sea side of Southwest Japan. A belt of schists, ranging in metamorphic grade from the glaucophane schist facies to the epidote amphibolite facies and called the Sangun Metamorphic Rocks, occurs to the south of the Hida metamorphic area.

The central area of the Inner Zone is occupied by the extensive Chichibu Devonian (mostly Carboniferous-Permian) to Triassic System and Sangun Metamorphic Rocks.

The zone just north of the Median Tectonic Line is characterized by concordant and discordant batholithic intrusives of granodioritic rocks, including banded gneisses and schists named the Ryoke Metamorphic Rocks.

The original rocks of Hida, Sangun and Ryoke Metamorphic Complexes would be Chichibu Paleozoic-Triassic sedimentary rocks.

Hida metamorphic area: The Hida Metamorphic Rocks occupies the northernmost belt of the basement of Southwest Japan (Fig. I–1). They are composed of amphibolite, hornblende and biotite gneiss (injection gneiss), staurolite-kyanite-biotite schist and granitic rocks. The gneissic rocks are characterized by the presence of intercalated crystalline limestone. The granitic rocks occur as lenticular masses that are concordant with the structure of the metamorphic rocks and are believed to have been emplaced during the metamorphism. Another granite (Funatsu granite) intrudes the Hida Metamorphic Rocks and the adjacent Chichibu System at the Kamioka mining area, and is overlain by non-metamorphosed Middle or Later Jurassic and Cretaceous rocks of the Tetori Group. This granite,

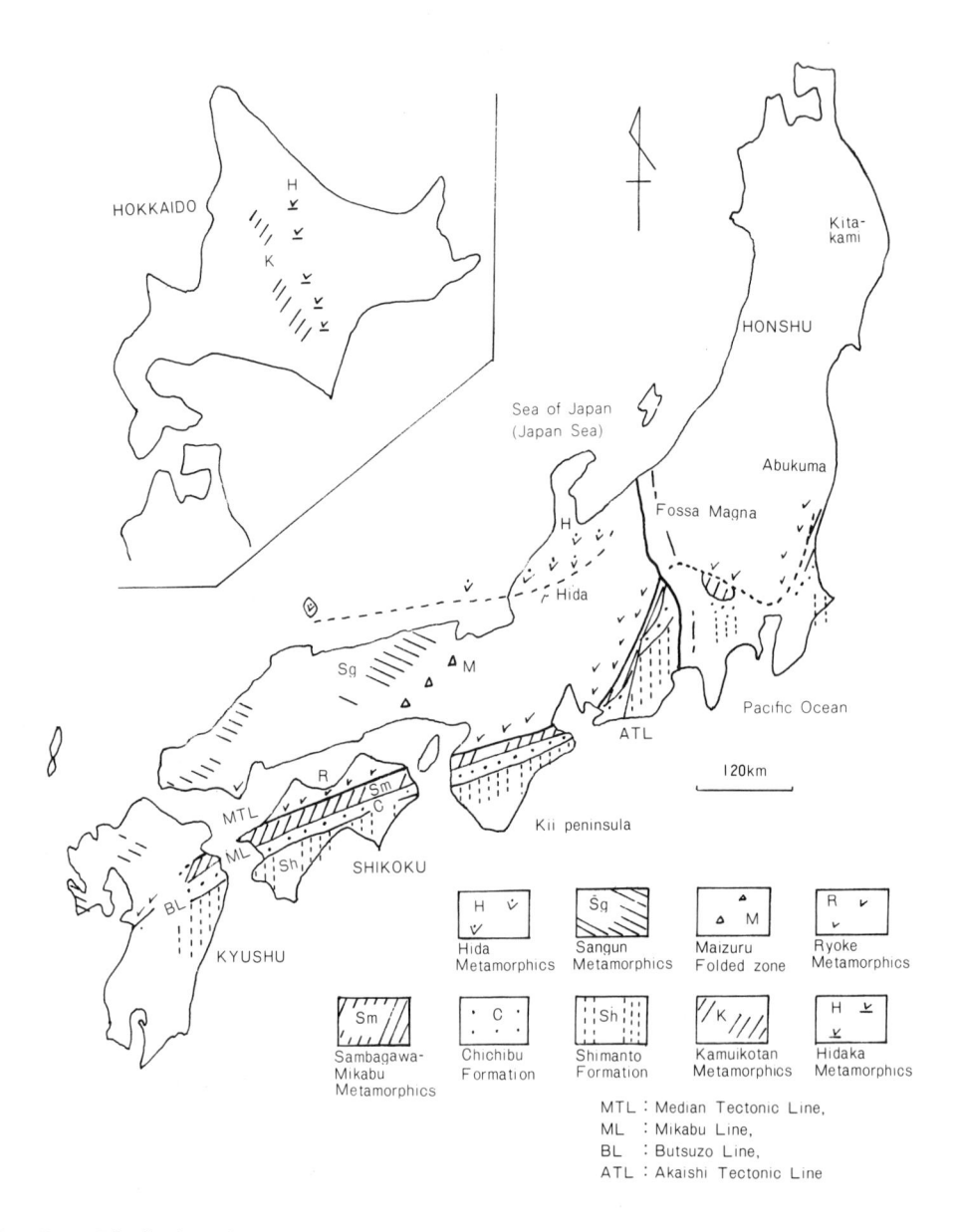

FIG. I–1. Distribution of metamorphic rocks in Japan.

therefore, is dated geologically as post-Hida Gneiss and pre-Middle Jurassic. The Liassic (Lower Jurassic) Kuruma Group adjacent to the Hida metamorphic area has conglomerates which contain granitic cobbles.

All these rocks are penetrated by later acidic igneous rocks (quartz porphyry, etc.) which intruded after the Tetori Epoch.

The writer does not recognize ore deposits genetically related to the gneissic rocks and affiliated granites. The deposit of the Kamioka mine is in the limestones intercalated in the gneissic rocks at the middle part of the Hida metamorphic area; the ore deposits may be genetically related to the later quartz porphyry.

Central area: Rocks formed by regional metamorphism crop out in patches from central Honshu to northern Kyushu (Fig. I–1). They are called the Sangun Metamorphic Rocks, composed of black schists, quartz schists and greenschists. In this area, they grade into the non-metamorphosed Chichibu System. So, the original rocks of these metamorphics are interpreted to be mostly Carboniferous-Triassic sedimentary rocks. Serpentinite and diabasic rocks are relatively abundant in these metamorphics, often intimately associated with greenschists.

The Sangun Metamorphic Group is unconformably covered by non-metamorphosed Triassic formations.

The folded zone called the Maizuru zone (Fig. I-1) runs NE-SW, trending apparently oblique to the nearly E-W arrangement of the Hida and Sangun belts. This zone is composed of amphibolite and other metamorphics, accompanied by basic and ultrabasic rocks such as diabase, gabbro and serpentinite. The original rocks of the metamorphics in this zone are also Chichibu Carboniferous-Triassic rocks.

From the absolute ages of these metamorphics (stated later), the Maizuru Folded zone would have been formed at the same period as the Sangun Metamorphism.

All these rocks in the central area of the Inner Zone are intruded by younger acidic rocks, which are genetically related to the ore deposits.

Ryoke metamorphic area: The Ryoke Metamorphic Rocks are distributed in a belt along the north side of the Median Tectonic Line extending from central Honshu to Kyushu (Fig. I-1). This belt is about 40 km to 70 km in width and 650 km in length. The metamorphic rocks comprise biotite schists (locally containing andalusite) and injection gneisses (locally containing sillimanite), accompanied by concordant and discordant granitic rocks. They were derived chiefly from sedimentary rocks, including clayslate, sandstone, chert, etc., of Carboniferous-Triassic age. Metamorphic grade as well as the degree of granitization generally increases toward the Median Tectonic Line. In this area, no ore deposits have been found.

Outer Zone of Southwest Japan

In the Outer Zone, highly metamorphosed rocks, named the Sambagawa System, border on the Median Tectonic Line. The system is composed of albite-epidote-chlorite schist, actinolite-chlorite schist, glaucophane schist, piedmontite-quartz schist, graphite-quartz schist, sericite-quartz schist and others. The schists are accompanied by sheet-like basic and ultrabasic rocks. Basic rocks are partly or entirely metamorphosed to rocks such as amphibolite. The strike of these rocks is nearly parallel to the direction of the Median Tectonic Line, i.e., NE-SW or E-W.

Towards the south or southeast, apart from the Median Tectonic Line, the rocks of the Sambagawa System change gradually to rocks of lower-grade metamorphism called the Mikabu System. The Mikabu System is composed of sericite-graphite-quartz phyllite (or semi-schist), epidote-actinolite-chlorite phyllite (or semi-schist) and others, accompanied by sheet-like metadiabases (or metagabbro) similar to those of the Sambagawa System.

Throughout the Sambagawa-Mikabu terrain, many deposits of cupriferous pyrite occur along planes of schistosity, commonly as lenticular or bedded deposits.

Farther to the south or southeast, sedimentary rocks of the Chichibu System (mostly Carboniferous-Triassic) occur. The Mikabu System is in fault contact with the Chichibu System at some places but grades into it at others. Most of the original rocks of the Sambagawa and Mikabu Systems are the sedimentaries of the Chichibu System. The Chichibu System is composed of sandstone, graywacke, clayslate, chert, limestone and schalstein; and its age ranges from mostly Carboniferous to Triassic. In some places it is intruded by sheet-like peridotite and serpentinite. Diabasic rocks also occur along the bedding planes.

To the south, the Chichibu System is thrust upon the Shimanto undifferentiated Mesozoic terrain along the Butsuzo Line. The Shimanto System is composed of non-fossiliferous sandstone and clayslate. Diabasic rocks like those in the Paleozoic area are also found in this terrain.

Since the early days of geologic research on the Sambagawa and Mikabu Systems, green rocks and greenschists were thought to be derivatives of basic tuffs and lava flows of submarine eruptions.

The problem of green rocks will be discussed in Chapter IV-2.

In the Chichibu Paleozoic and Shimanto Mesozoic Systems, cupriferous pyrite deposits similar to those in the Sambagawa-Mikabu terrain are present, though they are fewer in number and generally smaller in size than those in the Sambagawa-Mikabu Complex.

Ages of metamorphisms and igneous activities of Southwest Japan

In 1941, Kobayashi summarized the Paleozoic and Mesozoic geology of Japan and correlated the genetic history of the basement metamorphic zones with the crustal movements. He distinguished between the Akiyoshi Orogenic Cycle (Late Paleozoic-Early Mesozoic) and the Sakawa Orogenic Cycle (Middle to Late Mesozoic). Afterwards, Kobayashi (1957) revised the age of the Sakawa Orogenic Cycle to range from Mesozoic to Paleogene. According to Kobayashi (1941), the Sangun and Hida metamorphic areas represent the axial metamorphism of the Akiyoshi Orogenic Cycle, and the Ryoke and Sambagawa-Mikabu Systems were formed during the Sakawa Orogenic Cycle. The correlation between metamorphic rocks and crustal movements of the Akiyoshi and Sakawa Orogenic Cycles after Kobayashi is tabulated below.

Cretaceous	Ryoke Injection	Sambagawa-Mikabu Metamorphism
Jurassic		
Triassic	Hida Injection	Sangun Metamorphism
Permian		

He stated that the geosyncline and orogenesis throughout the Japan migrated towards the south with the lapse of geologic time.

Some geologists, however, have asserted that most metamorphic rocks of the Hida Metamorphics are a product of pre-Silurian (perhaps Pre-Cambrian) metamorphism, and also have insisted that metamorphic rocks and associated granitic intrusive rocks in the Ryoke metamorphic area and crystalline schists in both the Sangun and the Sambagawa-Mikabu Complexes were formed during a Late Paleozoic or Early Mesozoic orogenic phase (Minato et al., 1965).

The absolute ages of Hida Metamorphic Rocks as calculated by the K-Ar, Rb-Sr and U-Th-Pb methods are concentrated in 250–180 m. y., mostly in 180 m. y. (Kuno et al., 1960; Saito et al., 1961; Nozawa, 1968a; Ishizaka and Yamaguchi, 1969; Yamaguchi and Yanagi, 1970). The age of the granite associated with Hida Metamorphic Rocks is also mostly 180 m. y. (Nozawa, 1968a; Yamaguchi and Yanagi, 1970).

The absolute ages of the metamorphics from Sangun and Maizuru zones are 310–170 m. y. by the K-Ar method (Shibata and Nozawa, 1968; Shibata and Igi, 1969) and 330–215 m. y. by the Rb-Sr method and K-Ar method (Hayase and Ishizaka, 1967; Shibata, 1968). Data on the Sangun and Maizuru zones are very scarce.

The ages of Ryoke and Sambagawa Metamorphic Rocks are 110–65 m. y. calculated by means of the K-Ar, Rb-Sr and U-Pb methods (Miller et al., 1961; Ozima et al., 1967; Banno and Miller 1965; Hayase and Ishizaka, 1967; Kawano and Ueda, 1967a,b; Yamaguchi and Yanagi, 1970).

Miyashiro (1961) noticed paired metamorphic belts of injection gneiss type and greenschist type, i.e., a Hida-Sangun pair and a Ryoke-Sambagawa pair. Each pair was formed at the same geologic time, but the former pair is older than the latter one.

In the Inner Zone of Southwest Japan, widespread batholithic granitic rocks and their apophyses discordantly intruded the above metamorphic rocks and their original rocks, and are genetically related to many ore deposits. Their ages are 50–95 m. y. (Kawano and Ueda, 1967a,b). Those on the Japan Sea side are generally younger than those on the southern side.

Northeast Japan

The eastern extension of the Ryoke-Sambagawa belts is cut by the Fossa Magna. On the east side of the Fossa Magna, they are widely covered by Neogene volcanics and by the alluvial sediments. But it is certain that they change direction in arc and extend to the Abukuma plateau (Fig. I–1).

Abukuma plateau: The Abukuma plateau also has two metamorphic belts. The western

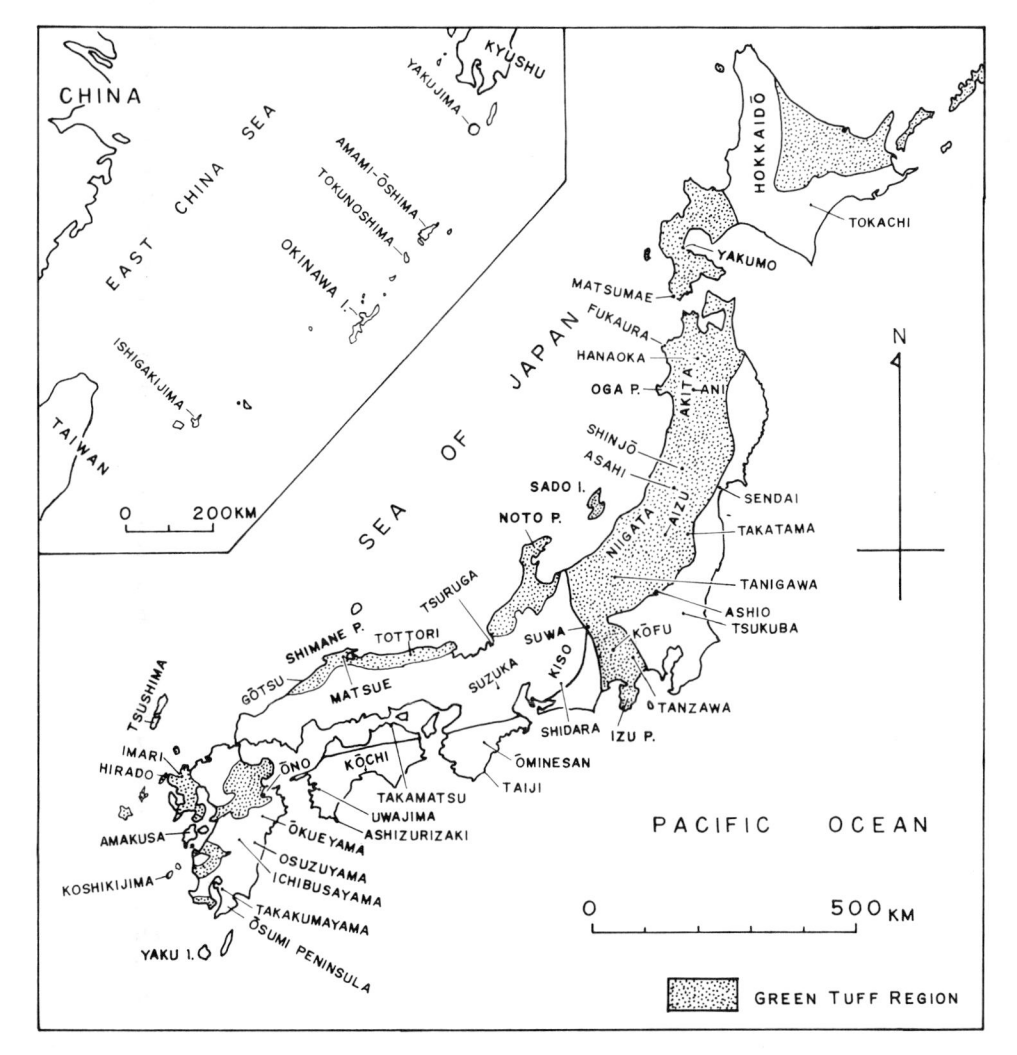

FIG. I–2. Index map showing distribution of green tuff and some of the locations referred to in the text where Tertiary radiometric datings have been done.

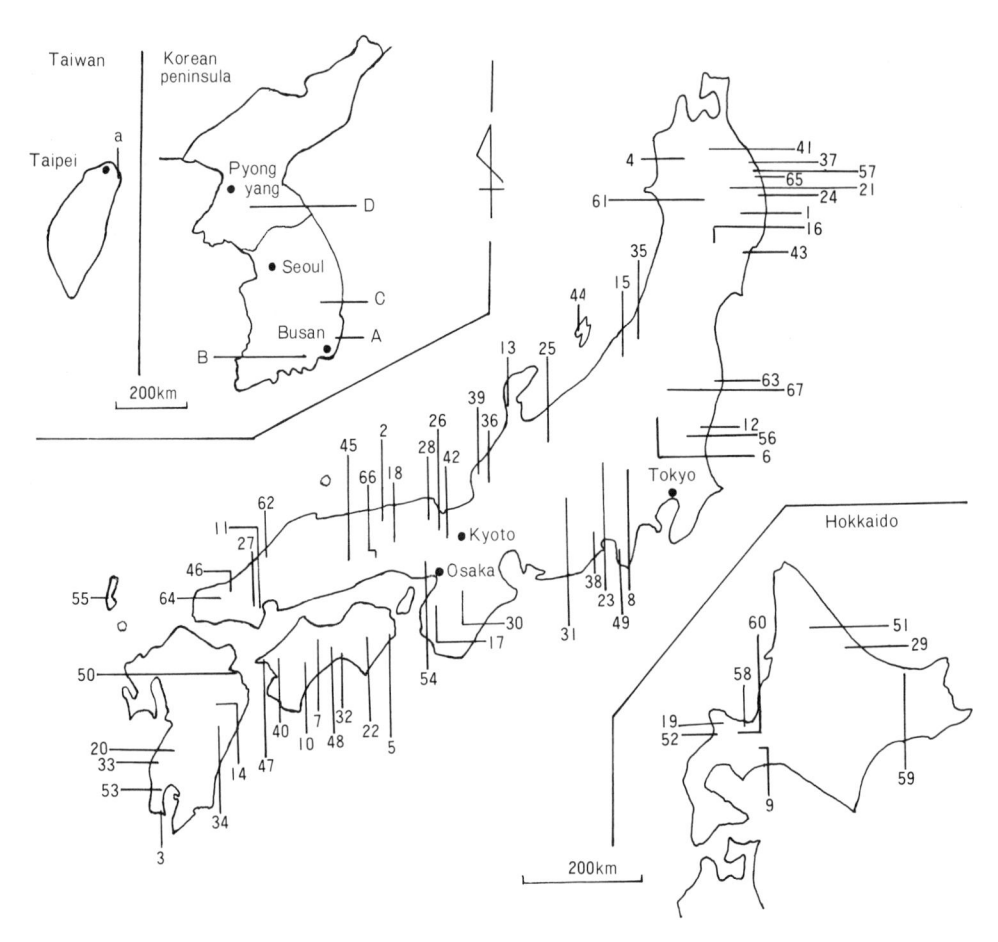

FIG. I-3. Index map of the mines and localities.

1. Akagane, 2. Akenobe, 3. Akeshi, 4. Ani, 5. Asakawa, 6. Ashio, 7. Besshi, 8. Chichibu, 9. Chitose, 10. Choja, 11. Fujigatani, 12. Hitachi, 13. Hiuchidani, 14. Hoei, 15. Hokuetsu, 16. Hosokura, 17. Iimori, 18. Ikuno, 19. Inakuraishi, 20. Iriki, 21. Isakozawa, 22. Jiro, 23. Kaize, 24. Kamaishi, 25. Kamioka, 26. Kaneuchi, 27. Kawayama, 11. Kiwada, 28. Komori, 29. Kohnomai, 30. Koya, 11. Kuga, 31. Kune, 32. Kunimiyama, 33. Kushikino, 34. Makimine, 23. Masutomi, 35. Mikawa, 31. Minenosawa-Nago, 14. Mitate, 36. Nakatatsu, 10. Nanogawa, 37. Noda-Tamagawa, 14. Obira, 38. Odake, 19. Oe, 39. Ogoya, 40. Okuki, 41. Osarizawa, 42. Ohtani, 43. Ohya, 44. Sado, 45. Sampo, 46. Sasagatani, 47. Sata, 48. Sazare, 49. Seikoshi-Toi, 50. Shibuki-Seki, 51. Shimokawa, 52. Suttu, 53. Suzuyama, 54. Tada, 55. Taishu, 56. Takatori, 57. Taro, 58. Teine, 59. Tokoro, 14. Toroku, 60. Toyoha, 61. Tsuchihata, 62. Tsumo, 63. Yaguki, 64. Yakuoji, 65. Yamaguchi, 66. Yanahara, 67. Yaso.

A. Ilkwang, B. Hanan, C. Sangdong, D. Suan, a. Chinkuashih.

belt is the main metamorphic belt of the plateau, and the eastern belt lies at the eastern margin of the plateau (Fig. I-1).

The main belt consists of biotite schist, gneiss (injection gneiss) and abundant mafic metamorphic rocks corresponding to the Ryoke belt. These rocks are accompanied by granitic rocks as in the Ryoke belt.

The belt lying at the eastern margin is made up of crystalline schists of glaucophane and actinolite types, corresponding to the Sambagawa-Mikabu type.

No fault is recognized between the two belts.

The ages of the gneisses and biotite schists in the main belt are 100–120 m. y. calculated by the K-Ar method (Ueda *et al.*, 1969). Those of the granitic rocks in the same area are 85–104 m.y. by the K-Ar method (Miller *et al.*, 1961; Kawano and Ueda, 1966a).

The ages of the muscovites from the crystalline schists in the eastern belt calculated by the K-Ar method also are 105–120 m.y. (Ueda *et al.*, 1969). These ages tell that the metamorphism and plutonism occurred in the late Mesozoic era, just as in the Ryoke-Sambagawa belts.

The Hitachi mine is situated at the southeastern margin of the plateau (Fig. I-3). It occurs in crystalline schists of actinolite type.

Kitakami plateau: This region is composed of mainly Carboniferous-Triassic sedimentary rocks, and granitic and basic igneous rocks (Fig. I-1).

The granites are widespread. Most of them intruded into the Carboniferous-Triassic formation during an interval between the Early Cretaceous (Ryoseki Epoch) and Middle Cretaceous (Monobegawa Epoch).

The absolute age of the uraninite from the Yamaguchi mine in Iwate Prefecture (Fig. I-3), a deposit which belongs to the pyrometasomatic type and is related to intrusion of the granitic rock, was determined by means of the U-Pb method by Imai, Saito *et al.* (1960). The result gives an age of 94 $\pm$ 8 m y. by the $U^{238} - Pb^{206}$ method, and 109 $\pm$ 11 m. y. by the $U^{235} - Pb^{207}$ method.

According to Kawano and Ueda (1965, 1966b), most of the granites in the Kitakami plateau show K-Ar ages of 128–107 m. y.

In the Kitakami plateau, there are many ore deposits genetically related to these granites. The largest is that of the Kamaishi mine (Fig. I-3), the largest magnetite deposit of pyrometasomatic type in Japan. The deposit of the Yamaguchi mine is the same type as that of the Kamaishi mine, though it is far smaller in scale. It occurs in the limestone of the roof-pendant Carboniferous formation on the granite. The ore minerals in Yamaguchi mine are chalcopyrite and pyrite, with small amounts of scheelite (bearing a powellite molecule), cubanite, molybdenite, sphalerite, ilmenite and uraninite. The occurrence of uraninite in the pyrometasomatic deposit is generally rare; the result of an age determination on this mineral is described above. The skarn and gangue minerals in the deposit are garnet, diopside, amphibole, titanite, apatite, allanite, chlorite, etc.

Hokkaido: In central Hokkaido, a belt of biotite schists and gneisses (injection gneiss) called Hidaka Matamorphics and granitic intrusives extends in a N-S direction (Fig. I-1). No ore deposits are found in this belt. To the west of the belt and separated by a thrust fault (?) from it, there is a glaucophane type metamorphic belt called the Kamuikotan belt. It is said that the original rocks of this belt belong to the Paleozoic or Mesozoic group. This belt is characterized by the association of numerous large bodies of ultrabasic and basic intrusives. Some bedded cupriferous pyrite deposits occur in this belt (for example, the Shimokawa mine in the Paleozoic sedimentary rocks).

The two belts are considered to have metamorphosed in the Late Mesozoic or Early Tertiary period, but not in the same age.

(B) Tertiary Period (Fig. I-2)

Precise igneous activity during the Paleogene Period in the Japanese Islands has gradually become known through accumulation of radiometric datings of rocks. During the Paleocene epoch (ca. 53–65 m. y.) granitic and rhyolitic rocks have their definite distributions as follows: The rhyolitic welded tuffs are distributed in the Asahi and Ashio Massives, northern Fossa Magna Region and southwestern part of Toyama Prefecture; and the granitic rocks in the Oga Peninsula, Asahi, Ashio, Yamizo and Tsukuba Massives of northeast Honshu, and in a continuous zone of the Kiso Mountainland-Suzuka Massif-Tsuruga-Okutango-

San'in District of southwest Honshu (Kawano and Ueda, 1964, 1966a,b; Shibata and Nozawa, 1966; Isomi and Kawada, 1968; Nishimura and Ishida, 1972; Shibata, 1973). The Early Eocene (ca. 40–53 m.y.) granitic intrusions took place in the northeastern part of Shimane Prefecture (the cities of Mitoya, Yokota, Hirose and Daito), the Late Eocene (ca. 35–40 m.y.) ones in Tottori Prefecture (Misasa area) and Shimane Prefecture (southeast of Gōtsu, and northeast of Masuda), and the Early Oligocene (ca. 30–35 m.y.) ones in the southern end of the Hidaka Massif in central Hokkaido (Kawano and Ueda, 1966a,b). It is to be noted that in the San'in District, a zone of Eocene granitic rocks separates the Paleocene zone from the Cretaceous zone; this fact means that the Paleocene granitic activity is not a mere continuation of the Cretaceous one.

Though the Japanese Islands were relatively quiet during the Eocene-Early Oligocene, the Green Tuff Movement began to take place in the Late Oligocene and continued throughout the Neogene period, and the Japanese Islands began to form the present island arcs. The movement was vigorous along the continental side of the Kurile, Honshu, Ryukyu and Izu-Bonin Arcs. The pyroclastic rocks thus formed are generally altered, show a light greenish appearance, and are called Green Tuff. Thus, the inner zones of the island arcs characterized by such green rocks are collectively called the Green Tuff Region, and the crustal movement is called the Green Tuff Movement. The Japanese standard of the "Miocene" stratigraphic subdivision was established in the Oga Peninsula, Akita Prefecture, as follows: Nishioga Stage (Upper Oligocene), Daijima Stage (Lower Miocene), Nishikurosawa Stage (Lower to Middle Miocene), Onnagawa Stage (Middle Miocene) and Funakawa Stage (Upper Miocene). These stages are primarily based upon lithostratigraphy and biostratigraphy of mollusks, plants and benthonic foraminifers. These stages have been revised by planktonic foraminiferal zonations, radiometric datings, and magnetic stratigraphy. However, the results are not necessarily good owing to a lack of due consideration of litho- and biostratigraphy, for example, subjective ultra-splitting of species by minor morphological changes of shells, direct application of tropical-subtropical deep-sea planktonic foraminiferal zonations to subarctic Japanese formations on the Japan Sea side, and misinterpretation of magnetic stratigraphy without radiometric datings.

Nishioga Stage (Upper Oligocene, ca. 25–30 m.y.): After the extensive Late Cretaceous-Paleocene upheaval of Honshu by intrusion of granitic rocks, local upheaval of land took place, probably owing to partial expansion and melting of the upper mantle and/or lower part of the crust, along zones nearly parallel to the island arcs. The upheaval caused deep-seated and high-angled fractures which crossed the island arcs obliquely or at right angles, and also caused collapsed basins, in which thin layers of clastic sediments (basal conglomerate, etc.) were deposited. Almost simultaneously vigorous volcanic activity took place, and thick pyroclastic layers and some lava flows of mostly rhyolite and andesite and sometimes basalt were deposited upon basal conglomerates, sandstones and/or mudstones containing plant fossils (Aniai-type flora) in the lacustrine basins. The pyroclastic rocks and lava flows (usually in the form of welded tuffs) received regional hydrothermal alteration (first-stage propylitization).

The Nishioga Stage at its type exposure covers the basement adamellite unconformably and consists, in its lower part, of lava flow and tuff breccia of andesite (propylite), and, in its upper part, of porphyritic andesite and rhyolite which are intruded by quartz porphyry, altered rhyolite, and olivine basalt. Mudstone below the Shinzan rhyolite contains the Aniai-type flora indicative of a cold climate. This stage is covered by the superjacent Daijima Stage unconformably. Thus the Nishioga Stage is defined litho- and biostratigraphically, but is not well defined chronostratigraphically owing to a lack of radiometric dating, and has long been considered to be Early Miocene in age. Only one radiometric dating was done on the uppermost part of the stage, namely the Shinzan rhyolite (26 m.y. by the fission-track method (Nishimura and Ishida, 1972)), but the age of the lower limit is unknown. Furthermore, there is some doubt about the accuracy of the fission-track method. These facts often cause serious difficulty in correlating strata to this stage. In general, local differences seem to be recognized in petrological characters of volcanic activity in this stage.

Daijima Stage (Lower Miocene, ca. 17–25 m.y.): Tectonic and igneous activities of this stage are similar to those of the Nishioga Stage, but this stage is more widely distributed and less local in petrological characters of volcanic activity than the Nishioga Stage. The Daijima Formation is covered unconformably by the superjacent Nishikurosawa Formation and consists mainly of massive green tuffs. The basal part is accompanied by lava flow (partly welded tuff) and agglomerate of pyroxene andesite and is called the Hokakejima dacite. The middle part of the formation is the Kanegasaki welded tuff, on which andesitic basalt lies. The Kanegasaki and Hokakejima welded tuffs are 25 and 20 m.y. in age, respectively, by the fission-track method (Nishimura and Ishida, 1972). The Kitaoguni Formation (rhyolitic tuff breccia, tuff and welded tuff) of the Oguni Basin, Yamagata Prefecture, is 22 m.y. in age by the K-Ar method (Ueda *et al.*, 1973). In the Sendai area, fission-track ages of the Tsukinoki Formation, covering the Paleozoic, etc., unconformably and containing the warm "Daijima flora," and of the Takadate volcanics are 22 and 24 m.y., respectively (Tamanyu, 1975). Radiometric datings have been carried out on the Minamisawa welded tuff (Shinjo Basin), Yatani Formation (welded tuff, Asahi Massif), the rhyolite of the Takadate Formation (Sendai) and moonstone rhyolites of the Hokuriku Group (originally erroneously reported as the uppermost part of the Futomiyama Group), and their ages are 22–25 m.y. (Kawano and Ueda, 1964; Yamazaki and Miyajima, 1970; Ueda, unpublished, quoted by Taguchi, 1973). Paleontologically, the Daijima Stage is characterized by the warm "Daijima Flora" and by shallow, warm tropical-subtropical "*Lepidocyclina-Miogypsina* fauna." On the other hand, three peaks (14, 21 and 55 m.y.) are recognized in the distribution of radiometric ages of granitic rocks in the Outer Zone of Southwest Japan (including the Ryukyu Islands) (Nozawa, 1968b). Therefore, the granitic rocks whose ages are ca. 21 m.y. belong to the Daijima Stage. They are the Osumi granitic rocks, Ishigaki granitic rocks, Kunimiya granodiorite, all in Kagoshima Prefecture, and a granodiorite from Amakusashimojima, Kumamoto Prefecture. Generally speaking, volcanic activities during the Daijima Age were characterized by dacite-rhyolite, and sometimes by basalt and dolerite. As can be understood from the foregoing description, sedimentation during the Nishioga and Daijima Ages were epeirogenic in nature and are much influenced by geologic structures of the basement complex of the pre-Tertiary period, while those of the Nishikurosawa, Onnagawa and Funakawa Ages are of geosynclinal nature, accumulating thick sediments rapidly.

Nishikurosawa Stage (Lower to Middle Miocene, ca. 12–17 m.y.): During the Nishikurosawa Age, subsidence took place in all areas of the Green Tuff Region, to which marine transgression proceeded. The areas of volcanic activity during the Nishioga and Daijima Ages upheaved in Northeast Japan. Simultaneously with the subsidence, fractures were formed along the boundary between subsiding and upheaving areas, and basalts were extruded from the fractures in Akita and Yamagata Prefectures (e.g. Horozuki, Sunakobuchi, Aozawa and Onisakatoge Formations). The basaltic eruption, in turn, caused further subsidence, and, during the following Onnagawa Age, dolerites intruded along the fractures which were formed in the margins of the subsiding areas. At the same time, rhyolitic to dacitic volcanic activities took place in the marginal areas of the sedimentary basins and in newly subsiding areas during the Nishikurosawa and Onnagawa Ages (e.g. Okuzu, Hanaoka, Kawajiri, Yoshino, Osawa and Oshio Formations in the backbone range of Northeast Honshu). Such bimodal volcanism of basalt-rhyolite association (Konda and Taguchi, 1976) is the characteristic feature of the Nishikurosawa and Onnagawa Ages. These acidic volcanic activities are closely related to the formation of the "Kuroko" (Black Ore) deposits.

These pyroclastic rocks were altered extensively to green rocks in the Nishikurosawa Age and locally in the Onnagawa Age. Shimazu (1973) and Shimazu *et al.* (1976) also reported multiple dikes of rhyolite-dolerite in the Otanigawa Formation (green tuff of the Nishikurosawa Age) in Tsugawa-Aizu Province.

The Nishikurosawa Formation at its type exposure and its vicinity consists of basal conglomerate and calcareous sandstone characterized by the warm "*Operculina-Miogypsina* fauna." The formation is covered by hard siliceous shale of the superjacent Onnagawa Stage conformably and partly unconformably. The Nishikurosawa tuff of the Nishikurosawa Formation is 16 m.y. in age by the fission-track method (Nishimura and Ishida, 1972). The upper part of the Iwaine Formation (andesite) in Toyama Prefecture is 16 m.y. in age by the K-Ar method (Shibata, 1973). The Nishikurosawa Stage is widely distributed, and two types of formations are recognized by lithological character. One group consists mostly of green tuff, tuff breccia with small amounts of sandstone and mudstone (e.g. Sunakobuchi Formation of Akita Prefecture and Tsugawa Formation of Niigata Prefecture), and the other group consists mostly of mudstone sometimes intercalated with thin layers of green tuff (e.g. Uyashinai Formation of Akita Prefecture and Nanatani Formation of Niigata Prefecture). The two are isochronous and interfinger with each other, but the latter usually covers the former in the field, because the latter is an offshore deposit.

Granitic rocks of the Nishikurosawa Age occur in Southwest Japan as follows: the Yakujima,* Osumi, Takakuyama, Nomadake, Mukaeyama, Koshikijima, and Shibisan granitic rocks (Kagoshima); Ichibusayama granitic rocks (Kumamoto); Osuzuyama acidic rocks (rhyolite and granodiorite, Miyazaki); Okinoshima granitic rocks (Tosashimizu, Kochi), Ashizurizaki granitic rocks (Kochi); Uwajima and Omogo granitic rocks (Ehime); Kumano acidic rocks and granite porphyry from Ugui and Taiji (Wakayama); Ominesan granodiorite (Nara) and granodiorites in parts of the Kofu Basin (Yamanashi). Their K-Ar ages are within a range of 12–17 m.y. (Kawano and Ueda, 1966b; Shibata and Nozawa, 1968a, b,c,d,e; Miller *et al.*, 1962; Kawai and Hirooka, 1966). The K-Ar age of the Tsushima granite** is 12 m.y. (Kawano and Ueda, 1967a). These granitic or acidic rocks represent igneous activities in the orogenic stage, but they did not intrude deeply into the axis of the orogenic zone, and transitions from plutonic to volcanic facies are observed in them. They are parts of a volcano-plutonic complex. Radiometric ages (by the K-Ar method, with one exception by the fission-track method) of the following volcanic rocks of the "First Setouchi Series" are 12–16 m.y.: altered rhyolite (Amakusashimojima, Kumamoto), Ono volcanics (Oita), andesite from Takamatsu and Sakaide (Kagawa), Nijo volcanics (Osaka-Nara boundary), Mikasayama (Nara), Muroo volcanics at Nabari (Mie), Shidara volcanics (Aichi) (Kawano and Ueda, 1964, 1967b; Kawai and Hirooka, 1966; Shibata and Ono, 1974; Shibata and Togashi, 1975). The K-Ar date of the Okawa rhyolitic welded tuff (Tokachi, Hokkaido) is 15.1 m.y. (Shibata *et al.*, 1975).

Onnagawa Stage (Middle Miocene, ca. 9–12 m.y.): Two glauconite samples from the Onnagawa Formation at Unozaki, Oga Peninsula, are 8.7 and 8.8 m.y. and glauconites from the Onnagawa Formation at Yumoto and Nishikurosawa, Oga Peninsula, are 9.0–9.8 and 8.4 m.y. by the K-Ar method (Ueda and Suzuki, 1973). Glauconite ages of some formations considered to belong to this stage are as follows: Najimi mudstone at Suzu, Noto Peninsula (14.8 m.y.); Odoji Formation in the Fukaura area, Aomori Prefecture (9.7 m.y.); Otaki Formation of the Odate-Hanawa area, Akita Prefecture (7.5 and ca. 10 m.y.); Yakumo Formation of Matsumae and Oshima, southwest Hokkaido (9.6 and 11.8 m.y.); Onnagawa Formation at Iwasaki, Aomori Prefecture (9.8 m.y.) (Ueda and Suzuki, 1973).

---

* Igneous rock genetically related to the wolframite-quartz veins on Yakujima Island.
** Igneous rock genetically related to the vein-type deposit at the Taishu mine (refer to Chapter II-6).

Volcanic activity of the lower part of the Onnagawa Stage is represented by the bimodal volcanism of basalt-rhyolite, as mentioned in the previous paragraph, though andesitic volcanism is also known. However, the bimodal volcanism is limited to the northern Akita and Tsugawa-Aizu districts, where dolerite intrusion is common. Andesite is known in Shinjo Basin, and dacite-rhyolite are common in the Sakunami area (west of Sendai). In the vicinity of Shimane Peninsula, the Onnagawa Stage is represented by the Kuri (Josoji) Formation in the lower part and the Omori (Ushigiri) Formation in the upper part. These formations consist mostly of black shale, andesite, dacite, rhyolite and their pyroclastic rocks, and they were intruded by many sheets of intermediate to basic intrusive rocks (dolerite, fine grained-gabbro and fine-grained diorite) at the end of the Onnagawa Age. These intrusions are closely related to the genesis of metallic ore deposits such as the Wanibuchi, Udo and Iwami mines (Kuroko deposits). These intrusive rocks are one of the so-called Tertiary granites intruded during the Nishikurosawa, Onnagawa and Funakawa Ages in many areas sporadically, not only in Northeast Japan and the Fossa Magna Region but also in the Outer Zone of Southwest Japan. The radiometric ages of many of them are in the range of 12–14 m.y., and they are listed in the Nishikurosawa Stage. The following granitic rocks are also transitional between the Nishikurosawa and Onnagawa, but are rather closer to the Onnagawa Age: quartz porphyry in Nago City (Okinawa), 11.6 m.y. by the K-Ar method (Shibata, 1975); biotite and orthoclase in granodiorite, Ani mine (refer to Chapter II–10), 12–15 m.y. (Kawano and Ueda, 1966b), and adularia a in vein of ore deposit of the same mine, 11 m.y. (Yamaoka and Ueda, 1974); part of the Ichibusayama granitic rocks (Kumamoto), 11 m.y. by the K-Ar method (Kawano and Ueda, 1967a). They are parts of a volcano-plutonic complex.

Funakawa Stage (Upper Miocene, ca. 6–9 m.y.): Fission-track ages of the Anzenji and Minamihirasawa tuffs of the Funakawa Formation at their type exposures, Oga Peninsula, are 12 and 13 m.y., respectively (Nishimura and Ishida, 1972). However, this method seems to give a little older age to analyzed rock than the true age, and the resulting age cannot always be trusted. The K-Ar ages of a glauconite of the Fujikotogawa Formation in the Takanosu Basin (Akita) are 5.7 and 6.1 m.y. (Ueda and Suzuki, 1973). The K-Ar age of an adularia in a gold-bearing quartz vein of the Shiraishi Formation from the Takatama mine (Fukushima) is 8.4 m.y. (Yamaoka and Ueda, 1974). Granite intruding into the Paleozoic Chichibu System in the Chichibu mine (Saitama) is ca. 8 m.y. in age (Ueno *et al.*, in preparation). The K-Ar ages of the following granitic rocks in the southern Fossa Magna region are within a range of 7.3–8.6 m.y.: granodiorite at Ishiwariyama (Yamanashi) in the Tanzawa Massif; granodiorite and quartz-diorite at Sakeishi and Somayama (Yamanashi) in parts of the Kofu Basin; granodiorite at Chino (Nagano) in the vicinity of Suwa Lake. Quartz-diorite at Doai of Minakami (Gunma) in the Tanigawadake area is 5.9 m.y. in age by the K-Ar method. These datings were reported by Kawano and Ueda (1966b). These rocks are parts of a volcano-plutonic complex. The Wakurayama andesite (Shimane) covering the Matsue Formation is 6.34 m.y. in age by the K-Ar method (Kawai and Hirooka, 1966).

In general, dacitic volcanic activity is very common, and andesitic activity is also common in the Funakawa Stage. Yoshitani *et al.* (1976) and Shimazu (1976) reported alkaline basalts from the Matsue and Misasa Formations in the San'in District.

From the standpoint of ore genesis, the Tertiary mineralization occurred predominantly in Nishikurosawa Stage, but, its remains still continued in the Onnagawa and Funakawa Stages, and further in the Pliocene epoch. The K-Ar age of an adularia from the gold-silver quartz vein at the Seikoshi mine, Izu Peninsula is 3.7 m.y. (Yamaoka and Ueda, 1974).

In Quaternary volcanic areas, sulfur deposits are distributed.

PART II
VEIN-TYPE DEPOSITS

A. GENERAL PROBLEMS

II-1. Formation of Fissures and Their Mineralization (Imai, 1963a, 1966a)

H. Imai

Introduction

In Japan, vein-type deposits were important as gold, silver, copper, lead-zinc, tungsten and tin producers. They are classified into two types according to the geological age of their formation:

(1) Vein-type deposits genetically related to the granitic intrusions of the Late Cretaceous to Early Tertiary epoch (older type veins). They include hypothermal, mesothermal and xenothermal types.

(2) Vein-type deposits genetically related to the volcanic activities of the Miocene epoch, i.e., rhyolitic and andesitic (propylitic) activites (younger type veins). They include epithermal and xenothermal types. They may be called subvolcanic types.

However, some of the deposits which are related to the Miocene volcanic activities have close relations to the Miocene granitic intrusion, as stated in the chapter on the Ani mine. The problem of igneous rocks relating to subvolcanic deposits will be discussed in later chapters.

Formation of Vein Fissures

The writer has recognized the following types of vein fissures:
(1) Fissures due to magmatic intrusion
  (a) Fissures due to magmatic upheaval
  (b) Fissures due to subsidence (including cauldron subsidence) after magmatic intrusion
  (c) Joints in igneous rocks due to solidification of magma
(2) Fissures due to lateral force (horizontal compressive force or couple)
(3) Fissures revealing multiple movements or displacements due to more than two different stresses

(1) Fissures Due to Magmatic Intrusion

Emmons (1933, 1940), Hulin (1929, 1948) and others have discussed the fissures related to igneous intrusion which were afterwards mineralized. By magmatic intrusion many kinds of fissures are formed not only in the surrounding rocks (the "roof" proposed by Emmons (1940)) but also in the marginal part (the "hood" proposed by Emmons) of the intruding igneous rock which solidified somewhat earlier than the "core" part.

(a) Vein fissures due to magmatic upheaval: These fissures correspond to the diapiric structure discussed by De Waard (1949) and De Sitter (1956).

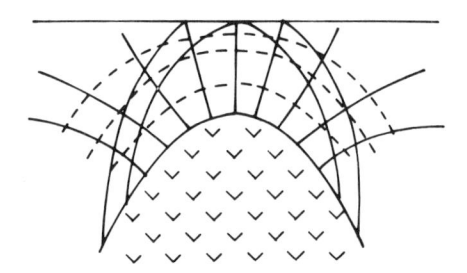

FIG. II–1–1. Stress diagram showing the supposed model of formation of cone sheets (parallel to solid lines with inward inclination) and ring dikes (parallel to solid lines with outward inclination). The broken lines represent the tension fractures at the time of magmatic shrinkage. (after Anderson 1936, 1937).

Tension cracks, perpendicular to the wall of the igneous mass (Fig.II–1–1), are formed by magmatic upheaval. They are apt to concentrate in the area just above the cupola of the igneous mass. The dike filling this kind of fissure is called cone-sheets, whose formation mechanism was discussed by Anderson (1936, 1937). In some cases, these fissures are mineralized. The Ohya mine is a vein-type deposit, filling the tension cracks in the roof just above the cupola of the granitic rock.

Also, in some cases network ore bodies occur in the roof just above, or in the hood of, the cupola of the igneous stock by magmatic upheaval (Emmons, 1938; Bryner, 1961). The writer recognized ore deposits of this type in the Ani mine and Tsuchihata mine and in the Ilkwang mine, South Korea.

Occasionally, normal faults resulting from pushing up by magmatic upheaval are recognized around the intrusive igneous mass (Imai, 1963a). They occur not only in the roof rocks surrounding the igneous intrusives but also extend into the igneous mass (hood) (Fig.II–1–2). It is noticeable that they dip away from the igneous mass. They are synthetic[*] normal faults produced at the time of igneous intrusion which is often accompanied by folding and warping of the surrounding rocks. Faults of this kind are similar to the cracks shown in the model of the experiments by Riedel (1929) carried out in Nadai's laboratory

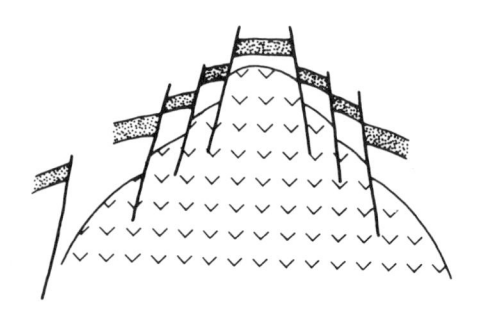

FIG. II–1–2. Pushing up of the roof and hood of magma by synthetic normal faults.

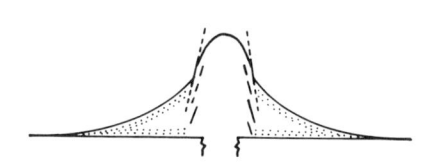

FIG. II–1–3. Bysmalith (After Iddings, 1898).

---

[*] "Synthetic" is used after the meaning of Cloos (1928), Balk (1936) and Wisser (1936). It means smaller movement that is associated with larger movements but has the same tendency as the larger movements.

(Nadai, 1963). De Waard (1949) recognized faults similar in the formation of the diapiric structures.

The blocks sandwiched between such faults are upheaved like horst. The writer has observed some deposits where synthetic normal faults of this kind were mineralized to copper, zinc and lead veins. In the later chapters, the writer discusses some examples of these veins, i.e., the veins in the Mikawa mine, Ani mine, Osarizawa mine and Akenobe mine. These fissures are similar to those accompanied by the intrusive mass of bysmalith type in the mechanism of fissure formation. A bysmalith is a variety of laccolith, the roof of which has been uplifted along the cylindrical faults (Fig.II–1–3) (Iddings, 1898).

Wisser (1960) discussed many antithetic* fault veins accompanied by folding and doming in the North American Cordillera. Balk (1936) reported flat-lying normal faults filled with vein materials at the flat part of the top of the igneous mass, which were caused by magmatic upheaval. Balk (1936) recognized marginal fissures filled with vein materials in the steep walls of the igneous mass, which were due to magmatic intrusion. The writer has not yet found veins of these kinds in Japan. Wisser (1960) described the radiated vein fissures around the igneous mass formed by magmatic intrusion. The writer has observed radiated veins at the Teine mine, Hokkaido.

(b) Fissures due to subsidence (including cauldron subsidence) after magmatic intrusion: The subsidence of the blocks occurs along the fissures after the intrusion of the magma. This would be caused by shrinkage of the magma due to solidification, to partial outflow of the magma from the reservoir or to other causes. Some of these fissures would have been previously formed, for example, by pushing up of the magma, and later subsidence would occur along these fissures. Others would be formed directly by collapse of the block at the time of subsidence. The cauldron subsidence might also include the both kinds. The fissures filled by ring dikes have an outward inclination (Fig.II–1–2) (Anderson, 1936, 1937) or vertical dip (Billings, 1945). In some cases, these fissures are mineralized into veins. According to Wallace *et. al.* (1957), the molybdenum deposit of the Climax mine belongs to the ring-like fracture mineralization. Takenouchi (1962a) studied the ring fractures mineralized by copper, zinc and lead at the Yaso mine. In the Ashio mine, the faults due to cauldron subsidence are the important paths of the ore-forming fluid. (Imai *et al.* 1975)

Also, Anderson (1936, 1937) assumed that the tension fractures would be formed parallel to the wall of the magma reservoir due to magmatic shrinkage (Fig. II–1–2). The vein fissures with a gentle dip or flatly-lying veins at Kaneuchi and Takatori mines would belong to this type.

Walker and Walker (1956) and Perry (1961) recognized the mineralized breccia pipes caused by subsidence due to magmatic withdrawal.

(c) Joints in the igneous rocks due to solidification of magma: Many joint systems are produced in the igneous mass during its solidification. Some vein-type deposits occur in these joints. Fujiki (1964a) studied the mineralized onion cracks at the Komori mine.

(2) Fissures Due to Lateral Force (Horizontal Compressive Force or Couple)

The writer has observed many vein-type deposits which occur in the fissures of wrench faults (strike-slip faults). McKinstry (1941, 1953) studied the wrench faults and their branches (shears of second order) which were mineralized. The writer (1956b) also has observed mineralized wrench faults and branching shears and tension cracks of the second order at the Hosokura mining area. They are epithermal or subvolcanic zinc and lead veins in the Miocene volcanic and pyroclastic rocks and sediments.

In the Inakuraishi-Oe mining area the wrench faults are filled with sphalerite, galena,

---

* Antithetic has a tendency opposite to synthetic.

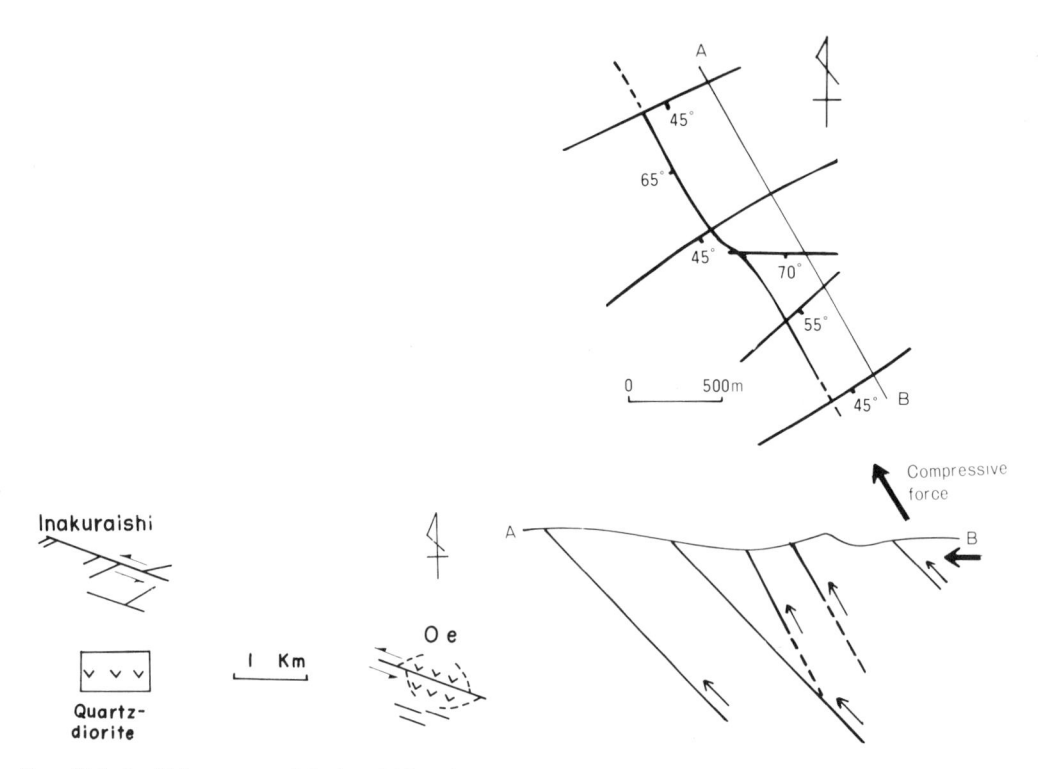

Fig. II-1-4. Vein system of Inakuraishi and Oe mines. Arrows show the displacements of the blocks.

Fig. II-1-5. Vein system of the Kushikino mine.

rhodochrosite, etc., forming the epithermal veins (Fig. II-1-4) (Imai, 1966b). In the Taishu mine, zinc and lead veins of a similar kind are developed (Imai, 1973).

The existence of mineralized reverse faults is naturally expected. The flat-lying vein group filling the bedding reverse faults in the Taishu mine is an example of this type (Imai, 1973). At the Kushikino mine in Kyushu, the reverse fault striking NE with a low-angle SE dip diverges the branch faults of the same strike, showing imbricated structure (Fig. II-1-5). They intersect a tension crack striking NW with a steep dip nearly at a right angle. They are all mineralized by gold-quartz veins. The country rocks are Miocene propylites and andesites. Also, the writer has observed veins filling the thrust faults in the Komori mine (Fujiki, 1964a; Imai, 1966b). Newhouse (1940) reported the same kind of veins in Alleghany, California.

The writer *et al.* (1959) recognized tension cracks perpendicular to the folding axis which are mineralized into the fissure veins at the Takatori mine. They ascribed the tension cracks to the lateral pressure which was probably exerted perpendicular to the folding axis. The tin-tungsten deposits mineralized in fissures of this kind are described in the chapters on the Takatori, Akenobe, Kuga, Ohtani and Kaneuchi mines. The same kind of fissure veins of copper will be described in the chapter on the Ani mine.

Examples of veins filling tension cracks which interesct the folding axes at right angles were described in the tin veins of the Quisma Cruz region, Bolivia, by Turneaure and Welker (1947), in the zinc-lead veins of Westfalen, Germany, by Hesemann and Pilger (1951) and in the zinc-lead veins of Oberharz, Germany, by Jacobsen (1950). In many cases, tension cracks of this kind would be formed by lateral pressure, combined with the force due to magmatic upheaval, as discussed in the chapter on the Kaneuchi mine.

(3) Fissures Revealing Multiple Movements or Displacements Due to more than Two Different Stresses

After the formation of a fissure accompanied by displacement or opening, it receives the other kind of stress and reveals a different movement. The writer discusses, in later chapters, the Heiko Fault vein in the Sado mine, the Tajima vein in the Toyoha mine and the Tenju vein in the Ikuno mine as examples of this type.

From the above discussion, it can be seen that the kinds of fissures are quite variable even under the same kind of stress. As an example, fissures due to magmatic upheaval are different in each case. It is responsible for many factors such as shape of magma reservoir, property of magma, speed of upheaval, kind of country rock, geologic depth of the magma, etc., which characterize each area.

Leading Types of Mineralization of the Vein Fissures in Japan

The leading kinds of vein-type deposits in Japan are as follows:
(1) Au-Ag veins
(2) Cu-Pb-Zn (-Au-Ag) veins including straight Cu veins and Pb-Zn veins
(3) Sn-W-Cu (-Pb-Zn-Au-Ag) veins

(1) Au-Ag Veins

Some of these are genetically related to the Late Cretaceous or Early Tertiary igneous activities belonging to the older type, while others are related to the Oligocene or Miocene volcanic activities belonging to the younger type. The former are hypothermal, mesothermal or xenothermal deposits, and the latter are epithermal or subvolcanic deposits and are far more prevalent than the former. The Ohya gold mine belongs to the older type. The Au-Ag veins of the younger type contain three to fifty times as much silver as gold, whereas those of the older type contain nearly the same amount of silver as of gold. The older type veins are quartz veins containing gold grains of high fineness, with extremely small amounts of sulphide minerals, while the younger type veins are generally quartz veins containing electrum and silver sulphide minerals such as argentite, polybasite, pearceite, pyrargyrite, etc., accompanied by chalcopyrite, galena, sphalerite, adularia, chlorite, etc. Alteration of the country rocks of the younger type veins is characterized by kaolinization, sericitization, chloritization and silicification.

The Kohnomai and Chitose mines in Hokkaido, Sado and Seikoshi mines in central Japan and Kushikino mine in Kyushu are representative of the gold-silver veins of the younger type. They are accompanied with either rhyolite or andesite (propylite) or both.

(2) Cu-Pb-Zn(-Au-Ag) Veins Including Straight Cu Veins and Pb-Zn Veins

The content ratio of the above metals varies with the deposit. Some of the veins are worked as copper mines, and others are worked as lead-zinc mines.

Some of these veins belong to the older type (hypothermal or mesothermal), and others to the younger type (epithermal or subvolcanic).

The hypothermal veins containing copper, lead and zinc are distributed in the regions of the granitic rocks, especially in western Japan. Tourmaline copper veins are found in the Yakuoji mine, Yamaguchi Prefecture (Kato, 1912). They are developed chiefly in the boss of granodiorite near the contact with the Permo-Carboniferous formations. The Ilkwang (Nikko) mine, South Korea, is similar to the Yakuoji mine. Both are situated in the same geologic province, the "Tsushima Basin."

Mesothermal veins are also distributed in the western part of Japan, and are related to the granitic rocks of the Late Cretaceous or Tertiary age. They are quartz veins bearing chalcopyrite, sphalerite, galena and pyrrhotite. They generally have no characteristic minerals indicating high-temperature formation except for pyrrhotite. The Pb-Zn veins of the Taishu mine belong to the mesothermal type.

Xenothermal veins are distributed in the areas where acidic intrusives coexist with the rhyolitic rocks or equivalent pyroclastics of the same age. Most of them belong to Late Mesozoic or Early Tertiary epoch.

The epithermal or subvolcanic veins are found in the terrains of the Oligocene or Miocene formations with predominant tuff and tuff-breccia accompanied by rhyolite and andesite (propylite).

A petrogenetical interpretation of the bimodal coexistence of rhyolite, andesite and propylite in areas of epithermal or subvolcanic deposits has not yet been made. A possible interpretation might be that they are the result of liquid immiscibility of magma (Roedder and Weiblen, 1970; Roedder, 1971d; Nakamura, 1974) or of fractional melting of the lower crust or upper mantle (Yoder, 1973).

According to Kato (1928), the effusive rocks and the associated ore deposits are evidently consanguineous or hold a relation of sisterhood, and the ore-feeder is concealed far beneath. He called effusive rocks of this kind sister rocks.

This kind of deposit is mostly quartzose veins with quartz as the chief gangue mineral, and chalcopyrite, galena and sphalerite as the principal ore minerals. In some cases, they change their metal contents from the lower level to the upper level. Generally, from the lower level towards the upper level, there occur (1) a copper zone, (2) a zinc zone, and (3) a lead zone, displaying zonal distribution. Some of them become gold-silver quartz veins at the higher levels, as in the case of the Washinosu deposit in the Tsuchihata mine.

Veins containing enargite and luzonite occur in terrains predominating in andesite or andesitic tuff. They also belong to epithermal or subvolcanic type. But, they are different from the chalcopyrite veins.

The Osarizawa and Tsuchihata mines, described later, were representative of the copper mines of the younger type in Japan, while the Toyoha and Hosokura mines are typical of zinc and lead veins of the same type.

In fluid inclusion studies, it is recognized that the minerals from epithermal or subvolcanic veins do not necessarily reveal low filling-(homogenization) and low decrepitation* temperatures. There are some differences in the temperatures between hypothermal, mesothermal and epithermal (subvolcanic) veins. Generally speaking, each kind of vein has the tendency to decline from lower level to upper in filling and decrepitation temperature.

The epithermal or subvolcanic veins described above include, in some cases, the higher-

---

* As to the decrepitation method, there are problems with the reliability of the data (Boyle, 1954; Hayakawa et al., 1972). Fundamentally, it is a question whether decrepitation really depends on the bursting of fluid inclusions. In the writer's laboratory, some experiments were carried out on this problem by Kashiwagi et al. (1955) and Takenouchi (1962b). It has not yet been fully solved.

The writer thinks at present that most of the decrepitation is due to the bursting of the inclusions, and that the values of the decrepitation temperatures are closely related to the filling temperatures.

But in minerals with predominating cleavage, lineage or fracture, such as sphalerite, galena, fluorite, calcite, etc., the measured decrepitation temperatures are not connected with the filling temperatures.

Boyle (1954) discussed the decrepitation temperatures of quartz as related to the state of crushing in the crystallization process.

Besides, there are problems of size effect in host minerals and in inclusions (Takenouchi, 1962b), and of contents of CO_2 and other gases in the inclusions (Takenouchi, 1971a).

The writer measured the filling or homogenization temperatures by a heating microscope and decrepitation temperatures by the electronic method and regards them as values related to the formation temperatures. For the sake of convenience, the writer did not make any pressure correction.

temperature mineral assemblages, such as magnetite, pyrrhotite, bismuthinite, etc., showing some xenothermal-type characteristics.

(3) Sn-W-Cu(-Zn-Pb-Ag-) Veins

These belong to hypothermal and xenothermal types. The hypothermal tin-tungsten-copper veins genetically related to Late Mesozoic granitic activity occur in the Permo-Carboniferous* Chichibu System. The veins of the Takatori, Ohtani and Kaneuchi mines are examples. The xenothermal veins contain polymetallic ores such as Sn-W-Cu-Zn-Pb-Ag minerals. The Akenobe, Tada, Ikuno and Ashio deposits are representative xenothermal types. Xenothermal-type veins such as the Ikuno and Akenobe deposits occur in regions where the rhyolitic rocks of the Mesozoic or Tertiary epoch are widely developed. From this point, they belong to the subvolcanic type. But granitic rocks belonging to the same igneous activity are also found.

In some cases, the fluid inclusions in a specimen are characterized by having various filling degrees (liquid-rich inclusions and gas-rich inclusions), indicating their formation from heterogeneous mixtures of liquid and gaseous phases. Perhaps boiling would occur in this case. The filling temperatures of these fluid inclusions under the microscope are variable over a wide range, but the lowest value of the range would be close to the trapping temperature of fluid (Smith and Little, 1959). Consequently, the temperature measurements were carried out on liquid inclusions having the highest filling degree in the same inclusion group.

Following this chapter the writer discusses the geology and ore deposit of each vein-type deposit. But many of the vein-type deposits in Japan have been closed in the past ten years. Among those discussed in the following chapters, the Ohya, Komori, Taishu Mikawa, Yaso, Ashio, Ikuno, Tada and Obira mines have been closed.

---

* Including Triassic period.

B. PEGMATITIC, HYPOTHERMAL AND MESOTHERMAL VEINS

II-2. Ohya Mine, Miyagi Prefecture   (Imai, 1961, 1966b)

H. Imai

The Ohya mine, located in the southern part of the Kitakami mountainous district, Northeast Japan, has been a well-known gold producer since ancient times. A large batholith of granodiorite and quartz diorite occupies the southern part of the Kitakami mountainous district. The age of the intrusion of these plutonic rocks is Middle or Late Cretaceous period. They intruded into the clayslate, sandstone, limestone and quartzite of the Permo-Carboniferous and Triassic periods. Around the southern margin of the batholith, many ore deposits (vein-type deposits) are distributed. The Ohya mine is one of them.

In the mining area of Ohya, the quartz diorite batholith projects its cupola toward the south into the Triassic clayslate and sandstone. These sedimentaries display a basin structure in the mining area and become hornfelsic. The projected cupola is hidden beneath the Triassic formation toward the south (Fig. II-2-1), and small bosses of dioritic rocks, which may be offshoots of the elongated cupola, occur along this zone.

The vein fissures occur mainly in the Triassic formation just above the southward projection of the quartz diorite (Fig. II-2-1). They strike N-S, dipping steeply to the west: that is, they stretch in the direction of the elongation of the cupola. Lodes parallel to the longer axis of an associated cupola were described by Emmons (1933, 1935). In some cases, the fissure veins in the Ohya mine penetrate the bosses of quartz diorite offshoots, so it is certain that the mineralization was later than the solidification of the marginal part of the dioritic magma.

The maximum strike length of the veins is 600m, the vertical distance between the uppermost level and the lowest levels is 500m, and the vein width is 20-30cm on the average.

The fissures containing vein materials are faults which have apparent displacements of 1-2m in the plan (Fig. II-2-2). The eastern blocks of the N-S faults moved laterally southward, and the western blocks moved north. As shown in Fig. II-2-3, in the southern extension of the Honmyaku vein, a group of veinlets occur *en echélon*, striking N 35°E. Though their widths are 2-3cm, the veinlets are gathered so closely that they were mined together with the country rocks as a massive deposit. The fissures of the veinlets may be tension cracks. This kind of *en echélon* structure may have been formed by right-handed shearing stress, as shown in Fig. II-2-3. This interpretation also explains with the displacement of the fissure of the Honmyaku vein.

The veins occur just above the the top of the cupola. According to the interpretation of Anderson (1936, 1937), these vein fissures would belong to the tension cracks (Fig. II-1-1). But, as described above, they are faults having small displacements. So it is certain that the shearing stress was exerted along fissures which were mineralized later. These facts may be consistent with the assumption of the asymmetrical shape of the batholith in horizontal and vertical planes, as shown by Emmons (1940) (Fig. II-2-4).

The Triassic sandstone and clayslate surrounding the veins suffered skarnization, becoming green or dark green massive rocks. Under the microscope, skarn minerals such as scapolite,* diopside, garnet, titanite, actinolite, epidote, etc., are observed in these rocks (Fig. II-2-5(a)). The skarnization is confined to areas just along the vein fissures.

---

* Dipyre: Uniaxial (negative), $\omega = 1.549$, $\varepsilon = 1.536$, $\omega - \varepsilon = 0.013$.

Part II: Vein-type Deposits

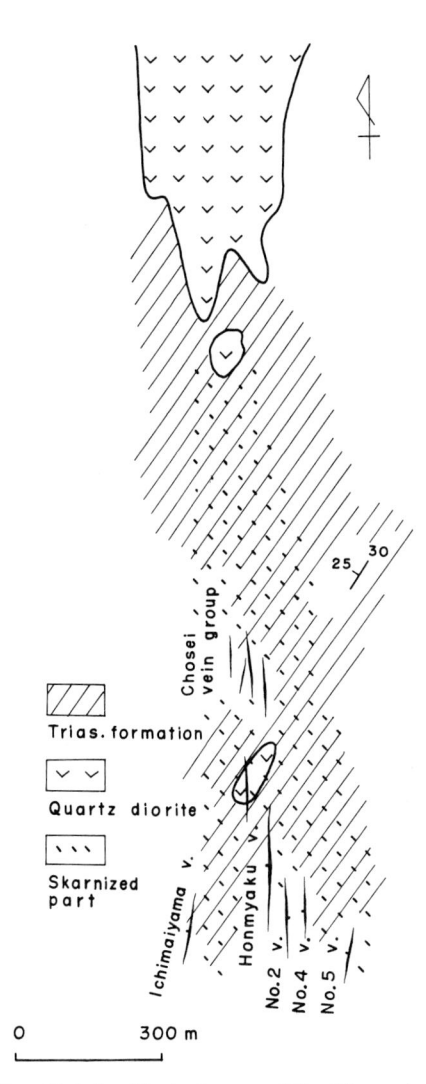

FIG. II-2-1. Geologic map of the Ohya mine.

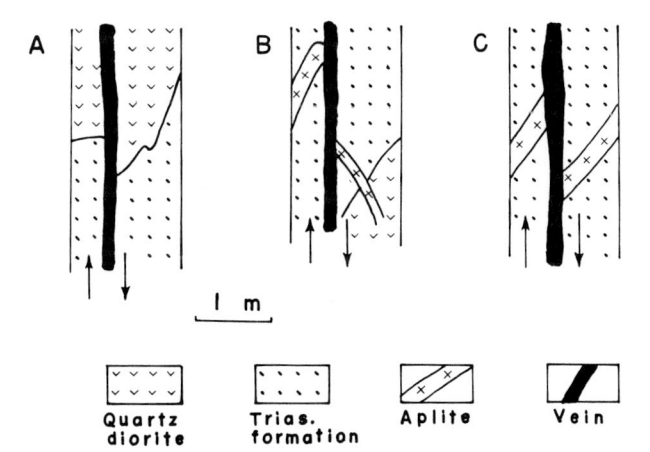

FIG. II-2-2. Sketches of the veins at the roof of the adits. The upper edge is north.

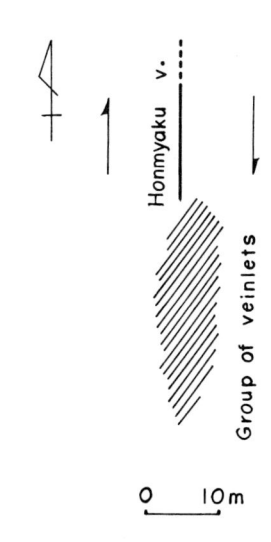

FIG. II-2-3. Group of the veinlets to the south of the Honmyaku vein, showing *en echélon* structure.

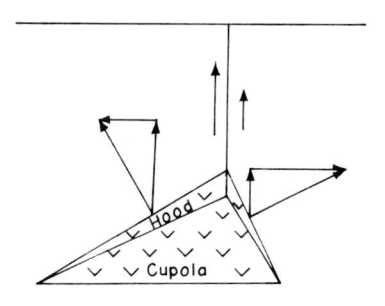

FIG. II-2-4. Diagram showing how a sheared zone may form on the unsymmetrical walls of an elongated cupola (after Emmons, 1940).

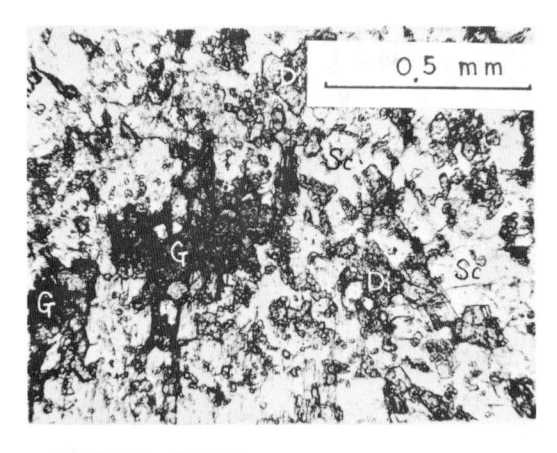

(a) Thin section. Skarnized country rock. One nicol.
Sc: scapolite, Di: diopside, G: garnet.

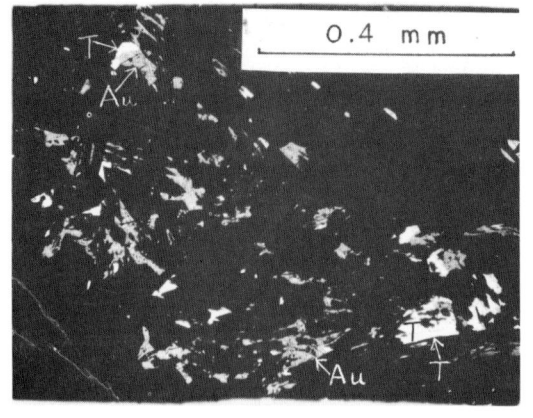

(b) Polished section. One nicol. Au: native gold, T: tellurobismuthite, Black part: skarn minerals.

Fig. II-2-5. Photomicrographs.

The vein-forming minerals are quartz, native gold, arsenopyrite, pyrrhotite, chalcopyrite, tellurobismuthite, etc. (Fig. II-2-5(b)). The gold content in the Chosei Main vein (tenor of gold × vein width) is shown in the longitudinal section of the vein (Fig. II-2-6). In this figure, it is seen that the distribution of the gold content plunges to the north. As this tendency is not recognized in the vein width, it is due to the tenor of the gold. The formation temperature of quartz from the same vein was studied by the decrepitation method. The results are shown in the longitudinal section of the vein (Fig. II-2-7(a)). The gradient of the decrepitation temperatures plunges to the south, which would indicate the path of the mineralizing solution. In view of the genesis of the ore deposit, it is interesting that the path of the mineralizing solution was nearly perpendicular to the equi-grade lines representing gold content. In the east-west section (Fig. II-2-7(b)), the decrepitation temperatures of the same level are high in the samples from the vein in the middle part and become lower in those from the marginal veins to the east or west. This fact tells that the middle part of the vein group was the center of the mineralization.

Part II: Vein-type Deposits

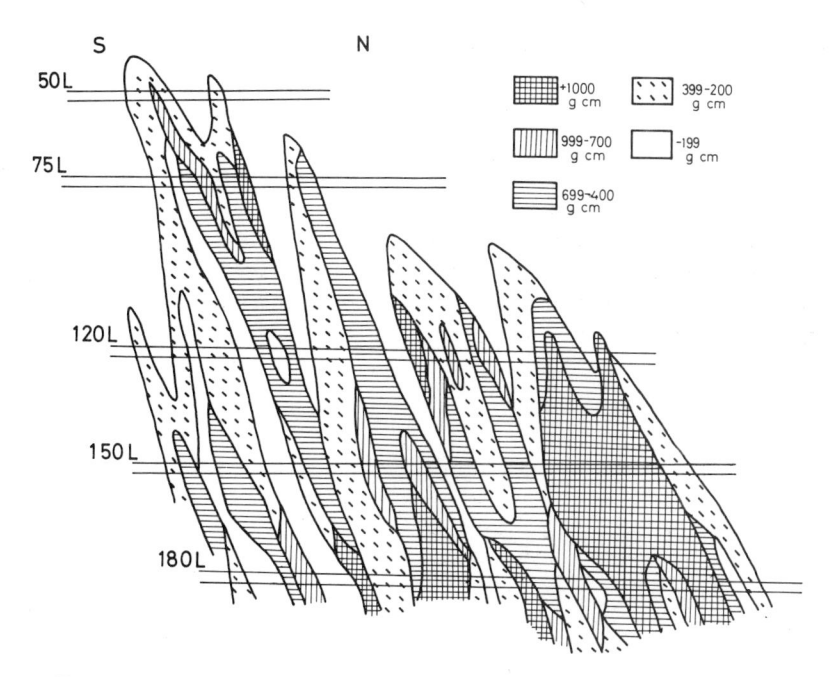

FIG. II-2-6. Distribution of gold content (tenor (g/t) × vein width (cm)) in the longitudinal section of the Chosei Main vein. (The data are obtained from the mine). Numerals represent the height differences from the adit level in meters.

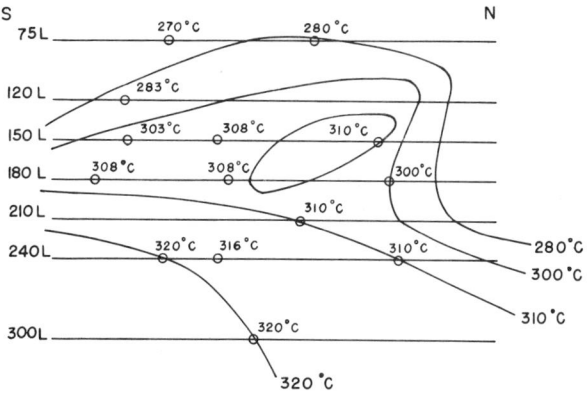

(a) Distribution of the decrepitation temperatures in the longitudinal section of the Chosei Main vein.

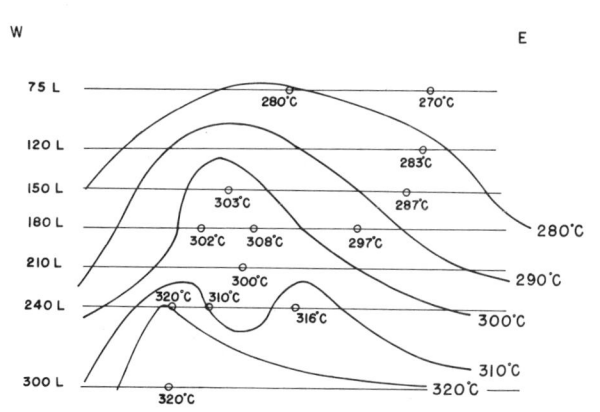

(b) Ditto in the transverse section of the Chosei vein group.

FIG. II-2-7.

II-3. Takatori Mine, Ibaragi Prefecture (Imai and Hayashi, 1959; Imai, 1966b)

H. Imai, S. Hayashi and T. Kashiwagi

This mine is situated about 110 km NNE of Tokyo. This district is geologically composed of Paleozoic sandstone, clayslate and quartzite (including Triassic), which generally strike N45°E, dipping 30° – 40° to NW. Limestone is scarce, occurring in a small area about 9 km south of the mine. Granitic rock occurs about 4 km northeast of the mine (Fig. II-3-1).

The mine produces tungsten and copper. The deposit of this mine belongs to the hypothermal quartz veins, accompanied by wolframite, cassiterite, chalcopyrite, arsenopyrite, pyr-

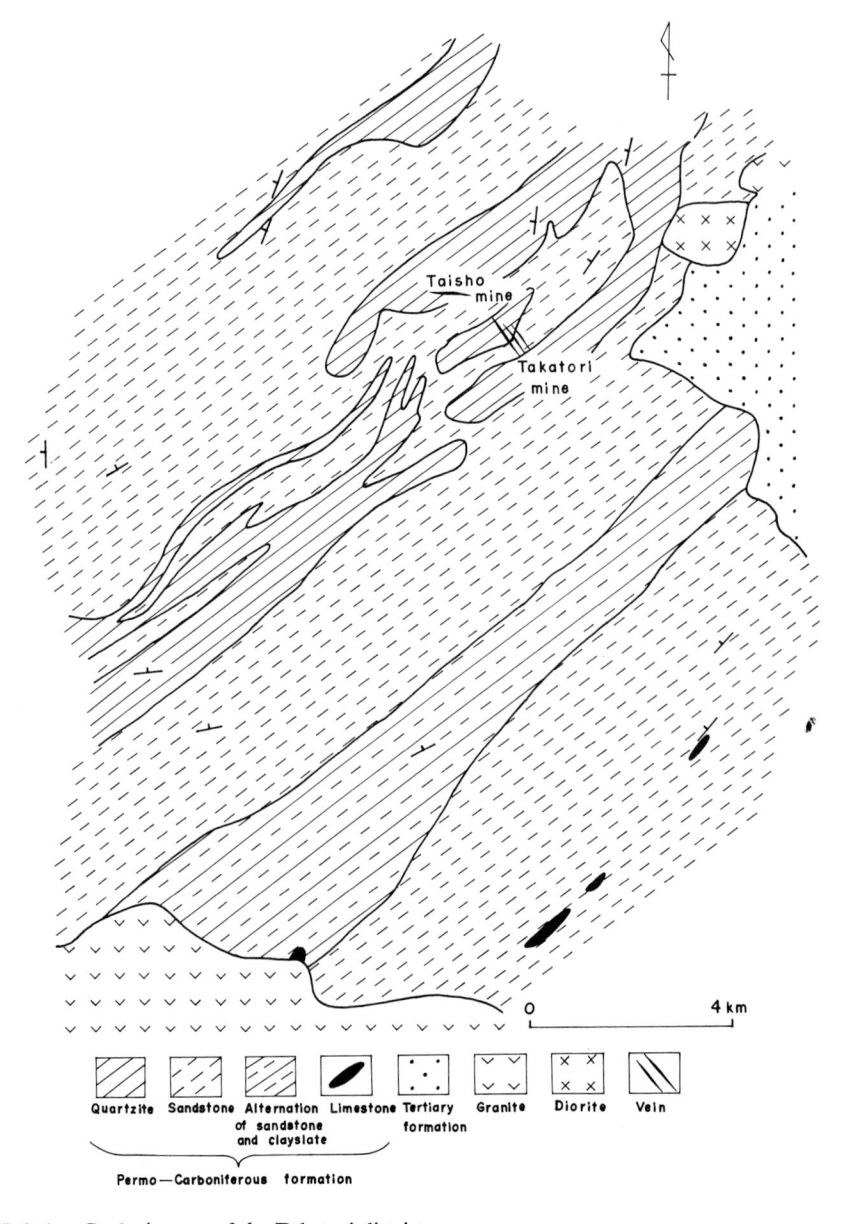

FIG. II-3-1. Geologic map of the Takatori district.

Part II: Vein-type Deposits

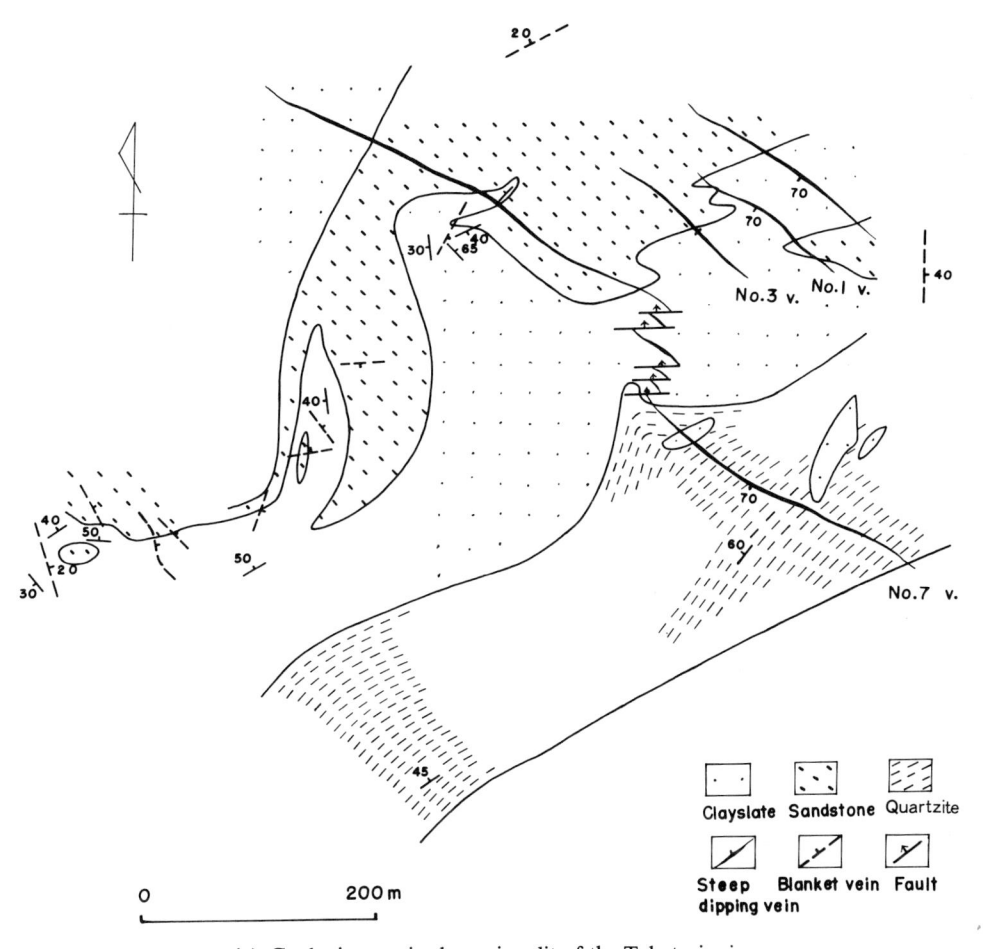

(a) Geologic map in the main adit of the Takatori mine.

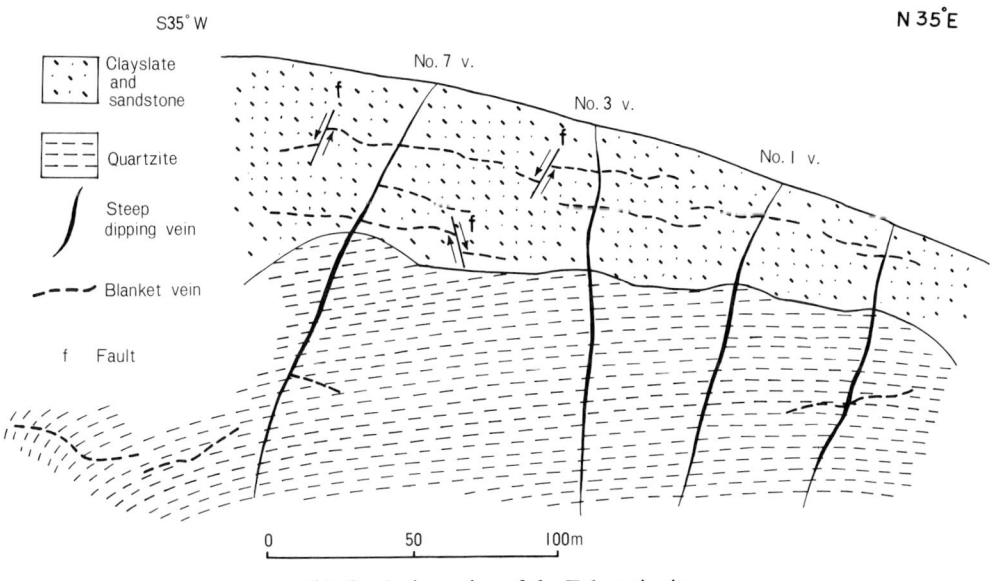

(b) Geologic section of the Takatori mine.

FIG. II–3–2.

rhotite, pyrite, sphalerite, galena, stannite, bismuthinite, kobellite, lithia mica (zinnwaldite), monazite, fluorite, topaz, rhodochrosite, calcite, sericite, chlorite and others.

The veins are classified by attitude into the following two types (Fig. II–3–2):

(a) Steeply dipping veins with strike N45°W, dipping 70°NE ∼ 70°SW. The maximum strike length of the veins is 700m, and the maximum width is 2m. They are mined as deep as 150m from the surface.

(b) Flatly-lying veins (blanket veins). They are well developed but small in scale. They dip less than 40°, and the directions of their strikes are variable. Even in a certain vein, strike and dip are irregular. But they are often intercalated between the strata. In some cases they link the veins belonging to (a) (Fig. II–3–2(b)).

Each fissure of the two types displays no displacement in either wall. The shapes of the hanging wall and foot wall of the fissures are apt to match. Generally the fissures have no fault clay or striation showing displacement in their walls. In Fig. II–3–3, it is seen that the two type veins (type (a) and type (b)) intersecting nearly at 90° show no displacement of each other. These facts indicate that these vein fissures are tension cracks.

The country rocks become whitish close to the veins, suffering sericitization and silicification. In some places, tourmaline, fluorite and topaz are recognized in country rocks.

The strikes of the veins of type (a) are nearly perpendicular to the general strike of the Paleozoic sediments. So, when writers admit a NW-SE lateral pressure as a cause of the

FIG. II–3–3. The photograph in the adit, showing the relation between type (a) and type (b) veins.

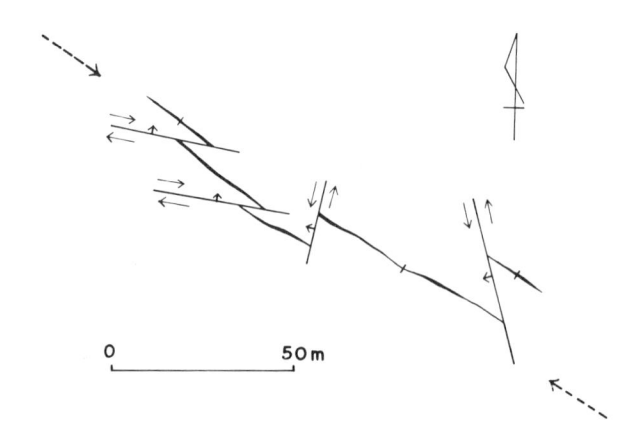

FIG. II–3–4. Conjugate faults in No. 3 vein. Broken arrows show the direction of lateral pressure.

folding of this area, the vein fissures of type (a) would correspond to the tension cracks. As shown in Fig. II–3–4, No. 3 vein belonging to type (a) is displaced by a group of conjugate faults striking nearly E-W and N-S. These faults are poorly mineralized, while No. 3 vein becomes wider forming ore shoots at the contact part with the above conjugated faults. It is certain that these faults were formed before the mineralization. The displacements by these faults may be explained by assuming that the wrench faults were due to the lateral pressure of the NW-SE direction which caused folding in this area and also formed the type (a) fissures. A similar set of faults is recognized also in No. 7 vein (Fig. II–3–2 (a)).

As already stated, the strike of the Paleozoic sedimentaries in this area is generally N45°E, with NW dip. But the writers recognize that the strikes of the sedimentaries gradually change close to veins of type (a), forming small dome structures (Fig. II–3–2). The dome structure of the sedimentaries is especially conspicuous in the western part of the mining area (Fig. II–3–2). It may be due to the pushing up of the roof by magmatic upheaval. Judging from the dome structure of the sedimentaries near the vein, a cupola of the cryptobatholith is inferred to occur underneath the area of the vein. It is generally believed that tin-bearing veins are apt to occur just above the cupola of the batholith (Emmons, 1933, 1935). So the formation of the veins of type (a) was related to the NW-SE lateral pressure as well as the pushing up of the acidic magma, as the writer will discuss in the next chapter.

The vein fissures of the flat-lying veins (type (b)) are tension cracks, as stated above. They are formed by the depression of the cupola of the cryptobatholith accompanied by magmatic solidification, as discussed in Chapter II–1.

II–4. Geologic Structure and Fluid Inclusion Study at the Ohtani and Kaneuchi Mines, Kyoto Prefecture (Imai *et al.*, 1972; Kim *et al.*, 1972)

H. Imai, M. Y. Kim, Y. Fujiki and S. Takenouchi

The Ohtani and Kaneuchi mines are situated in the western district of Kyoto City, which is occupied by the Permo-Carboniferous Tamba Formation and Late Cretaceous or Early Tertiary granitic rock. About 80 km northwest of Kyoto, there occurs a large mass of granitic rocks (Fig. II–15–2). According to Kawano and Ueda (1967a,b), the absolute age of this granitic mass is 55–68 m.y. by the K-Ar method, corresponding to the Late Cretaceous or Early Tertiary period. Shibata and Ishihara (1974) determined the absolute age of the granodiorite (biotite) from the Ohtani mine to be 93 ±3.7 m.y. by the K-Ar method. Ishizaka (1971) studied the absolute age of the Ibaragi granitic complex about 10 km to the south of the Ohtani mine by the Rb-Sr method applied to whole rock, feldspar and biotite (Fig. II–15–2). The result was 96 m.y., 76 m.y. and 76 m.y., respectively.

As is discussed in a later chapter, the absolute ages of the acidic rocks in the Ikuno-Akenobe mining area to the west of this district give a range of 55–80 m.y. All these districts belong to the Western Kinki Metallogenetic Province proposed by the writers, which they discuss in the chapter on xenothermal veins.

In the western part of this district, the Maizuru Folded zone runs from NE to SW, accompanied by basic plutonic rocks.

Geology and Ore Deposit of the Ohtani Mine

This area is composed of sandstone, clayslate and chert, belonging to the Tamba (Chichibu) Formation (Fig. II–4–1). The formation displays an anticlinal structure with an axis of E-W, where the granodiorite occurs in the core part (Fig. II–4–1). The granodiorite would have intruded nearly at the same time as the folding. In the area about 500 m from the contact with the granodiorite, the sedimentary rocks change into hornfels bearing cordierite.

The hypothermal tungsten veins exist in the granodiorite stock (Fig. II–4–2). Generally, they are arranged in parallel, striking NE-SW. Each vein is composed of small units. Each unit extends within tens of meters, and is classified into two types. Namely, the one strikes N40°E, dipping 75°–85° NW, and the other strikes N20°E, dipping 75°-85° NW. The

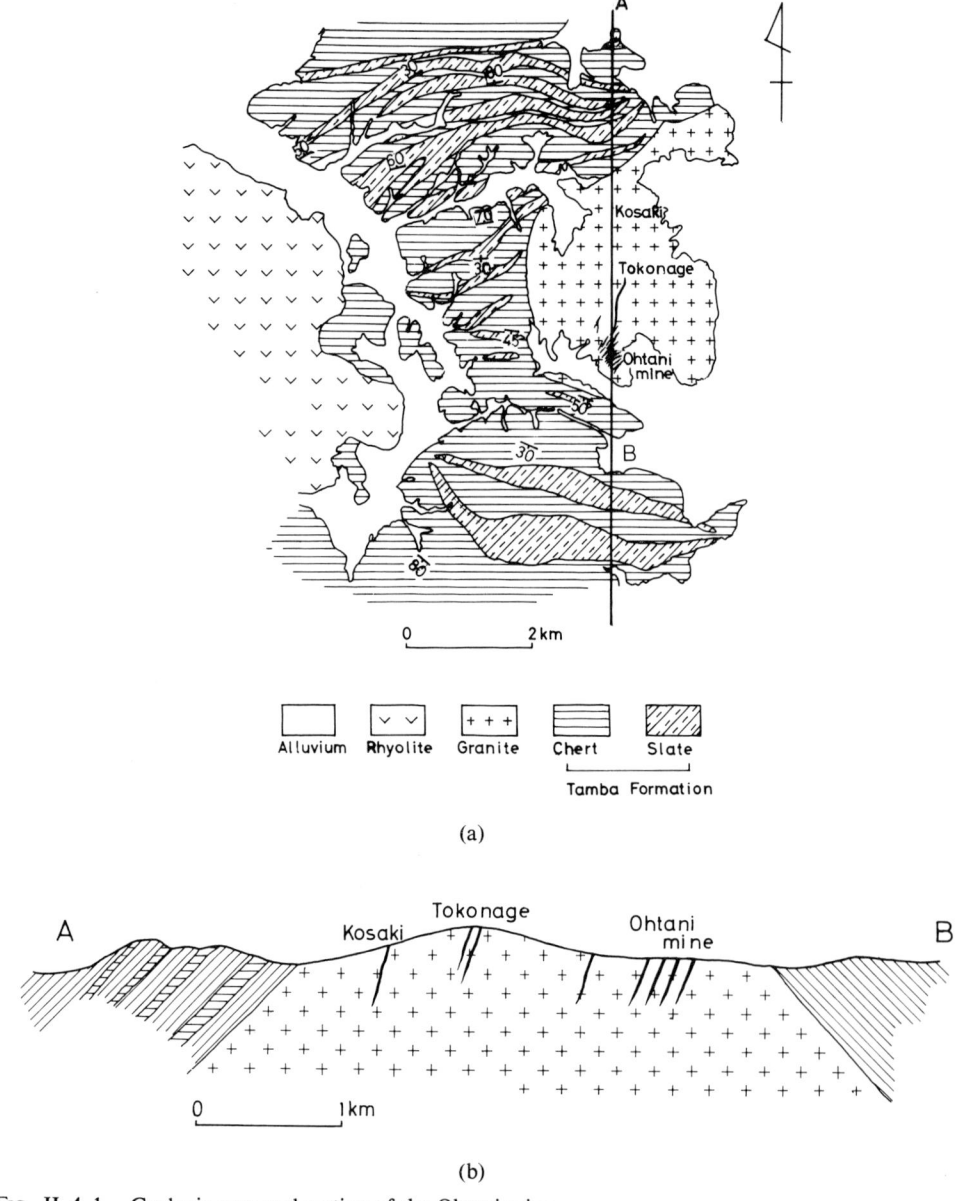

B. Pegmatitic, Hypothermal and Mesothermal Veins

FIG. II-4-1. Geologic map and section of the Ohtani mine.

vein width of both types is 50–100 cm. The fissure of the former type is wrench fault, having the fault gouge along the boundary plane with the country rock. It has horizontal striation on the slickenside of the boundary wall. In some places the unit is linked with the next one by a small connecting veinlet, or is transferred into the next by curved shingles (Fig. II–4–3(a)). The fissure of the latter is tension crack without fault gouge and striation. In some cases, the cracks are arranged *en echélon* (Fig. II–4–3(b)), stretching N40°E as a whole. In a certain part, the aplite dike is displaced about 50 cm along the fissure of the latter (Fig. II–4–3(c)). This fissure is classified as a tension crack, having a small component of displacement.

The two types of fissures are combined with each other, forming one vein.

FIG. II-4-2. Vein pattern of the Ohtani mine.   V: vein

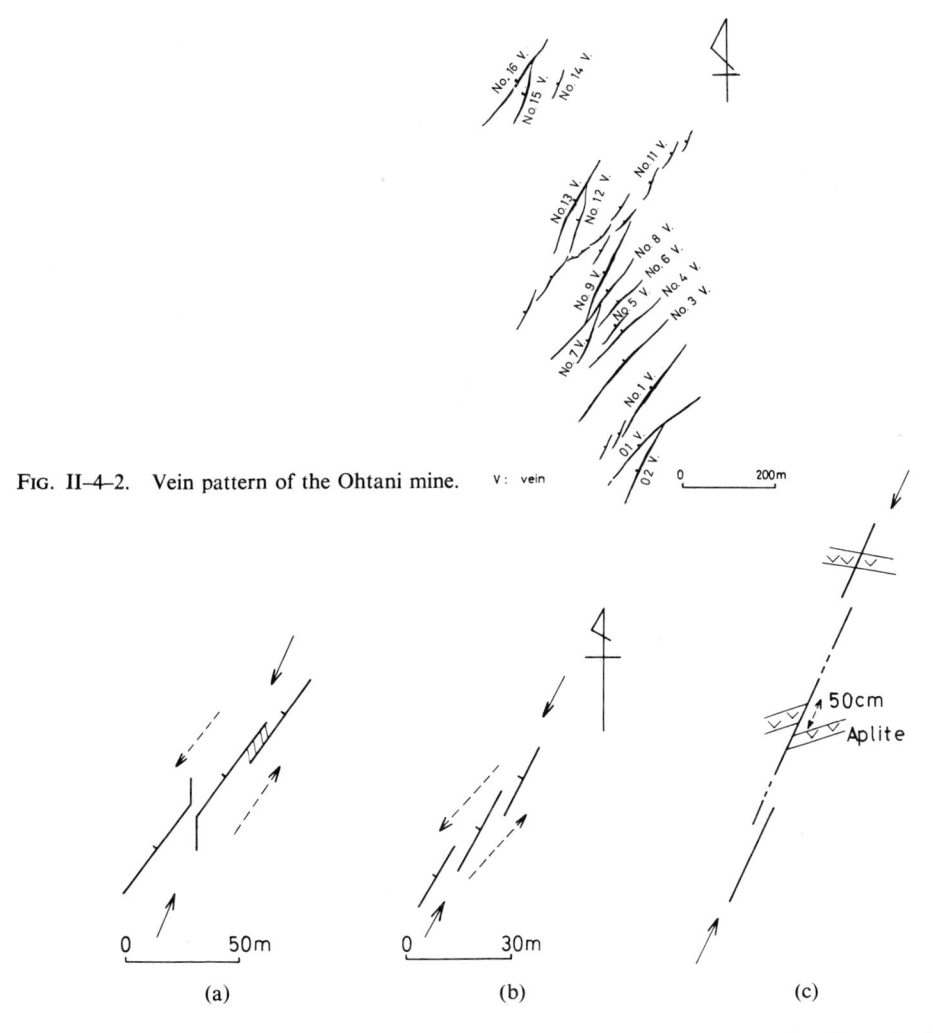

FIG. II-4-3. Fissure formation at the Ohtani mine. Solid line arrow: compressive force, broken line arrow: shearing stress.

These fissures would have been produced by the lateral pressure from N20°E-S20°W, combined with upheaval pressure accompanied by the intrusion of granodiorite, as discussed later. The direction of the lateral pressure coincides generally with that which caused the folded structure in this district.

The host rock of the veins, i.e., granodiorite, suffered greisenization about 50–100 cm along with the veins. Mineralogically, the greisen is composed of quartz, feldspars (orthoclase, microcline and plagioclase), muscovite, fluorite, tourmaline, apatite, allanite, zircon, etc., with sporadic occurrence of pyrite and arsenopyrite. Locally it suffered chloritization in the later stage. According to Shibata and Ishihara (1974), the absolute ages of the muscovite from the greisen envelope and of the muscovite from the vein calculated by the K-Ar method are 90 ± 3.6 m.y. and 91.4 ± 3.7 m.y., respectively.

The vein-forming minerals are quartz and scheelite, accompanied by cassiterite, pyrrhotite, arsenopyrite, pyrite, chalcopyrite, sphalerite, cubanite, mackinawite, stannite, bismuthinite, native bismuth, muscovite, fluorite, calcite, etc.

It is characteristic that wolframite is lacking. Scheelite is the only tungsten ore mineral in this mine. The muscovite occurs along the boundary of the vein with the host rock.

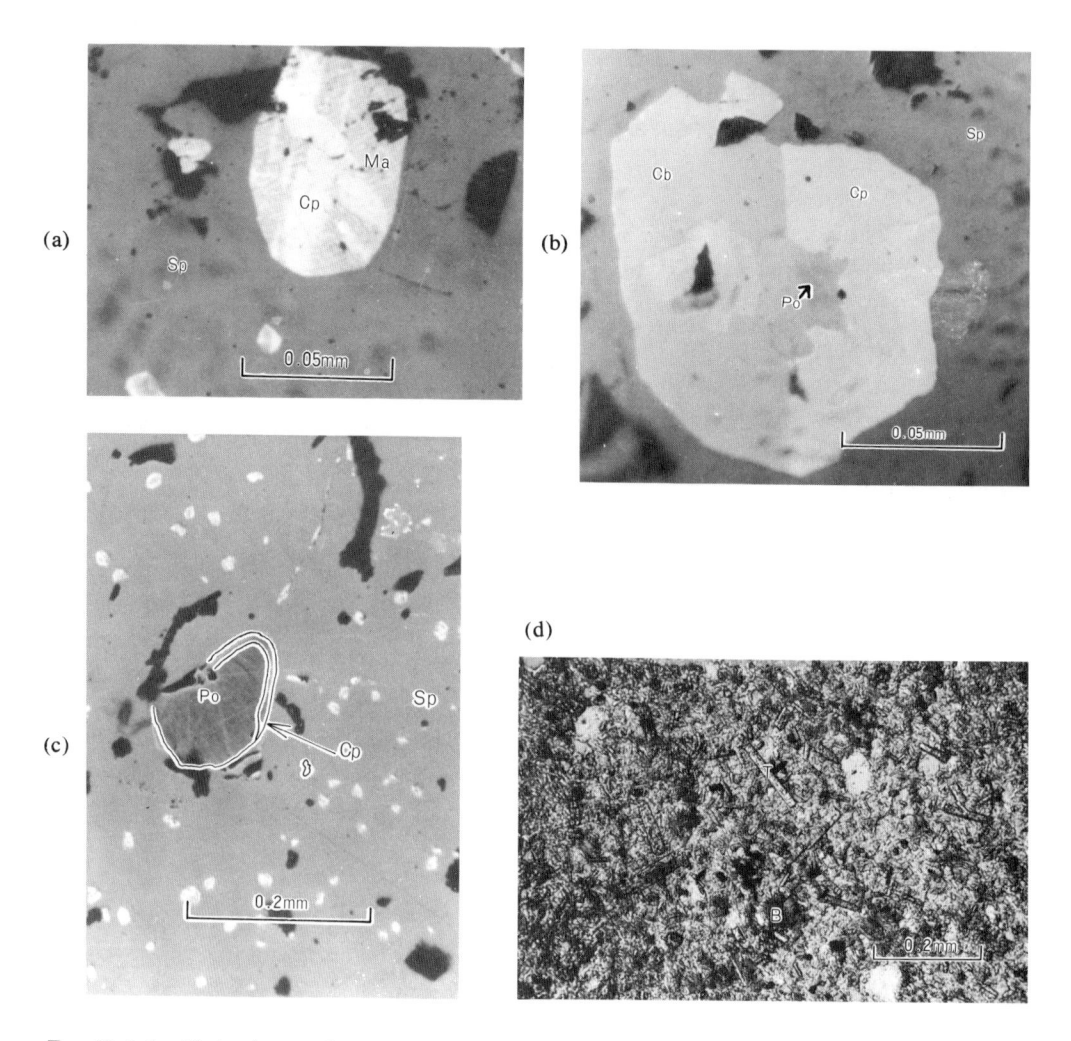

Fig. II-4-4. Photomicrographs.
  (a) Polished section (Ohtani mine). One nicol. Exsolution texture in two stages. Sp: sphalerite, Cp: chalcopyrite, Ma: mackinawite.
  (b) Polished section (Ohtani mine). One nicol. Exsolution texture in three stages. Po: pyrrhotite, Cb: cubanite.
  (c) Polished section (Ohtani mine). One nicol. Cell structure of pyrrhotite. Exsolution texture in two stages.
  (d) Thin section (Kaneuchi mine). One nicol. Tourmaline hornfels. B: biotite, T: tourmaline.

It is recognized under the microscope that drop-like aggregates composed of chalcopyrite and mackinawite (Fig. II-4-4(a)) and of pyrrhotite, chalcopyrite and cubanite (Fig. II-4-4(b)) occur in the sphalerite matrix. They are the results of exsolution, successively occurring in two or three different stages. The same would be true in a drop of pyrrhotite with membrane-like chalcopyrite film as a cell structure in a sphalerite matrix (Fig. II-4-4(c)). Also, drop-like stannite is observed in the sphalerite matrix, perhaps due to exsolution.

The basalt dikes in granodiorite penetrate the veins.

Geology and Ore Deposit of the Kaneuchi Mine

As already stated, the Tamba Formation strikes nearly EW, displaying repeated folded structure. The formation is composed of slate, sandstone, chert and schalstein. In this area, the formation shows a dome-like structure, and underwent thermal metamorphism, which might reflect the existence of the underlying cupola of the granitic batholith (Fig. II-4-5). By the effect of thermal metamorphism, biotite and tourmaline were produced in the sedimentary rocks. The area which underwent thermal metamorphism nearly corresponds to the area displaying the dome structure (Fig. II-4-5), which is about 2 km in diameter.

Tourmaline occurs sporadically in the area of dome structure suffering thermal metamorphism (Fig. II-4-4(d)).

The veins being worked at the present time strike N10°-30°E, dipping 60°-70°E. They occur at the west flank of the dome structure. The veins are divided into three groups, i.e., from east to west, Gessei (eastern) group, Suei (middle) group and Yomei (western) group (Fig. II-4-6). The maximum strike length is 500 m, and the mean width of the veins is 30-50 cm. They are tension cracks, revealing no displacement of either wall along the vein fissures(Fig. II-4-5(b)).

In the central part of the dome structure, there exist vertical veins, also striking N10°E. In the eastern flank of the dome, veins having the same strike dip to the west (Fig. II-4-5).

In the central part of the dome, the lineation attitude observed in the sedimentary rocks is E-W and horizontal, while in the western flank it plunges to the west (Fig. II-4-7) and in the eastern flank it plunges to the east.

The vein fissures were formed by lateral pressure from N10°-30°E which caused the folded structure in this district, accompanied by upheaval pressure due to the doming-up of cryptobatholith. The strike direction of the veins depends on the direction of the lateral pressure, but the attitude of the dip would be related to the stress from underneath caused by the upheaval pressure of the magma. Doming up of the cryptobatholith exerts pressure at right angles to the boundary plane of the magma chamber with the surrounding rocks (Imai et al., 1972). These pressures correspond to two principal stresses, i.e., maximum principal stress (σ_1) and intermediate principal stress (σ_2) (Fig. II-4-7). The vein fissures belong to the tension crack perpendicular to the boundary plane between the cryptobatholith of the granitic rock and the surrounding Tamba Formation. The tension cracks include the directions of maximum and intermediate principal stresses, i.e., they are perpendicular to the minimum principal stress (σ_3) (tensile stress). The formation mechanics of the vein fissures in the Ohtani mine as described above would be similar to those in this mine.

They are also coincident with the veins described in the Takatori and Kuga mines.

Also in the southern part of the Kaneuchi mining area, there exist veins striking from N60°E to E-W, with a gentle dip to N (Fig. II-4-5). The veins of the Funai and Takao adits are of this type. The fissures of these veins are tension cracks. The same kind of flat-lying veins exist in the Takatori mine, as described before. They would have been produced by the subsidence or depression due to the consolidation of the magma, as discussed in Chapter II-1.

The host rock surrounding the veins was subject to sericitization, but not intensively. Tourmaline is found sporadically throughout the hornfels area, and more often in the vicinity of the veins.

The vein-forming minerals are wolframite,* scheelite, cassiterite, molybdenite, arsenopy-

---

* The values Fe/Fe + Mn in the wolframites from this mine are 0.43–0.67, mostly 0.55–0.65, from electron microprobe analyses, and from d(200) values by X-ray diffraction charts, using the data of Berman and Cambell (1957) and Sasaki (1959).

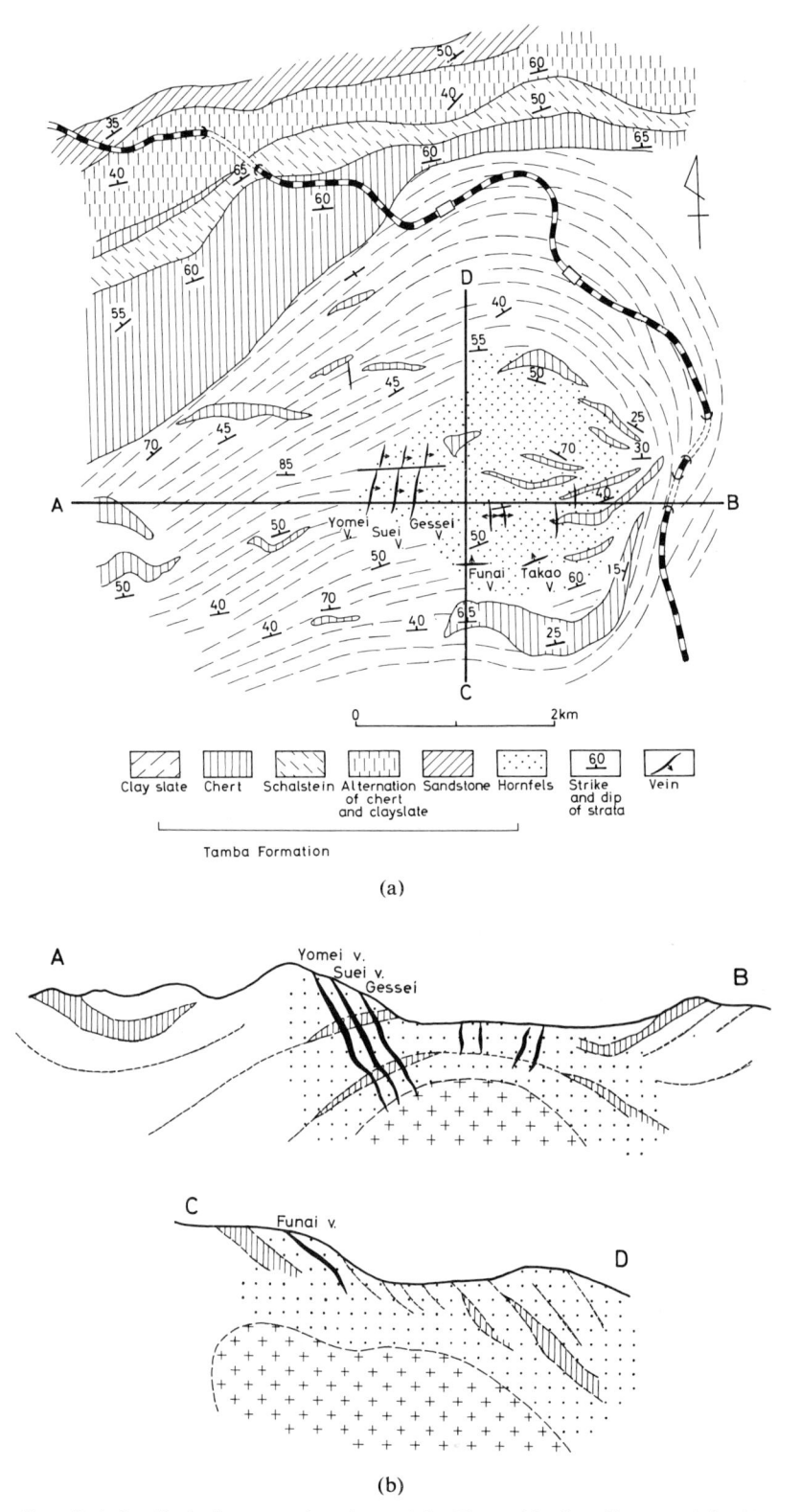

B. Pegmatitic, Hypothermal and Mesothermal Veins

Fig. II-4-5. Geologic map and sections of the Kaneuchi mine. Cross mark in (b) represents the acid intrusive.

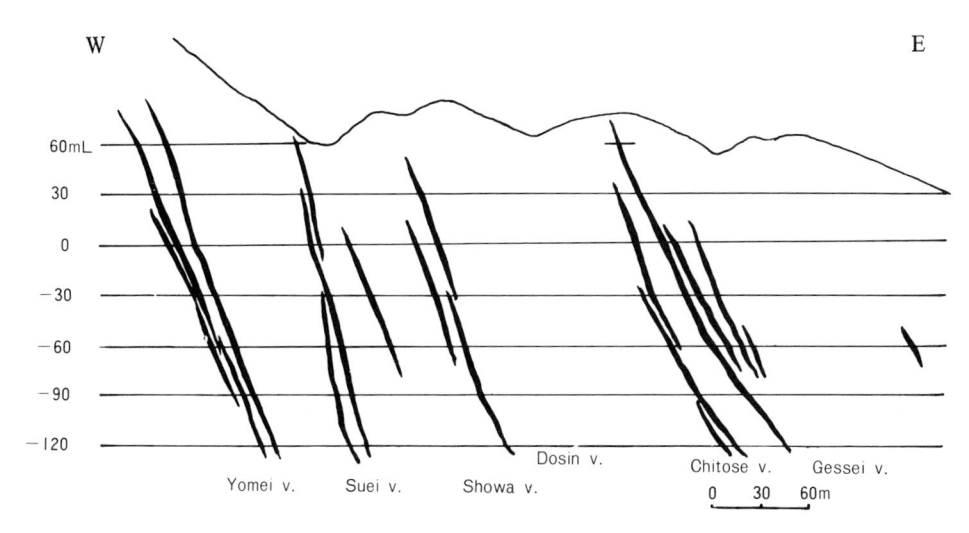

FIG. II–4–6. E-W vertical section of the Kaneuchi mine.

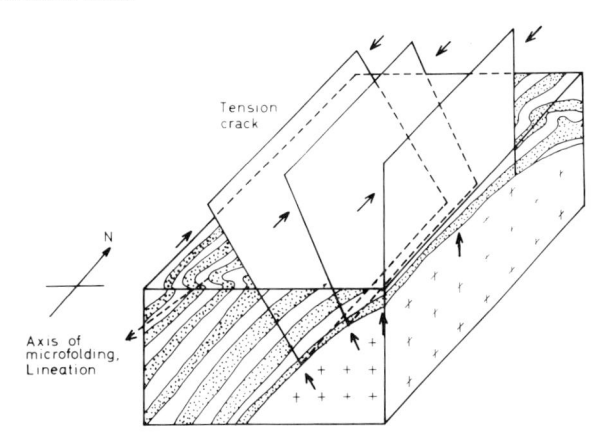

FIG. II–4–7. Tension cracks due to the maximum (σ_1) and intermediate (σ_2) principal stresses. Lateral pressure: σ_1. Upheaval pressure of magma: σ_2.

rite, pyrrhotite, pyrite, mackinawite, chalcopyrite, sphalerite, galena, stannite, bismuthinite, native bismuth, etc., accompanied by quartz, orthoclase, microcline, muscovite, apatite, beryl, tourmaline, etc.

The veins are of pegmatitic or hypothermal type. Wolframite, muscovite or beryl occurs adjacent to both hanging and foot walls of the veins, while the sulfide minerals occupy the median part of the veins.

Under the microscope, drop-like chalcopyrite, stannite and pyrrhotite are observed in sphalerite as exsolution products. At the 60m L. in the adits of the mining area (Fig. II–4–6), the Gessei vein of the eastern group is rich in scheelite and cassiterite, while the Yomei vein of the western group is rich in pyrite and chalcopyrite. The Suei vein in the middle group shows an intermediate character rich in pyrrhotite. This is due to the zonal distribution of minerals in this area. This fact reflects the geologic occurrence of the granitic cryptobatholith to the east of the mining area (Fig. II–4–5). It is also coincident with the results of the decrepitation experiment, as described below.

The porphyrite dikes penetrate the veins in the same way as in the Ohtani mine.

Yamaoka and Ueda (1974) determined that the absolute ages of orthoclase and muscovite from the veins at the Kaneuchi mine are 73 and 93 m.y., respectively, by the K-Ar method. Shibata and Ishihara (1974) obtained a similar value of 91.2 m.y. by the K-Ar method for vein muscovite from this mine.

Fluid Inclusion Studies

The writers (1972) studied the filling temperatures and NaCl equivalent concentrations of the fluid inclusions in minerals from the Ohtani and Kaneuchi mines by means of the heating-stage- and freezing (cooling)-stage-microscope and decrepitation method. The minerals studied were quartz, cassiterite, wolframite and scheelite. The quartz in the vein is generally milky in appearance to the naked eye. The inclusions in this quartz are very small, i.e., smaller than 10 microns, so it is almost impossible to study the filling and freezing temperatures of the fluid inclusions in this kind of quartz.

However, in this kind of quartz, transparent quartz crystal is sporadically seen. In the transparent quartz, comparatively large-sized fluid inclusions exist, generally below 50 microns and up to 200 microns. This transparent quartz would belong to later-stage mineralization, perhaps to post-mineralization of tungsten and tin minerals.

Most of the fluid inclusions are composed of liquid (solution) and gas (vapor), i.e., they are two-phase inclusions. But, in rarer cases, three-phase inclusions composed of liquid (solution), liquid (CO_2 liquid) and gas (vapor) are found in the drusy quartz. The same is true in the Takatori tungsten mine (Takenouchi and Imai, 1971). Also, in the Taishu mine, the same kind of three-phase fluid inclusion is recognized in quartz occurring at the marginal part of the deposit.

Sushchevskaya and Ivanova (1967) reported this kind of three-phase inclusion in a tungsten deposit in the Transbaikal region. According to Takenouchi (1971a), CO_2 and NaCl concentrations in the three-phase inclusions from the Ohtani mine are 11–20 wt. % and 0.9–3.5 wt. %, respectively. The filling temperatures are 220°–280°C. He assumed that the external pressure would be more than 250 times the atmospheric pressure during the ascending of the fluid when CO_2 boils (Takenouchi, 1971a).

Filling temperatures of the fluid inclusions: The filling temperatures of the fluid inclusions of some samples from these two deposits are shown in Figs. II-4-8 and -9. The data shown in each line represent the results of the measurement of a hand specimen. It is recognized that in some cases the filling temperatures of a hand specimen have a range of up to 100° C.

As to the Ohtani mine, the filling temperatures are 375°–225°C in quartz, 337°–262° C in scheelite and 345°–297° C in cassiterite. As shown in this figure, the maximum filling temperature of the inclusion in quartz, i.e., 375° C, is exceptionally high. The temperatures are generally below 350° C and predominantly below 300° C.

The filling temperatures of the quartz in greisen and in granodiorite just adjacent to the greisen envelope are 336°–312° C, and 355°–287° C, respectively (Fig. II-4-8). These values are nearly the same as or a little higher than the filling temperatures in the vein minerals. These inclusions would be formed by the vein-forming fluid. The filling temperatures of the liquid inclusions in topaz, quartz, fluorite and albite from the greisen in Kamennye Mogily granite were 430°–320° C (Zatsikha, 1968). The decrepitation temperatures of the greisen accompanied by ferberite veins at Maidental were 700°–420° C (Elinson et al., 1969).

As to the Kaneuchi mine, the filling temperatures are 308°–231° C in quartz, 318°–276° C in scheelite and 337°–286° C in wolframite. The filling temperatures of both mines are similar, but those from the Ohtani mine are somewhat higher than those from the Kaneuchi mine.

As shown in both figures, the filling temperatures in quartz are generally lower than those in cassiterite, scheelite and wolframite. This is due to the fact that the quartz in which the filling temperatures of the inclusions were measurable belongs to a later stage than the tungsten-tin mineralization, as stated above.

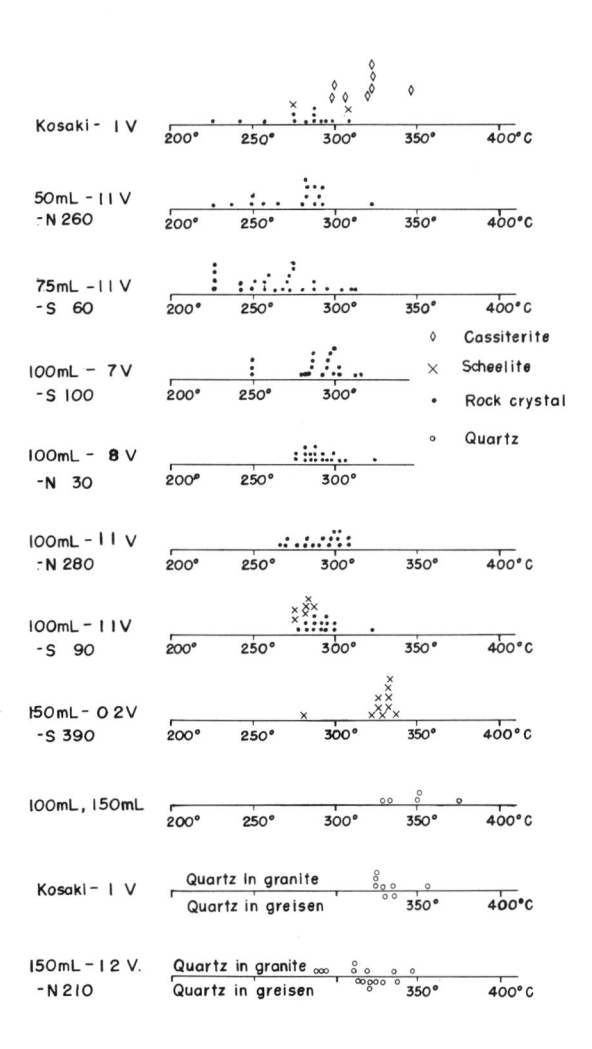

Fig. II-4-8. Filling temperatures of the fluid inclusions in the minerals from the Ohtani mine.

NaCl equivalent concentrations of the fluid inclusions: The NaCl equivalent concentrations of the inclusions in both mines are shown in Figs. II-4-10 and -11. As in the filling temperatures, the data in each line represent the results for a hand specimen.

At the Ohtani mine, the NaCl equivalent concentrations of the inclusions in vein minerals are 7.4–4.0 wt.% in quartz, 8.2–6.1 wt.% in scheelite and 8.7–6.1 wt.% in cassiterite. At the Kaneuchi mine, they are 8.2–3.7 wt.% in quartz, 8.6–6.4 wt.% in scheelite and 8.4–8.1 wt.% in wolframite. The results show that there is no difference between the two deposits.

The relation between filling temperature and NaCl equivalent concentration, which were measurable in the same inclusion, is shown in Figs. II-4-12 and -13. These values show that the two have a positive relation, i.e., the higher the filling temperatures, the higher the NaCl equivalent concentrations. This would be due to the dilution of the ore-forming fluid by ground water or connate water in the course of ascending. In the case of the Kaneuchi mine, the NaCl equivalent concentration rises abruptly at 280°–290° C, which would

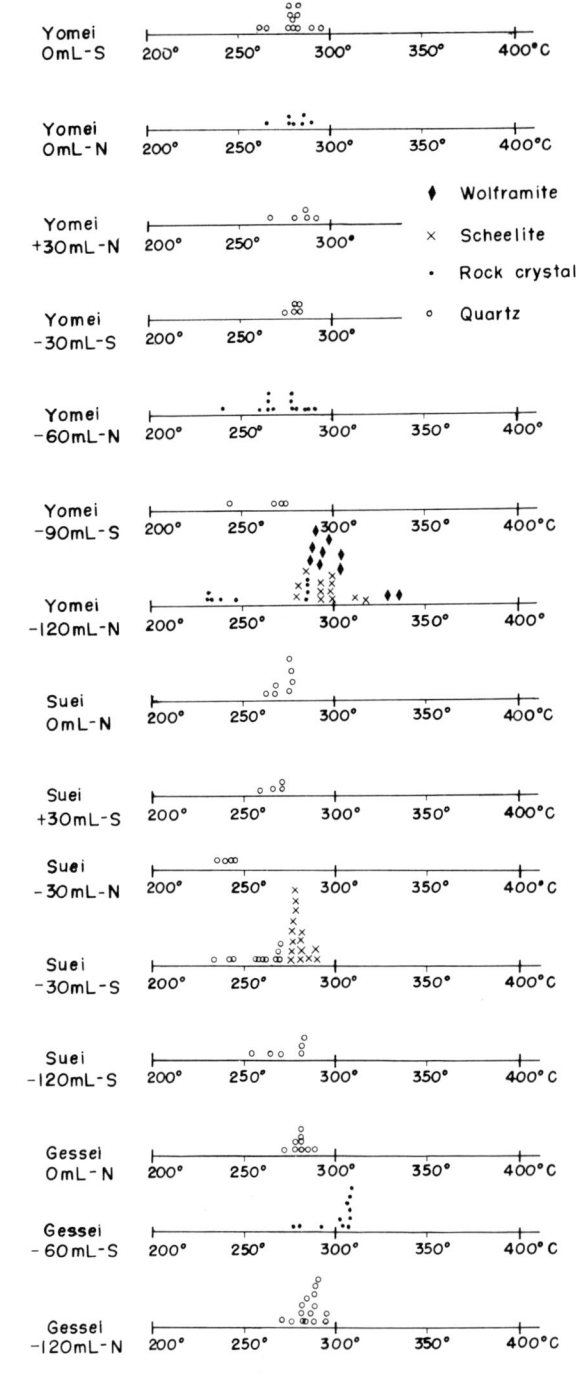

B. Pegmatitic, Hypothermal and Mesothermal Veins

FIG. II-4-9. Filling temperatures of the fluid inclusions in the minerals from the Kaneuchi mine.

Part II: Vein-type Deposits

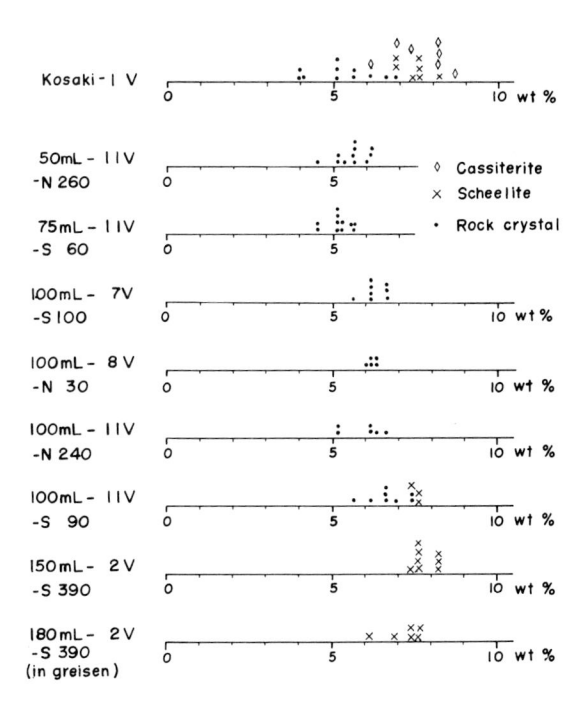

Fig. II–4–10. NaCl equivalent concentrations of the fluid inclusions in the minerals from the Ohtani mine.

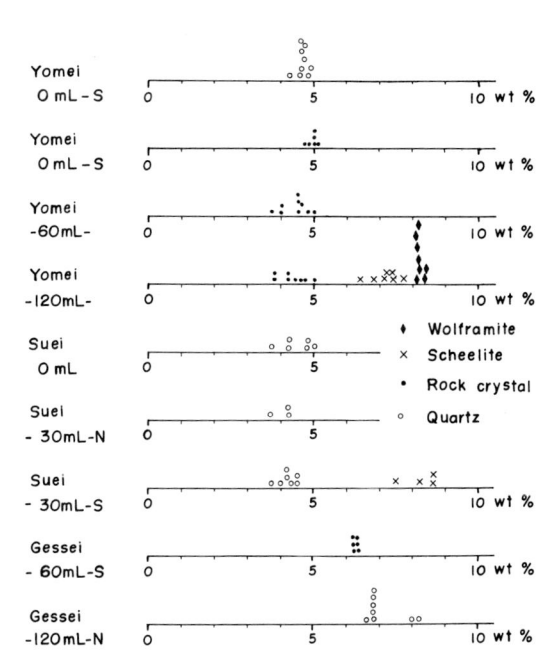

Fig. II–4–11. NaCl equivalent cocentrations of the fluid inclusions in the minerals from the Kaneuchi mine.

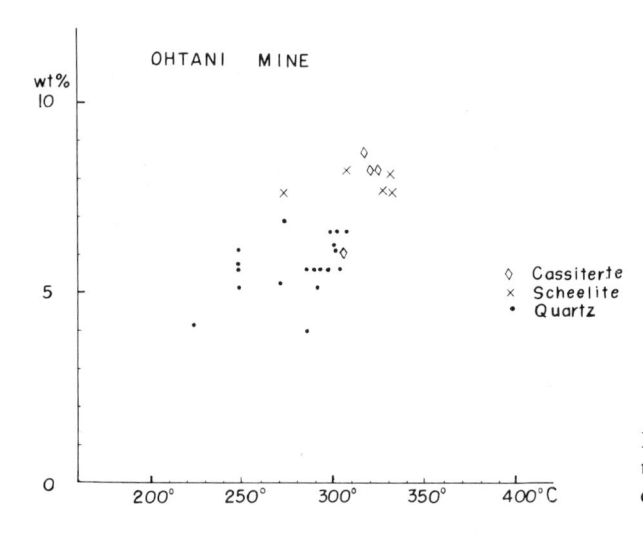

B. Pegmatitic, Hypothermal and Mesothermal Veins

FIG. II–4–12. Relation between filling temperatures and NaCl equivalent concentrations at the Ohtani mine.

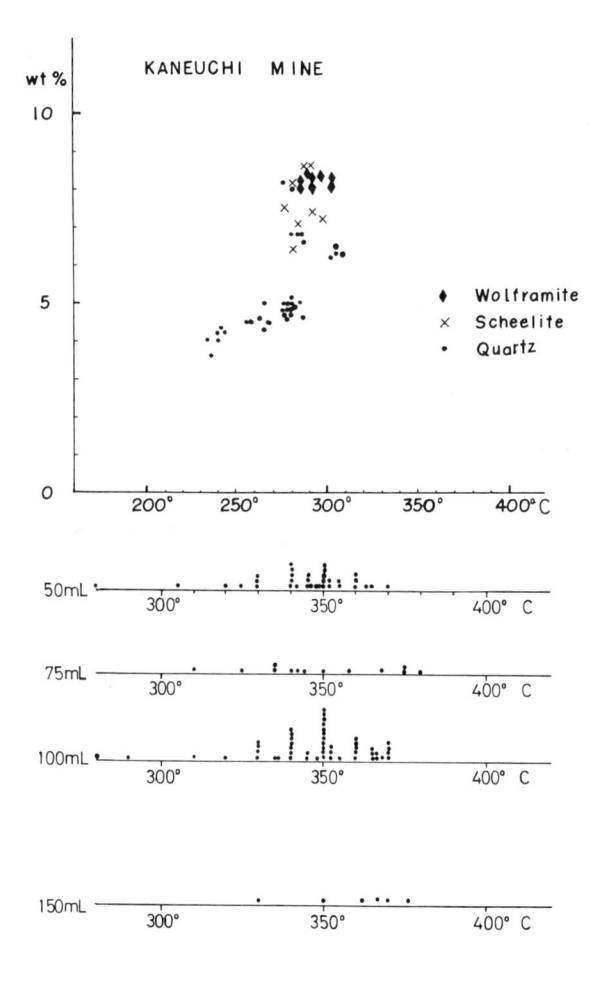

FIG. II–4–13. Relation between filling temperatures and NaCl equivalent concentrations at the Kaneuchi mine.

FIG. II–4–14. Decrepitation temperatures of the minerals from the Ohtani mine.

be interpreted to be due to mixing of the heated groundwater or connate water. The results of the decrepitation experiments on 100 samples of quartz from the Ohtani mine are shown in Fig. II–4–14.

Decrepitation temperatures: The decrepitation temperatures of the quartz samples from the Ohtani mine range from 382°C to 280°C. This range is 30°–50°C higher than the filling temperatures measured by means of a heating-stage microscope. It might be due to "overshoot," as discussed by Peach (1951) and Takenouchi (1962). Some inclusions in the quartz of the later stage deposition contain CO_2 liquid. They might show smaller "overshoot" than those with no CO_2 liquid.

This also might be considered to be due to the stage difference of the deposition of quartz. As discussed in the previous section, the quartz in which filling temperature was measured by means of the heating-stage microscope would belong to a later stage than the mineralization of tungsten and tin, while the quartz for decrepitation experiments necessarily includes quartz accompanying the mineralization of tungsten and tin.

The decrepitation temperatures of the quartz samples from +60 mL to −120 mL in the Gessei vein of the Kaneuchi mine were 376°–275° C. Those from +60 mL to −120 mL in the Yomei vein of the same mine were 366°–255°C. The number of measurements was 124. In both veins, the decrepitation temperatures increase in a deeper and southward direction, which intersects at right angles with the plunge of the ore-shoots of both veins. This temperature gradient would indicate the ascending paths of the ore-forming fluid. The decrepitation temperatures of the Gessei vein are somewhat higher than those of the Yomei vein. This fact is consistent with the geologic occurrence of the acidic plutonic emplacement. Hayakawa *et al.* (1975) also recognized that the filling temperatures measured by means of a heating stage microscope increase from the veins of the western group towards those of the eastern group. The decrepitation temperatures of the Kaneuchi mine are somewhat lower than those of the Ohtani mine, which reflects the geologic occurrences in the two mines.

Discussions and Conclusions

(1) The vein-type deposits of the Ohtani and Kaneuchi mines are geologically controlled by the anticlinal (or domed) structure due to the regional-scale folding and the synchronous intrusion of acidic plutonic rock at the core part of the structure.

(2) The fissures of both deposits were formed by the lateral pressure which caused the regional-scale folded structure, accompanied by upheaval pressure due to doming up of the batholith or cryptobatholith. These pressures correspond to two principal stresses, i.e., maximum principal stress (σ_1) and intermediate principal stress (σ_2).

(3) The deposits are genetically related to Late Cretaceous granitic activity. According to Shibata and Ishihara (1974), the age of the biotite from the granodiorite in the Ohtani mine is 93 m.y. The muscovite separated from the greisen envelope in this mine has an age of 90 m.y. Vein muscovite from the same mine gives an age of 91.4 m.y. Similar values of 91.2 and 93 m.y. are obtained for vein muscovite from the Kaneuchi mine.

(4) It is characteristic that the tungsten mineral in the Ohtani mine is only scheelite, while in the Kaneuchi mine both wolframite and scheelite are produced. Generally speaking, the tungsten mineral in the vein-type deposit is wolframite. But the vein in the limestone area contains scheelite, instead of wolframite, as described in the chapter on the Kuga mine. The writers interpret this to mean that the limestone of the Tamba Formation exists underneath the present granodiorite mass.

According to Takimoto (1944) and Ishihara (1971), wolframite veins are genetically related to granite, and scheelite veins to granodiorite. The writers interpret this to mean that granodiorite would be formed as the result of contamination of granitic magma with limestone.

In the constituent quartz of greisen and granodiorite in the Ohtani mine, the writers observed fluid inclusions composed of gas-liquid phases, which have nearly the same filling temperatures and the same NaCl equivalent concentrations as those in the vein quartz. Imai, et al. (1971) found inclusions of high filling temperatures and high NaCl equivalent concentrations in the constituent quartz of the granite at the Taishu mine. They recognized that this granite would be the source of the ore-forming fluids which are represented by these inclusions. Examples of original ore-forming fluids having high NaCl equivalent concentrations were discussed by Roedder (1971a) at some porphyry copper deposits in the United States and Canada, by Kelly and Turneaure (1970) at tin deposits in Bolivia, by Stollery et al. (1971) at the Providencia igneous mass in the Zacatecas lead and zinc mine in Mexico, and by Sillitoe and Sawkins (1971) at some porphyry copper deposits in Chile.

From these facts, the writers conclude that the original ore-forming fluid of the Ohtani mine was concentrated in the granitic magma underneath the present country rock, i.e., granodiorite, which would have been formed by the contamination of the granitic magma with the calcareous rock at the time of intrusion. Scheelite would have been formed by the calcareous rock through which the ore-forming fluid flowed upwards. So the present country rock of the veins, granodiorite, was not the direct source of the mineralizing fluid.

(5) The vein minerals in the Ohtani mine are scheelite, cassiterite, pyrrhotite, arsenopyrite, pyrite, chalcopyrite, sphalerite, cubanite, mackinawite, stannite, bismuthinite, native bismuth, etc., accompanied by quartz, muscovite, fluorite and calcite. The exsolution texture of chalcopyrite, pyrrhotite and mackinawite in the sphalerite matrix can be observed under the microscope. They might exsolve in two or three different stages. The drop-like stannite in sphalerite would be due to the exsolution. The granodiorite underwent greisenization around the veins. The deposit of the Ohtani mine is the hypothermal vein type. The vein minerals in the Kaneuchi mine are wolframite, scheelite, cassiterite, molybdenite, arsenopyrite, pyrrhotite, pyrite, mackinawite, chalcopyrite, sphalerite, galena, stannite, bismuthinite, native bismuth, etc., accompanied by quartz, muscovite, orthoclase, microcline, tourmaline, apatite, beryl, etc., and are of the pegmatitic or hypothermal vein type.

(6) The filling temperatures of the vein minerals under the heating-stage microscope are 375°–225°C in the Ohtani mine and 337°–231°C in the Kaneuchi mine. The decrepitation temperatures are 382°–280°C in the Ohtani mine and 376°–255°C in the Kaneuchi mine.

(7) The NaCl equivalent concentrations in the inclusions are 8.7–4.0 wt% in the Ohtani mine and 8.6–3.7 wt% in the Kaneuchi mine. The filling temperatures increase with the increase of the NaCl equivalent concentrations. This would be due to the dilution of the ore-forming fluid by underground water or connate water at the time it was ascending.

II–5. Komori Mine, Kyoto Prefecture (Imai and Fujiki, 1963; Fujiki, 1963, 1964a,b)

H. Imai and Y. Fujiki

This mine is situated about 60 km west of Kyoto. Geologically, it occurs in the Maizuru Folded zone, and from the point of ore genesis, in the Western Kinki Metallogenetic Province which the writers discuss in Chapter II–15 (Fig. II–15–2).

This region is geologically composed of the Carboniferous-Triassic formation, intermediate-basic plutonic rocks (consisting of diorite and gabbro), ultrabasic rocks (consisting of peridotite, serpentinite and pyroxenite) and younger granite (Fig. II–5–1). The basic and ultrabasic rocks, often associated with each other, are called Yakuno basic intrusives. The

Part II: Vein-type Deposits

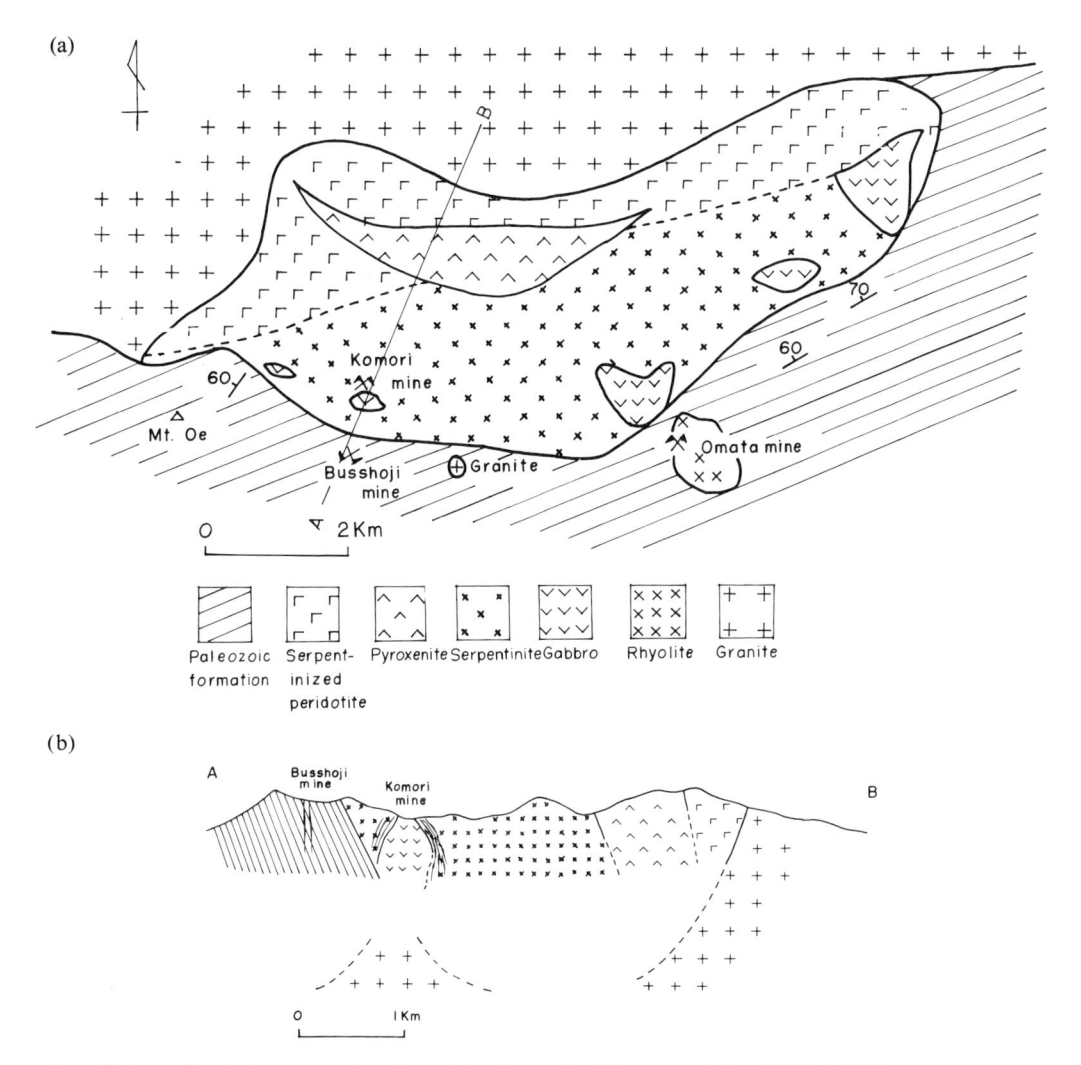

FIG. II-5-1. Geologic map and section of the Komori mining area. (after Fujiki 1964a) Notice the sheared zones around the gabbro in (b).

ultrabasics intrude the Carboniferous formation and are penetrated by the basic rocks. They occur in the Maizuru Folded zone (Fig. II-15-2). They are closely related to the folded structure. In this folded zone, metamorphic rocks such as biotite schist and biotite amphibolite were produced, related to the folding. The absolute ages of the metamorphic rocks were determined to be 215–217 ± 17 m.y. by the K-Ar method (Shibata and Igi, 1966) and 332 m.y., 306 m.y. and 269 m.y. by the Rb-Sr method (Hayase and Ishizaka, 1967).

They are intruded by the younger granite, the so-called Miyazu granite (Fig. II-5-1), of the Late Cretaceous epoch. The ore deposit of the Komori mine is related genetically to this granite.

The ore deposits are classified into two types, the mesothermal vein type and the impregnated type. The vein-type deposits occur along sheared zones in the serpentinite around or near the contacts with the gabbro and diorite stocks intruding the serpentinite (Fig. II-5-1). Due to lateral pressure, the sheared zones were formed in the serpentinite around the rigid gabbro or diorite stocks. Consequently, the vein-type deposits surround the gabbro or diorite mass. The

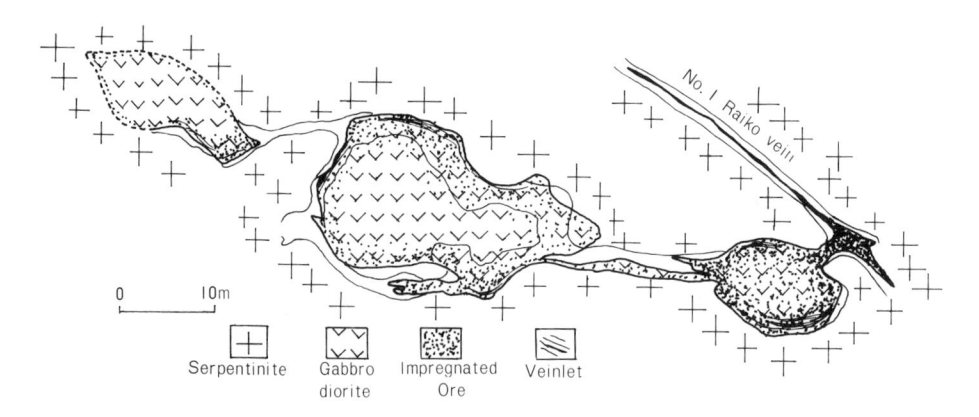

B. Pegmatitic, Hypothermal and Mesothermal Veins

FIG. II-5-2. Impregnated ore bodies in the dioritic rocks of the Komori mine (after Fujiki, 1964a).

direction of displacement of a sheared zone at a site depends on the shape of the gabbro or diorite mass of that site. So, some sheared zones are wrench faults and others are thrust faults.

The impregnated-type deposits exist in the gabbro or diorite stocks in serpentinite, especially in the marginal parts of the stocks (Fig. II-5-2). Some ore deposits of this type occur in the onion cracks in the marginal parts of the stocks, which were formed at the time of magma solidification.

The granite is most widely developed in the northern part of the mining area, but a small outcrop is found to the east of the Komori mine. It is inferred that a cupola of cryptobatholithic granite exists underneath the deposit of the Komori mine (Fig. II-5-1).

Ore minerals in the vein-type deposit are pyrrhotite, chalcopyrite, sphalerite, cubanite, nickel-bearing mackinawite (Imai and Fujiki, 1963), cobalt-bearing pentlandite, molybdenite, star-shaped sphalerite, galena, native bismuth, bismuthinite, polybasite (?), arsenopyrite, magnetite, pyrite, etc., with such gangue minerals as chlorite, talc, quartz and carbonate minerals; ore minerals in the impregnated type consist of pyrrhotite, chalcopyrite, sphalerite, ilmenite, etc., with gangue minerals such as chlorite, quartz, sphene, talc, amphibole, apatite, plagioclase, and carbonate minerals.

In the ore from this mine, Fujiki (1963) described the complex exsolution textures of chalcopyrite, cubanite, pyrrhotite, Co-bearing pentlandite and Ni-Co-bearing mackinawite (Imai and Fujiki, 1963), as described later.

As for the minor elements, relatively high contents of Sn, Mn, In, Ag, Ga, etc., are detected in the sphalerite of this mine (Fujiki, 1964b). Also, high Ni and Co contents are recognized in the pyrrhotite of this mine. According to Shazly et al. (1957) and Muta (1958), the high contents of Sn, Mn, In, Bi, etc., are characteristic in sphalerite from the high-temperature deposit genetically related to granitic rocks.

Molybdenite is found in the sulphide ore, though it is very rare (Fig. II-5-3).*

As already stated, it is inferred that a cupola of granite exists beneath the area of the ore deposits.

From these facts the writers believe that the ore deposits were formed by the activity of the granite magma, though it is highly probable that the serpentinite played a part in, or had some influence on, the ore deposition, especially on the formation of cubanite, as discussed in Chapter III-10.

---

* The molybdenite-quartz veins of the Busshoji mine (now abandoned) occur at an adjacent area just to the south of the Komori mine (Fig. II-5-1).

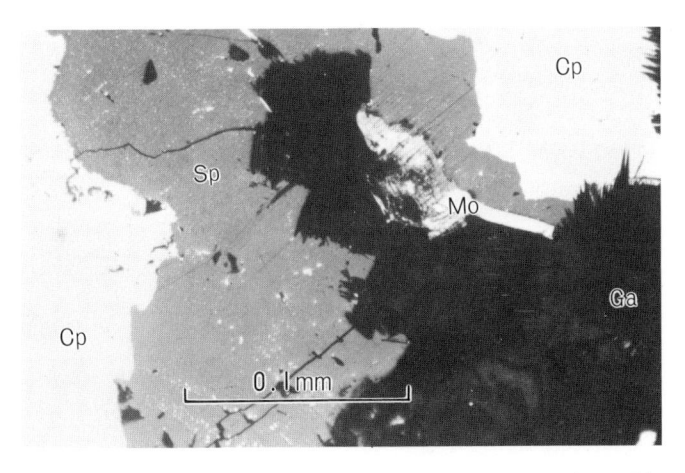

FIG. II-5-3. Photomicrograph. Polished section. One nicol. Molybdenite in sulfide ore. Mo: molybdenite, Cp: chalcopyrite, Sp: sphalerite, Ga: gangue mineral.

II-6. Geologic Structure and Fluid Inclusion Studies at the Taishu Mine, Nagasaki Prefecture (Takenouchi, 1962b,c; Imai *et al.*, 1971; Imai, 1973)

H. Imai, S. Takenouchi and T. Kihara

The Taishu mine, a zinc and lead deposit of mesothermal veins, is situated in Tsushima Island, which lies between Kyushu and the Korean Peninsula (Fig. I-3). In 1972, this mine produced 23,000 t of crude ore per month. Grades were 4.45 % zinc, 2.7% lead and 58 g/t silver. But since 1973 it has been abandoned. The total production of the crude ore was 3.6 million tons, yielding about 260 thousand tons of zinc, 151 thousand tons of lead and 325 tons of silver.

Geologic Setting

The major part of Tsushima Island consists of the shale and sandstone of the Taishu Formation belonging to the Paleogene epoch. The total thickness of the formation amounts to several thousand meters, which is divided into three units, Lower, Middle and Upper Formations.

In the southern part of the island, there is a granite mass occupying an area of 12 km². The mass extends NE-SW. The veins of the Taishu mine occur along the western side of this granite mass (Fig. II-6-1), and are genetically related to it.

In this mining area, the Taishu Formation is folded with an axis of N30°E, plunging at 10°–15°. The granite mass occurs at the major anticlinorium part of the folding, perhaps indicating synchronous intrusion.

Sheets and dikes of quartz porphyry or porphyrite, cut by the granite, are observed throughout the island (Fig. II-6-2). As stated above, granite occurs at the anticlinorium part of the folding, and extends in the same direction as the plunge of the folding axis, i.e., N30°E 10°–15°.

Sheets of quartz porphyry are folded concordantly with the strata. However, along the major synclinal axis of the folding just to the west of the above stated, major anticlinorium dikes of quartz porphyry are sporadically recognized. Many of the dikes strike N30°E, i.e., the direction of folding axis, dipping vertically. This fact is observed at the surface, as well

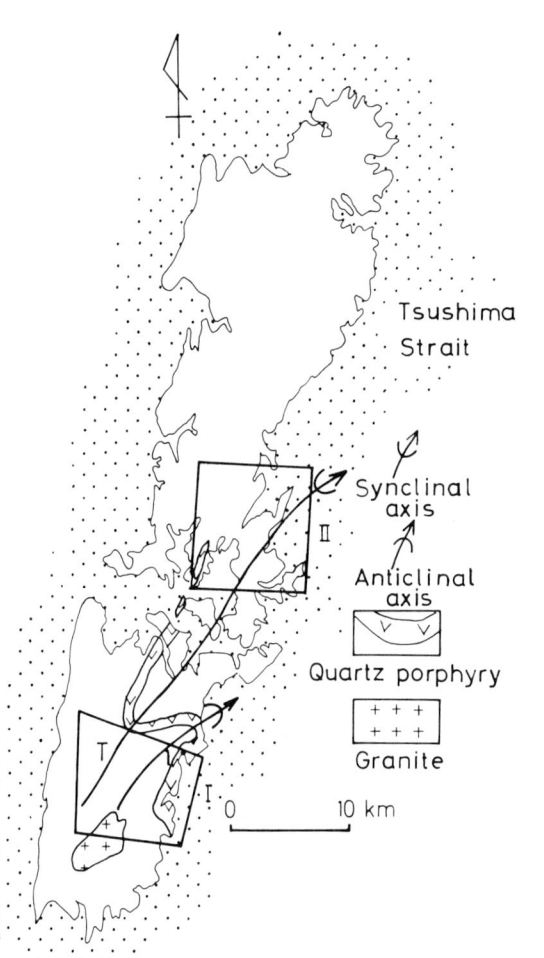

B. Pegmatitic, Hypothermal and Mesothermal Veins

FIG. II-6-1. Tsushima Island.
I: Taishu mine area, II: Nezumi Island area, T: Taishu mine.

as in the adits (Fig. II-6-2) where the major synclinal axis runs. Especially in the Nezumi Island area (Fig. II-6-1), about 25 km northeast of the Taishu mine site, where the major synclinal axis submerges in Tsushima Strait (Fig. II-6-3), small stocks, dikes and sheets of quartz porphyry and porphyrite occur. Some of the dikes are composite or multiple dikes composed of quartz porphyry and porphyrite. In these cases, porphyrite preceded quartz porphyry.

The writers' interpretation is that the Taishu Formation in this zone was sagged or dragged into the deeper part at the initial stage of folding. The quartz porphyry magma intruded as dikes or stocks at the sagged part, and then it spread laterally between the beds, forming sheets. This was followed by continued folding, resulting in folded sheets of quartz porphyry or porphyrite (Fig. II-6-1). This interpretation is similar to that made of the formation of a certain lopolith by Billings (1972).

In the northern part of Tsushima Island, the lineation or linear structure is developed in the shale, which is parallel to the folding axis. Controlled by this structure, the sample of shale is split into a columnar or long prismatic shape (Fig. II-6-4). It is due to the intersection of two schistosities or cleavages. The photomicrograph of the thin section perpendicular to the lineation (Fig. II-6-5) indicates bedding (plane) schistosity and flow cleavage (slaty cleavage). Both were formed by lateral pressure.

Granite intruded into the major anticlinorium in the course of the folding of sedimentary rocks and porphyritic rocks.

Part II: Vein-type Deposits

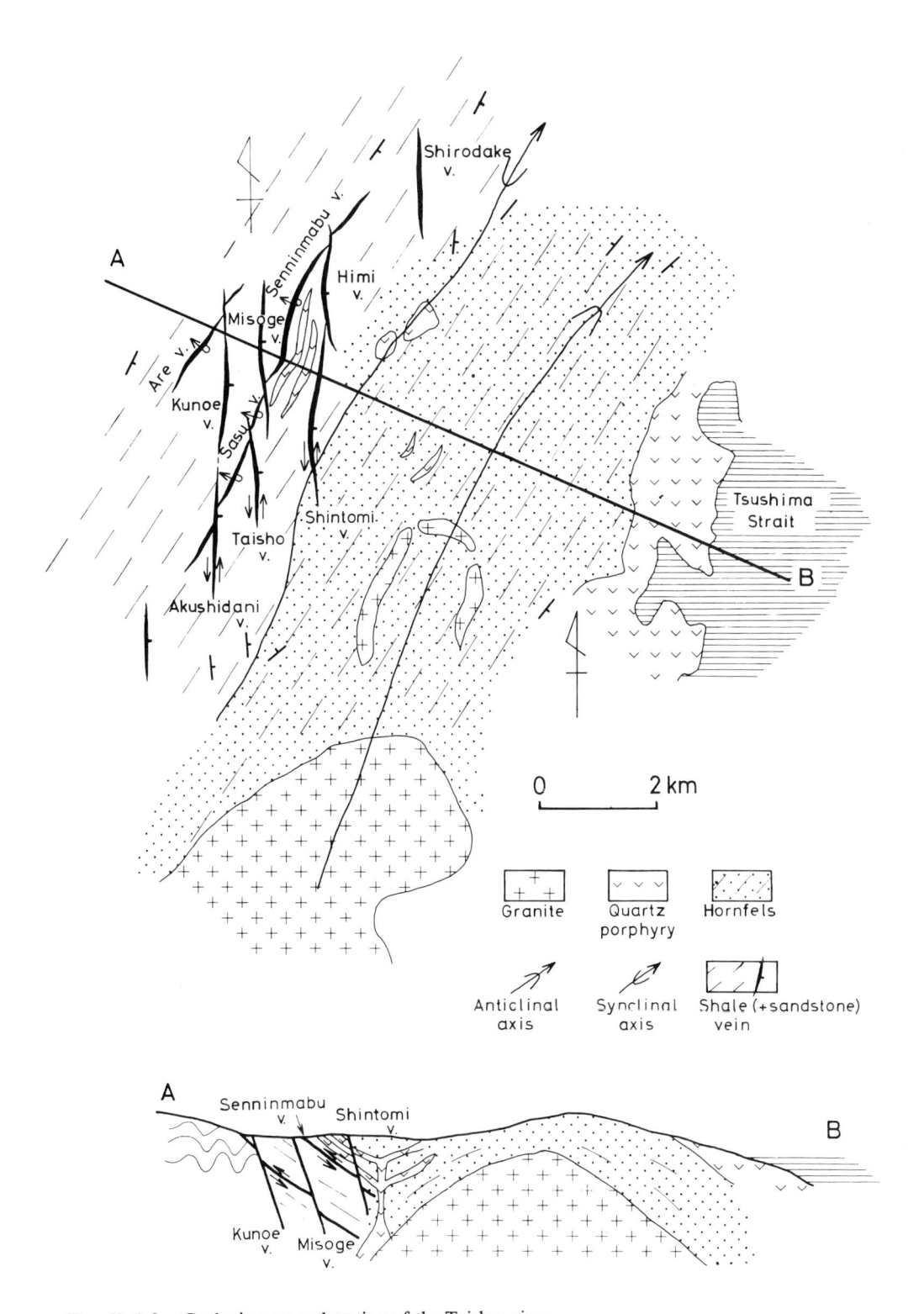

FIG. II-6-2. Geologic map and section of the Taishu mine.

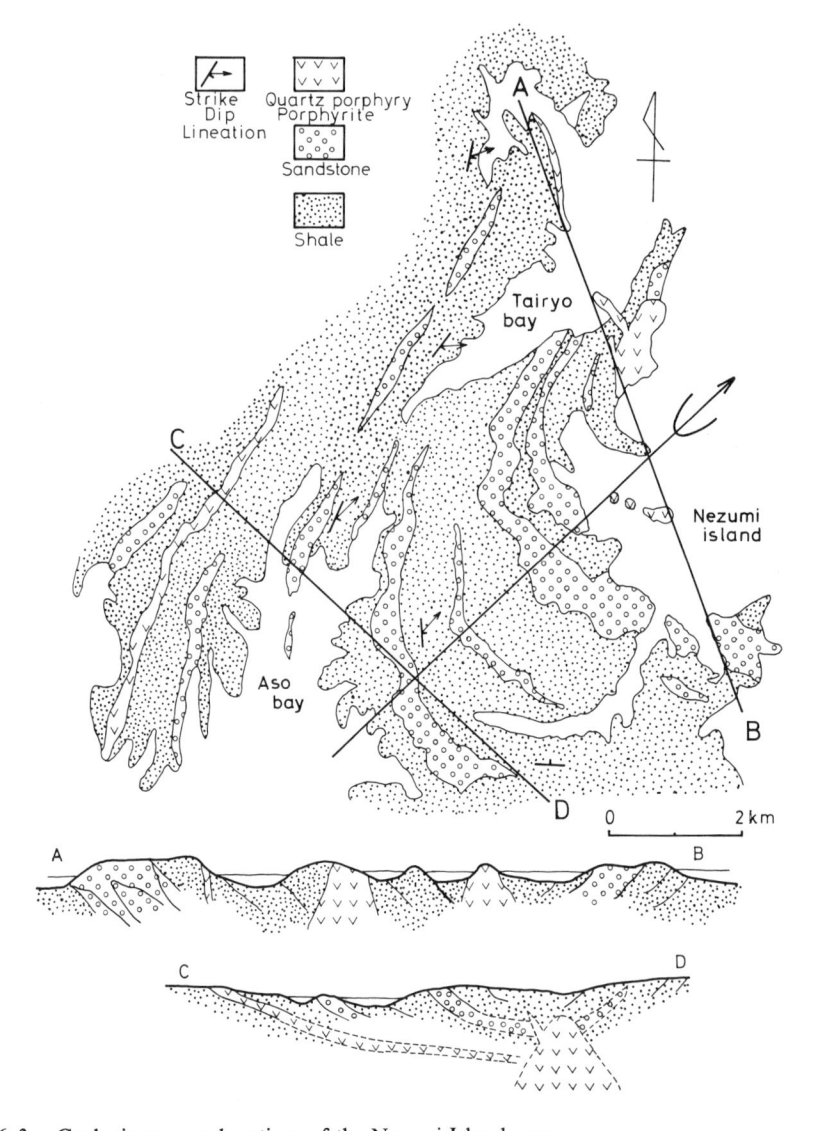

FIG. II-6-3. Geologic map and sections of the Nezumi Island area.

FIG. II-6-4. Lineation in the shale, showing two kinds of schistosities or cleavages. The plane in which sample number is written represents the bedding schistosity, and the upper plane represents the flow cleavage. The intersection of both planes corresponds to the lineation.

As the conclusion of geologic events, lateral pressure in a NW-SE direction caused the folding of Taishu Formation. At the initial stage of the folding, the dikes and stocks of quartz porphyry or pophyrite were intruded along the major synclinal axes, from which sheets of the same rocks were formed. The sheets were folded with the continuation of the folding, while granite intruded at the major anticlinorium. The mineralization at the Taishu mine would be genetically related to this granite. According to Kawano and Ueda (1967a), the absolute age of the granite was determined by the K-Ar method to be 12 m.y., corresponding to the Miocene epoch.

Vein fissures: The vein fissures are classified into the following two types (Fig. II–6–2). Both would have been formed by NW-SE lateral pressure.

(1) Gently dipping bedding reverse faults: They strike N25°–45°E and dip 20°–40° SE.

Drag folds on the bedding fault planes are recognized (Fig. II–6–6). The angle (α) between the axial plane of the drag fold and the bedding fault plane in an upward direction reveals an acute angle. This fact tells that the hanging wall of the fault thrust up on the fault plane. By measurement of the directions of the axes of these drag folds, the direction of the thrusting up might be determined, which is perpendicular to the axes. The axes of the drag folds on the fault plane are N50°E10° (−)*, so the direction of the thrust is perpendicular to this, i.e., N30°W30° (+).*

FIG. II–6–5. Photomicrograph. The thin section perpendicular to the lineation. One nicol. The plane represented by a solid line corresponds to the bedding schistosity. The bar line represents the flow cleavage.

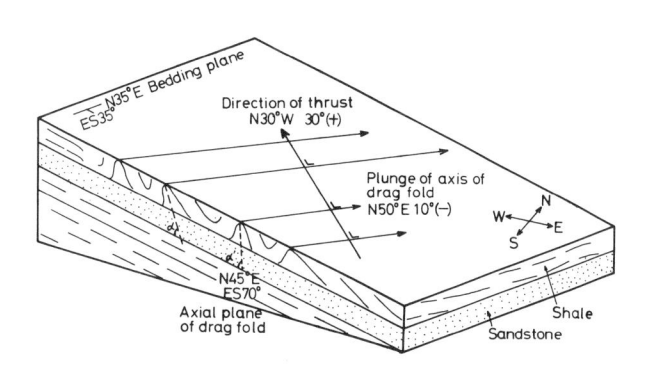

FIG. II–6–6. The drag folds on the bedding reverse fault.

There are several veins of this kind in this mining area. The Senninmabu vein is the largest (Fig. II–6–2). This kind of vein fissure was formed by the lateral pressure producing the folded structure in this district.

These vein fissures are cut by the following wrench fault striking N-S.

In some places of bedding reverse fault planes near the intersections with the wrench faults,

---

* The upward direction from the horizontal plane is represented by (+) and the downward by (−).

drag folds whose axes occur in the dip direction of the bedding fault plane, i.e., up and down on the fault plane, are observed. This would be due to the effect of the displacement of the wrench fault, described in the following.

(2) Wrench faults striking N-S: They strike N20°E-N15°W and dip 50°–70°E. On the fault planes, horizontal striation is often recognized. Sheets of quartz porphyry and bedding reverse faults are displaced by the wrench faults. The eastern blocks of these wrench faults were displaced towards the north, and the western blocks towards the south. In other words, this kind of fault is a left-handed (left-lateral) fault (Fig. II–6–2).

Along the Taisho vein fissure, the shale of the Taishu Formation shows a drag fold, shown in Fig. II–6–7. The folding axes of the drag folds are N40°–80°E 0°–30°, or S60°–80°W 10°–20°. These drag folds might be interpreted as being due to the lateral displacement on the wrench fault, judging from the strain ellipsoid shown in Fig. II–6–7. They correspond to the second order drag fold proposed by Moody (1956), or H. V. Faltung (Faltung durch Horizontalverschiebung) by Paroni (1961a,b).

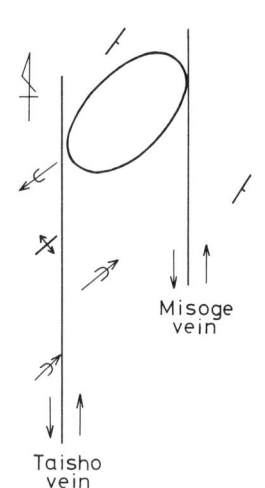

FIG. II–6–7. The second order drag folds along Misoge and Taisho veins.

Wrench faults might be also produced by the lateral pressure in a NW-SE direction which caused the regional-scale folded structure described above.

Mineralization at the Taishu Mine

The vein-forming minerals are pyrrhotite, chalcopyrite, sphalerite, galena and calcite with a small amount of quartz. Microscopically tetrahedrite can be seen.

The ore-forming fluid originated from the granitic magma. This might be interpreted from the state of zonal distribution of minerals described below. It ascended along the bedding reverse faults or dikes and sheets of quartz porphyry, and then poured into the wrench faults (Fig. II–6–2(b)).

In this mine, the sequence of the zones from the lower part upwards is barren quartz-pyrrhotite (locally containing chalcopyrite)-sphalerite-galena-barren calcite.

The major feeder of the ore-forming fluid from the granitic magma would have been the Senninmabu bedding reverse fault vein (Fig. II–6–2(b)). The above zonal distribution of minerals is clearly recognized in the Senninmabu vein, as well as in the Shintomi wrench fault vein shown in Fig. II–6–8. In parts where both types of faults intersect, the transition of the assemblage is from bedding reverse faults to wrench faults (Fig. II–6–9), indicating the flow direction of the mineralizing fluid.

Also, in the Himi wrench fault vein, the ore shoots are developed from the intersection

Part II: Vein-type Deposits

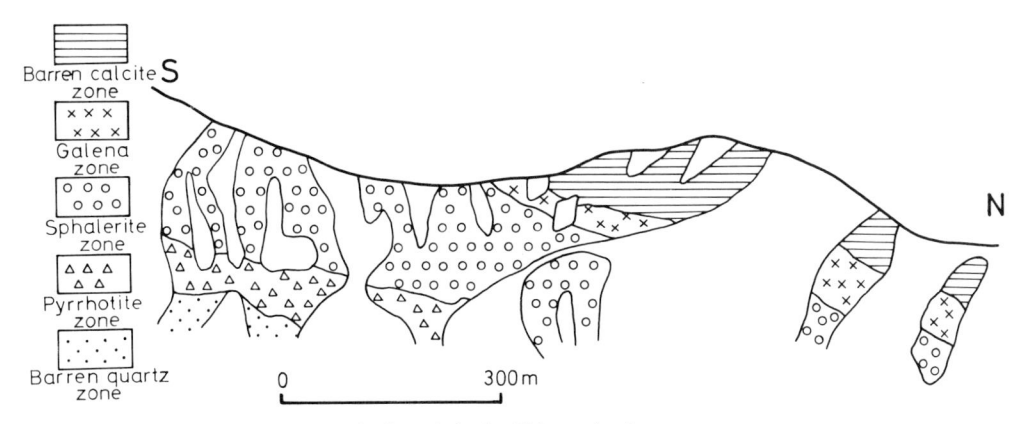

FIG. II-6-8. Zonal distribution of minerals in the Shintomi vein.

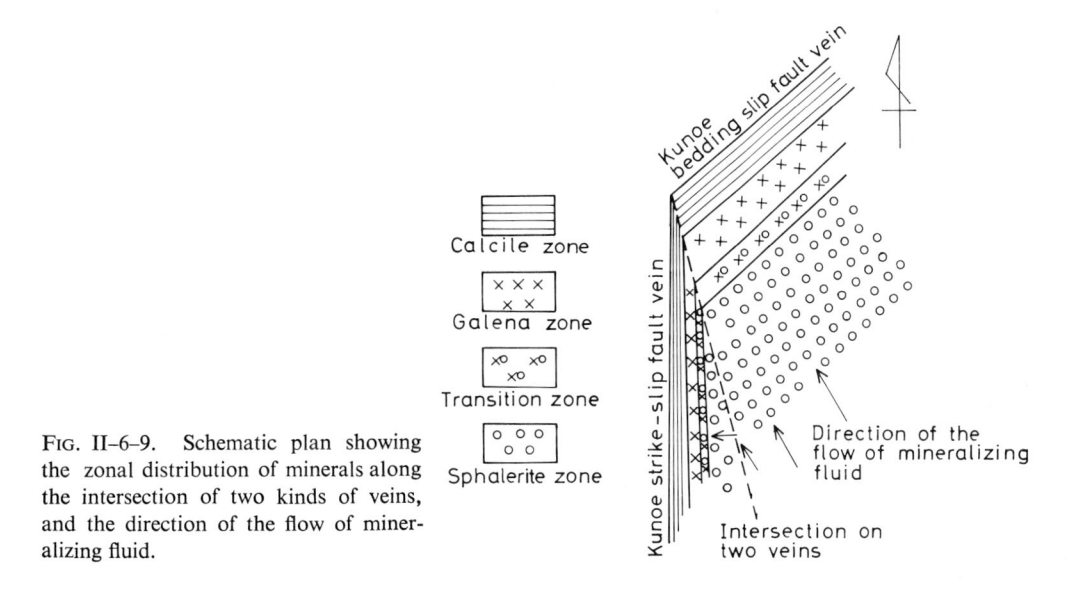

FIG. II-6-9. Schematic plan showing the zonal distribution of minerals along the intersection of two kinds of veins, and the direction of the flow of mineralizing fluid.

FIG. II-6-10. Zonal distribution of ores in the Himi vein. The dotted line represents the intersection of Himi and Senninmabu veins.

of this vein with the Senninmabu vein (Fig. II-6-10), indicating the flow direction of the ore-forming fluid. In some wrench faults (for example, the Misoge vein), where more than two bedding faults or sheets of quartz porphyry intersect, there is disorder in the zonal distribution. This would be expected from the multiple paths of the ore-forming fluid into the wrench fault.

Fluid Inclusion Studies

The writers studied inclusions in quartz crystals from the veins as well as from the granite, and found the following types of inclusions:

(1) liquid water, vapor and solids (Figs. II-6-11, -12, -13);
(2) liquid water and vapor (liquid-rich and vapor-rich) (Fig. II-6-14);
(3) liquid water, liquid CO_2 and vapor (Fig. II-6-15).

The quartz from the granite contains inclusions of types (1) and (2). In type (1) inclusions in the quartz from the granite, halite, a transparent birefringent mineral (perhaps calcite), two unknown transparent minerals and an opaque mineral are observed (Fig. II-6-11). But all of these solids are not necessarily observed in every inclusion. It is highly improbable that these solids existed at the time of trapping of the inclusions. They probably crystallized in the course of cooling after trapping. By volumetric calculation under the microscope, the NaCl equivalent concentration of inclusion of type (1) from the granite was 31–50 wt. % at the time of trapping. Although the solubility of halite in water as a function of pressure and temperature are still lacking, the data of Keevil (1942) and Sourirajan and Kennedy (1962) show that the salinity corresponds to the temperature of deposition above 250°–450°C, respectively. The coexistence of types (1) and (2) in quartz from the granite is due to separation of the residual fluid by boiling in the magmatic chamber, or to a difference in the stage of trapping of the inclusion. On the heating microscope stage, the vapor phase of type (1) in many cases fades out earlier than the halite crystals. In this case the halite crystals disappear at 250°–490°C. Similarly, these data show that the salinity (as NaCl equivalent concentration) is 34–56 wt. %. In some cases, halite crystals disappear earlier than the vapor phase, which fades out at 350°–440°C.

The vein quartz contains inclusions of types (1), (2) and (3). Takenouchi (1962a) already recognized the liquid water-vapor-solid inclusions (type 1) (Fig. II-6-12) as well as the liquid water-vapor ones (type 2) (Fig. II-6-14) in some barren quartz from the lowest level of the Shintomi vein (Figs. II-6-2 and II-6-8), which is near the granite mass. He presumed that the concentration of NaCl in an inclusion is 256 gr/l by volumetric calculation. He

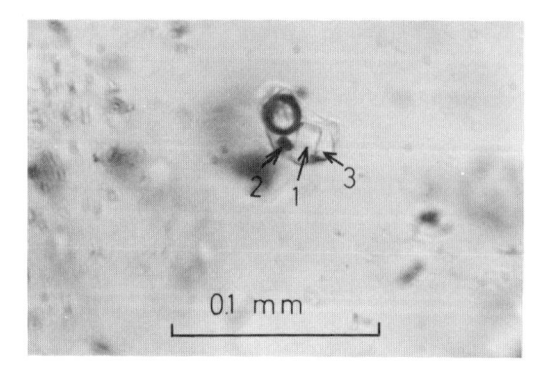

FIG. II-6-11. Fluid inclusion in quartz from the granite. 1: halite, 2: calcite (?), 3: undetermined transparent mineral.

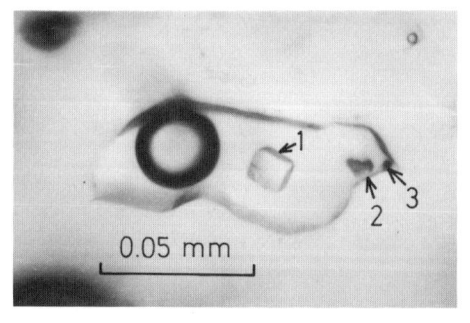

FIG. II-6-12. Fluid inclusion in quartz from the Shintomi vein. −70 mL. 1: halite, 2: calcite (?), 3: opaque mineral. Three phases (liquid water, vapor, solids) (after Takenouchi, 1962c).

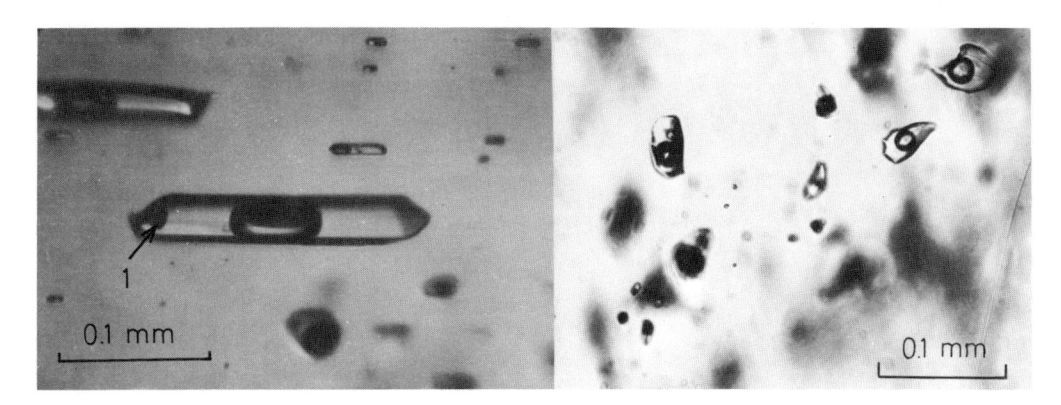

FIG. II-6-13. Fluid inclusions in quartz from the Shintomi vein. −70 mL. 1: calcite (?). Three phases (liquid water, vapor, solid) (after Takenouchi, 1962c).

FIG. II-6-14. Fluid inclusions in quartz from the Shintomi vein. Two phases (liquid water, vapor).

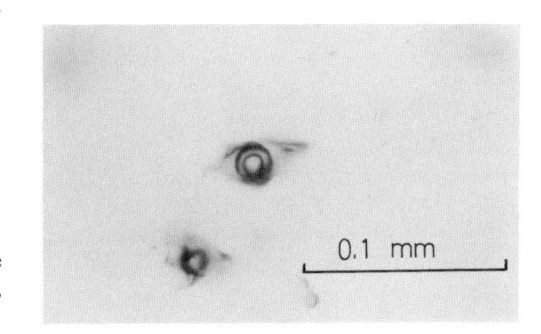

FIG. II-6-15. Fluid inclusions in quartz from the Are vein. −140 mL. Three phases (liquid water, liquid CO_2, vapor).

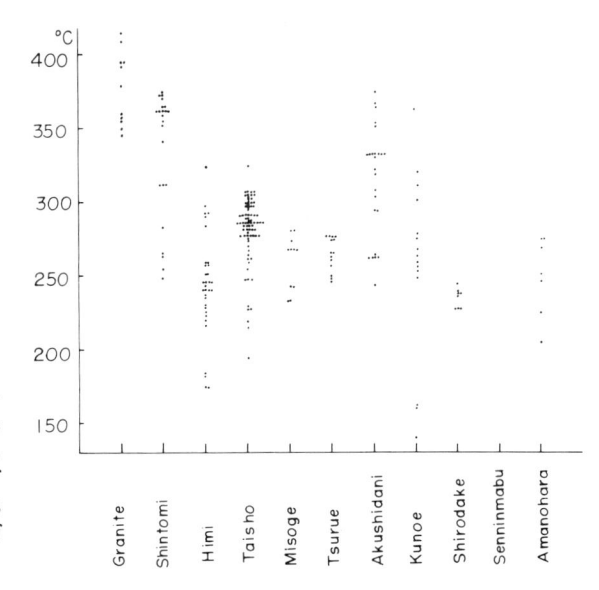

FIG. II-6-16. Homogenization temperatures of fluid inclusions in quartz crystals from the Taishu mine. Refer to Fig. II-6-2. The Tsurue vein is the northern extension of the Misoge vein. The Amanohara vein is situated about 1,000 m to the southwest of the Akushidani vein.

described halite, calcite (?) and an unknown opaque mineral in the inclusions. The same experiments were carried out by the writers. The NaCl concentration is about 30–33%. The homogenization temperatures of these inclusions are 350°–370°C.

Inclusions of type (2) in the same quartz have homogenization (filling) temperatures in

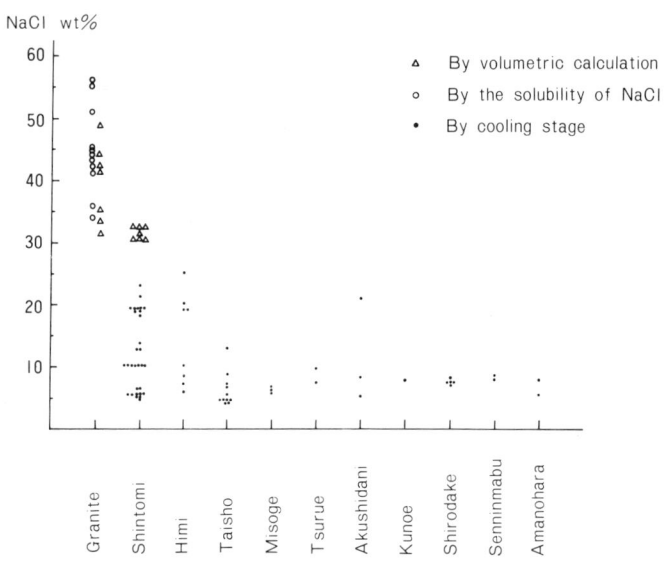

B. Pegmatitic, Hypothermal and Mesothermal Veins

FIG. II–6–17. NaCl equivalent concentration in fluid inusions in quartz crystal from the Taishu mine.

the range of 250°–310°C and have NaCl equivalent concentrations of 5–20 wt. % determined from lowering the freezing point. The coexistence of the two types of inclusions in the same quartz crystal would be due to different stages of mineralization. Inclusion type (1) would be primary and type (2) would be secondary. Towards the outside and upwards from the granite mass, type (1) loses the halite crystal (Fig. II–6–13).

The inclusions in vein quartz further away from the granite belong to type (2) (Fig. II–6–14).

In Figs. II–6–16 and –17, both homogenization temperatures by volumetric calculation and by heating stage, and NaCl equivalent concentrations by volumetric calculation, by solubility of NaCl and by cooling stage are shown collectively, and indicate that both the maximum values drop gradually outwards and upwards from the granite mass. This could be due to mixing of the ascending ore-forming fluid with meteoric or connate water.

It is recognized that some of the inclusions in the quartz collected from the Are vein and the Shigekuma vein consist of three phases, liquid water, liquid carbon dioxide and vapor (Fig. II–6–15). The Are vein is remote from the granite mass (Fig. II–6–2). The Shigekuma vein is situated about 40 km north of the Taishu mine. This vein is the same type

TABLE II–6–1 Volumetric analyses of the CO_2 and N_2 gas in the fluid inclusions from the Taishu and Ohtani mines.

Mineral	No.	Filling temperature	Vapor volume under 1 atm / Vapor volume in inclusion	CO_2 volume %	N_2 volume %
Quartz (Taishu mine, Shintomi vein — 54 mL)	1	305°C	17.2	76.2%	23.8%
	2	360°C	2.6	83.3	16.7
	3	305°	11.5	85.7	14.3
Quartz (Ohtani mine, No. 11 vein — 100 mL)	1	265°	1.7	82.2	17.8
	2	320°	3.2	71.5	28.5
	3	315°	1.4	73.5	26.5
	4	320°	1.5	72.8	27.2

Analyzed by N. A. Shugurova (Takenouchi, 1971a).

as the veins in the Taishu mine. The sample from the Shigekuma vein was collected at the part farthest from the granite. Such fluid inclusions containing liquid CO_2 are often formed at a low-temperature stage. The reason why the inclusions containing liquid CO_2 occur at the marginal part of the deposit is unknown, but similar phenomena are often observed in other examples. For example, inclusions containing liquid CO_2 are found in quartz of the late stage in some deposits (Takenouchi and Imai, 1971) (refer to Chapter VII-3), and mofettes, CO_2-bearing fumaroles, belong to the later stage of volcanic activities. Takenouchi (1971a) showed the results of analyses of the CO_2-bearing fluid inclusions from this mine and the Ohtani mine (refer to Chapter II-4), as shown in Table II-6-1.

Conclusions

The Taishu mine is a zinc and lead deposit, consisting of mesothermal veins. Veins occur in the folded Paleogene strata around the granite mass which was intruded at the core part of the anticlinorium of the strata. Sheets of quartz porphyry which preceded the granitic intrusion are folded concordantly with the beds. Dikes or stocks of quartz prophyry are found at the synclinorium part. The quartz porphyry magma intruded as dike or stock at the sagged part in the initial stage of the folding, and then it spread laterally between the beds forming sheets at the same time as the folding.

The vein fissures are classified into two types: (1) gently dipping bedding reverse faults, and (2) wrench faults. These two kinds of faults are believed to have been formed by the lateral pressure which produced the folded structure in the district. The former were displaced by the latter, indicating the age relationship.

The ore-forming fluid originated from the granitic magma. It would ascend along the bedding reverse faults or dikes and sheets of quartz porphyry, and then pour into the wrench faults. This process is indicated by the zonal distribution of ore minerals around the granite mass and the distribution of the ore shoots in the veins, as well as by fluid inclusion studies.

Fluid inclusions in quartz crystals from the veins as well as from the granite in the mining area were studied by means of heating tage, freezing stage microscope and volumetric calculation. The writers recognized the following types of inclusions:
(1) liquid water, vapor and solids
(2) liquid water and vapor
(3) liquid water, liquid CO_2 and vapor

Both homogenization temperatures and the salinities of the inclusions drop gradually upward and outward from the granite mass. This could be due to mixing of the ore-forming fluid with meteoric or connate water. The liquid CO_2-bearing inclusion is found at the marginal part of the deposit.

From the [34]S exchange fractionation factor for the coexisting sphalerite-galena pair, Kiyosu and Nakai (1977) have estimated the formation temperatures of these minerals in the Taishu mine to be 350°–300°C. [Kiyosu, Y. (1977) Sulfur isotope ratios of ores and chemical environment of ore deposition in the Taishu Pb-Zn sulfide deposits, Japan: Geochem. J. **11**, 91–99.]

C. EPITHERMAL (SUBVOLCANIC) VEINS

II-7. Sado Mine, Niigata Prefecture

H. Imai and M. Bunno

The Sado mine is situated in the northwestern part of Sado Island in the Japan Sea, central Japan. It was discovered in 1542. The Sado mine has produced 15 million tons of crude ore, yielding 77 t of gold, 2,300 t of silver and 5,400 t of copper. It is the largest mine in Japan in total production of gold and silver.

The northwest part of Sado Island is geologically composed of Miocene and Pliocene volcanic and pyroclastic rocks, and shale, generally stretching NE-SW in the anticlinal structure.

The veins occur in the Miocene formation as follows (Sakai and Ohba, 1970). The basement is rhyolitic mass (Ohdate Formation) followed conformably by rhyolitic tuff (Ohgiri Formation), black shale (Hidarisawa Formation) and andesitic tuff and tuff breccia (Kohshinzuka Formation) in succession (Fig. II–7–1). Dikes of rhyolite and propylite penetrate these rocks. The basement rhyolite would upheave and be active up to the time after deposition of the above successive sedimentary and pyroclastic rocks. This movement was important from the point of formation of vein fissures, as is discussed below.

The champion veins run nearly E-W in two parallel groups. The Ohgiri-Torigoe vein fissure group and the Aoban (I and II)-Ohdate-Heiko Fault vein fissure group are two such examples (Fig. II–7–1).

The Ohgiri and Torigoe vein fissures originally belonged to the same fissure. This fissure was displaced by the Otsu Fault, stretching N-S with an easterly dip (Fig. II–7–1 (a)).

The Ohgiri-Torigoe vein fissure would have been a right-handed wrench fault, judging from the distribution of the surrounding rocks and the striations on the vein walls. On the walls of the Torigoe and Ohgiri veins, horizontal striations are often recognized.

Also, Heiko Fault was originally a wrench fault like the Ohgiri-Torigoe vein fissures. It is cut by Otsu Fault, and extends to Aoban II fissure (Fig. II–7–1 (a)) in the east of Otsu Fault.

The parallel fissures of Ohgiri-Torigoe fissures and Heiko Fault would have been formed by the lateral pressure exerted from NW-SE, which caused the anticlinal folding in Sado Island.

As to the Aoban I and Ohdate vein fissures, the writers will discuss them below.

In the block sandwiched by two parallel faults, the branching fissures linking them, such as Shichisuke, Nakadate and other veins, would have been produced by lateral shearing due to the displacement of the above two parallel faults (Fig. II–7–1 (a)).

In the wall of the Otsu Fault which strikes N-S dipping 60°E, striations stretching S35°E 40° are observed (Fig. II–7–1 (a)). From the distribution of the adjoining rocks along this fault and the striations on its walls, it appears that the western block (footwall) of the Otsu Fault was pushed up in the direction of the striation. Thus, this fault is an oblique normal fault formed in pre-ore time, and partly mineralized (Fig. II–7–1 (c)).

The Aoban I and Ohdate vein fissures running nearly E-W are perhaps tension cracks. They are branches of the Heiko Fault which also runs E-W dipping south (Figs. II–7–1 (b), and –2). From the distribution of the adjoining rocks along the Heiko Fault, from the attitudes of the fissures branching from this fault, such as Aoban I and Ohdate vein fissures,

Part II: Vein-type Deposits

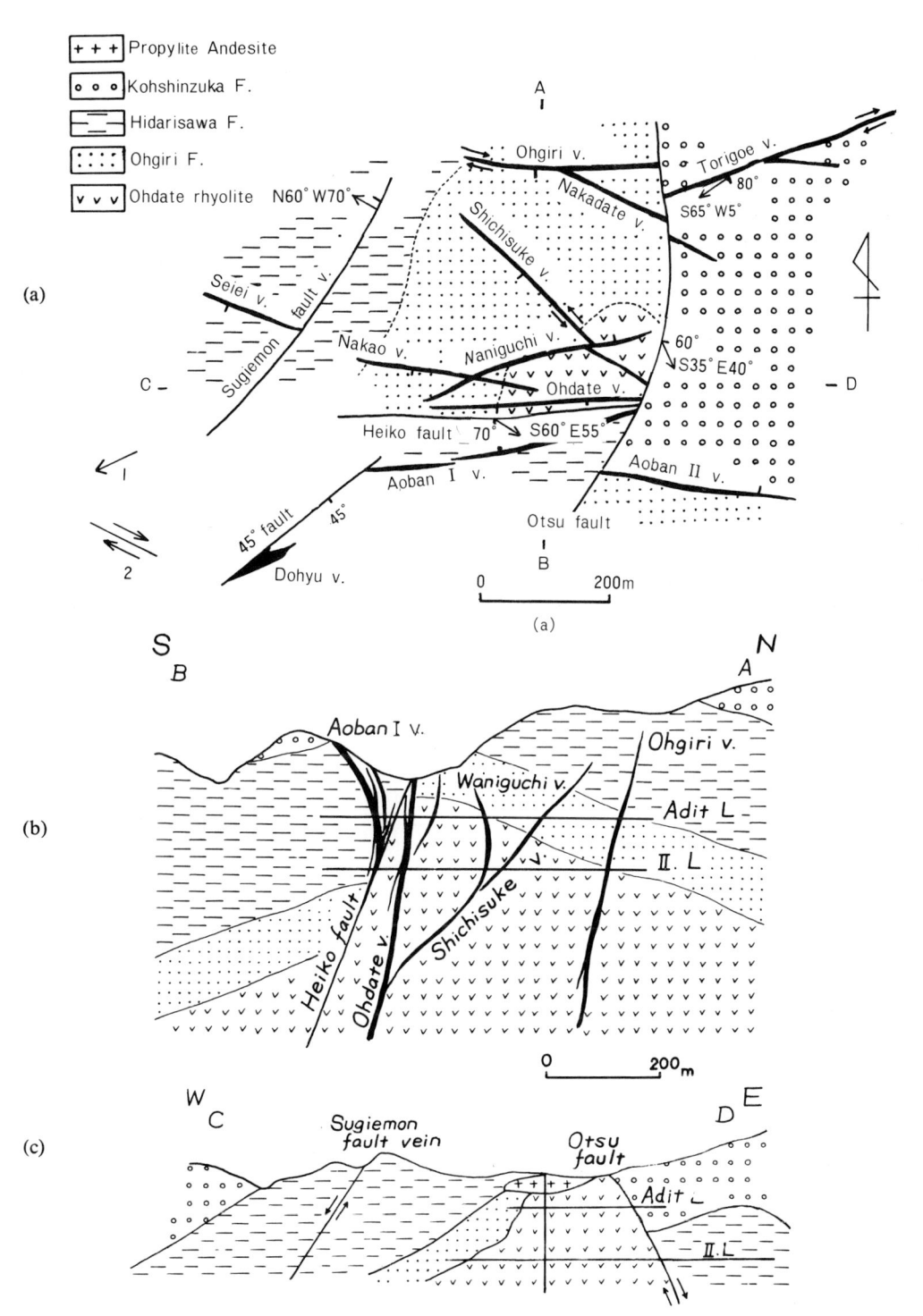

FIG. II-7-1. Geologic map and section of the Sado mine.
  (a) Geologic map of the adit level.
    1. Direction of striation on the fault plane.
    2. Direction of relative displacement along the fault.
  (b) Geologic section along A-B in (a).
  (c) Geologic section along C-D in (a).

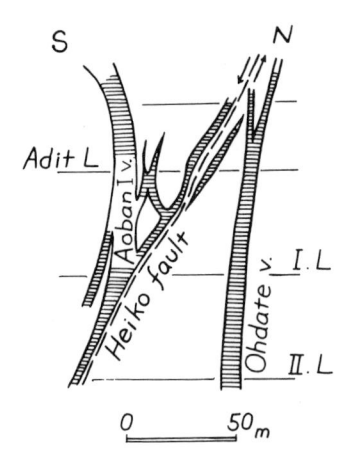

FIG. II–7–2. The relation of Heiko Fault vein with Aoban 1 and Ohdate veins.

and from the striation on the vein wall streching S60°E 55° (Fig. II–7–1 (a)), it appears that the northern block (footwall block) of Heiko Fault was pushed up as a normal fault (Fig II–7–1 (b)). So the Heiko Fault, which would have been formed originally as a right-handed wrench (right-lateral) fault at the initial stress condition, would afterwards move as a normal fault, producing tension cracks such as the Aoban I and Ohdate fissures.

The Sugiemon Fault in the west part of the mining area, which strikes N30°E dipping NW, is also a normal fault (Fig. II–7–1 (a), (c)).

The block surrounded by the Otsu, Heiko and Sugiemon faults was upheaved. This upheaval would have been caused by the pushing-up of the rhyolite (Ohdate rhyolite) (Fig. II–7–1).

So the Heiko Fault is a fissure revealing two kinds of displacement due to different stresses, as discussed in Chapter II–1.

The vein materials are composed of milky quartz, adularia, calcite, montmorillonite, sericite, chlorite, fluorite, rhodochrosite, rhodonite, etc., with ore minerals such as electrum, native silver, argentite, stromeyerite, pyrargyrite, proustite, polybasite, arsenopolybasite stephanite, pearceite, tetrahedrite, pyrite, marcasite, chalcopyrite, bornite, covellite, sphalerite, galena, hessite (Bunno, 1971), jalpaite accompanied by mckinstryite (Taguchi et al., 1974), etc.

II–8. Mikawa Mine, Niigata Prefecture   (Imai, 1963a, 1966b)

H. Imai

This mine is situated about 230 km north of Tokyo, in the central part of Japan.

In this area, middle Miocene sedimentaries and pyroclastics overlie the basement of granitic rocks and the Carboniferous formation. A stock of andesite (now altered to propylite) occurs in these rocks (Fig. II–8–1). Rhyolite is widely developed in this area. According to Nagasawa (1961), the activity of andesite was succeeded by that of rhyolite. As to the emplacement of the stock-like andesite (propylite mass), the following four cases might be considered:

   (1)   The andesite erupted as submarine volcanic flow at the same time as the sedimentation of the Miocene sedimentaries and pyroclastics.

   (2)   The andesite erupted as a volcanic island at the time of the sedimentation of the Miocene sedimentaries and pyroclastics, as the writer discusses in the case of the propylite at the Hosokura mine in a later chapter.

Part II: Vein-type Deposits

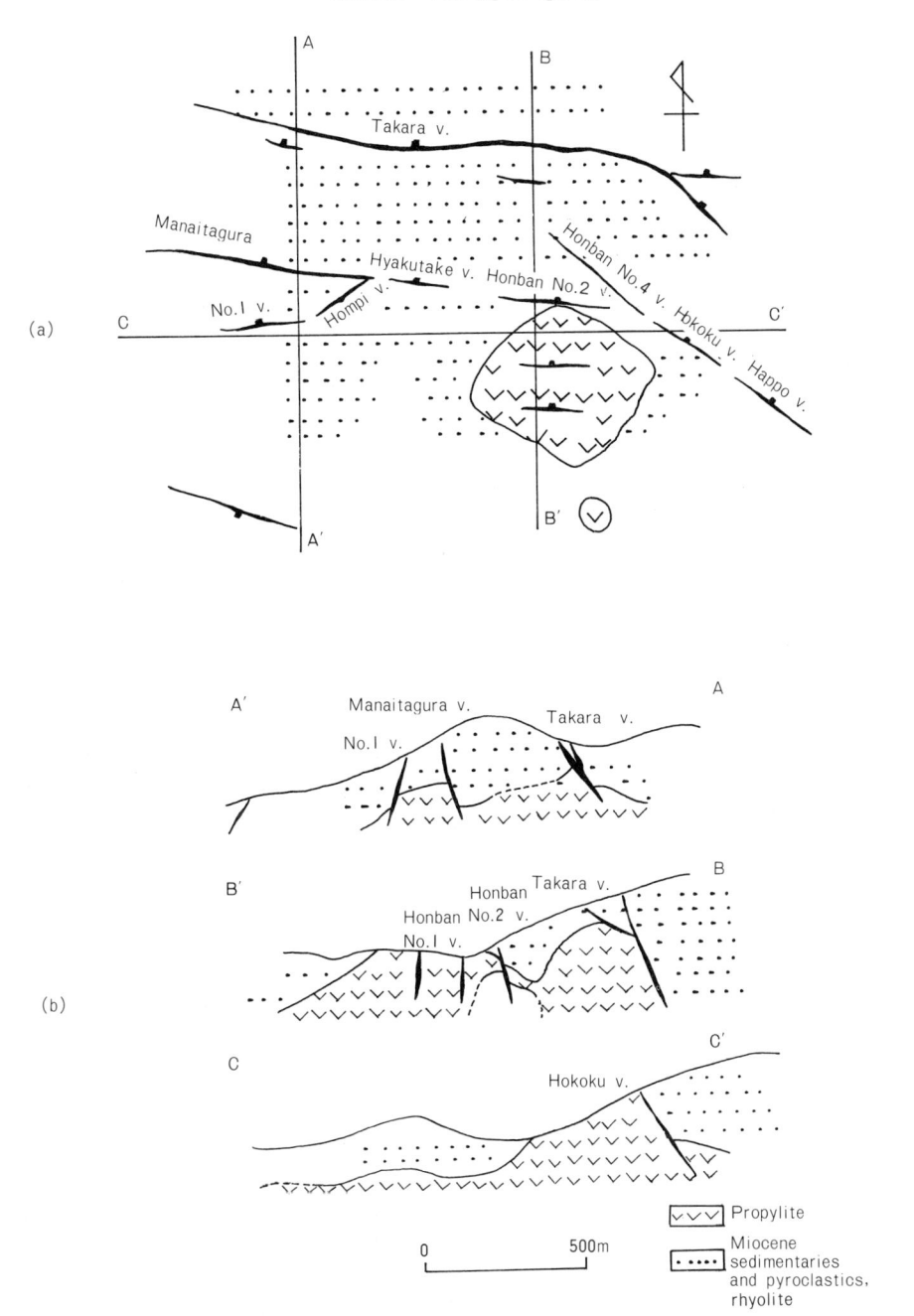

FIG. II-8-1. (a) Geologic map and sections of the Mikawa mining area.
(b) Geologic section of the Mikawa mine.

(3) The andesite intruded as a stock into the shallow zone at the time of sedimentation of the Miocene sedimentaries and pyroclastics.

(4) The andesite intruded into the Miocene sedimentaries and pyroclastics after their sedimentation, perhaps at the time of warping of these rocks.

The vein fissures surround the andesite (propylite) mass, dipping away from the andesite (Fig. II-8-1). They are normal faults, probably synthetic normal faults produced by the magmatic upheaval (Fig. II-1-2).

C. Epithermal Veins

These faults are developed not only in the andesite, but also in the sedimentaries and pyroclastics above the andesite. So, in any of the above four cases, magmatic upheaval of the andesite continued after sedimentation of the middle Miocene formation. In the previous chapter, the writers discussed a similar process in the doming up of rhyolite at the Sado mine. The petrographic properties of propylite are discussed in a later chapter.

Rhyolite is developed widely in this area. Whether it is a lava flow or a sheet is unknown, but the necks of the eruptions of the rhyolite are observed in places. In the adits of the Akatani mine just north of the Mikawa mine, necks of the rhyolite are found in the limestone of the Carboniferous formation. The occurrence of skarn minerals containing hematite ore is controlled by the shape of the rhyolite necks.

The ore and gangue minerals in the veins of the Mikawa mine are as follows (Nagasawa, 1961):

Ore minerals: chalcopyrite, sphalerite, pyrite, galena, argentite, polybasite, native gold, marcasite.

Gangue minerals: quartz, chlorite, dolomite-ankerite, siderite, hematite, magnetite, barite, adularia, kaolinite, sericite, calcite.

There are some examples of vertical zoning in the veins, namely, the upper part of the vein is rich in zinc and lead, while the lower part is rich in copper and iron sulfides.

The decrepitation temperatures of quartz samples from the Manaitagura vein (Fig. II–8–1) were measured. The result plotted in the longitudinal section of the vein reveals that the temperatures become higher towards the eastern lower part (Fig. II–8–2). From the geological map it is noticed that the center of the andesitic (propylitic) mass exists in the eastern part of this vein. From these facts, it may be suggested that the mineralizing solution migrated from the eastern deeper part into the Manaitagura vein.

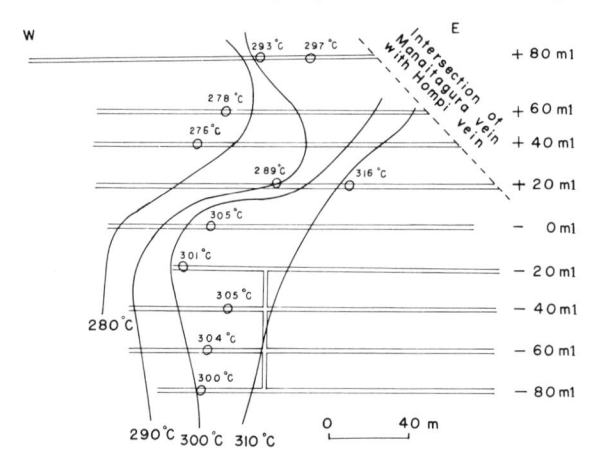

FIG. II–8–2. Decrepitation temperatures in the longitudinal section of the Manaitagura vein of the Mikawa mine (refer to Fig. II–8–1(a)).

II–9. Osarizawa Mine, Akita Prefecture (Imai, 1963a, 1966b)

H. Imai

The Osarizawa mine is situated in Akita Prefecture, northern Japan. It produced about 27 million tons of crude ore containing 340,000 t of copper up to 1975.

The area is composed mainly of Miocene formation, i.e., an alternation of green tuff and black siliceous shale, intruded by propylite and dacite (or plagiorhyolite). The alternation of green tuff and black siliceous shale strikes N30°–60°W and dips NE or SW at angles less than 30°, revealing an anticlinal structure. The propylite occurs at the core of the anticline and forms a domed structure (Fig. II–9–1). The strike of the boundary between

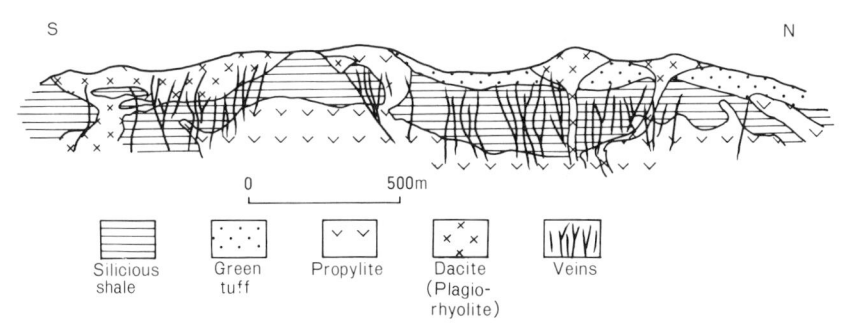

FIG. II–9–1. Geologic section of the Osarizawa mine (after Suga, 1952).

the propylite and the Miocene formation is generally parallel to the strike of the bedding plane of the latter. The writer believes that the propylite mass may have intruded into the core part of the folded structure nearly at the same time as the warping. Dacite erupted as necks or dikes after the emplacement of the propylite.

The deposit occurs in the anticlinal part of the folding. More than 300 veins are found in an area extending 2000 m from north to south and 1500 m from east to west.

The strikes of the vein fissures are predominantly E-W and NE-SW. These fissures are generally normal faults. According to Suga (1952), the strikes of many of the veins are parallel to the boundary between propylite and the Miocene sedimentaries. The veins dip mostly outward from the center of the propylite stock (Fig. II–9–1). From these facts, it may be suggested that many of the fissures containing vein materials were produced by the magmatic upheaval of the andesitic (propylite) magma. The veins are prevalent in the siliceous shale, but pinch in the propylite about 50 m in depth from the boundary with the siliceous shale.

The veins are classified into the following two types:
(1) copper, zinc and lead veins;
(2) gold veins.

The former are composed of chalcopyrite, sphalerite, galena, pyrite, quartz and chlorite, with small amounts of hematite, gold, barite, rhodochrosite, etc. Copper predominates in the lower levels and zinc and lead in the upper levels, displaying a zonal distribution.

The latter is composed of quartz, hematite, chlorite and native gold, accompanied by small amounts of pyrite, chalcopyrite, sphalerite and galena.

The former is far more developed, but occasionally both types coexist, forming a composite vein.

II–10. Ani Mine, Akita Prefecture (Imai, 1963a, 1966b)

H. Imai

This mine is situated about 40 km northeast of Akita City, in northern Japan.

Geologically, the mine area is composed of sedimentary rocks, volcanic rocks and dioritic rocks, all belonging to the Miocene epoch. The sedimentary rocks including tuff, tuff breccia and coal-bearing shale, are folded with a NNE-SSW axis. The volcanic rocks, such as rhyolite, andesite and basalt, correspond to the Miocene volcanic activities in northeastern Japan. Both sedimentary rocks and basalt were intruded by the dioritic rocks. Outcrops of the dioritic rocks are very scarce on the surface, but in the underground adits they are widely developed (Fig. II–10–1). The dioritic mass presents a remarkable change in facies. From the core toward the marginal part, the facies changes as follows: quartz diorite-

C. Epithermal Veins

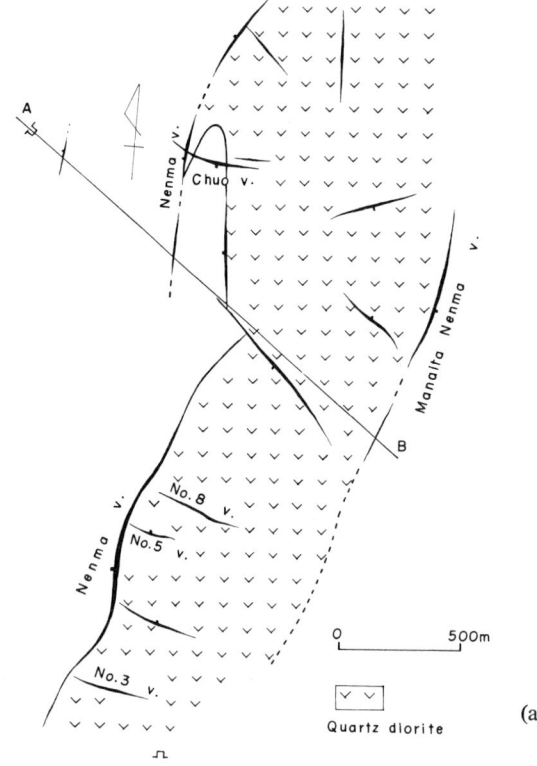

FIG. II–10–1.
(a) Geologic map of the Ani mine (adit level).
(b) Geologic section along A–B line in (a).

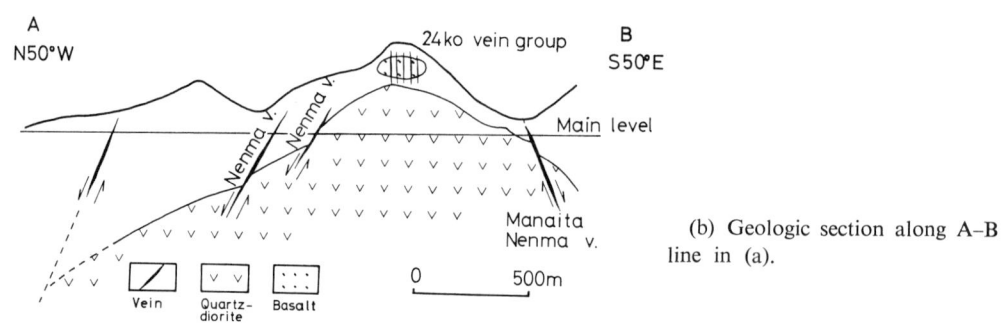

diorite-porphyrite. The intrusion of the dioritic rocks may have been accompanied by folding.

According to Kawano and Ueda (1966b), the absolute ages of orthoclase and biotite from the quartz diorite in this mine are 15–12 m.y. calculated by the K-Ar method. Yamaoka and Ueda (1974) determined the absolute age of adularia from the veins at the Ani mine to be 11 m.y. by the K-Ar method.

Along the elongated boundaries between the dioritic mass and the surrounding rocks, faults dipping away from the dioritic mass are found on both limbs of the dioritic mass (Fig. II–10–1). They stretch NNE-SSW. They are normal faults, partly mineralized. The fault vein on the west side of the dioritic mass is named the Nenma vein; the one on the east side of the dioritic mass is named the Manaita Nenma vein. The writer suggests that these normal faults were produced by the upheaval of the dioritic magma.

The Nenma vein is partly mineralized with galena, chalcopyrite, sphalerite and pyrite,

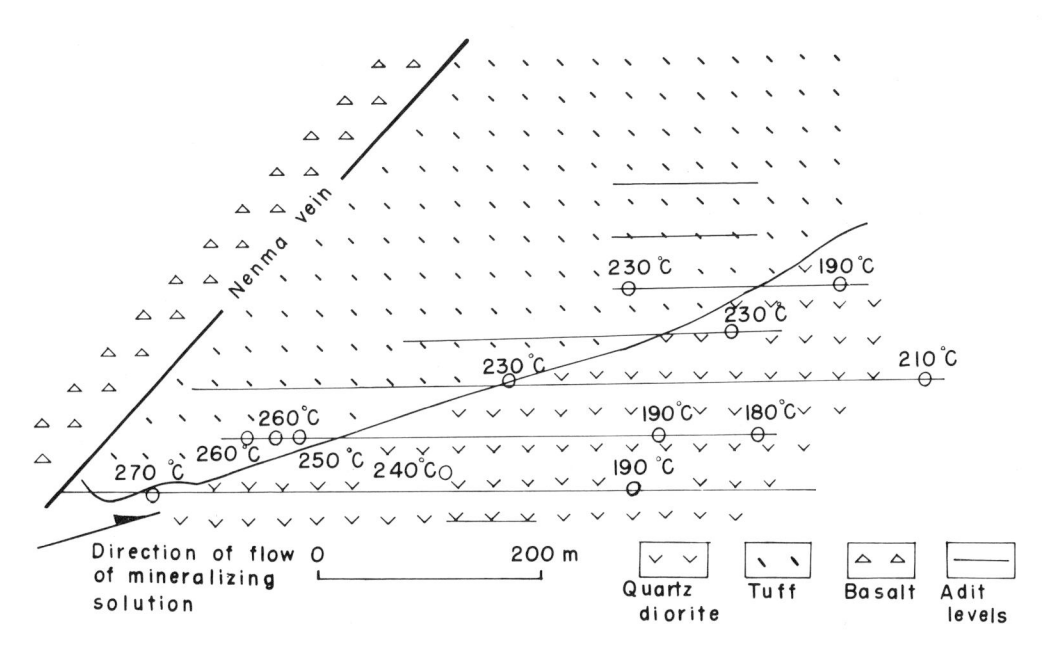

FIG. II–10–2. The distribution of the decrepitation temperatures in the longitudinal section of the Chuo vein of the Ani mine (refer to Fig. II–10–1 (a)).

accompanied by gangue minerals such as quartz and chlorite, and rich in clayey materials.

In the foot-wall side of the Nenma vein or across this vein are the fissures of the Chuo, No. 3, No. 5, and No. 8 veins, intersecting the Nenma vein at right angles (Fig. II–10–1). They occur mainly in the dioritic mass, having a strike of N70°–80°W and dipping 70°–80°NE. These fissures are tension cracks. No displacement is recognized in the walls of these vein fissures. The vein materials are almost the same as those of the Nenma vein. These fissures are nearly perpendicular to the anticlinal axis of the folding of this area. From these facts, the writer concludes that the fissures (tension cracks) were produced by lateral pressure causing the folded structure. As stated above, the intrusion of the dioritic rocks may have been contemporaneous with the folding. In the marginal part of the dioritic mass, the above tension cracks may have been produced by the lateral pressure which was exerted soon after the intrusion. These events probably occurred in the Middle or Later Miocene.

In the basaltic rock just above the cupola of the dioritic mass, two sets of veinlets with EW and NW-SE trends occur displaying a network pattern (24 Ko vein). They are tension cracks produced by doming up of the dioritic magam, and filled with gold-bearing chlorite-hematite-quartz (Fig. II–10–1(b)).

Using quartz separated from the other vein materials, the writer measured the formation temperatures of the Chuo vein (Fig. II–10–1(a)) by the decrepitation method, and plotted the obtained values in the longitudinal section of the vein, as shown in Fig. II–10–2.

From this result, it is recognized that the temperature rises towards the lower intersection of this vein with the Nenma vein. It is supposed that the mineralizing solution migrated along the Nenma vein and flowed into the fissure of the Chuo vein. As described above, the fissure of the Nenma vein is a fault vein rich in clay, while the fissure of the Chuo vein is a tension crack rich in open space. So it is considered that the adiabatic expansion occurred when the mineralizing solution from the Nenma fissure flowed into the Chuo fissure, causing a drop in temperature.

II-11. Tsuchihata Mine, Iwate Prefecture (Imai, 1964, 1966b)

H. Imai

The Tsuchihata mine is situated about 120 km north of Sendai in the northern part of Japan.

This area is geologically composed of Miocene pyroclastics, volcanics and sedimentary rocks. Rhyolite is intercalated between the underlying tuff and the overlying shale. It poured out of many vents or necks which stretch downward in the underlying tuff.

The ore bodies occur in the rhyolite above the vents (Fig. II–11–1). Each ore body is a group of networks of narrow veinlets of chalcopyrite and quartz, densely penetrating rhyolite. It is an oval-cylinder in shape, resembling a breccia pipe.

In this area, the rhyolitic rock was favorable for the formation of ore deposits, probably due to the availability of fissures in the rock. The fissures of the predominant veinlets belong to the tension cracks. The fissures may have been formed by the pushing up of the rhyolitic magma. In the Shiratsuchi ore body, it is observed that the ore body is bordered by a "Curved Fault" (Fig. II–11–2). The overlying shale is partly pushed up by the ore body bordered by the "Curved Fault." The "Curved Fault" may have been produced by pushing up due to magmatic upheaval.

The horizontal sections of the vents or necks of the rhyolite underneath ore bodies must be elongated N-S or NE-SW, in relation to the longer axes of the ore bodies. The longer axis of the Hatabira ore body is NE-SW, and the longer axes of the other ore bodies are N-S (Fig. II–11–2).

The ore bodies are arranged generally in a N-S direction (Fig. II–11–2), which would correspond to the trend of distribution of the rhyolite vents. The primary ore minerals are chalcopyrite and pyrite, with subordinate amounts of sphalerite, galena, enargite, luzonite, a tetrahedrite group mineral, bismuthinite and others. The main gangue mineral is quartz.

In some ore bodies of this mine (for example, the Washinosu deposit), the upper part of the deposit reveals a gradual decrease in copper content and an increase in gold and silver, which becomes a gold-silver network deposit.

Just above the network deposit at Shiratsuchi, a small bedded Kuroko ore body composed of sphalerite, chalcopyrite, galena, barite, etc., occurs underneath the shale bed.

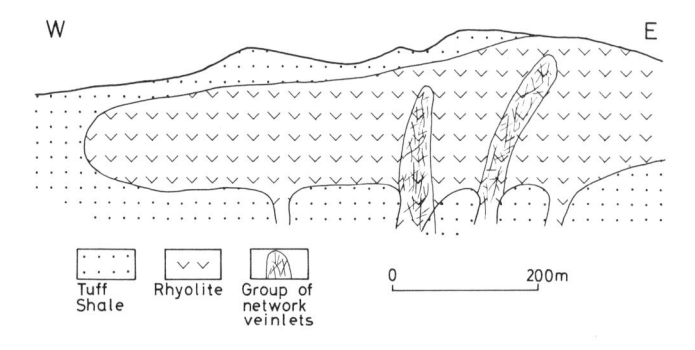

FIG. II–11–1. Geologic section of the Tsuchihata mine (Uenono area).

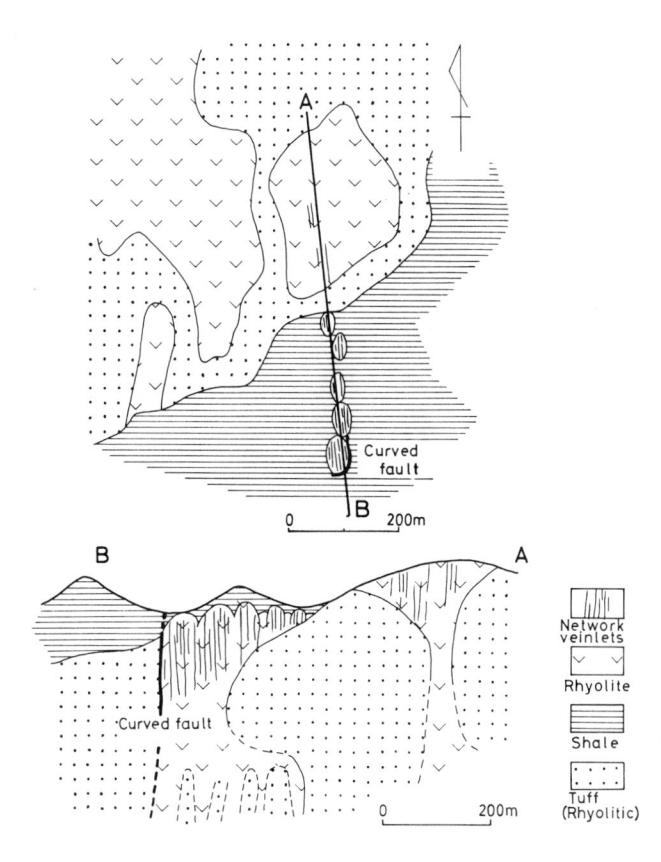

FIG. II-11-2. Geologic map and section of the Shiratsuchi area, the Tsuchihata mine. In this figure, the Kuroko ore body is omitted.

II-12. Yaso Mine, Fukushima Prefecture (Takenouchi, 1962a; Imai, 1966b)

S. Takenouchi and H. Imai

The Yaso mine is situated about 150 km north of Tokyo in the central part of Japan. It is a vein-type copper deposit of the subvolcanic type.

This region is geologically composed of Paleozoic (Permo-Carboniferous) rocks, Mesozoic (?) granite, Miocene pyroclastic and sedimentary rocks, Miocene rhyolitic and andesitic dike rocks and younger dacite lava flow.

Paleozoic slate, sandstone and quartzite, and granite form the basement of this region. The Miocene formation, consisting of basal conglomerate, tuffaceous sandstone and rhyolitic tuff, is widely developed, lying on the basement. These rocks are intruded by many dikes of lithoidite, rhyolite and nevaditic plagioclase rhyolite and by a few smaller dikes of propylitized andesite. In the northeastern part of the area the above-mentioned rocks are covered by a thick lava flow of dacite.

The rhyolitic dikes show a U-shape (Fig. II-12-1). They dip steeply toward the inner side. The block surrounded by the U-shaped dikes shows a cauldron subsidence. It can be said that the occurrence of the dikes corresponds to a ring dike in the meaning of Billings (1943, 1945) (Fig. II-12-2).

The mineralization took place along the U-shaped ring dike, particularly along the west wing. Consequently, it is inferred that the ore solution originated from the underground acidic magma which produced the rhyolitic ring dike.

C. Epithermal Veins

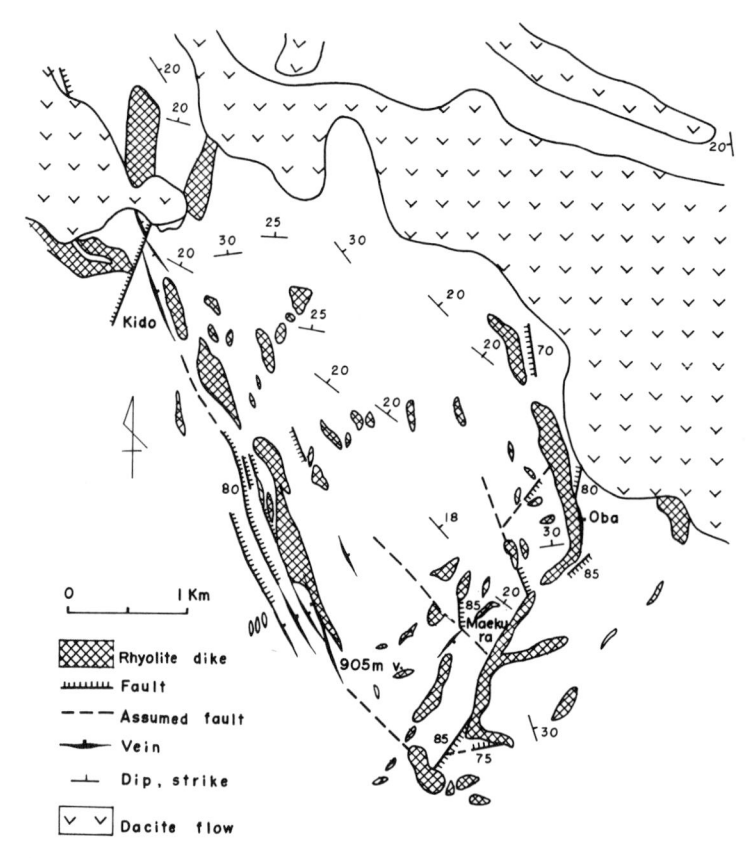

FIG. II–12–1. Geologic map of the Yaso mine (after Takenouchi, 1962).

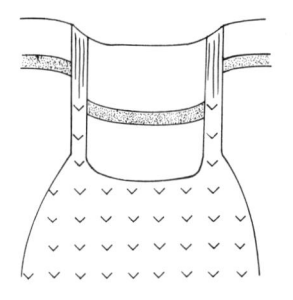

FIG. II–12–2. Ring dike (after Billings, 1945).

In the longitudinal section of the 905 M vein in the western wing of the U-shaped block (Fig. II–12–1), the traces of the boundaries between the Paleozoic formation and Miocene conglomerate in both hanging wall and foot wall sides are shown in Fig. II–12–3. The arrow in the figure indicates the direction of the relative displacement of the hanging wall at the time of the cauldron subsidence.

The deposits of this mine comprise typical chlorite-quartz-chalcopyrite veins. The largest vein is the 905 M vein, which extends as far as 1,200 m in the strike direction and as deep as 400 m.

The ore minerals are chalcopyrite, pyrite, sphalerite, galena, bismuthinite, etc., and the gangue minerals are chlorite, quartz and calcite. Though the zonal distribution of minerals is not so conspicuous, the upper parts of the veins are rich in sphalerite and galena, and the lower parts are rich in chalcopyrite. Bismuthinite has an intimate relation with chalcopyrite.

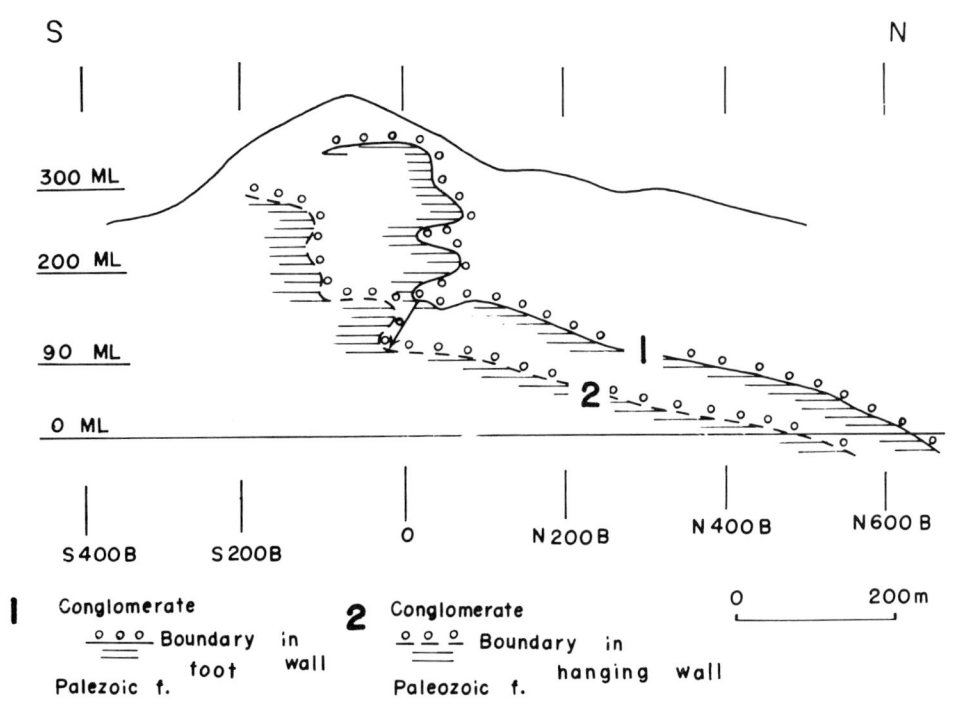

FIG. II-12-3. The boundaries between the Paleozoic formation and Miocene conglomerate intersecting the longitudinal section of 905 M vein (after Takenouchi, 1962a).
  1: The boundary in the hanging wall block of the vein.
  2: The boundary in the foot wall block of the vein. Arrow indicates the direction of slip of the hanging wall.

Wall rocks show the alteration of silicification and chloritization. The silicification is stronger in the upper parts of the deposit and the chloritization is striking in the lower.

II-13. Geology and Vein System of the Hosokura Mine, Miyagi Prefecture
(Imai, 1955, 1956b)

H. Imai

The Hosokura mine is situated about 60 km north of Sendai in northeastern Japan. This mine produced about 16 million tons of crude ore containing about 640,000 t of zinc and 235,000 t of lead up to 1975.

Geology of the Mining Area

The mine area is composed of propylite, green tuff (mostly andesitic tuff) and rhyolite, which belong to the Miocene formation and are extensively covered by dacitic pumice flow and younger sedimentary rocks. The dacitic pumice flow probably erupted in the Pliocene epoch. Most of the veins occur in propylite and some in green tuff. A mass of propylite occurs in the green tuff and crops out throughout an area of about 3.5 km². The distribution of propylite narrows towards the lower levels of the mine adits. This mode of occurrence indicates that the mass of propylite seems to have penetrated the surrounding green tuff as a volcanic neck (Fig. II-13-1). The green tuff containing boulders of propylite (volcanic conglomerate) lies on the fringing part of the propylite mass, as if it were in unconformable

C. Epithermal Veins

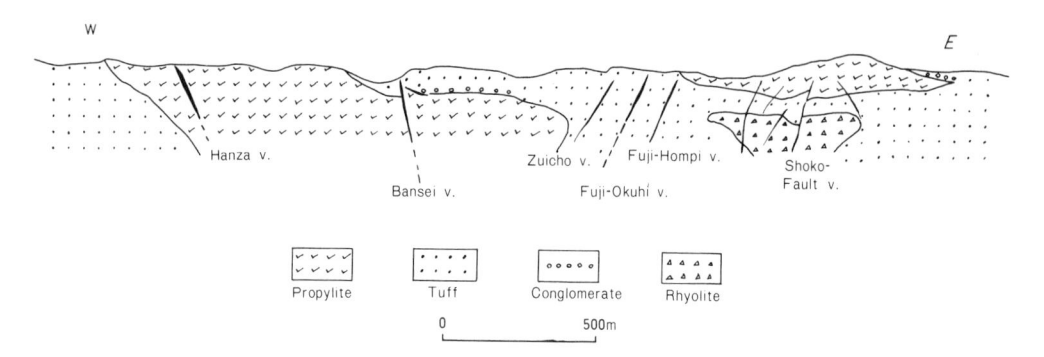

FIG. II-13-1. Geologic section of the Hosokura mine.

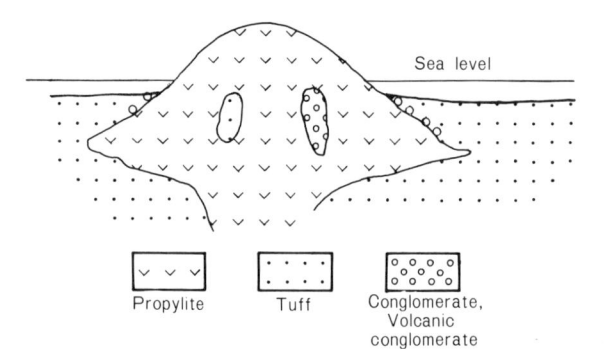

FIG. II-13-2. Geologic model of volcanic activities in the Hosokura area.

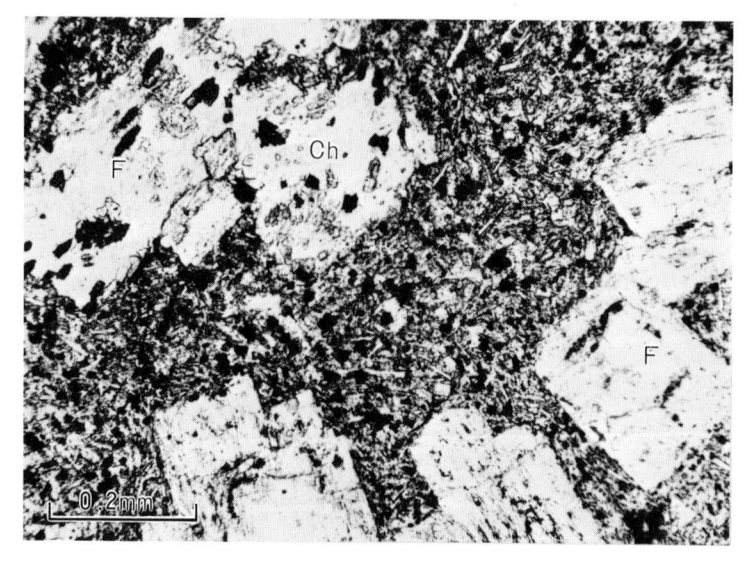

One nicol.
F: alkaline (potash) feldspar.
Ch: chlorite (bastite).

FIG. II-13-3. Photomicrograph of propylite in the Hosokura mine (sample no. Hosokura-294).

relation (Fig. II-13-1). It might be concluded that the original andesite (now propylite) erupted as an island during the period of sedimentation of the green tuff (Fig. II-13-2). So the geologic age of the andesite is the same as the deposition of the surrounding tuffaceous rock (green tuff). In the propylite mass are found small pockets of green tuff and volcanic conglomerate in places (Fig. II-13-2). From this occurrence it is thought that the original andesite mass was not formed by a single eruption but by repeated eruptions, and the green tuff and the volcanic conglomerate must have filled the depressions of the andesite during the intervals of eruption. The exact mode of occurrence of the rhyolite is still unknown.

The propylitization of andesite was not due to the vein-forming fluid; it had already taken place before the veins were formed, as described later. The ore deposits were formed at the vent of the andesite eruptions, and may be genetically related to the eruptions of andesite, andesitic tuff or rhyolite. The veins are covered unconformably in places by the dacitic pumice flow. Faults which displace the veins after the mineralization are very scarce.

The ore and gangue minerals are sphalerite, galena, "wurtzite" (it has been converted into sphalerite), pyrite, chalcopyrite, hematite, stibnite, silver minerals (pyragyrite, polybasite and tetrahedrite), quartz, chlorite, fluorite, calcite, kaolinite and other minerals.

In some places alteration of country rocks has been intense, while in other places almost no alteration is recognized. In the altered country rocks, the writer recognized quartz, kaolinite, sericite, montmorillonite, sepiolite (?), zunyite and chlorite.

TABLE II-13-1. Chemical analyses of propylite.

	Propylite[1]	Propylite[2]	Propylite[3]	Argillized propylite[4]
	Hosokura 294	Hosokura K. 43	Mikawa E. 9	Hosokura 95
SiO_2	55.31 %	54.72 %	60.57 %	59.74 %
TiO_2	0.61	0.84	0.62	0.48
Al_2O_3	17.45	16.08	16.23	16.20
Fe_2O_3	2.27	5.55	1.80	1.19
FeO	4.82	3.07	5.08	2.47
MnO	—	0.21	—	0.05
CaO	4.18	4.58	0.56	6.10
MgO	5.21	4.21	1.84	2.30
K_2O	4.62	3.64	6.92	1.40
Na_2O	0.84	2.51	2.13	0.95
CO_2	2.56	0.28	—	4.59
FeS_2	0.10	0.96	1.80	—
$H_2O(+)$	1.45	2.91	1.69	1.11
$H_2O(-)$	0.90	0.94	0.41	3.09
P_2O_5	—	tr	0.20	0.22
Total	100.32	100.50	99.85	99.89
Analyst	K. Iwase*	(Narita, 1961)	(Nagasawa, 1961)	K. Iwase*

(*Dept. of Mining, Univ. of Tokyo, unpublished data, 1952)

1) Hosokura 294: Hanza vein, 0 mL; 2) Hosokura K. 43: Kanoko vein, 0 mL; 3) Mikawa E. 9: Homban 0 mL; 4) Hosokura 95: Fuji vein 0 mL.

Propylite

It is characteristic that the andesites in the terrains of subvolcanic deposits, in most cases, underwent propylitization. Propylite is a dark green compact rock. Every stage of alteration from almost fresh rock to typical propylite can be observed in the field. Propylitization commenced with change of the mafic minerals to chlorite or epidote and albitization or the formation of potash feldspar from the basic plagioclase along the borders, cleavages and cracks showing network structure, and resulted finally in entire replacement by albite or potash feldspar (Fig. II-13-3). The alteration product of the plagioclase shows the original outline, but usually shows no twinning lamellae. Formerly, it was as-

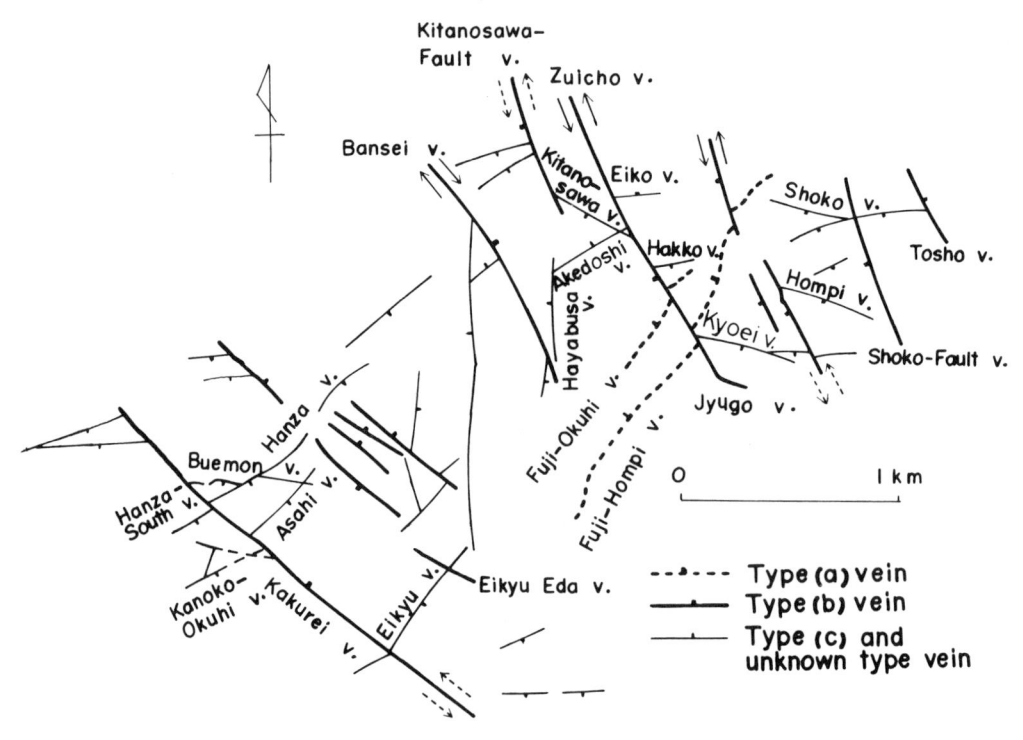

FIG. II-13-4. Vein map of the Hosokura mine.

sumed that the feldspar in the propylite was albite or albite-molecule-rich plagioclase (Kato, 1931). The chemical analyses of the propylite from the Hosokura mine are shown in Table II-13-1. The addition of potash is reflected by the formation of potash feldspar. This fact was anticipated by Tsuboya (1928) on some propylites from northeast Japan, and was ascertained by Nagasawa (1961) on the propylite from the Mikawa mine and by Narita (1961) in the Hosokura mine.

The distribution of propylite is not limited to the small area surrounding the ore deposit. The spatial relationship between the vein and the propylite suggests that propylitization was not due to the vein-forming solution. Observation in the Hosokura mine suggests that propylitization corresponds to the hydrothermal alteration antecedent to ore deposition as Tsuboya (1928), Schneiderhöhn (1928), Coats (1940), Imai (1956b) and Nagasawa (1961) have stated already. At some places in the Hosokura mine, propylite underwent argillic alteration. An example of the chemical composition of argillized propylite in this mine is shown in Table II-13-1. This is the kaolinized rock.

Vein System

Apparently, the pattern of fissures filled with vein materials in this district is complicated. The writer has classified these fissures into three types (Fig. II-13-4):

(a) NE-SW fissures (fault fissures); examples, Fuji-Hompi vein and Fuji-Okuhi vein.

(b) NW-SE fissures (fault fissures); typical examples, Tosho vein, Shoko-Fault vein*, Zuicho vein, Kitanosawa-Fault vein*, Bansei vein, and Kakurei vein. They dip toward east or west; some of them dip at comparatively low angles, i.e., 50°–60°.

(c) Branch fissures which diverge from (a) and (b); examples, Shoko vein, Hompi vein,

---

* As the Shoko-Fault vein intersects with the Shoko vein, it is named as such. The same is said for the Kitanosawa-Fault vein.

Part II: Vein-type Deposits

FIG. II-13-5. Main zone of Ninth Level, the Siscoe mine, Canada (after McKinstry, 1941).

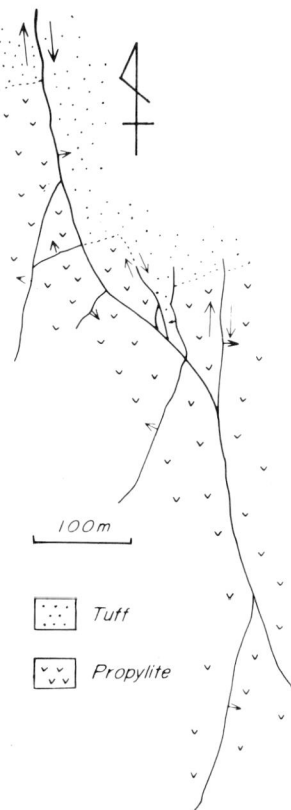

FIG. II-13-6. Bansei and its branch veins.

Kitanosawa vein, Akedoshi vein, Hakko vein, Hayabusa vein, Buemon vein, Hanza vein, Asahi vein, Eikyu vein, Hanza-South vein and Kanoko-Okuhi vein. Most of the veins now being worked belong to this type. They dip steeply toward north or south.

This classification was made on the basis of the mechanism of formation of the fissures, not on the basis of the fissure-filling materials.

The fissures of type (a) have not yet been sufficiently studied. They also originated in

FIG. II–13–7. Splay fault (after Anderson, 1951). a: master fault, b: splay fault.

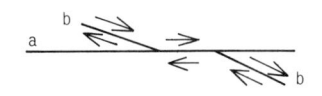

FIG. II–13–8. Shear of the second order (after McKinstry, 1953). a: master fault, b: shear of the second order.

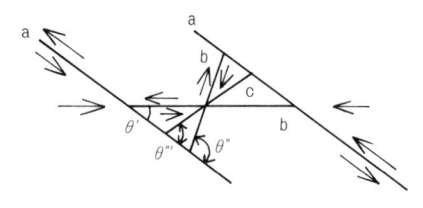

FIG. II–13–9. Formation of shears of the second order and tension crack (after Mckinstry, 1953).
a: master fault, b: shear of the second order, c: tension crack.

wrench (strike-slip) faults. Different kinds of propylite border along the Fuji-Hompi vein, which is a type (a) fissure. This indicates that propylitization had already occurred before the fissures were formed.

The veins which belong to type (b) are fault veins which contain much gouge and are rich in slicken-sides. The Zuicho vein and the Bansei vein are typical of this type.

The Fuji-Hompi vein and the Fuji-Okuhi vein, which belong to type (a), are displaced about 50 m by the fissure of the Zuicho vein (Fig. II–13–4). This is considered to be normal or wrench. If it is a normal fault, the western block of the Zuicho fissure must have been relatively downthrown. If it is a wrench fault, the eastern block would have been displaced relatively northward. Of course, an intermediate case between the above two may exist. The writer believes the fault is a wrench fault for the following reasons: The southeastern end of the Zuicho vein dies away in the Jyugo vein (Fig. II–13–4). This feature resembles to the terminal condition of the wrench fault described by McKinstry (1941) in the Siscoe mine in Canada (Fig. II–13–5). In the Hosokura area, the fissure veins such as the Akedoshi vein, the Hakko vein, etc., diverging from the Zuicho vein have the characteristics of tension cracks, and the Kyoei vein is a fault vein. The arrangement of these branch fissures indicates that the displacement by the Zuicho fissure is mainly in the direction of the strike, and they were formed by this movement. The amount of vertical displacement which was accompanied by horizontal displacement is not known.

The boundary between propylite and tuff is displaced about 170 m along the Bansei vein (Fig. II–13–6). The characteristics of the branch veins diverging from the Bansei vein are the same as those diverging from the Zuicho vein. So, interpreted in the same way, the Bansei fissure is a wrench fault, with relative displacement of the western block toward the north.

The ascertained directions of displacement along faults are shown by solid arrows in Fig. II–13–4.

In the Zuicho vein and the Bansei vein, the acute angles between the main veins and the branch fissures generally point in the direction of displacement of the block, as shown in Figs. II–13–4 and –6.

Generally speaking, the orientations of the branch fissures do not always indicate the relative movements between the displaced blocks. Anderson (1951) described the branch faults shown in Fig. II–13–7 and named them splay faults. An example of splay faulting can be seen in the dislocations of the Craven group in the West Riding of Yorkshire. Anderson (1951) explained the dynamics of this fracture pattern by ellipse coordinates. McKinstry (1953) discussed the branch fissures shown in Fig. II–13–8, and named them shears of the second order. According to him, they would make angles of roughly $\theta' = 45° - 1/2\, \varphi$, $\theta'' = 135° - 3/2\varphi$ with the fault (Fig. II–13–9). If tension cracks develop, they should make an angle $\theta''' = 90° - \varphi$ with the fault, where θ', θ'' and θ''' are the angles between the master fault and branch faults in the direction opposite to the displacement of this block along the master fault, and φ is the angle of internal friction of the rock. From these examples, it is seen that the orientation of the branch shear does not necessarily indicate the direction of displacement of the block. But the acute angle which the tension crack makes with the master fault points in the direction in which the block moved on its own side of the fault.*

The other fissures belonging to type (b) are also interpreted to be wrench faults of the same type, judging from the orientation of diverging branch fissures. However, no geologic indications such as key beds are recognized in the surrounding country rock. So the displacement of each wall along these faults has not been ascertained. The writer believes that these NW-SE fissures may have the same attitude as the Zuicho and Bansei fault fissures in the relation between orientations of the branch fissures and direction of displacement of the master fault. Since the country rock is the same, it might be expected that a similar fracture would be generated in the block by the dislocation of the master fault. Further, most of the acute angles which the branch veins filling tension cracks make with the NW-SE master fault veins would point to the direction in which the block moved on its own side of the master fault. In Fig. II–13–4, the inferred directions of displacement of NW-SE faults are marked by broken arrows. The branch fissures whose angles with the master faults are close to 90° have the characteristics of tension cracks. Fractures by tension are often gash-like, short and wide. Generally they are not accompanied by gouges or slicken-sides. The walls would fit with each other if the filling were removed. This fitting of the shapes of walls can be more distinctly recognized in small veinlets which are parallel to the main fractures. Also, in tension fractures, no displacement of the walls by shearing is recognized. Actually, they may often have opened by movements oblique to the plane of fractures—that is, a shearing component as well as a normal tensile component has been active in opening them. The Hakko vein, the Akedoshi vein, the Asahi vein, the Kanoko-Okuhi vein and the Hanza-South vein seem to belong to the tension fracture type (Fig. II–13–4). Also, in the eastern part of this mining district, mesh-like fissures are recognized as a type of tension fracture, which the writer discusses in a later section.

The branch fissures diverging with small angles, mostly smaller than 45°, in a direction opposite to the displacement of this block by the master fault, have the characteristics of shear fractures. They are rich in gouges and slicken-sides, and both of their walls are relatively displaced. The Kyoei vein diverging from the Zuicho vein is a typical example of this type (Fig. II–13–4). They are wrench faults which belong to shears of the second order. In Hanza (Fig. II–13–4) and other veins which belong to this type, a pattern of so-called curved shingles is observed, which would be caused by horizontal displacement (Fig. II–13–10). The existence of branch fissures, i.e., fissures of type (c) diverging from the master fault (b), depends on the actual displacement of the block on its own side at the time of faulting.**

Kayaba and Ishii (1953) reported that in the Ogoya mine, Ishikawa Prefecture, branch veins which diverge from the master veins by 45° are predominant. These branch cracks

---

* E. Cloos concluded that the appearance of feather joints can be taken as valuable indicators of relative movements between blocks. In his paper, however, feather joints seem to include second order shears, tension joints, etc., so the opinion of Cloos can be applied in some cases but not in all cases (Cloos, E. (1932) Feather joints as indicators of direction of movements on faults, thrusts, joints and magmatic contacts: *Proc. Nat. Avad. Sci.*, **18**, 387–395).

** It goes without saying that the arrows indicating the opposite directions on both sides of the fault only indicate the relative displacement of the blocks.

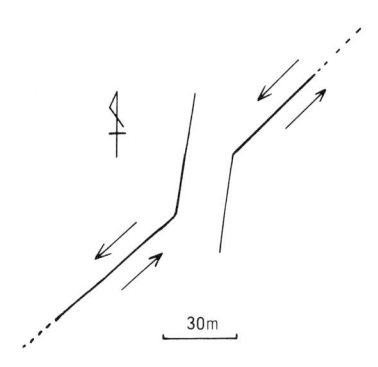

FIG. II-13-10. Curved shingles observed in the Hanza vein.

filled with vein materials would be produced by the horizontal displacement of the master faults. Also, the fracture patterns of the Yellowknife district described by Henderson and Brown (1949) and of the Darwin Hills mine, California, described by Wilson (1943), belong to this type, as pointed out by McKinstry (1953).

As determined above, the displacement along most of the fissures of type (b) is left-handed. But the displacement along the Bansei vein is right-handed (Fig. II-13-6). From this fact, it is impossible to conclude simply that the fracture pattern of this district was produced by E-W lateral pressure.

The three types of fissures described above are filled with vein materials. It is noticeable that type (b) fissures are low ore grade which resembles the veins of Darwin Hills described by Wilson (1943). The Zuicho vein, which belongs to type (b) (Fig. II-13-4), is poorly mineralized in general, and rich in gouges and slickensides, but in the vicinities of the junctions of the Hakko vein and the Fuji-Hompi vein, the ore grade of the Zuicho vein becomes higher. The vein materials of the Zuicho vein near the junction with the Fuji-Hompi vein resemble those of the latter, and they resemble those of the Hakko vein near the junction with the Hakko vein. Observations made at the junctions of the Zuicho vein with the Kyoei and Fuji-Hompi veins tell that the ore-forming solution flowed into the former vein out of the latter two veins (Fig. II-13-4).

Some Characteristics of the Branch Veins

Relations between shears of the second order and tension cracks of the second order: As stated before, the Buemon vein, the Hanza vein and the Asahi vein branch from the Kakurei vein (Fig. II-13-4). The former two are shears of the second order, and the Asahi fissure is a tension crack of the second order. The relation between the Buemon vein and the Hanza vein is shown in Fig. II-13-11. The Buemon vein was displaced about 50 m by the movement of the Hanza vein. That is, the formation of the Hanza fissure was later than that of the Buemon fissure. This direction of displacement of the Hanza fissure coincides with the attitude of the shear of the second order as shown in Fig. II-13-9. The Buemon vein and the Asahi vein intersect each other with no displacement at the 0 m level, though at the Lower II level, the former is displaced 1–2 m by the latter. As described above, the Asahi fissure is a tension crack. From the above relation it is suggested that the tension crack of the Asahi fissure was formed later than the shear of the second order of the Buemon fissure. The vein materials flowed into these fissures after the above movements. The junction of the Hanza and Buemon veins formed an ore shoot.

The same relation is observed in the Kitanosawa vein and the Akedoshi vein. Both are branch veins diverging from the Zuicho vein (Fig. II-13-12). The former is a shear of the second order, and the latter is a tension crack. The fact that Kitanosawa vein has been displaced 1–1.5 m by the Akedoshi vein, indicates that the Kitanosawa fissure, a shear of the

Part II: Vein-type Deposits

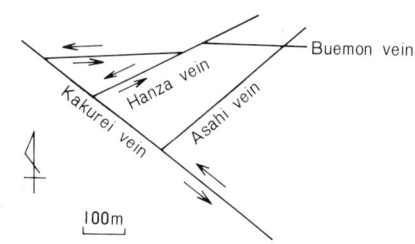

FIG. II–13–11. The relation between shears of the second order and tension crack (1).

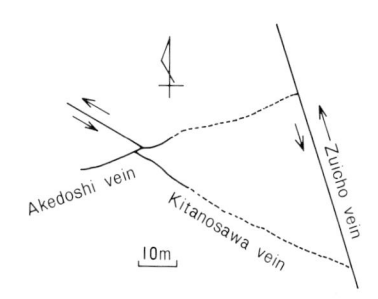

FIG. II–13–12. The relation between shear of the second order and tension crack (2).

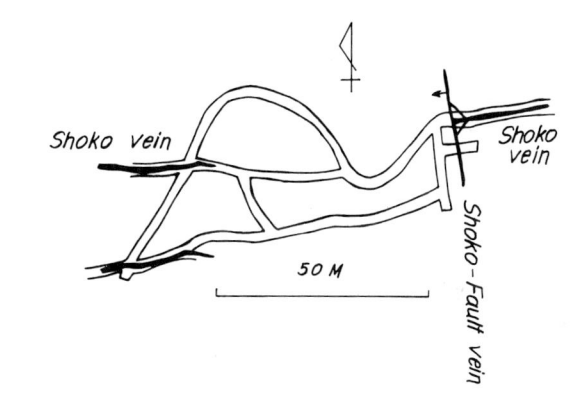

FIG. II–13–13. Vein map near the junction of the Shoko vein and the Shoko-Fault vein, −110m level.

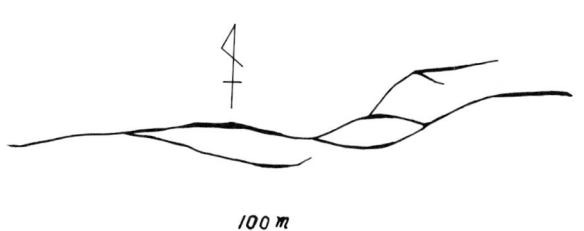

FIG. II–13–14. A part of the Shoko vein, −170m level.

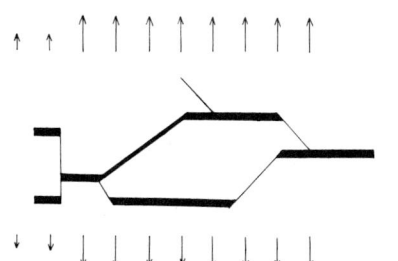

FIG. II–13–15. Formation mechanism of mesh-like veins by tensile force.

second order, was formed earlier than the Akedoshi fissure, a tension crack. As already mentioned, in a tension crack of Akedoshi fissure a shearing component has been active.

Mesh-like veins of the eastern area of the Hosokura mine: In the eastern area of the Hosokura mine the Shoko-Fault vein belonging to type (b) fissures strikes NNW (Fig. II–13–4). In Fig. II–13–4, no displacement is observed in the Shoko vein or in the Shoko-Fault vein at their junction. However, from the plan of the junction on a larger scale shown in Fig. II–13–13, it is thought that the so-called Shoko vein fissures on each side of the Shoko-Fault vein are different, having been produced by the displacement of Shoko Fault. That the Shoko vein fissures look like one vein crossing the Shoko-Fault vein is a casual result of two different fissures starting from nearly the same position on both sides of the Shoko-Fault vein.

The Shoko vein, Hompi vein and other branch veins in the eastern area (Fig. II–13–4) are of mesh-like type. For example, a part of the Shoko veins is shown in Fig. II–13–14. Mesh-like vein patterns are characteristic of an area where tensile stress worked. The mechanism of mesh-like pattern formation by tensile stress is shown diagrammatically in Fig. II–13–15. It is composed of pure tension cracks, pure shear cracks and fissures which have both characters. It is conceivable that the ore shoots occur in the parts of pure tension cracks in which the width of the fissure becomes large.

Summary and Conclusions

The writer has described the vein fissures of the Hosokura mining district as belonging to the following three types:
(a) NE-SW fissures probably related to wrench faults;
(b) NW-SE fissures related to wrench faults;
(c) Branch fissures from type (a) and (b).

After the displacement of the fissures of type (a), wrench faults of type (b) were formed. Due to this displacement, the branch fissures of type (c) were produced. The fissures of type (c) are composed of shears of the second order and tension cracks. Most of the branch fissures have attitudes such that the acute angles which they make with their master faults point to the directions of displacement—that is, to the directions in which their own walls of the master shears were moving.

The characters of the vein materials are not necessarily coincident with the type of fissure. The characters might depend on the origins of the ore-forming fluid, its characteristics, its flow directions and other factors. It has been observed that the veins of type (b) are generally poor in ores.

From the above-mentioned fissure pattern, it is concluded that the deposit should be prospected along the NW-SE fissures now occupied by the vein materials. Thus, the branch veins filling second order shears and tension cracks would be caught. Further, by following these branch veins, fissure veins of the third order would be found.

In connection with the experimental studies of the cracks (Hirata, 1931), it is probable that a close relationship exists between the distances of the fissures and their depth persistences. Further studies are needed not only from the point of crack formation in the earth crust but also from that of prospecting the ore deposits. It is generally said that epithermal veins, particularly those in Tertiary lavas, commonly occupy normal faults, whereas veins of the deeper zones commonly occupy reverse or wrench faults or faults that have slipped in direction between the dip and the strike (McKinstry, 1941). However, the writer has observed that the epithermal (subvolcanic) veins in Japan frequently occupy wrench faults.

II-14. Geologic Structure and Fluid Inclusion Study at the Toyoha Mine, Hokkaido
(Imai, 1956b; El Shatoury *et al.*, 1974, 1975)

H. M. El Shatoury, S. Takenouchi and H. Imai

The Toyoha mine is situated about 25 km west of Sapporo, Hokkaido. Since 1939, this mine has produced 8.5 million tons of crude ore containing Ag, 120 g/t; Zn, 6.5–7.0 %; Pb, 2.5–2.8 % and S, 13 %. Recent production is 35,000 t of crude ore per month. The deposit is now developed by underground workings reaching down to the 450 m level below 0 m level (adit level).

Geologic Setting

Rocks of Miocene age comprise pyroclastic formations intercalated with marine and epiclastic sedimentary members as well as extrusive and intrusive rocks including andesite, propylite, basaltic dikes, rhyolite and quartz porphyry. These rocks are categorized into five formations in ascending order as follows: Koyanagizawa (rhyolitic and andesitic lavas), Motoyama (sandstone, mudstone and conglomerate), Nagato (tuff, tuff breccia, andesitic lava and tuffaceous sandstone), Sanbonmata (dacitic lava and agglomerate) and Oeyama (andesitic lava and agglomerate). The main mineralization, which is supposed to be genetically related to the intrusion of quartz porphyry, occurs in the Motoyama and Nagato Formations.

Fissure Formation and Mineralization

The veins of the Toyoha mine have two main directional trends: E-W trend and NW-SE trend (Fig. II–14–1). The former system is characterized by the predominance of quartz as the gangue mineral, while calcite and/or mangano-calcite characterize the latter system. The characteristics of the vein material are be due to the stage of the mineralization.

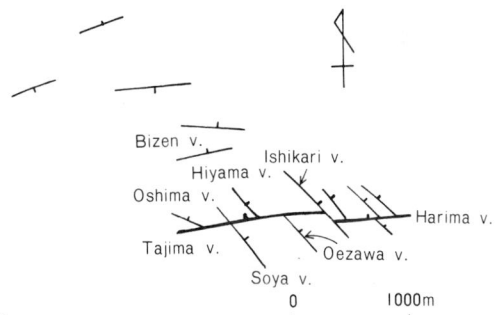

FIG. II–14–1. Vein system of the Toyoha mine.

The Tajima and Harima veins run E-W and dip 60° N. The Harima vein is the eastern extension of the Tajima vein. Some of the veins trending NW with NE dips diverge from the Tajima and Harima veins, forming a horse-tail structure. The Hiyama, Oezawa, and Oshima veins correspond to the veins branch from the master faults, i.e., the Tajima and Harima veins.

The vein pattern resembling a horse-tail structure with respect to the master fault zone would indicate wrench fault movement. But the emplacement of quartz porphyry shows that the downward dip-slip of the hanging-wall along the master fault took place by 40–50 m, as shown in Fig. II–14–2(b).

C. Epithermal Veins

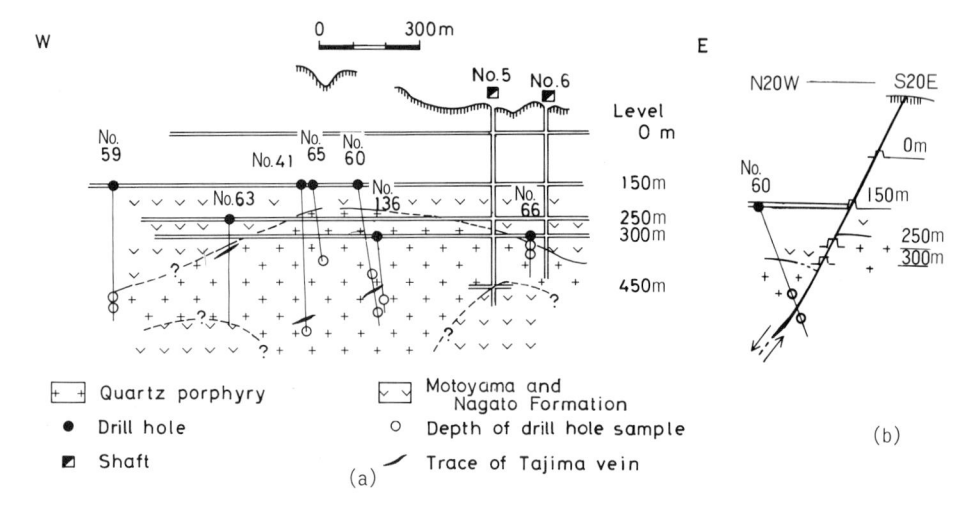

FIG. II–14–2. Configuration of the outline of the quartz porphyry along the Tajima vein inferred from drill holes and underground workings, and the location of core samples investigated for fluid inclusions. Rock boundaries in (a) is on the footwall of the vein.

From these facts, the writers suggest that the master fault which had been formed as a wrench type by the lateral pressure, moved afterwards as a normal type by the upward pressure due to the pushing up of the quartz porphyry. This interpretation is similar to Bryner's at the Mankayan mine (Bryner, 1969). Namely, it corresponds to case (3) of fissure formation, as described in Chapter II–1.

In some of the NW trending veins, e. g. the Soya vein, the vein fissure and material cut across the E-W vein, e. g. the Tajima vein, without displacement of both veins, indicating that this NW-trending vein fissure is a tension crack which was formed and mineralized after the mineralization of the E-W vein (Akome and Haraguchi, 1967).

The ore minerals are sphalerite, galena, chalcopyrite, pyrite, pyrargyrite, argentite, native silver, and rarely pyrrhotite, magnetite, hematite, arsenopyrite, alabandite, tetrahedrite group mineral, marcasite, cassiterite and stannite (Yajima, 1977), canfieldite and berthierite (Yajima, 1977)*. The gangue minerals are quartz, chlorite, calcite and manganocalcite, rarely accompanied by rhodochrosite, inesite, barite, etc.

Shikazono (1973, 1974, 1975) discussed f_{S_2}-f_{O_2} conditions in the course of the mineralization of the Soya vein from the mineral paragenesis and FeS content in sphalerite.

Fluid Inclusion Study of the Quartz Porphyry

Sampling: Samples used in this study were collected from the quartz porphyry exposed in underground workings at the 150, 200, 250, 300, and 450 m levels (Fig. II–14–2). The samples from the 250, 300, and 450 m levels represent the main body of the intrusion, whereas the samples from the 150 and 200 m levels represent an apophysis of the quartz porphyry main body. Besides these samples, some drill core samples representing depths equivalent to the 425, 500, 550, and 600 m levels were also collected. Fig. II–14–2(a) shows the depth of samples taken from drill holes as well as the shape (on foot wall) of the quartz porphyry intrusion configurated from extensive underground working and drill holes at the Toyoha mine. Another set of samples was collected from underground workings representing the mineralized quartz veins as well as the post-ore quartz veins.

---

* Oral communication.

Part II: Vein-type Deposits

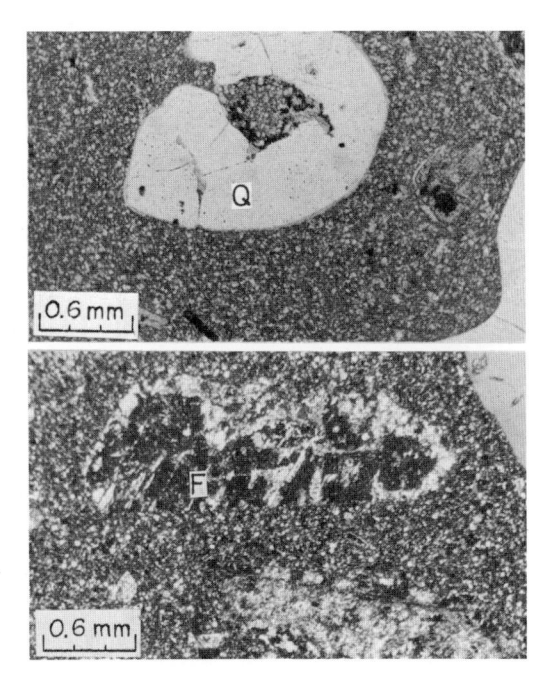

FIG. II-14-3. Photomicrographs of thin section of quartz prophyry. Crossed nicols.
Q: resorbed quartz phenocryst;
F: altered plagioclase phenocryst.

Quartz porphyry exposed in the Toyoha mine is a leucocratic rock composed essentially of quartz and altered plagioclase phenocrysts in an aphanitic groundmass of sericitized orthoclase and quartz. The quartz phenocrysts are generally resorbed (Fig. II–14–3) and vary in diameter from less than 1 mm to about 2 mm. The plagioclase shows intense argillic alteration, but the lath shape of the phenocrysts is preserved. Magnetite and apatite occur in accessory amounts.

General statement: The shape of fluid inclusions in quartz phenocrysts of quartz porphyry varies from negative crystal to irregular shapes. It is considered that the fluid inclusions in phenocrysts of porphyry would be secondary in origin (refer to Chapter VII–3).

The fluid inclusions in some parts of the quartz porphyry intrusion are characterized by various filling degrees, indicating their formation from heterogeneous mixtures of liquid and gaseous phases. The filling temperatures of these fluid inclusions vary over a wide range, but the lowest value would be close to the trapping temperature of fluid (Smith and Little, 1959). Consequently, temperature measurements were carried out on liquid inclusions having the highest filling degree in the same inclusion group.

Although careful selection of liquid inclusions for temperature measurements was maintained, the writers cannot exclude certain elements of uncertainty in the proper selection of inclusions which trapped only the liquid phase from those which trapped some gaseous phase with liquid or those highly recrystallized after their formation. The uncertainty, anyhow, is reflected in the variation of filling temperatures of inclusions.

Types of inclusions: Microscopic examination of thin polished plates of the quartz porphyry samples revealed the presence of different type of fluid inclusions as well as of glass inclusions (Roedder, 1971a; Roedder and Coombs, 1967) (Fig. II–14–4). Systematic description of different samples representing different levels of the underground workings at the Toyoha mine indicates a remarkable change in the type of inclusions from liquid-rich (liquid inclusions) to gas-rich (gaseous inclusions) type. This change is rather gradational from the 450 m level and reaches a maximum predominance of gaseous inclusions in the 300 m and 250 m levels. It is noteworthy that the 250 m level approximately marks the up-

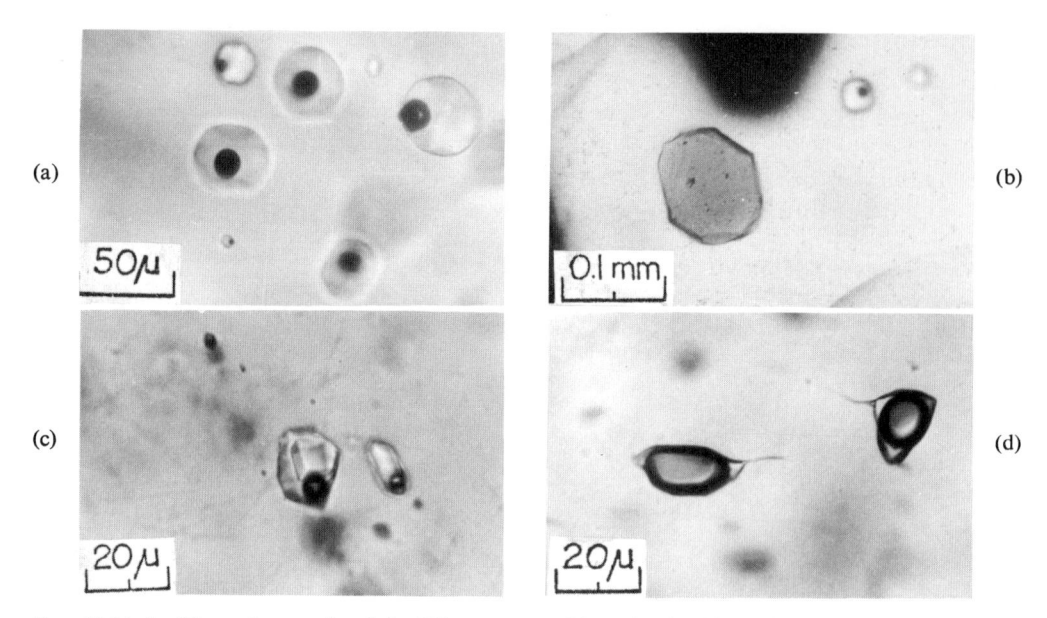

FIG. II–14–4. Photomicrographs of the different types of inclusion in thin sections of quartz porphyry.
 (a): Diamond-shaped transparent glass inclusions with vacuoles.
 (b): Gray partially devitrified glass inclusion without vacuole. Opaque: glass inclusion.
 (c): Polyphase inclusion of type 1. A negative crystal composed of liquid, gas and a halite cube (Drill-hole 41 at a depth corresponding to the 600 m level).
 (d): Gaseous inclusion of type 3 (300 m level).

per portion of the quartz porphyry intrusion (Fig. II–14–2). Gaseous inclusions are also observed in samples representing quartz porphyry apophysis exposed in the 200 and 150 m levels.

Besides the gaseous and liquid inclusions, halite- and/or sylvite-bearing inclusions are also recognized in the samples representing the lower levels of the mine and occasionally in some samples from the upper levels.

Fluid inclusions, in general, are abundant in samples representing the central part of the intrusion. These fluid inclusions coexist with glass inclusions which vary greatly in shape and degree of devitrification (Fig. II–14–4(a)(b)). The western margin of the intrusion (represented by two samples from drill hole No. 59), however, is very poor in fluid inclusions. Extremely few two-phase (liquid and gas) fluid inclusions were recognized in drill core samples representing the 500 m level (Fig. II–14–2). Another sample from the same drill hole representing the 530 m level was found devoid of any fluid inclusions.

The following is a description of the different types of fluid inclusions in quartz porphyry:

Type 1. Polyphase inclusions: Polyphase inclusions are those composed of liquid, gas and one or more solid crystals. The solid crystal is either halite or sylvite (identified by cubic form and isotropism). Sylvite is distinguished from halite during heating measurements. Sylvite generally dissolves faster at relatively lower temperatures than halite, which withstands heating except for a rounding of its corners. In some fluid inclusions sylvite dissolves completely at temperatures between 94°C and 182°C, while halite cubes remain at temperatures above 250°C with only partial dissolution of their corners. Other solid crystals in the polyphase inclusions, although they occur only rarely, are very small grains of dark brown color (hematite?) or birefringent unidentified minute prismatic crystals or flakes. This latter mineral is observed in fluid inclusions from the upper levels of the mine.

The occurrence of this type of fluid inclusion is frequent in samples from the 600 m level

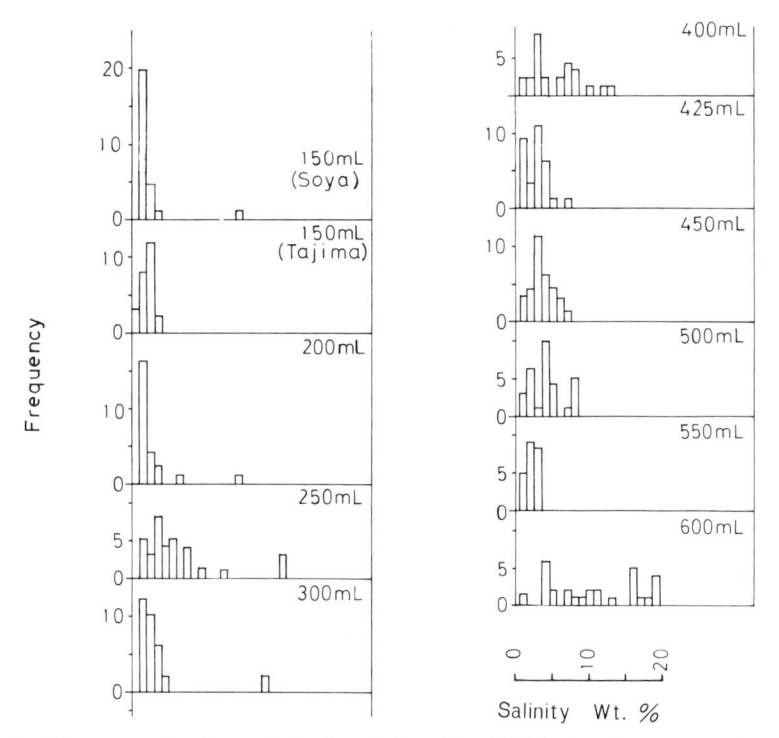

FIG. II–14–5. Histograms showing variation in salinity of liquid inclusions in quartz porphyry with depth in the Toyoha mine.

and occasionally present in samples from the upper levels, probably because of upgrading of the salt concentration during boiling (Fig. II–14–4(c)).

Type 2. Liquid inclusions: This type of inclusion is common in all samples representing the different levels of the mine. These are characterized by a high filling degree of liquid, and all homogenize to the liquid phase at different temperatures. The salinity range of these inclusions varies greatly from as low as 1 to over 20 wt. % NaCl equivalent concentration. Some of these inclusions contain a small volume of liquid CO_2 upon cooling.

Type 3. Gaseous inclusions: This type of inclusion occurs sporadically in samples from the different levels of the mine but is characteristically abundant in samples representing the upper part of the intrusion. These inclusions are characterized by having filling degrees of about 5 % or even less (Fig. II–14–4(d)). The gas-liquid ratio of these inclusions varies greatly, and consequently they homogenize at different temperatures characteristically higher than those of the liquid-rich type.

Because of the dark border of their gas bubbles, it is often difficult to recognize the presence of liquid CO_2. However, in some of these inclusions it was possible to recognize it upon cooling.

Type 4. Monophase (liquid) inclusions: Monophase inclusions are scarce, compared with the above types of inclusions. These inclusions are evidently the product of necking-down of two-phase inclusions.

Salinity (NaCl Equivalent Concentration) of Fluid Inclusions in Quartz Porphyry

Type 2 liquid inclusions described above were measured for salinity by freezing-stage microscope. The salinity of the halite-bearing inclusions, however, is estimated visually from the volume of salt crystals.

The range of salinity obtained from 316 fluid inclusions from quartz phenocrysts lies between that of nearly fresh water and over 20 wt. % NaCl equivalent concentration. It is noteworthy that few of the inclusions that were run under the freezing-stage microscope showed peculiar behavior upon freezing and defrosting. Upon freezing, the gas-bubble of these inclusions completely disappears. This is different from normal behavior where the gas-bubble shows different degrees of deformation due to the expansion of ice. On defrosting, the bubbles of these peculiar inclusions appear very suddenly at $+4.0°C$. Reruns on such inclusions do not give the same temperature at which the bubble appears but averages around $+4.0°C$, indicating a state of disequilibrium. Roedder (1971b, c) discussed similar conditions and assumed that such inclusions are under negative pressure.

The appearance of liquid CO_2 or CO_2-gas hydrate is observed in some inclusions while they are cooling.

Variation in salinity with depth: The results of salinity measurements of fluid inclusions in specimens of quartz porphyry with a freezing-stage microscope are shown in Fig. II–14–5. This gives the frequency distribution of the measured salinity exclusive of halite-bearing inclusions. This Figure illustrates the frequency distribution of salinity of liquid inclusions in the different levels of the mine and points out the general trend of decreasing salinity of solutions from the 600 m level to the 150 m level. The data, combined with visual microscopic investigation of inclusions characteristic of each level, explains some of the irregularities and discrepancies in the salinity of ore-forming fluids.

The 600 m level is characterized by fluid inclusions of all types mentioned earlier, but the proportions of liquid and polyphase inclusions are greatly in excess of those in gaseous inclusions. The range of salinity of liquid inclusions at this level is wide, between 1.5 and 18.5 wt. %. Visual volumetric measurement of the salinity of the halite-bearing type was estimated to be about 40 wt. %.

The 550 m level is also characterized by having all types of fluid inclusions in different proportions to each other. Visual volumetric estimation of the salinity of halite-bearing inclusions lies between 30 and 40 wt. %. The range of salinity of inclusions at the 500 m level falls between 1.5 and 8.7 wt. %, and the maximum frequency is around 5 wt. %.

The 450 m and 425 m levels, on the other hand, are characterized by higher proportions of gaseous inclusions than that estimated in the lower levels. The salinity of liquid inclusions covers a narrow range between 1.5% and 7.5 wt. %.

The salinity of inclusions from the 400 m level is characterized by a wider range, 1.5 to 13 wt. %, and by the scarce presence of halite-bearing inclusions coexisting with gaseous and liquid inclusions.

A wider range of salinity is also seen in inclusions from the 300 m and 250 m levels, which show ranges of 1.5 to 18.5 wt. % and 1.5 to 20 wt. %, respectively.

It should be mentioned that the 250 m level cuts across the upper part of the intrusive body of quartz porphyry and that a remarkable abundance of gaseous inclusions is recognized in samples from this level.

The remarkable change in the proportion of gaseous inclusions from the 450 m level and their predominance in the upper part of the intrusive body indicates that "boiling" was manifested within this part of the intrusion, which might explain the discrepancy in the salinity of the circulating ore-forming fluids.

The quartz porphyry exposed in the underground working in the 200 m level and 150 m level is an apophysis of the main intrusion. The salinity of liquid inclusions from these levels ranges from 1 to 6.3 wt. % and from 1 to 3.5 wt. %, respectively, although one inclusion with 14.8 wt. % was measured from the 200 m and 150 m levels. This wider distribution might also be accounted for by the enrichment of salts by boiling, which is indicated by the coexistence of gaseous and polyphase inclusions.

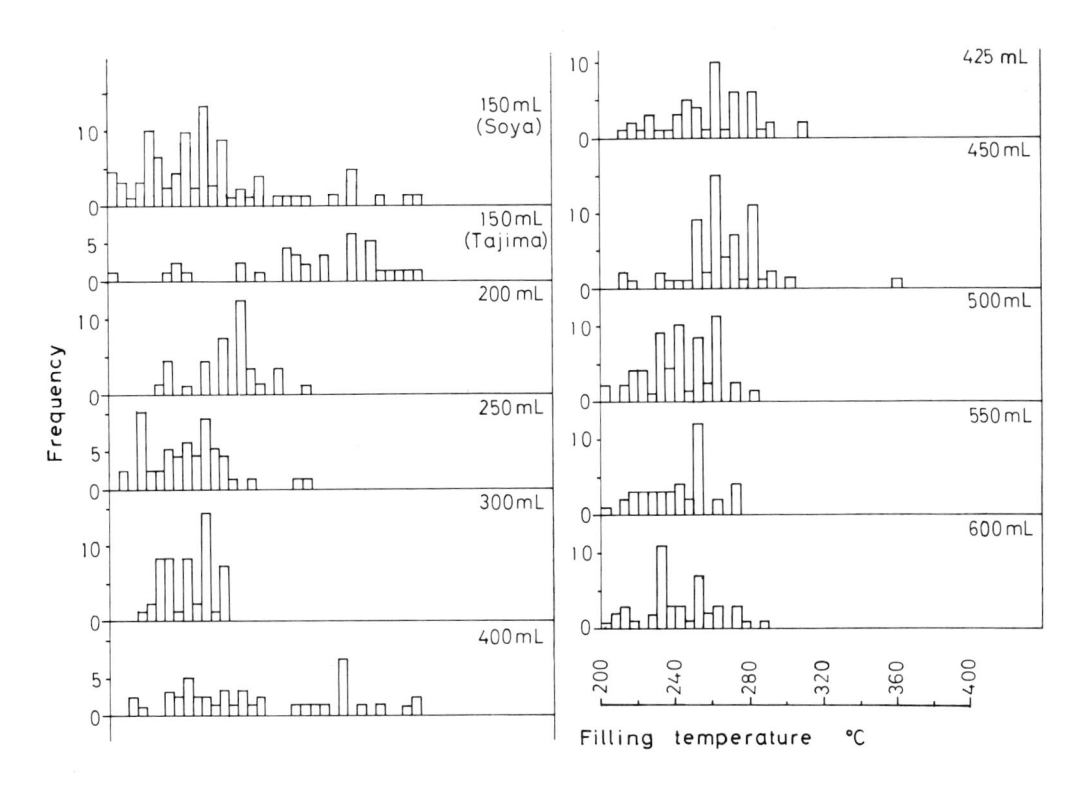

FIG. II-14-6. Histograms showing variation with depth in filling temperatures of liquid inclusions in quartz porphyry in the Toyoha mine.

Filling Temperatures of Fluid Inclusions in Quartz Porphyry

The filling temperatures of some 583 liquid inclusions from quartz porphyry were measured. The range of filling temperatures of inclusions from quartz porphyry fall between 200°C and 360°C. The normal distribution of these measurements shows a maximum frequency representing about 53% of the measurements in the range of 240°C to 280°C.

Variation in temperature with depth: The histograms shown in Fig. II-14-6 illustrate the variation in filling temperatures of liquid inclusions among the different levels of the mine. Analysis of this Figure shows that both ranges and frequencies of the filling temperatures vary from the lowest to the uppermost levels of the mine.

The filling temperatures of the liquid inclusions from the 600 m level range between 200°C and 290°C. Inclusions in samples representing the 550 m and 500 m levels have a similar range, between 200°C and 285°C.

Inclusions from the 450 m level, however, are characterized by a wider range of filling temperatures between 210°C and 360°C.

The filling temperatures of inclusions from the 425 m level range between 210°C and 315°C, but the majority of measurements are on the order of 260°C to 265°C. Meanwhile, some inclusions in samples from the 400 m level are characterized by higher filling temperatures, and their range falls between 210°C and 370°C.

A drop in values of the filling temperatures is noticed for inclusions from the 300 and 250 m levels. In both levels the majority of measurements fall between 210°C and 255°C.

Inclusions in the quartz porphyry dike exposed in the underground workings at the 200 and 150 m levels also show a wide range of filling temperatures. At the 200 m level, liquid

82   C. Epithermal Veins

inclusions homogenize over a wide range between 225°C to 310°C, with maximum frequency at 270°C to 275°C.

Three samples of the quartz porphyry forming the footwall of the Tajima and Soya veins at the 150 m level are also shown in the histograms of Fig. II–14–6.

Relation between Filling Temperature and Salinity of Inclusions

The relation between filling temperature and salinity of fluid inclusions offers some valuable information about the thermal history and changes in salinity of the ore-forming fluids during its path and course of mineralization. It is, then, appropriate to examine the relation between filling temperature and salinity of fluid inclusions for each level of the mine (Fig.

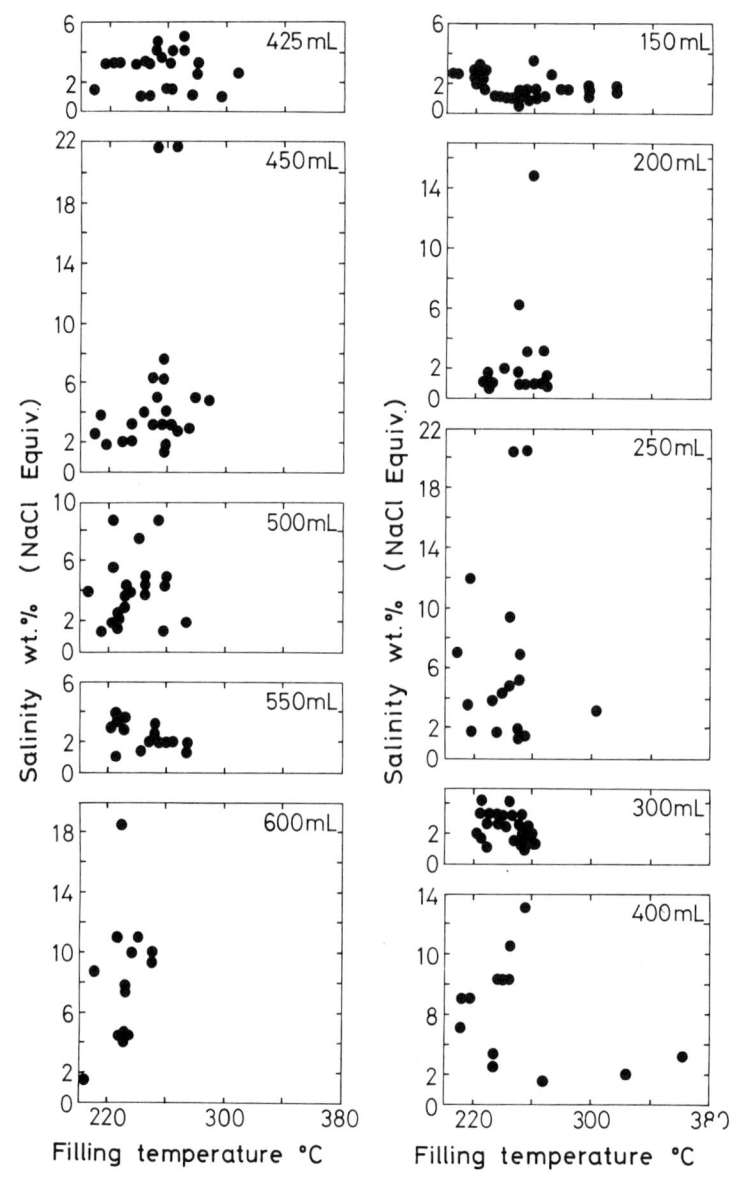

FIG. II–14–7. Histograms showing variation in salinity with filling temperatures of liquid inclusions in quartz porphyry in different levels of the Toyoha mine.

II–14–7). Although the number of measurements from some levels was not large enough to warrant a solid conclusion, some trends can still be discerned from the present data.

With few exceptions, the filling temperatures of liquid inclusions lie within a limited range of about 200°C to 270°C. The fluid inclusions with higher filling temperatures, however, probably represent those trapping the excess of the gas phase with the liquid during boiling of the solutions.

The change in salinity of fluid inclusions in samples from the lower levels of plotted on the filling temperature-salinity diagram of Fig. II–14–7 shows a nearly vertical or slightly inclined trend to the temperature axis. This trend can probably be explained by the mixing up of saline fluids with preheated dilute meteoric water. This trend generally holds up to the 300 m level where "boiling" is assumed to have been active. The relation between salinity and filling temperature of inclusions from this level shows a grouping of points representing a low range of salinity and a limited range of filling temperature. A rather vertical relation between salinity and filling temperature is also displayed for inclusions from the 250 and (partially) the 200 m levels. However, fluid inclusions from the 150 m level show a distinct parallel distribution to the temperature axis. Takenouchi (1970) described such a variation trend in quartz druses collected from some localities in Japan. This trend can be explained as due to the trapping of already diluted solutions at different temperatures.

Correlation with Data Obtained from Fluid Inclusions in Ore and Post-Ore Quartz Veins

The writers (El Shatoury *et al.*, 1974) studied the salinity and filling temperatures of fluid inclusions of the vein-forming minerals from the Toyoha mine. The filling temperatures are 260°–150°C, and the salinity is 4.8–0 wt. %. The data on temperatures are coincident with those reported by other authors. Temperature measurements have also been given for quartz and sphalerite from the Toyoha mine by Tokunaga (1970), Miyazawa *et al.* (1971) and Yajima and Okabe (1971).

Figure II–14–8 shows the present data on the temperature and salinity gradient of the fluid inclusions from quartz porphyry, quartz veins cutting the quartz porphyry, sphalerite, quartz associated with sphalerite and post-ore quartz veins. The following remarks can be made:

(1) The temperature gradient of the ore-forming fluids from quartz porphyry to the post-ore quartz veins is fairly gentle (Fig. II–14–8(a)). The difference in temperatures of the ore-forming fluids inferred from inclusions in quartz porphyry and those in sphalerite amounts to about 80°C. A drop of about 20°C is inferred between the deposition of sphalerite and of the later-stage quartz constituting the matrix of the ore.

From microscopic examination and the established paragenetic sequence of mineralization at the Toyoha mine, quartz was evidently deposited prior to and simultaneously with sphalerite, but it continued to crystallize after the completion of sphalerite deposition. This is, in fact, reflected in the range of filling temperatures assumed for fluid inclusions in quartz. This range overlaps that shown for inclusions in sphalerite, and in part extends beyond it to lower and higher temperatures by a few tens of degrees.

Accordingly, the temperature ranges represented by fluid inclusions reflect remarkably well the paragenetic sequence of ore and gangue minerals.

(2) The salinity values of the ore-forming solution are generally low. Although fluid inclusions in quartz porphyry are characterized by a wide range of salinity from nearly fresh water to a saturated solution of salts at room temperature, the maximum frequency of salinity measurements lies between 1 and 5 wt. %.

The salinity gradient (Fig. II–14–8(b)) from quartz porphyry to post-ore quartz veins is very gentle and is of almost constant value between the quartz veins in porphyry and the post-ore veins.

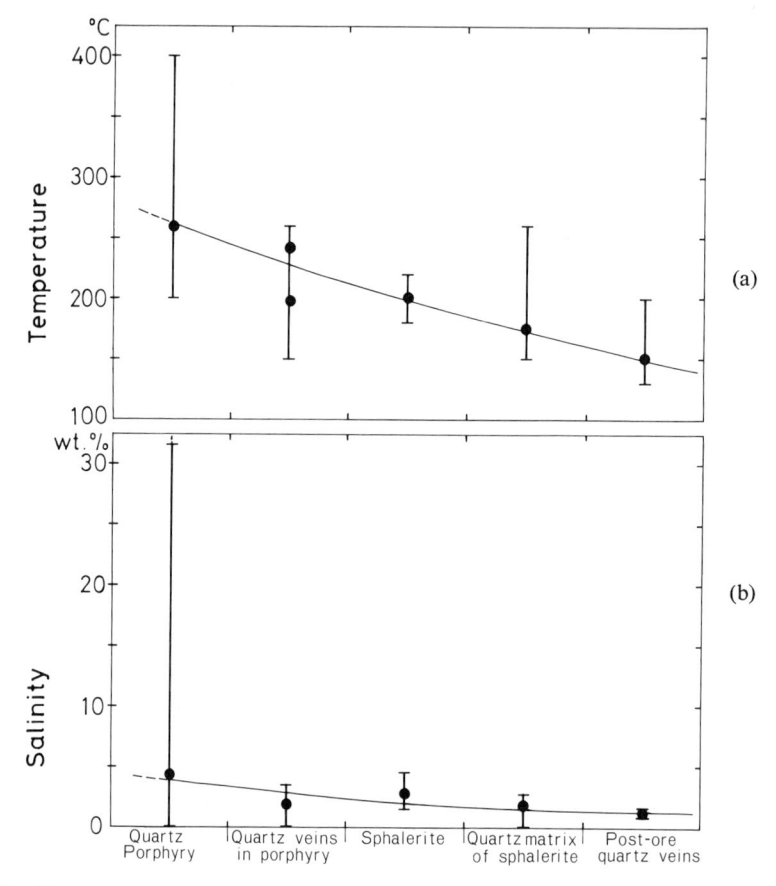

FIG. II-14-8. Temperature-salinity gradient curves of the ore-forming fluids in inclusions in quartz porphyry and vein minerals of the Toyoha mine.
(a) Temperature gradient curve—Solid line: range of filling temperatures. Solid circle: mean value of the maximum frequency of filling temperatures.
(b) Salinity (NaCl equivalent concentration) gradient curve—Solid line: range of salinity. Solid circle: mean value of the maximum frequency of salinity.

These two figures would inform us of the relation in time and space between the temperature and salinity of ore-forming fluids at the Toyoha ore deposit. Polyphase fluid inclusions in quartz phenocrysts of porphyry may represent post-magmatic brines, which would have played an important role in mineralization at the upper levels.

Conclusions

The Toyoha mine is a typical example of epithermal or subvolcanic vein-type deposits (xenothermal to some extent) in the "Green Tuff" province of Japan. The deposit is intimately related to an intrusive body of quartz porphyry.

A vein pattern resembling a horse-tail structure in relation to the master fault zone indicates wrench fault movement. But the master faults afterwards move like normal faults, perhaps through upward pressure due to the intrusion of quartz porphyry.

Filling temperature and salinity (NaCl equivalent concentration) of fluid inclusions were measured in quartz from veins and quartz porphyry, which are related to mineralization. The following is a summary of the results:

(1) Fluid inclusions are categorized into four main types: polyphase, liquid, gaseous and

monophase (liquid) inclusions. This is in addition to the presence of many glass inclusions of various shapes and different degrees of devitrification.

(2) Systematic microscopic examination of quartz porphyry revealed a remarkable change in proportion and type of inclusions from liquid-rich to gas-rich in the central upper part of the intrusion. The gaseous inclusions are characterized by having various low filling degrees. This indicates their formation from a heterogeneous mixture of liquid and gaseous phases and signifies that boiling occurred in the upper part of the intrusion.

(3) A remarkable decrease in salinity of the ore-forming fluids is observed from measurements of liquid inclusions from the lower 600 m level to the upper 150 m level. An increase in salinity in some liquid inclusions, however, is shown in areas of active boiling.

(4) The filling temperatures of fluid inclusions from the different levels of the mine fall within a rather limited range. However, in areas where "boiling" was prevalent, fluid inclusions with higher filling temperatures exist. This fluctuation is probably due to the trapping of the gas phase under heterogeneous conditions caused by boiling.

(5) The relationship between filling temperature and salinity of fluid inclusions indicates the possibility of dilution of the ore-forming fluids by preheated underground waters. Inclusions from the upper level of the mine, however, show constant low salinity and variable filling temperatures. This is explained by the trapping of already diluted solutions under different temperatures.

(6) The salinity of the fluids in inclusions from quartz porphyry, sphalerite, quartz associated with sphalerite as well as post-ore quartz veins indicates that the deposition of ore occurred from low-saline solutions and that the salinity gradient curve is very gentle.

(7) The filling temperatures of inclusions from quartz porphyry, ore, quartz associated with the ore and the post-ore quartz veins indicate that the temperature gradient of fluids is fairly gentle. The difference between the filling temperatures of inclusions in the ore and those in quartz porphyry is about 80°C. The deposition of sphalerite and part of the quartz associated with it occurred at around 200°C. Quartz, however, is revealed to belong to more than one generation. The earlier quartz was deposited simultaneously with or slightly prior to sphalerite; the later stage quartz, on the other hand, probably crystallized at a slightly lower temperature of about 180°C.

From the $^{34}S/^{32}S$ ratios of coexisting sphalerite and galena from the Toyoha mine, Kiyosu and Nakai (1977) have estimated the formation temperatures of these minerals in this mine to be 250°–200°C.

D. XENOTHERMAL VEINS

II-15. Geologic Structure and Mineralization of Polymetallic Xenothermal Vein-type Deposits in Japan (Imai *et al.*, 1975)

H. Imai, M. S. Lee, S. Takenouchi, Y. Fujiki, K. Iida, T. Sakimoto and S. Tsukagoshi

The classic works of Kato (1920, 1927b) on the geology of Sn-W-Cu-Zn-Pb-Ag-bearing vein-type deposits of the Akenobe and Ikuno mines have been referred to in such textbooks as those by Lindgren (1933), Schneiderhöhn (1941) and Park and MacDiarmid (1964) (Fig. II-15-1). Buddington (1935) also cited Kato's paper on the Akenobe deposit when he proposed the term "xenothermal."

The Ashio mine was the largest copper producer in Japan. This mine consists of two kinds of deposit, the one a fissure-filling vein-type deposit in rhyolitic rocks and the other a replacement deposit mainly in chert (Kato, 1926b). Nakamura (1951) recognized cassiterite and wolframite in the vein deposits of the Ashio mine and therefore classified the deposits as the xenothermal type.

In the Akenobe and Ikuno districts, rhyolitic rocks are widely developed. They were formerly considered to be Miocene in age, but by the early 1960's, the entire formation, including the rhyolitic rocks in the Akenobe and Ikuno districts, was found to be Late Cretaceous or Early Tertiary (Ikebe *et al.*, 1961).

The same kinds of rhyolitic rocks are developed in places throughout Japan (Fig. II-15-1). Sn-W-Cu-Zn-Pb-Ag-bearing veins occur with or in close proximity to these rhyolitic rocks. Imai (1966b) and Imai *et al.* (1967) proposed that the Akenobe and Ikuno deposits were genetically related to the granitic rocks which intrude the rhyolitic rocks and that the mineralization epoch was Late Cretaceous or Early Tertiary.

In 1939, Yamaguchi described "brown stannite" in ore from the Ikuno mine. Nakamura (1954) recognized "brown stannite" and "brownish-orange stannite" in ore from the Ashio mine and later Nakamura (1970) revised them to be stannoidite and mawsonite. Previously Ramdohr (1960) named the "brown stannite" hexastannite, and Markham and Lawrence (1965) named "brownish-orange stannite" mawsonite.

Imai *et al.* (1967) observed hexastannite and mawsonite in ore from the Tada mine in the same district, and they classified the deposit as the same type as the Ikuno deposit, i.e., xenothermal. Kato (1969) studied hexastannite from the Konjo mine (Fig. II-15-2) to the west of the Akenobe mine, determined the chemical composition to be $Cu_5(Fe, Zn)_2SnS_8$, and named it stannoidite. Kato and Fujiki (1969) described the occurrence and paragenesis of stannoidite from xenothermal vein deposits of the Akenobe, Ikuno, Tada, and Fukoku mines in the Akenobe-Ikuno district. In 1973, the chemical composition of stannoidite was revised by Petruk (1973) to be $Cu_8(Fe, Zn)_3Sn_2S_{12}$. The present writers have studied the hydrothermal syntheses of stannoidite, mawsonite, bornite and chalcopyrite to investigate the paragenesis and formation condition of these minerals, as described later.

Some pegmatitic or hypothermal tungsten-tin veins and mesothermal chalcopyrite and pyrrhotite deposits occur in the areas composed of Permo-Carboniferous sedimentary rocks in this district. These rocks are intruded by Late Cretaceous or Early Tertiary granitic rocks. Examples of pegmatitic or hypothermal tungsten-tin veins which occur in the area adjacent to xenothermal vein-type deposits are the Ohtani and Kaneuchi mines (Chapter II-4)(Fig. II-15-2). The Komori mine consists of mesothermal copper veins containing chalcopyrite,

Part II: Vein-type Deposits

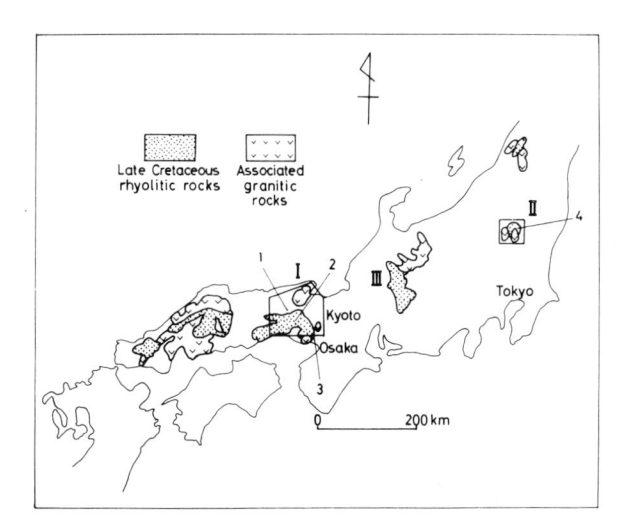

FIG. II–15–1. Index map.
I: Western Kinki district, II: Ashio-Nikko district, III: Nohi district. 1: Akenobe mine, 2: Ikuno mine, 3: Tada mine, 4: Ashio mine.

cubanite and pyrrhotite, and occurs in the same metallogenetic province (Fig. II–15–2) as the Kaneuchi and Ohtani mines (Chapter II–5). The mineralization epochs of these deposits are also Late Cretaceous or Early Tertiary.

Mineralized areas of the above xenothermal, pegmatitic or hypothermal and mesothermal deposits to the west of Kyoto constitute the Western Kinki Metallogenetic Province (Fig. II–15–2) (Imai et al., 1967; Imai, 1970).

Western Kinki Metallogenetic Province

Geologic Outline of the Western Kinki District

In the district to the west of Kyoto, that is, the western Kinki district, Permo-Carboniferous formations (Tamba Formation) form the basement. The general trend of the formation is E-W. Granitic rocks are intrusive into the formations (Ikebe et al., 1961), and Kawano and Ueda (1967 a,b) determined the ages of some of the granitic rocks by the K-Ar method. The ages range from 55 to 80 m.y., falling within the Late Cretaceous and Early Tertiary epochs.

On the southeastern margin of the district, the Ibaragi granitic complex intrudes the Tamba Formation (Fig. II–15–2). The absolute age of this complex is 73.8 to 75.6 ± 3 m.y. by the K-Ar method (Shibata, 1971). According to Ishizaka (1971), whole rock (feldspar-biotite) isochrons by the Rb-Sr method give an age of 76 to 96 m.y. Arguing that the initial $^{87}Sr/^{86}Sr$ ratio of the isochron of the whole rock is 0.7060 (± 0.0001), Ishizaka asserted that the complex was formed by fractionation of a single parental dioritic magma generated in the lower crust or in the upper mantle.

In the middle and southern parts of the district, the Permo-Carboniferous basement is overlain by the formation of volcanic, pyroclastic and sedimentary rocks known as the Ikuno (or Arima) Formation. Though it includes andesitic and basaltic rocks, rhyolitic volcanics and pyroclastics are predominant in the formation. Stratigraphic evidence suggests that the formation is Late Cretaceous or Early Tertiary. The formation is intruded by granitic rocks. As discussed later, these granitic rocks have the same age as the above-mentioned granitic

D. Xenothermal Veins

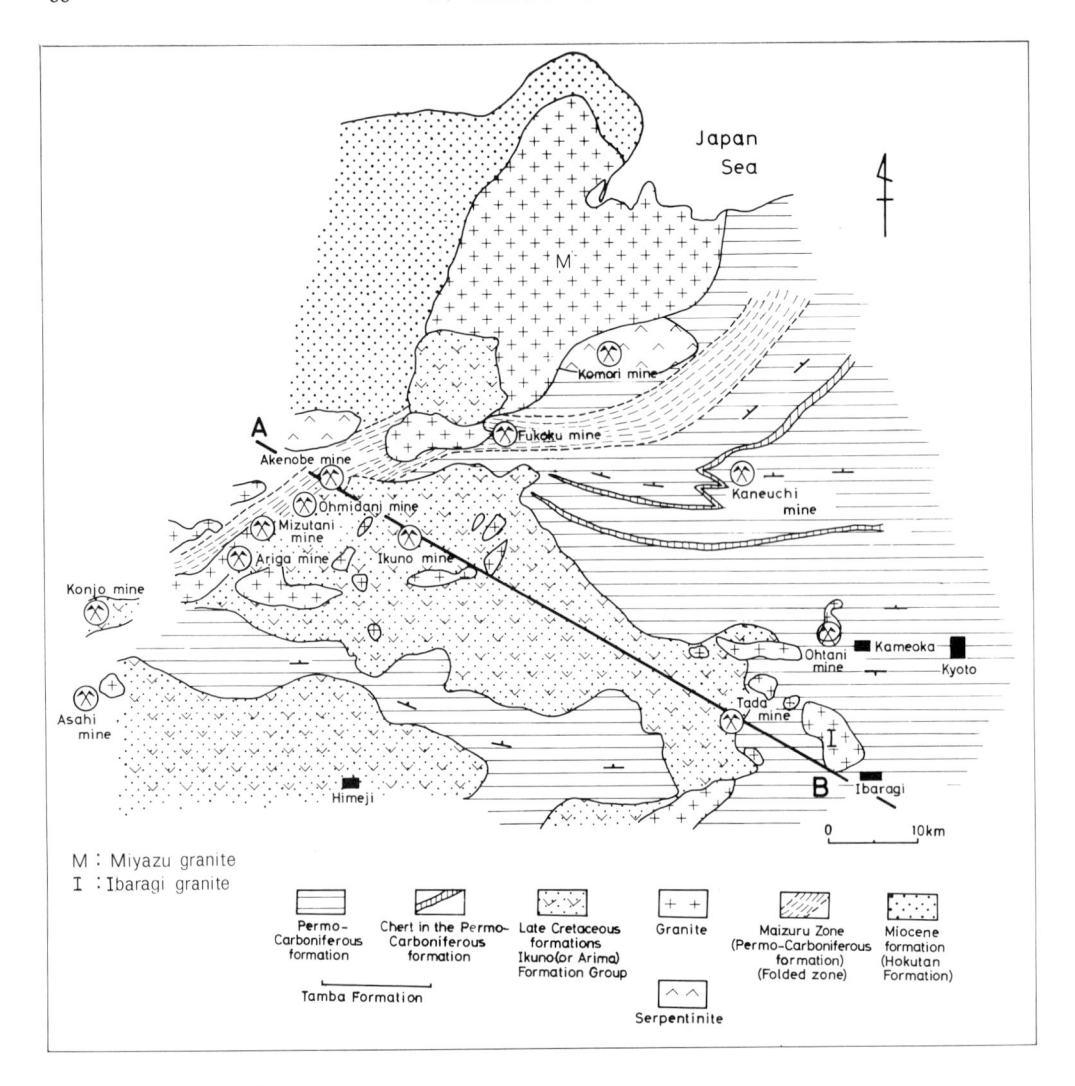

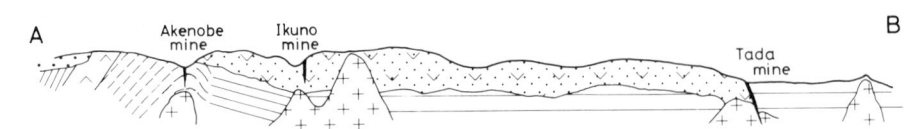

FIG. II–15–2. Geologic map and section of the western Kinki district.

rocks in the area of the Permo-Carboniferous formation. The writers believe that the granitic rocks intruding the Ikuno (or Arima) Formation are products of the same igneous activity as the volcanic rocks in this formation.

In the northern part of the western Kinki district, the Hokutan Formation of Miocene sediments with some volcanics and pyroclastics is developed. It unconformably overlies the Permo-Carboniferous formations, granites, basic intrusives and the Ikuno (Arima) Formation (Fig. II–15–2).

As shown in Fig. II–15–2, the district contains many vein-type deposits of tin, tungsten, copper, lead, zinc and silver. The writers described in the foregoing chapters the pegmatitic or hypothermal tungsten veins of the Ohtani and Kaneuchi mines and the mesothermal copper veins of Komori mine. It is believed that these deposits are genetically related to the granitic rocks of the Late Mesozoic or Early Tertiary period.

The deposits of the Akenobe, Tada and Ikuno mines are xenothermal vein-type deposits containing various minerals of tin, tungsten, copper, zinc, lead and silver. The Akenobe deposit occurs in the Permo-Carboniferous formation, while the Ikuno and Tada deposits are in the Ikuno or Arima Formation.

Kato (1920) had originally believed that the Akenobe deposits were genetically related to the granitic activity of the Mesozoic age. However, after his study of the geology of the area around the Ikuno mine, Kato suggested in 1927 that both the Ikuno and Akenobe deposits were formed by the Miocene rhyolitic activity and proposed the name "Ikuno-Akenobe Metallogenetic Province" (Kato 1927b).

The age of many copper, zinc and lead vein deposits in northern Japan is considered by Japanese geologists to be Middle Tertiary. Sekine (1959), Nakamura (1970), Tatsumi *et al.* (1970) and others classified the xenothermal deposits as Tertiary subvolcanic type. However, as mentioned above, recent stratigraphic studies indicate that the age of the Ikuno Formation is probably Late Cretaceous or Early Tertiary. Besides, Prof. M. Ozima[*] and the writers have obtained a K-Ar age of 77 m.y. for granite that intrudes the Ikuno Formation in the Ikuno mining area (Imai *et al.*, 1967).

Based upon the distributions of the granite and the ore deposits as well as on the absolute ages of the granite, the writers believe that the Akenobe and Ikuno deposits are genetically related to the granitic activity of Late Mesozoic or Early Tertiary which was intruded soon after the rhyolitic eruption. Buddington (1959) noticed that in the Cascade Mountains, Oregon, granite coexists with volcanic rocks which are presumably the products of the same igneous activity as the granite. It seems probable that the Ikuno and Akenobe deposits were formed by the same igneous activity that formed the deposits of the Ohtani, Kaneuchi and Komori mines.

The veins of the Ohmidani mine to the southwest of the Akenobe mine are situated in the silver zone of the zonal distribution in the Akenobe mine (Figs. II–15–2, II–15–3). The Asahi mine in the southwestern margin of the western Kinki district is the same type as the Ohmidani mine (Fig. II–15–2).

Ishihara and Shibata (1972) determined the absolute ages of the dike rocks from the Akenobe and Ikuno mines by the K-Ar method. The post-ore felsite (granophyre) dikes from the Akenobe mine are dated 57.8 ± 2.9 m.y. and 52.6 ± 2.1 m.y., while the pre-ore rhyolite (presumably from the same igneous activity as the mineralization) is 72.8 ± 2.9 m.y. Accordingly, Imai *et al.* (1967) proposed the "Western Kinki Metallogenetic Province" of Late Cretaceous or Early Tertiary age, including the Ohtani, Kaneuchi, Komori, Akenobe, Tada and Ikuno mines.

As stated before, the absolute ages of granodiorite and greisen in the Ohtani mine are 91–93 m.y. So the igneous activity of the Late Cretaceous and Early Tertiary epochs in the Western Kinki Metallogenetic Province ranges from 93 to 50 m.y. In the Basin and Range Province of the U. S. A. the igneous activity ranges 80–50 m.y. (Damon, 1968; Gilmour, 1972).

Ore deposits of the Akenobe Mine, Hyogo Prefecture

The Akenobe mine is located about 100 km to the northwest of Osaka City, and it is believed that the mine was discovered more than 1,000 years ago. Monthly production of

---

[*] Dept. of Geophysics, University of Tokyo.

D. Xenothermal Veins

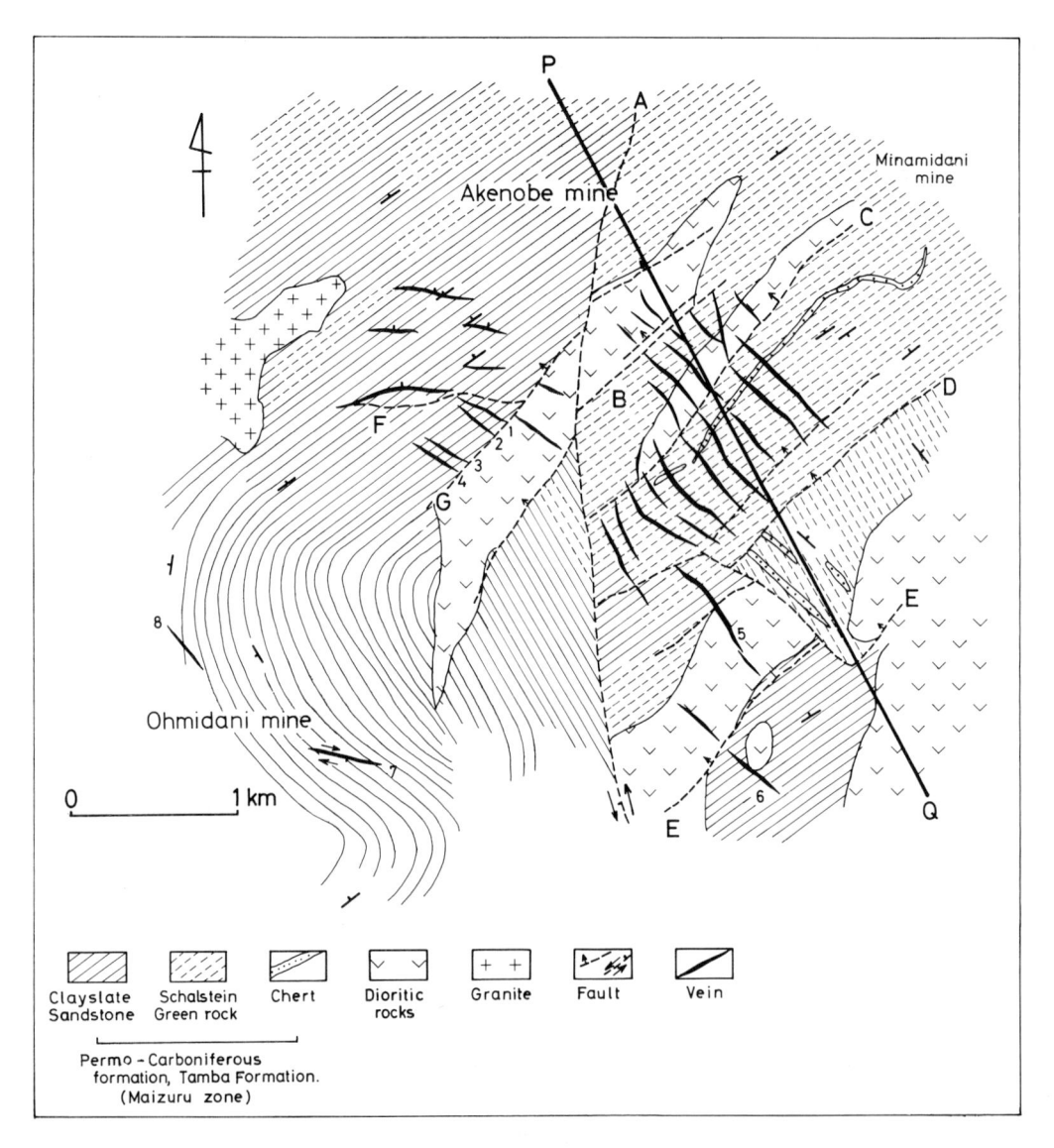

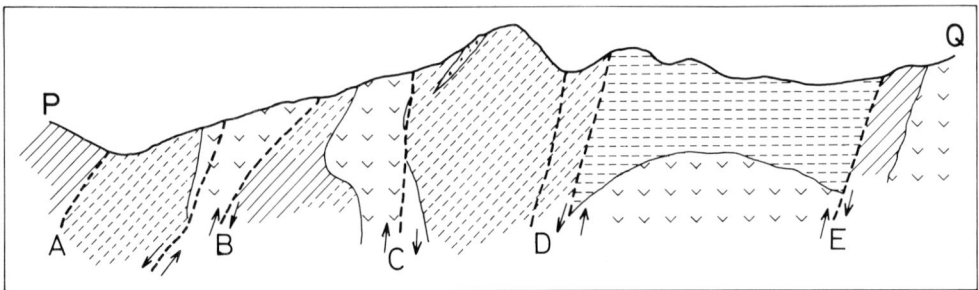

FIG. II-15-3. Geologic map and section of the Akenobe and Ohmidani mines.
  Faults—A: Akenobe Fault, B: Seiei Fault, C: Komine Fault, D: No. 25 Fault, E: Kanakidani Fault, F: No. 4 Vein Fault, G: Sedani Fault.
  Veins—1: Sedani No. 4 vein, 2: Sedani No. 7 vein, 3: Nansei vein, 4: Fudono vein, 5: Ryusei vein, 6: Ginsei vein, 7: Fusei vein, 8: Ohdate vein.

crude ore in 1977 was 24,000 t. Grades were 1.40% copper, 2.55% zinc, 0.29% tin, and 18g/t of silver. The total production up to the present is 15 million tons.

Geologic setting: The area is mainly covered by Permo-Carboniferous sediments rich in phyllites and green rocks (schalstein) and by intermediate to basic intrusives (Hirokawa *et al.*, 1954). Some of the green rocks might be considered "ophiolites," and it is generally believed that they were derived from basic tuff and volcanics. The green rocks are similar to the greenschist in Outer Zone of southwest Japan. As described in Chapter IV–2, Imai (1960) concluded that the green rocks were derived from Mg-Fe metasomatism of the sediments at the time of the intrusion of basic rocks, accompanied by metamorphism.

The Permo-Carboniferous sediments strike N40°–60°E, dipping generally to the NW at 30°–70° but occasionally dipping to the SE. They are folded along the NE-SW axes, indicating that lateral pressure from a NW-SE direction was exerted. Dioritic rocks occur as intrusive sheets or dikes in the Permo-Carboniferous formation. They include gabbro and diabase in addition to diorite, and occur nearly parallel to the strike of the Permo-Carboniferous formation (Fig. II–15–3).

In the eastern part of the mining area, the green rocks were developed while in the western half black slate facies predominates on the surface but grades into green rocks in deeper parts. The Akenobe Fault, running N-S with a dip of 60°–70°W, divides the mining area approximately in half. By this fault a boundary between diorite and sediments is found to have displaced horizontally about 300 m, as shown in Fig. II–15–3. It is also known that the bottom level of the mineralization is about 150 m deeper in the western area than in the eastern area; i.e., the Akenobe Fault is perhaps obliquely displaced, i.e. a left-handed wrench fault with a normal fault component. However, it is also possible that this fault was formed as a left-handed wrench fault at the initial stage and later moved as a normal fault, as discussed in Chapters II–1, –7 and –20 and in the Tenju vein (Ikuno mine) of this chapter. The displacement on this fault before the diorite intrusion is not known. The faulting is related to the NW-SE lateral pressure mentioned above.

The Ikuno Formation is developed in the southeast part of the area. Granite crops out near the Fukoku mine to the northeast of Akenobe and also in the Mizutani-Ariga area to the southwest, presumably extending beneath the Akenobe mining area as a cryptobatholith (Fig. II–15–2). This presumption is supported by gravity and airborne magnetic survey

FIG. II–15–4. Pushing-up of the green phyllitic rock by the intrusion of diorite. Di: diorite, Phy: phyllite.

D. Xenothermal Veins

maps. The low gravity* and negative anomaly in the magnetic survey extend along this NE-SW zone (Ministry of International Trade and Industry of Japan, 1973).

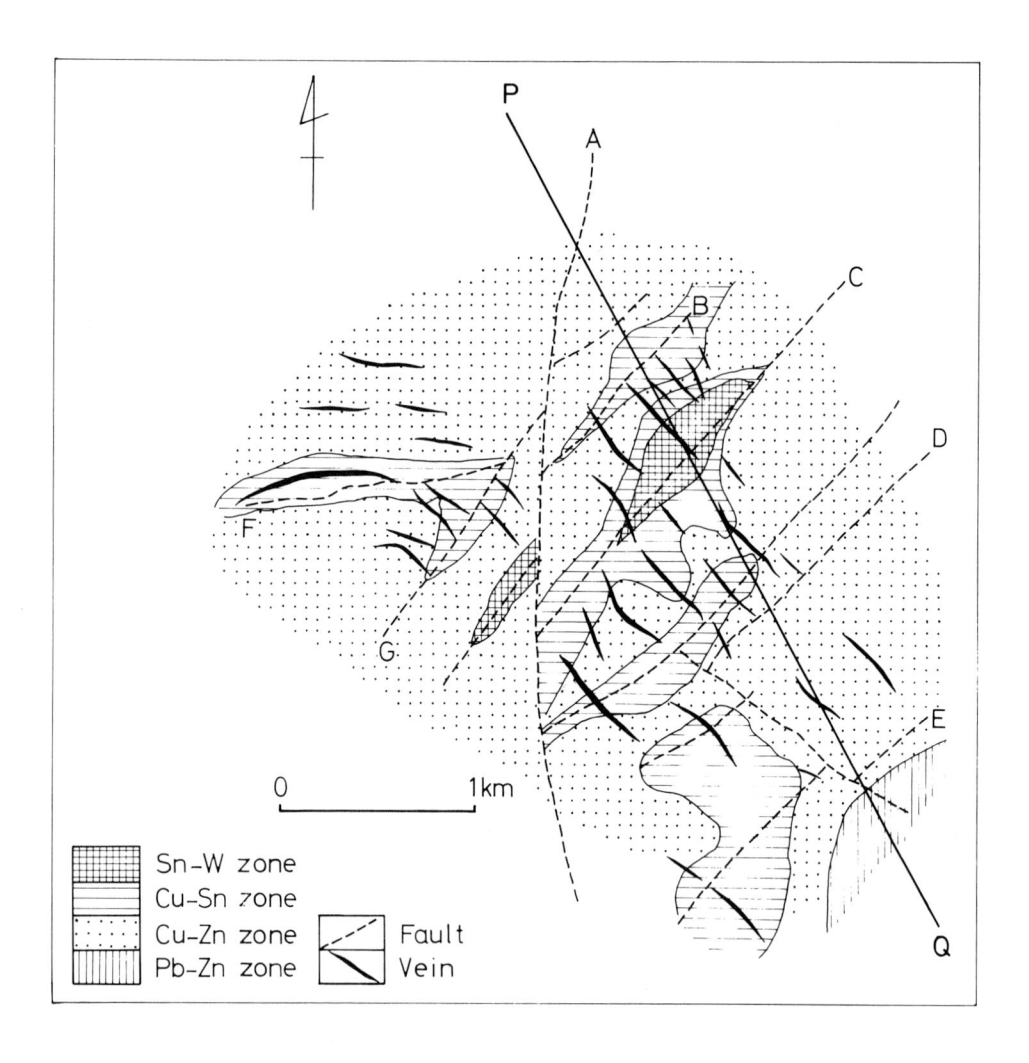

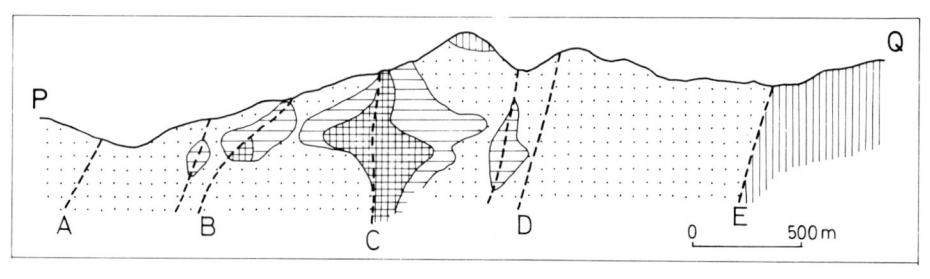

FIG. II-15-5. Zonal distribution map of the Akenobe mine (after Abe, 1963). Plan: 0m level. Refer to the names of faults in Fig. II-15-3.

---

* In this district, the low gravity anomaly has been verified by Nabetani et al. (1972) in the area of the Ibaragi granite complex (Fig. II-15-2).

Vein-fissure system: Along the boundaries between the elongated dioritic masses and the Permo-Carboniferous sediments, faults often occur (Fig. II–15–3). The writers propose that the faults were caused by a pushing-up of dioritic magma intruded as sheets or dikes (Imai, 1966b). As shown in Fig. II–15–3, the block sandwiched between the No. 25 Fault and the Kanakidani Fault was pushed up, and consequently the dip and strike of the strata in this block are different from the general trend. They strike mostly NW, dipping NE. Underneath the disturbed block, a dioritic body occurs (Fig. II–15–3). At a surface outcrop in this block, it is seen that phyllite was pushed up by the diorite (Fig. II–15–4).

The fault fissures (NE faults) are occasionally mineralized with tin, tungsten or magnetite veins. The Seiei fault is the most remarkable in having tin-tungsten ore (Fig. II–15–5).

Fissures of the leading veins in this district are tension cracks striking N30°–50°W with dips of 60°–90°NE or SW (NW tension cracks). No displacement of the host rocks across the fissures is recognized. The strike orientation of the fissures is found to be nearly perpendicular to the folding axes of the sediments. The folded structure and tension cracks may have been formed by lateral pressure exerted from a NW-SE direction. In general, no displacement is observed in the NW veins where they intersect the NE fault veins (Fig. II–15–3). This indicates that the NE faults were formed earlier than the NW tension cracks. However, in some parts they show small displacements along the NE faults. This may be explained in three possible ways: (1) the NE faults of these parts were formed later than the NW tension cracks; (2) the pre-existing NE faults were partly displaced by the lateral pressure that produced the NW tension cracks and continued to prevail in the area; (3) the upheaval of dioritic bodies that formed the NE faults or the withdrawal of diorite bodies continued after the formation of the NW tension cracks.

In some cases, tension cracks produced by NW-SE lateral pressure are found only in blocks sandwiched by two faults, but do not extend across the faults. For example, Sedani No. 7 vein, Fudono vein and Nansei vein exist only in the block sandwiched by the Sedani Fault and the No. 4 Vein Fault (Fig. II–15–3). This may be due to differences in the mechanical properties of rocks from those of the adjoining blocks.

As stated above, the displacement of the Akenobe Fault, presumably a product of NW-SE lateral pressure, continued to post-mineralization, though it is weakly mineralized (Fig. II–15–3).

Dikes of andesite, rhyolite and granophyre (felsite) are scattered throughout the mining area. Some of them cut across the ore veins, while some are impregnated by tin-copper-zinc ores.

Mineralization and zonal distribution: The distribution of ore minerals in the Akenobe deposit shows a remarkable zoning (Fig. II–15–5). Five zones are recognized, namely: (1) tin-tungsten, (2) tin-copper, (3) copper-zinc, (4) zinc-lead and (5) silver. The mineral assemblage of each zone is as follows (* indicates presence in small amounts):

(1) Tin-tungsten zone: cassiterite, wolframite, scheelite*, chalcopyrite, lamellar magnetite* (Sekine, 1959), marmatite*, galena*, bornite*, pyrite*, native bismuth*, bismuthinite*, molybdenite*, arsenopyrite*, quartz, fluorite, topaz*, apatite*, calcite, chlorite, epidote.

(2) Tin-copper zone: chalcopyrite, cassiterite, bornite, lamellar magnetite*, stannite*, wolframite*, scheelite*, sphalerite, galena*, stannoidite*, mawsonite*, tennantite*, arsenopyrite*, pyrite*, molybdenite*, native bismuth*, bismuthinite*, chalcocite*, fluorite, calcite, apatite, chlorite.

(3) Copper-zinc zone: chalcopyrite, sphalerite, lamellar magnetite*, galena*, cassiterite*, arsenopyrite*, bornite*, tennantite*, roquesite* (Kato and Shinohara, 1968), wolframite*, scheelite*, quartz, calcite, chlorite, fluorite*.

(4) Zinc-lead zone: sphalerite, galena, arsenopyrite*, chalcopyrite*, lamellar magnetite*, pyrite*, cassiterite*, quartz, calcite, chlorite, fluorite*.

D. Xenothermal Veins

FIG. II–15–6. Sedani No. 4 vein of the Akenobe mine, 700-ft level. 1: wall rock, 2: chalcopyrite-sphalerite zone, 3: cassiterite-wolframite zone, 4: fluorite zone, 5: barren quartz zone.

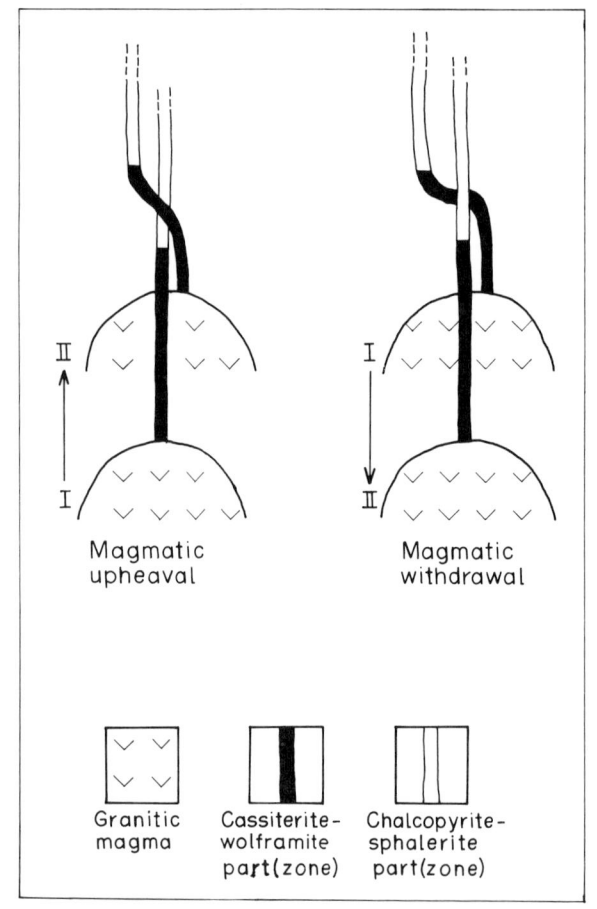

FIG. II–15–7. Magmatic movement (pulsation) and zonal distribution of minerals.

(5) Silver zone: argentite, pyrargyrite*, stephanite*, native silver, chalcopyrite, pyrite, sphalerite*, galena*, electrum*, quartz, adularia, calcite, chlorite, rhodochrosite*.

In the Akenobe mine lamellar magnetite characteristically occurs with copper minerals such as bornite and chalcopyrite, and an exsolution texture of bornite and chalcopyrite is often observed. Also bornite is generally transformed into digenite along the later calcite veinlets. From the shape and habit of magnetite under the microscope, it is believed that it was originally deposited as hematite and was later transformed into magnetite at the time of copper mineralization, probably by a decrease in oxygen fugacity.

The mineral assemblage of chalcopyrite, stannoidite, bornite and mawsonite is less common in the Akenobe mine than it is in the Tada mine.

The zonal distribution of minerals according to Abe (1963) is shown in Fig. II-15-5. From the zonation pattern it is inferred that the mineralizing fluids may have ascended through the NE faults and flowed into the NW tension cracks, depositing the minerals during the migration. The mineral assemblage in the NW veins thus changes successively with lateral distance from the intersection with the NE faults, forming the pattern of zoning described above. The same trend in mineral zoning has been confirmed in vertical section (Fig. II-15-5). No sharp boundaries exist between the zones, and generally one zone changes gradually to another.

From the above description, one may consider the zoning of the deposit to be of the monoascendant type, and the distribution of zones to coincide generally with the classic zoning pattern described by Emmons (1924). On the other hand, the banded structure observed in the veins indicates the possibility of polyascendant (pulsation) zoning (Smirnov, 1960; Kutina, 1965). Kato (1920) recognized five successive stages of mineralization in these veins; they are as follows, from the earliest to the latest stage: (1) deposition of main cassiterite ore, (2) deposition of wolframite-cassiterite ore, (3) deposition of chalcopyrite, (4) deposition of sphalerite and (5) deposition of barren quartz with a small amount of chalcopyrite.

In the Sedani vein, however, Saigusa (1958) mentioned that chalcopyrite-sphalerite-bearing quartz occurs in contact with both the hanging wall and the footwall of the vein and that cassiterite-wolframite-bearing quartz occupies the median part of the vein (Fig. II-15-6), indicating that at least in this part of the deposit the chalcopyrite-sphalerite mineralization preceded the cassiterite-wolframite mineralization. The writers observed the same kind of banded structure in many parts of the veins, especially in the western area of the mine. Berg (1928), Kutina (1957) and Sekine (1959) introduced the term "rejuvenation" to explain such a phenomenon.

A possible explanation of this apparent contradiction may be that the mineral deposition was drastically modified by the movement of the source magma during the mineralization. As indicated in Fig. II-15-7, the upheaval (advance) of the magmatic body or of the source of the mineralizing fluids would explain the apparently reversed succession of mineralization at any particular point. Through upheaval, the early-formed chalcopyrite-sphalerite part (zone) could have been cut by the cassiterite-wolframite part (zone), and thus the zonal distribution would be partly out of sequence. As already mentioned, there appears to be evidence of diapiric movement of intrusive bodies in this mining field. By the withdrawal (retreat) of the magmatic body the zonal distribution also would be out of sequence. However, the banded structure reveals normal (direct) succession (Fig. II-15-7). This case is described later in the section on the Ashio mine.

Some geochemical features: Yamamoto (1974b) noticed that $\delta^{34}S$ values of sulfides mainly from the Ryusei vein have negative values with respect to the meteorite standard (Table II-15-1). The range of variation is relatively small, varying from -0.4 to -4.4 per mil in 45 determinations of chalcopyrite and sphalerite. These values are smaller, for example, than those from the Shakanai mine (Kuroko deposit) reported by Kajiwara (1971). In a state of isotopic equilibrium between chalcopyrite and sphalerite (in coexistence), the sulfur of sphalerite is generally heavier than that of chalcopyrite (Sakai, 1968; Kajiwara and Krouse, 1971; Yamamoto, 1974a). However, samples from the Akenobe mine show a reverse relationship. This may be due to the fact that the two minerals were not deposited simultaneously.

D. Xenothermal Veins

TABLE II–15–1. δ^{34} S values and Se/S ratios of sulfides from the Akenobe mine (after Yamamoto, 1974b).

Sample no.		δ^{34}S (‰)	Se/S ($\times 10^{-5}$)	Sample no.		δ^{34}S (‰)	Se/S ($\times 10^{-5}$)
AR101	Cp	−4.2	20.4	AR0203	Cp	−0.9	18.9
	Sp	−4.2	8.3	AR0205	Cp	−0.6	11.3
AR102–1	Sp	−0.8		AR0207	Cp	−3.0	29.3
−2	Cp	−0.8	16.5	AR0208	Cp	−2.2	51.9
−3	Cp	−2.8	18.8		Sp	−3.8	2.66
−4	Cp	−3.1	13.2	AR0303	Cp	−1.3	24.6
AR103–1	Cp	−0.9	26.5	AR0305–1	Bn	−0.8	43.7
−3	Cp	−2.8	17.8	AR0307	Cp	−2.1	31.1
AR001	Cp	−3.7	13.0	AR0310	Cp	−2.5	31.7
AR002–1	Cp	−1.2	6.7	AR0401–1	Bn	−2.4	40.3
−2	Cp	−1.2	35.4	AR0401–2	Cp	−2.3	44.4
AR005	Cp	−2.7	48.7		Bn	−2.4	41.8
	Sp	−3.5	51.9	AR0405	Cp	−3.6	21.0
AR006	Cp	−1.7	13.4	AR0406	Cp	−2.7	19.8
	Sp	−4.4	62.2	AR0408	Cp	−1.2	25.9
AR0102	Cp	−1.6			Sp	−2.4	17.6
	Sp	−1.2		AR0409	Sp	−1.2	32.8
AR0104	Cp	−2.5	26.5	AR0603	Cp	−1.2	44.5
AR0107	Cp	−0.4	10.9	AG0202	Cp	−2.3	
	Sp	−0.5	14.6		Sp	−3.0	
AR0110	Cp	−2.8	23.4	AS1201	Cp	−1.6	
AR0201	Cp	−2.7	20.7		Sp	−2.4	
					Ga	−2.9	

Bn: bornite.
Cp: chalcopyrite.
Ga: galena.
Sp: sphalerite.

Se/S: unpublished data.

According to H. Sakai[*] and M. Yamamoto (personal communication), the selenium content of sulfide minerals in the Akenobe mine (Table II–15–1) is high (refer to Fleischer, 1955). The writers observed that indium, scandium and iron are higher in cassiterite from xenothermal deposits such as Akenobe and Ikuno than in those from some hypothermal deposits. On the other hand, tantalum and niobium contents are very low in cassiterite from the Akenobe and Ikuno deposits (Table II–15–2).

It is worthy of mention that Kato and Shinohara (1968) recognized roquesite ($CuInS_2$) showing exsolution textures with chalcopyrite and sphalerite. This suggests that indium might be rather rich in xenothermal mineralization.

Decrepitation temperatures: The writers (Takenouchi and Imai) studied the decrepitation temperatures of the minerals from No. 4 and No. 7 veins of Sedani (Takahashi *et al.*, 1955). Measurements were carried out on about 70 samples collected from various places between the 400-ft. and 1,000-ft. levels. The temperatures range from 355° to 155°C. The data indicate that the highest decrepitation temperatures in a level have a tendency to increase with depth, but they do not show any clear distinction between tin-tungsten and copper-zinc mineralization.

Igneous rock genetically related to mineralization: The ore deposit of the Akenobe mine may be genetically related to the Late Cretaceous or Early Tertiary acidic igneous activity. In the Akenobe mining district granitic rocks occur in a small area, but it is most probable that a cryptobatholith of acidic rock exists underneath the deposit. From a reconnaissance

---
[*] Institute for Thermal Spring Research, Okayama University.

TABLE II–15–2. Emission spectrochemical analyses of cassiterites.

	Localities	Fe	In	Mn	Nb	Sc	Ta	Ti	V	Zr
Hypothermal type	Ohtani, Tokonage area (Japan)	0.07	—	VVW	0.8	—	—	0.4	0.01	0.04
	Kaneuchi, Gessei vein (Japan)	0.1	—	W	0.08	—	—	0.4	0.03	0.025
	Takatori mine (Japan)	0.2	*	*	0.01	—	*	0.4	0.02	0.04
	Naegi district (Japan)	*	VW	VW	*	*	—	*	*	*
	Foot mine (U.S.A.)	*	—	W	*	*	VVW	*	*	*
	Ulchin mine (Korea)	0.4	VW	W	2.n	—	VW	0.2	0.01	0.1
	Aberfoyle mine (Australia)	0.07	—	W	0.3	—	—	0.2	0.005	0.002
	Walwa mine (Australia)	0.3	VW	W	3.n	—	VW	0.1	0.02	0.1
Xenothermal type	Unknown mine (Bolivia)	3.n	W	VVW	0.01	—	—	0.7	0.02	0.3
	Ikuno mine (Japan)	5.n	M	W	0.01	0.03	—	0.2	0.02	0.08
	Akenobe mine (Japan)	3.n	M	W	0.01	0.03	—	0.2	0.02	0.01
	Akenobe mine (Japan)	3.n	M	VW	0.01	0.02	—	0.1	0.02	0.01

* Not analyzed. —: not detected.
Quantitative analyses by Mr. K. Ikeda, Geological Survey of Japan; qualitative analyses by Mr. S. Oda, Dept. of Applied Chemistry, University of Tokyo. M: moderate, W: weak, VW: very weak, VVW: very very weak. Quantitative analyses in wt. %.

study of the adjacent areas it is believed that an elongated cupola of the cryptobatholith runs from NE to SW underneath the mining area (Imai, 1966b, 1970). It is noticed that many andesite, rhyolite and granophyre (felsite) dikes are developed in the mining area and that they were intruded just before and after the mineralization. They would be related to the same igneous activity as the granitic intrusion.

Deposits in the areas adjacent to the Akenobe mine: There are some mines which are developed to the southwest of the Akenobe mine along the outcrop of the granitic batholith. A brief description of one of these mines follows.

The Ohmidani mine is located to the southwest of the Akenobe mine (Fig. II–15–3). This vein-type, silver-producing mine is located in the silver zone of the Akenobe zonal pattern. The Fusei vein strikes N70°W while the Ohdate vein strikes N45°W, but both veins dip steeply to the north. The Fusei vein is presumably a right-handed wrench fault, with branching shears and tension cracks of the second order. The Ohdate vein, on the other hand, is a tension crack. The fissure pattern in this area might also be the result of lateral pressure exerted from a NW-SE direction, as described earlier.

The veins of the Ohmidani mine are filled with quartz, which is accompanied by orthoclase (adularia)*, argentite, native silver, pyrargyrite, polybasite, stephanite, tetrahedrite, pyrite, chalcopyrite, sphalerite, galena, etc. The tenor of the ore is 1.8 g/t gold and 480 g/t silver on the average (Sumita, 1969).

In the Ohmidani mine area, the Permo-Carboniferous sediments show a dome-like structure (Fig. II-15-3) which might be an indication of an underlying cupola of the granitic batholith. This viewpoint was also postulated by Kojima and Asada (1973).

Ore Deposits of the Tada mine, Hyogo Prefecture

The Tada mine is situated about 25 km northwest of Osaka. Monthly production of the mine prior to its closing in the spring of 1973 was approximately 1,000 tons of crude ore assayed at 2.6 % copper and 400 g/t silver.

Geologic setting: The mining area is composed essentially of the Permo-Carboniferous Chichibu Formation, which is known locally as the Tamba Formation. This formation is un-

---

* According to Yamaoka and Ueda (1974), the absolute age of orthoclase from the vein was 66–88 m.y., dated by the K-Ar method.

D. Xenothermal Veins

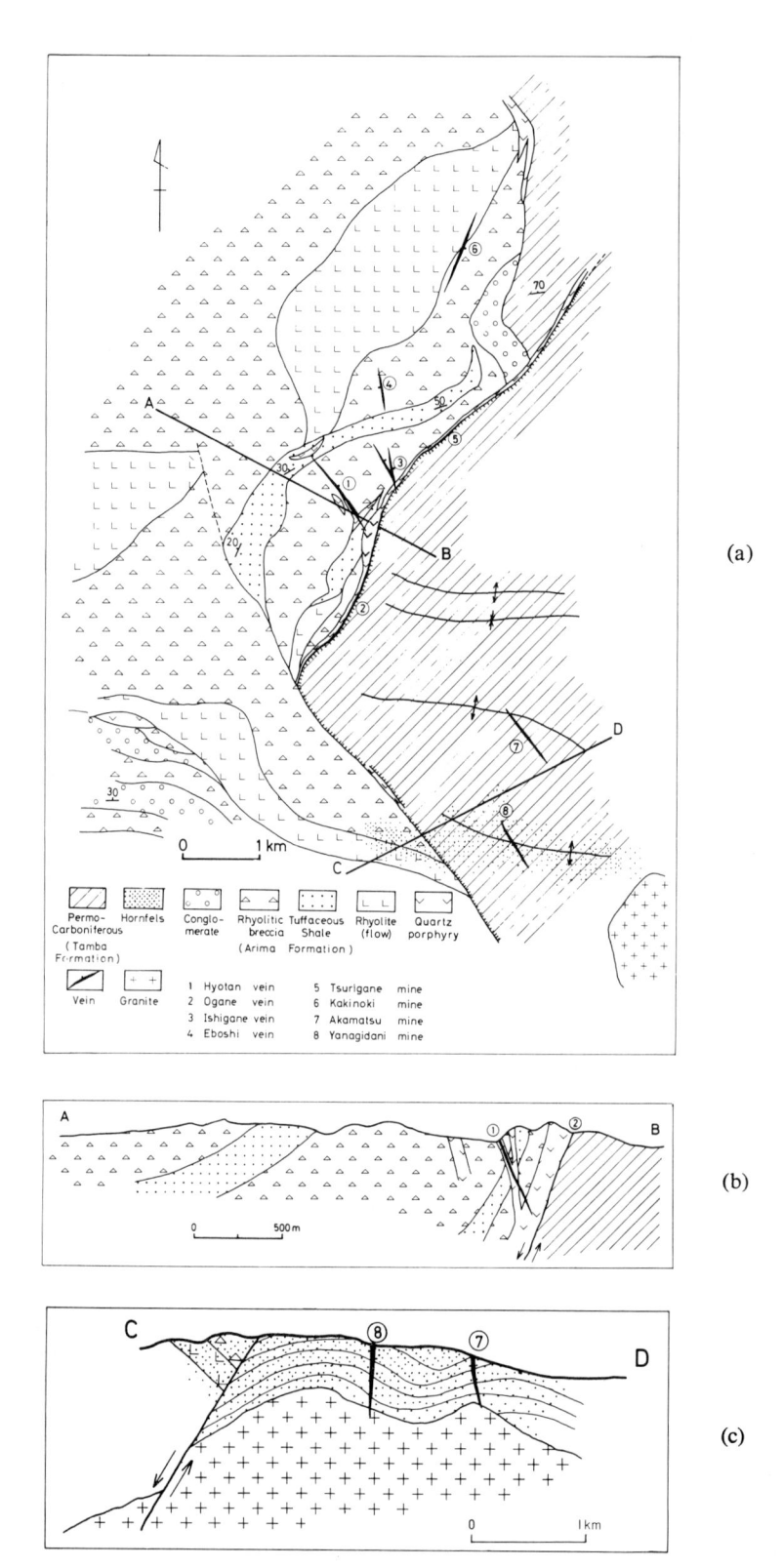

FIG. II-15-8. Geologic map and sections of the Tada mine.

conformably overlain by the Late Cretaceous or Early Tertiary Arima Formation (Figs. II–15-2, and -8).

The Tamba Formation consists mainly of sandstone and slate striking in a WNW-ESE or E-W direction and dipping steeply N or S. The Arima Formation, on the other hand, consists of rhyolite and rhyolitic tuff intercalated with shale, sandstone and andesitic rocks that generally strike in a NE direction and dip NW. Rhyolitic and andesitic dikes frequently occur in the vicinity of the mine; similar kinds of dikes are described in the Akenobe mine.

Some parts of the Tamba Formation are thermally metamorphosed to hornfels by a granitic intrusion which also had a feeble thermal effect on some andesitic dikes in the Arima Formation (Fig. II–15-8).

The contact between the Tamba and Arima Formations in this mining area is a normal fault which strikes NE and dips steeply NW. The strike direction of the Arima Formation is parallel to that of the fault but dips rather steeply to the NW near it, and gently away from it (Fig. II–15-8). The steep dip of the Arima Formation near the normal fault is caused by the fault dragging. In the southern part of the mining area, the Tamba and Arima Formations are in contact along another normal fault striking NW and dipping SW (Fig. II–15-8).

The spatial distribution of the granite which crops out only in the area covered by the Tamba Formation in the vicinity of the Tada mine, and the hornfelsic aureole, as well as the structural setting of the Arima Formation, suggest that the Tamba Formation, which covers the eastern part of the mining area, was pushed up by granite developing at depth underneath the area (Fig. II–15-8). The two normal faults bounding the Tamba and Arima Formations might have been developed as a result of the upheaval of the granitic magma.

Ore deposits: The deposits of the Tada mining area are mainly of two groups. The first group comprises veins occurring in the Arima Formation of the xenothermal type, very similar to those of the Akenobe, Ikuno and Ashio mines. Ore minerals of this group are bornite, cassiterite, chalcopyrite, stannite, stannoidite, mawsonite, sphalerite, galena, native silver etc., in a matrix of quartz. The second group comprises those veins occurring in the Tamba Formation which are mesothermal vein-type deposits corresponding to the Kajika-type deposits in the Ashio mine described later. Ore minerals of this group are chalcopyrite, pyrrhotite, pyrite and sphalerite, in a matrix of quartz. The Akamatsu and the Yanagidani mines are typical mesothermal-type deposits (Fig. II–15-8).

The xenothermal vein-type deposits in the Tada mine are represented by many veins, the most important being the Hyotan, Gochaku and Ohgane veins (Figs. II–15-8, and -9). The Hyotan and Gochaku veins occur in the rhyolitic rocks of the Arima Formation. These are fissure veins filling up tension cracks and striking generally in a NW-SE direction perpendicular to the NE-trending normal faults (the contact between the Tamba and Arima Formations). These tension cracks were formed by lateral pressure caused by the upheaval due to the granitic intrusion, which also produced the normal NE-trending faults occupied by the mineralized Ohgane vein (Fig. II–15-8). The mineralized veins are characteristically accompanied by quartz porphyry and andesite dikes and frequently by composite or multiple dikes of both rocks (Fig. II–15-9). In some places the quartz porphyry dike is recognized as preceding the andesite dike.

The alteration zones in the country rocks are classified into three successive zones outward from the vein: silicification zone, sericitization zone and chloritization zone. This pattern of alteration zoning is similar to that in the Ashio mine which is described later.

Mineralization of the deposit: The deposit was developed at the Hyotan and Gochaku veins down to the 270 m level from the top of the shaft. The ore minerals are composed of pyrite, cassiterite, chalcopyrite, stannite, bornite, stannoidite, mawsonite, tetrahedrite, sphalerite, galena, chalcocite (digenite), stromeyerite, mckinstryite (?) or jalpaite (?) and

D. Xenothermal Veins

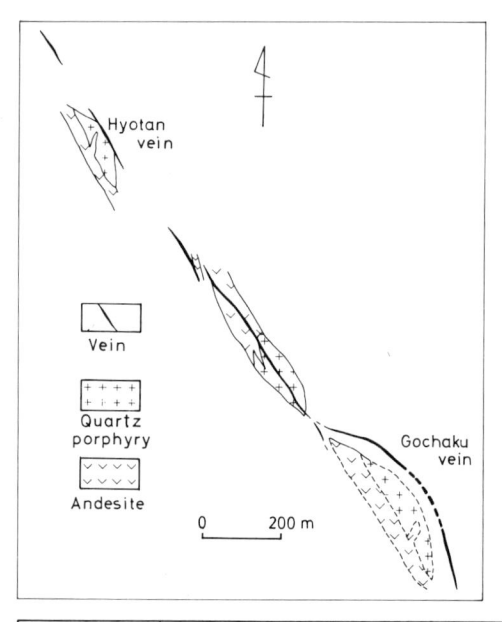

FIG. II-15-9. Vein map of the Hyotan and Gochaku veins.

FIG. II-15-10. Paragenesis of ore minerals in the Hyotan vein.

native silver in a matrix of quartz, calcite and fluorite. Bornite, chalcopyrite, native silver and stromeyerite are the important ore minerals.

At the Hyotan vein, pyrite, cassiterite and chalcopyrite tend to increase at the lower levels, while bornite and native silver concentrate at the upper levels. The paragenetic sequence of mineralization is shown in Fig. II-15-10. It is observed under the microscope that both cassiterite and sphalerite are replaced by stannoidite, which in turn is replaced by mawsonite (Fig. II-15-11(a)). Kato and Fujiki (1969) analyzed the chemical composition of stannoidite and mawsonite from this mine by electron microprobe and indicated that the zinc content of stannoidite reaches 4.5 %, while mawsonite is zinc-free. The above texture might not be due to the reaction of cassiterite with the ascending materials which deposited mawsonite. If stannoidite was formed by the reaction of cassiterite and the surrounding mawsonite which is virtually zinc-free, it is, then, expected that the resultant stannoidite should be poor in zinc. Stannite is sporadically the exsolution product from the chalcopyrite. The exsolution texture of chalcopyrite and bornite also is observed (Fig. II-15-11(b)).

Mawsonite usually occurs in very small and irregular grains and is usually associated with chalcopyrite, stannoidite and bornite, and sometimes with galena and tetrahedrite. Mawsonite always replaces the margin of stannoidite or a part of the grain boundary of stannoidite, bornite and chalcopyrite (Fig. II-15-11(c)). From these facts, it is certain that mawsonite could have been formed by the reaction of the ore-forming fluids with chalcopyrite, stannoidite and bornite.

Part II: Vein-type Deposits

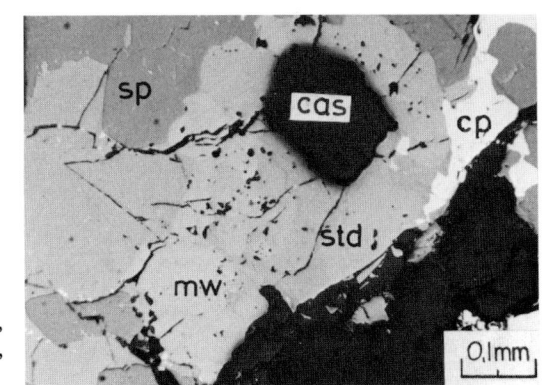

(a) Polished section. One nicol. cas: cassiterite, std: stannoidite, cp: chalcopyrite, sp: sphalerite, mw: mawsonite.

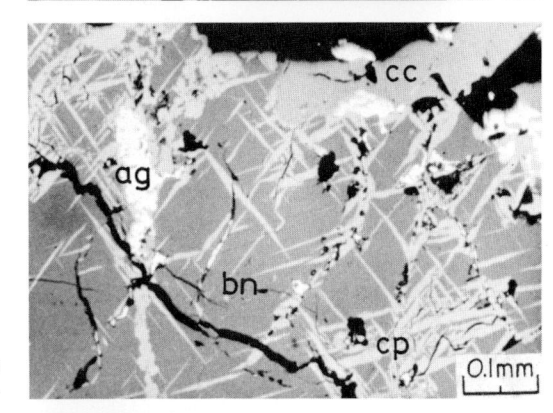

(b) Polished section. One nicol. bn: bornite, cp: chalcopyrite, ag: native silver, cc: chalcocite.

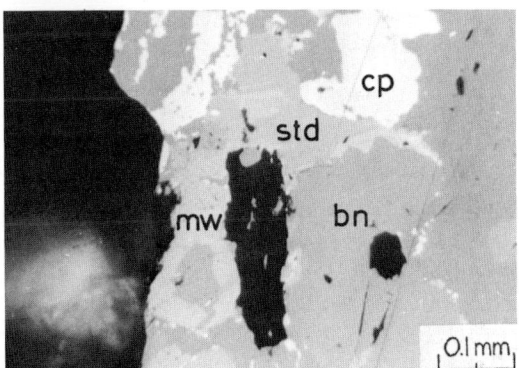

(c) Polished section. Parallel nicols. mw: mawsonite, cp: chalcopyrite, std: stannoidite, bn: bornite.

FIG. II–15–11. Photomicrographs.

From the fugacity-temperature studies of the writers (Lee *et al.*, 1975), the univariant log f_{S_2}-T curve of mawsonite — (stannoidite + bornite + chalcopyrite) in a Cu-Fe-Sn-S system is located between those of (bornite + pyrite) — chalcopyrite and of pyrite — pyrrhotite* (Fig. II–21–7).

Lee *et al.* (1974) studied the stability relations of stannoidite and mawsonite. The natural zincian stannoidite decomposes into bornite, stannite and chalcopyrite above 500°C, while zinc-free stannoidite is stable even at 800°C. Mawsonite, however, decomposes above 390°C. The writers also pointed out that mawsonite is not formed from the original composition $Cu_6Fe_2SnS_8$ by dry synthesis above 500°C, but instead a mixture of bornite + stannoidite

---

* Refer to Chapter II–21.

D. Xenothermal Veins

+ chalcopyrite is formed. By hydrothermal and flux methods, mawsonite can be formed at 200° to 380°C accompanied by bornite, stannoidite and chalcopyrite.

The hydrothermal synthesis of mawsonite, bornite and stannoidite is apparently favored in NaHCO$_3$ aqueous alkaline solutions. Based on the results of synthesis and hydrothermal alterations of the country rocks described above, the mineralization in this mine has possibly proceeded under alkaline conditions, though it is not certain.

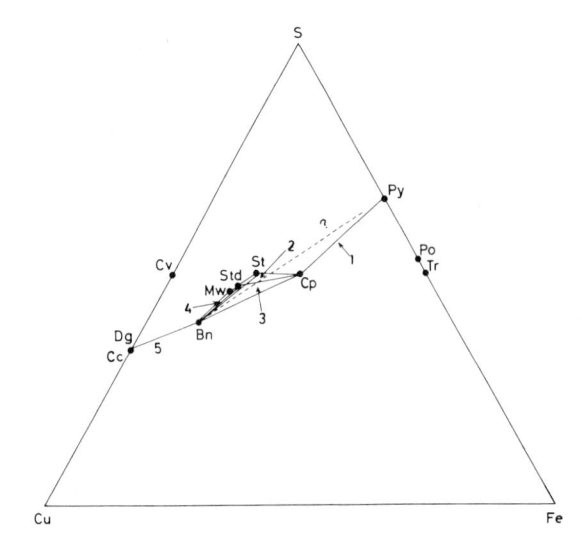

FIG: II-15-12. Diagram showing the succession of the Cu-Fe-Sn-S mineral paragenesis in the Hyotan vein. The diagram is projected from Sn to the Cu-Fe-S plane.

Py: pyrite, Po: pyrrhotite, Tr: troilite, Cp: chalcopyrite, St: stannite, Std: stannoidite, Mw: mawsonite, Bn: bornite, Cc: chalcocite, Dg: digenite, Cv: covellite.

The paragenetic relation of Cu-Fe-Sn-S minerals is expressed as shown in Fig. II-15-12. The succession of Cu-Fe-Sn-S minerals from the earlier stage to the later stage of mineralization shows a sequence of 1 to 5 as displayed in the figure.

TABLE II-15-3. Cu-Fe-Sn-S minerals in the Hyotan vein.

Mineral	Composition	Cu:Sn	Metal: Sulfur
Chalcopyrite	CuFeS$_2$	—	1:1
Stannite	Cu$_2$FeSnS$_4$	2:1	1:1
Stannoidite	Cu$_8$Fe$_3$Sn$_2$S$_{12}$	4:1	13:12
Mawsonite	Cu$_6$Fe$_2$SnS$_8$	6:1	9:8
Bornite	Cu$_5$FeS$_4$	—	6:4

Table II-15-3 shows the Cu/Sn and metal/sulfur ratios in the mineral paragenesis described above, revealing the successive increases of Cu: Sn and metal: sulfur. The writers (Lee et al., 1975) have recognized that mawsonite is more stable under conditions of high sulfur fugacity than is the assemblage bornite + stannoidite + chalcopyrite. Similarly, it is considered that stannoidite would be stable under higher sulfur fugacity than stannite + bornite + chalcopyrite*. So it seems as if the mineralization proceeded with an increasing sulfur fugacity. The same process was postulated by Sales and Meyer (1949), McKinstry and Kennedy (1957), McKinstry (1957), Gustafson (1963) and others, when they stated that fugacity of sulfur increases toward the later stage of sulfide mineralization.

But the boundary line of the stable areas of mawsonite and stannoidite in the f$_{S_2}$-T diagram would decline with a decrease in temperature (Lee et al., 1975) (Fig. II-15-13)*. Similarly, it is considered that the boundary line between the stannoidite and stannite areas

---

* Refer to Chapter II-21.

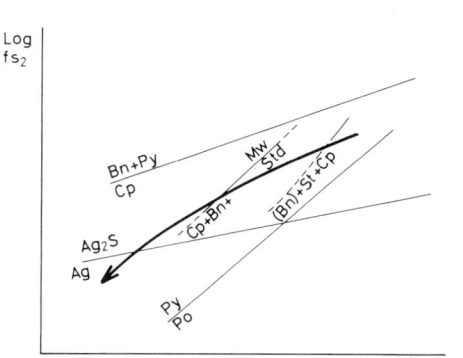

FIG. II–15–13. Sequential mineralization at the Tada deposit (arrow direction) in log f_{S_2} — T(1/T) diagram. Abbreviations are the same as in Fig. II–15–12.

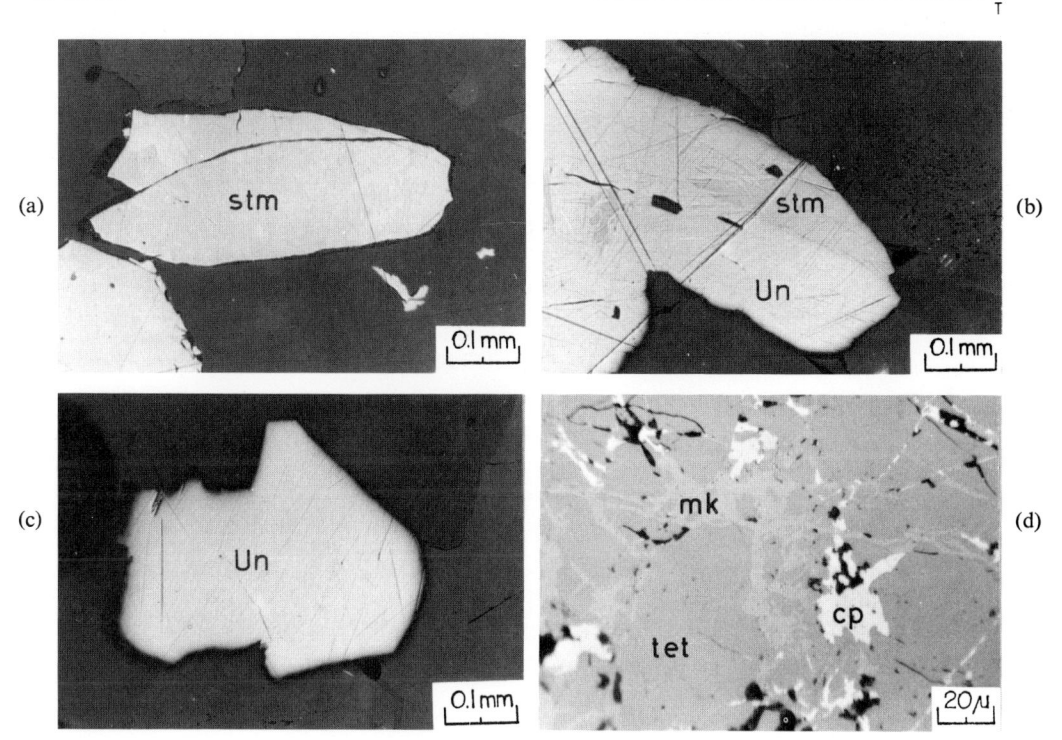

FIG. II–15–14. Photomicrographs. Polished section. One nicol.
stm: stromeyerite, Un: unknown mineral, mk: mckinstryite (?) or jalpaite (?), tet: tetrahedrite, cp: chalcopyrite.

would decline with a decrease in temperature. The boundary line between stannoidite and (stannite + bornite + chalcopyrite) is tentatively drawn as in Fig. II–15–13. So, the mineralization at the Tada mine would proceed towards the lower fugacity and lower temperature, crossing the boundary lines of the respective stability areas of stannite, stannoidite and mawsonite in the f_{S_2}-T diagram (Fig. II–15–13). But bornite accompanied by stannite has not been found in the ore deposit.

Silver minerals such as tetrahedrite, stromeyerite, mckinstryite (or jalpaite?) and native silver are recognized under the microscope, accompanied by bornite, chalcocite and digenite; they belong to the later stages of mineralization (Fig. II–15–14). It is worthy of mention that all these silver minerals occur predominantly in the upper levels of the mine.

The intimate association of stromeyerite and an unknown Ag mineral with bornite, galena, chalcopyrite and tetrahedrite is revealed by ore microscope as well as by electron microprobe. The results of analyses of stromeyerite in bornite (Type I) and in calcite vein

D. Xenothermal Veins

TABLE II-15-4. Analyses of Cu-Ag-S minerals by electron microprobe.

Type	Minerals	Beam diam. (μm)	Ideal comp. (wt. %) Cu	Ag	S	Total	Analyzed comp. (wt. %)* Cu	Ag	S	Total	Mol. % Cu$_2$S	Ag$_2$S
I	Stromeyerite in bornite	2	31.2	53.0	15.8	100	33.7	50.3	16.0	100	53.0	47.0
II	Mckinstryite or jalpaite in tetrahedrite	2					16.3	68.3	15.4	100	28.9	71.1
III-1	Unknown Ag mineral (Ag-poor domain)	20 2					77.4 78.4	2.8 1.9	19.8 19.7	100 100	97.6 98.4	2.4 1.6
	Unknown Ag mineral (Ag-rich domain)	20 2					76.2 61.3	3.6 20.7	20.2 18.0	100 100	97.5 83.5	2.5 16.5
	Unknown Ag mineral with stromeyerite lamella	2					74.5	5.3	20.2	100	95.9	4.1
III-2	Stromeyerite lamella in unknown Ag mineral	2	31.2	53.0	15.8	100	26.8	54.2	19.0	100	45.7	54.3
	Stromeyerite	20					32.6	51.4	16.0	100	51.5	48.5

* Recalculated after correction.
Take-off angle, 40°; specimen current, 0.02 μA on ZrO$_2$; accelerating voltage, 15 kV.

TABLE II-15-5. X-ray powder diffraction data for an unknown Ag-bearing mineral.

Cu Kα radiation, Ni filtered
35 kV, 15 mA

Unknown mineral I	2θ	d	Chalcocite I/I$_0$	d	Stromeyerite I/I$_0$	d	Calcite I/I$_0$	d
w	20.98	4.230			10	3.98		
m	24.70	3.601	10	3.60				
s	26.63	3.348			80	3.33		
vw	28.10	3.173	20	3.21				
s	29.42	3.033			60	3.07	100	3.04
s	31.28	2.857	20	2.88			20	2.85
s	34.08	2.629			100	2.61		
w	35.04	2.559	10	2.54	60	2.55		
w	36.65	2.450	20	2.47			40	2.51
s	38.83	2.314					50	2.29
vw	42.13	2.144	10	2.14				
vw	42.70	2.116			20	2.10		
vw	43.23	2.092			80	2.07		
vw	44.80	2.021			70	1.99		
vw	45.83	1.979	80	1.969				
vs	46.85	1.938	5	1.937			70	1.93
vw	47.88	1.898			60	1.89		
vs	48.98	1.858	100	1.870			60	1.87
vw	50.93	1.792	5	1.787				
vw	51.75	1.765			50	1.75		
m	54.73	1.704			30	1.69		
m	56.48	1.628	20	1.645	10	1.63		
w	61.83	1.493	20	1.514				
vw	66.83	1.399			60	1.42		
w	70.23	1.339	10	1.351				
w	74.80	1.267	30	1.278				

w: weak, vw: very weak, m: moderate, s: strong.

(Type III-2) by the probe are shown in Table II–15–4 (Fig. II–15–14(a)). The exsolution texture of the unknown Ag mineral and stromeyerite is shown in Fig. II–15–14(b). The unknown Ag mineral in the calcite vein of the latest stage (Fig. II–15–14(b), (c)) shows different domains of silver (Table II–15–4). Using a 20 μm dimameter beam, a silver-poor composition is obtained, while a 2 μm diameter beam indicates a rather silver-rich as well as a silver-poor domain (Tyep III-1) This indicates a heterogeneous aggregate of minerals that vary in their silver content. X-ray powder diffraction analysis of this mineral revealed a mixture of chalcocite and stromeyerite (Table II–15–5). This explains that this unknown Ag mineral is an ultramicroscopic exsolution product of these two minerals (Imai and Lee, 1972)*. Stromeyerite in Fig. II–15–14(b) represents the larger exsolved part in the chalcocite and stromeyerite aggregate. According to Skinner (1966), this texture might be formed at 67°C.

The mineral described as mckinstryite (or jalpaite?) (Type II) occurs as a fine network in a lump of tetrahedrite (Fig. II–15–14(d)). It reveals a pale grayish color, weak reflection pleochroism and strong anisotropism. Its properties under the microscope are the same as those of stromeyerite. From electron microprobe analysis (Table II–15–4), the writers tentatively distinguish it from stromeyerite. Native silver occurs as veinlets in bornite, chalcocite and stromeyerite, and belongs to the latest stage of mineralization (Fig. II–15–11(b)). The fugacity of sulfur decreases towards the later stage, depositing native silver.

Ore Deposits of the Ikuno Mine, Hyogo Prefecture

The Ikuno mine is located about 17 km southeast of the Akenobe mine (Fig. II–15–2). The vein-type deposit produced tin, tungsten, copper, zinc, lead and silver ores. The total output of the mine from 1886 to 1973 when the workings were closed, amounted to 17 million tons of crude ore, producing 165,000 t of copper.

Geologic setting: The area is covered by the Ikuno Formation, which is composed of rhyolite, rhyolitic tuff, sandstone, shale, andesite and basalt of Late Cretaceous age. This formation is intruded by Late Cretaceous granite, which was presumably the product of the same volcanic activity as the Ikuno Formation (Fig. II–15–15). The Ikuno Formation forms a basin structure elongated in a NW-SE direction (Fig. II–15–16), representing the depressed part of the formation in the vicinity of the mining area.

Dikes of rhyolite, andesite and basalt are well developed in the mining area. These dikes penetrated the Ikuno Formation before and after the mineralization. It is noticeable that in many places the veins run along the pre-ore basalt dikes. It is also observed that some dikes of andesite and basalt form sheets of the same rock in the upper levels of the adit (Fig. II–15–17).

In the rhyolitic rocks of the Ikuno Formation in this area, there are several nonmetallic deposits of replacement origin, containing pyrophyllite, kaolinite and small amounts of alunite, diaspore, pyrite, etc. The relation between the xenothermal veins and these nonmetallic deposits is uncertain.

Vein-fissure pattern: The directions of strike of the leading veins are N0°-20°W and N40°-50°W, dipping steeply NE. The N0°-20°W fissures are tension cracks filled by vein materials. The Senju vein, which is the champion vein of the mine, belongs to this type (Fig. II–15–16). The N40°-50°W fissures would originally have been formed as wrench faults, judging from the pattern of the branch fissures. The Tenju vein is a typical example of this type (Fig. II–15–18). Both types of fissures would be formed by lateral pressure exerted along a NNW-SSE direction.

---

* It is uncertain whether this phase corresponds to the stromeyerite mineral group 3Cu$_2$S.Ag$_2$S described by Haranczyk and Jarosz (1966) and Jarosz (1966).

106   D. Xenothermal Veins

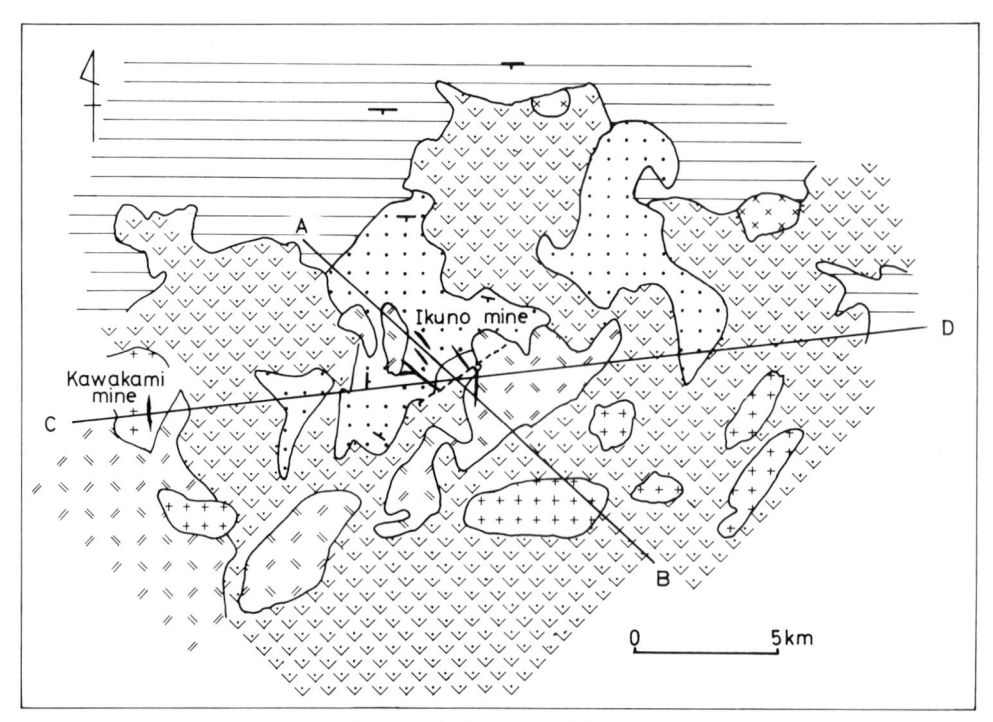

(a) Geologic map of the Ikuno mining area.

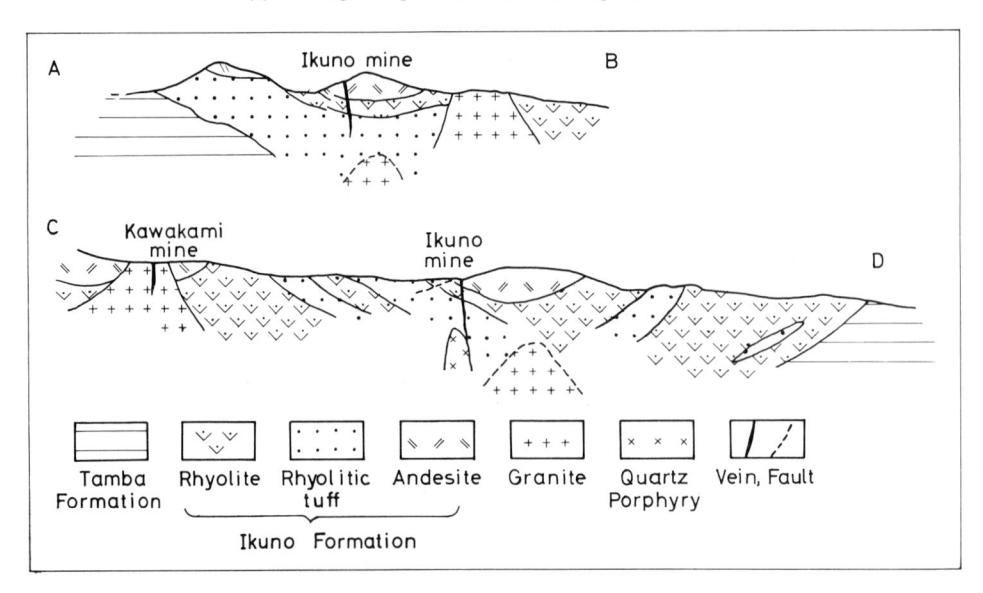

(b) Geologic sections of A–B and C–D in (a).

FIG. II–15–15. Geologic map and sections of the Ikuno mine.

The Senju vein is cut by many faults, such as the Nendo (Clay) Fault (Fig. II–15–16). They are classified into two types by strike direction, the N40°–50°W and the N40°–50°E. Both dip steeply to NE or NW. They reveal a conjugated pattern, and they were presumably formed by the lateral pressure discussed above. But in the fault plane, striation to the dip direction is observed. The Senju vein is displaced by these faults. Displacements due to the faults correspond to those of normal faults, judging from displacements of the vein and

Part II: Vein-type Deposits

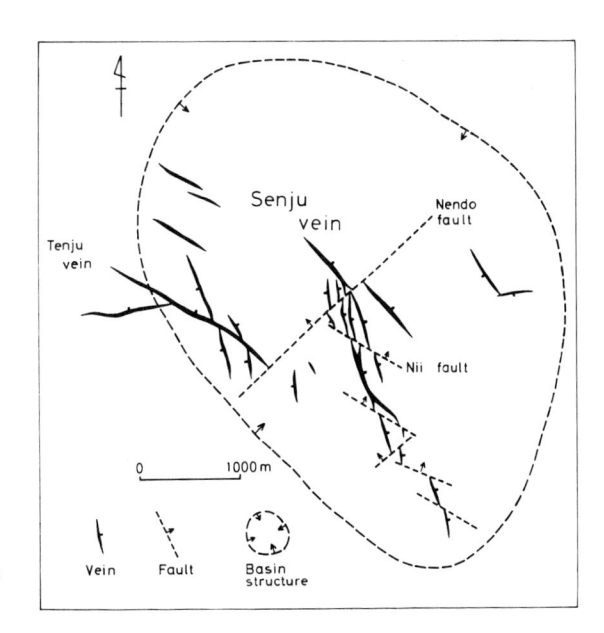

FIG. II–15–16. Basin structure of the Ikuno mining area.

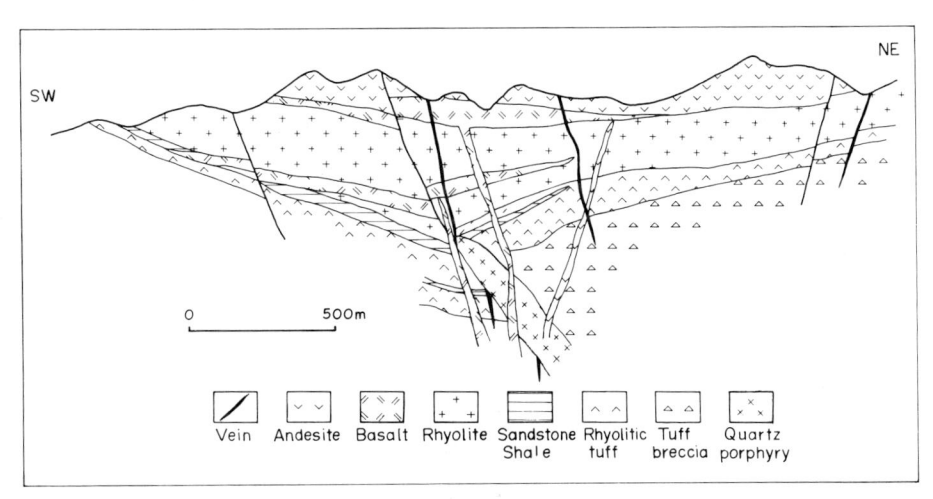

FIG. II–15–17. Geologic section of the Ikuno mine (after Maruyama, 1957).

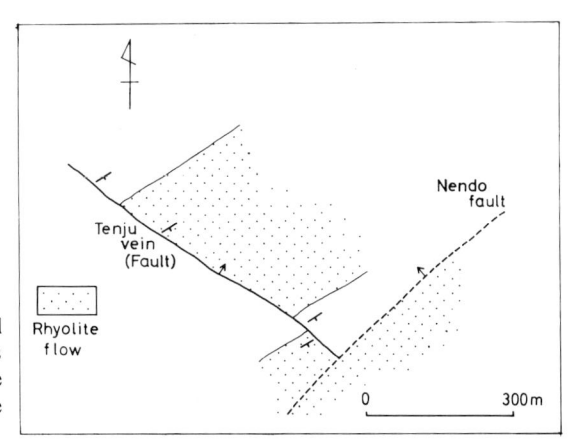

FIG. II–15–18. Displacement of the wall rocks by the Tenju Fault vein. Slickensides in the fault plane and distribution of rhyolite flows in both walls indicate the Fault to be normal.

D. Xenothermal Veins

striations in the fault plane. The writers presume that the fissures were originally formed as wrench faults by lateral pressure from a NNW-SSE direction, and that they were displaced afterward as normal faults. The blocks were depressed by these faults step by step to the north. These block depressions reflect the basin srtucture of the mining area. Weak mineralizations by a later hydrothermal solution are developed along these faults.

From the pattern of the Tenju vein (Fig. II–15–16), it is possible to assume that this fissure vein is a wrench fault. But, striation parallel to the dip and distribution of rhyolite in both walls reveals that the actual displacement of the host rocks is a normal fault. It might arise by the same processes as discussed in Chapter II-1.

The structural depression of the mining area corresponds to the caldera formation of volcano or cauldron subsidence. The center of the depression in this area is in the northern part of the Senju vein (Fig. II–15–16).

Mineralization: The Senju vein is the main vein of the mine. It extends about 1.5 km along its strike and more than 1 km down the dip, with an average width of 1 m. This vein is accompanied by a pre-ore basaltic dike.

The ore minerals include chalcopyrite, sphalerite and cassiterite, with small amounts of galena, ferberite, huebnerite, scheelite, arsenopyrite, pyrite, bornite, stannite, stannoidite, mawsonite, tetrahedrite-tennantite, bismuthinite, native bismuth, ikunolite ($Bi_4(S, Se)_3$; Kato, 1959), sakuraiite (($Cu, Zn, Fe, Ag)_3 (In, Sn)S_4$; Kato, 1965), argentite, pyrargyrite, proustite (Tanaka *et al.*, 1971), native silver, stromeyerite, native gold and orpiment.

The gangue minerals are quartz, calcite, fluorite and chlorite. Topaz was recognized in the lower level of the mine by Yamaguchi (1939). The relation between chalcopyrite, bornite, stannoidite and mawsonite is similar to that described at the Tada mine. Zonal distribution of the ore minerals both horizontally and vertically is recognized in this deposit (Maruyama, 1959). Zones from the lower part to the upper part and from the center to the margin are as follows: tin-tungsten, tin-copper, copper-zinc, zinc-lead and silver. The vertical zonal distribution in the Senju vein is shown in Fig. II–15–19. Each ore zone grades into adjacent zones without boundaries.

In certain limited areas, banded structures of the vein are typically manifested and reveal the composite nature of the vein. Kato (1927b) described the following succession of mineralization, from earlier to later: (1) cassiterite ore, (2) wolframite-cassiterite ore, (3) chalcopyrite ore, (4) quartz vein, sometimes with gold and silver ores. However, in a few cases, a

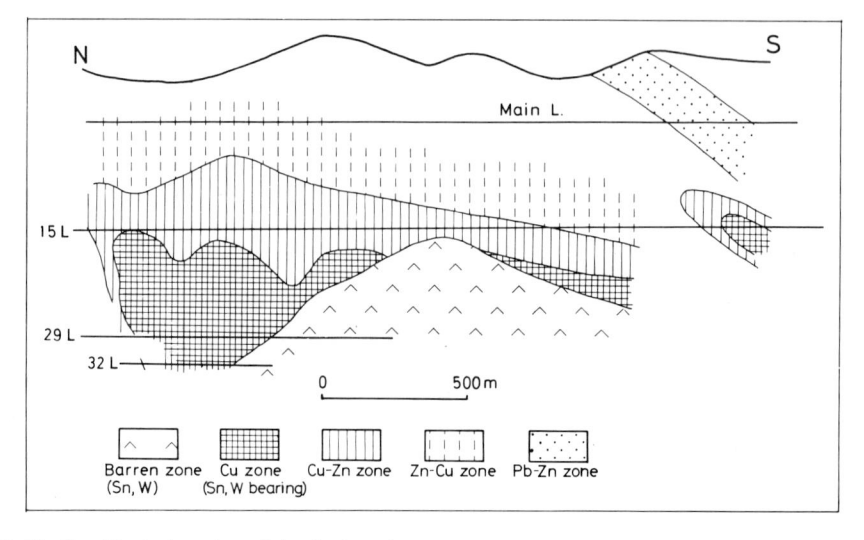

FIG. II–15–19. Vertical zoning of the Senju vein.

composite vein showing a reverse relation to that described in the Akenobe mine is observed. These observations might be explained as due to the advance and withdrawal (pulsation) of the magma as discussed in the section on the Akenobe and Ashio mines.

The writers believe that this deposit is genetically related to the granitic rocks which probably form a cryptobatholith underneath the mining area. The volcanism represented by the eruption of rhyolite and andesite would belong to the same igneous activity as the granitic intrusion. It should be noted that rhyolitic eruption was followed directly by andesitic activity (Fig. II–15–17). Small granitic bodies dispose sporadically in linear arrangement, running E-W in this district, indicating the probable E-W extension of the cryptobatholith. In and around these granitic stocks some copper-bearing veins are distributed (Fig. II–15–15). The west extension of this row of granitic bodies joins with the granitic stock in the area of the Ariga mine, extending to the southwest of the Akenobe mine (Figs. II–15–2, and–15).

Ashio-Nikko Metallogenetic Province

The Ashio-Nikko district is situated about 110 km north of Tokyo (Fig. II–15–1).

Geologic Outline of the Ashio-Nikko District

The basement rocks in the Ashio-Nikko district are Permo-Carboniferous sedimentary rocks known as the Chichibu Formation. They include slate, sandstone, chert, limestone and dolomitic limestone, and schalstein (Fig. II–15–20). The formation strikes NE-SW and dips steeply to NW or SE, forming a complex folded structure. The Chichibu Formation is overlain by Irohazaka rhyolitic welded tuff and rhyolite flows. Both formations are intruded by granodiorite and granodiorite porphyry. According to Yanai (1972), the ages of these granitic rocks determined by the K-Ar method are from 116 to 44 m.y. Formerly the Irohazaka welded tuff and rhyolite were considered to be of Miocene age on account of their similarities to the widely developed Miocene rocks in northeastern Japan. Kawada (1966) speculated that these volcanic rocks are Late Cretaceous or Early Tertiary from his studies on the Nohi Rhyolitic Rocks in central Japan (Fig. II–15–1). According to Yanai's isotopic age determination, it is evident that the Irohazaka welded tuff and rhyolite are pre-Early Tertiary, as they are intruded by the above granitic rocks. The present writers believe that these rocks are related to the same igneous activities as those which are developed in the western Kinki district.

The rhyolite flows and the rhyolitic tuff are widely developed to the east and north of the Ashio district and are known locally as the Katashina-Kinugawa Formation (Fig. II-15-20). This formation is stratigraphically identified as Miocene and is known as the host rock of several epithermal gold-silver-copper veins and several xenothermal veins. All the above-listed rocks are overlain by andesitic lavas and their equivalent pyroclastics of the Quaternary volcanoes.

In the Ashio mining district, the rhyolite mass occurs on the Permo-Carboniferous formation as a funnel-shaped mass (Fig. II–15–21) with a longest diameter of 4.4 km. The mass is called the Ashio Rhyolite Mass. The contact between it and the underlying Permo-Carboniferous sedimentary rocks is often marked by the presence of basal (talus) breccia (Figs. II–15–21 and–24) consisting of slate and sandstone blocks and, rarely, rhyolitic blocks. The maximum thickness is 80 m. From the occurrence of this basal breccia, it has been considered that the Ashio Rhyolite Mass erupted at the depressed part of the Permo-Carboniferous formation (Kusanagi, 1955). The writers also suggest that it is a result of cauldron subsidence, as discussed later.

The geologic age of the Ashio Rhyolite Mass is not yet settled, and the present writers

D. Xenothermal Veins

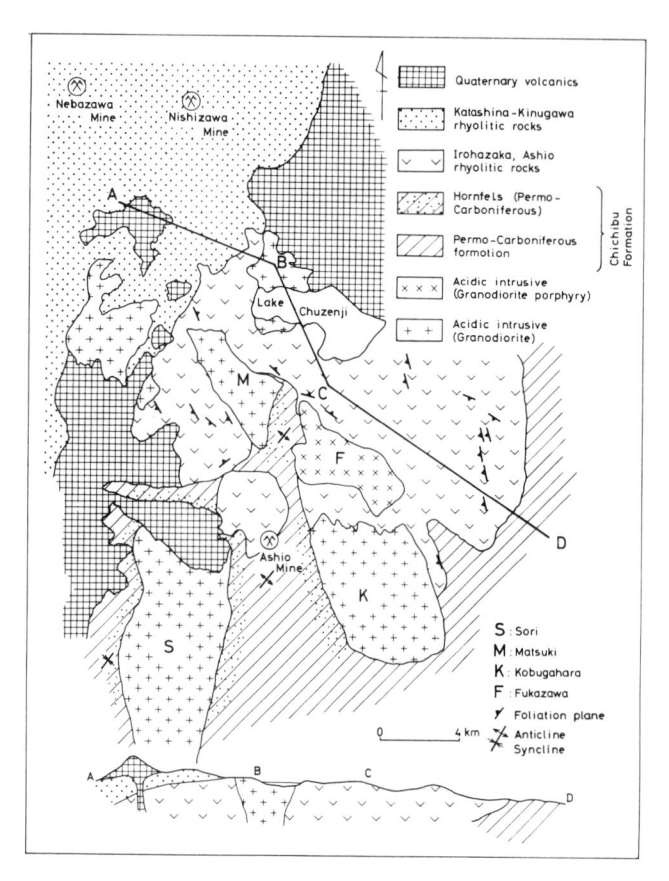

FIG. II-15-20. Geological map of the Ashio-Nikko district (after Yanai, 1972; revised by Imai). 1: Foliation plane of welded tuff. 2: Anticline and syncline.

speculate that it is probably the same age as the Irohazaka welded tuff, i.e., Late Cretaceous or Early Tertiary. But, Shibata and Ishihara (1974) determined the absolute age of sericitized rhyolitic rock at the Ashio mine by the K-Ar method. It was dated at 14.8 m.y., telling that the mineralization probably occurred in Middle Miocene epoch.

Granodioritic rocks are widely developed around the Ashio mine (Fig. II-15-20). Yanai (1972) gave some K-Ar isotopic ages for these plutons. The granodiorite to the east is known locally as the Kobugahara granodiorite and Fukazawa granodiorite porphyry, with absolute ages of 64 to 63 and 44 m.y., respectively. The Matsuki granodiorite which intrudes the Irohazaka Fomation to the north is 116 m.y., while the Sori granodiorite to the west is 87 m.y. This latter (Sori) granodiorite intrudes the Permo-Carboniferous formation with the development of a hornfelsic aureole (Fig. II-15-20).

The direct relation between the Ashio Rhyolite Mass and the above-described granodioritic masses is not yet established, as they are not in direct contact with each other. It is speculated, however, that a granodiorite or granodiorite porphyry cryptobatholith intrudes the Ashio Rhyolite Mass. The crystobatholith must have a genetic relationship with the ore deposits at the Ashio mine (Fig. II-15-21).

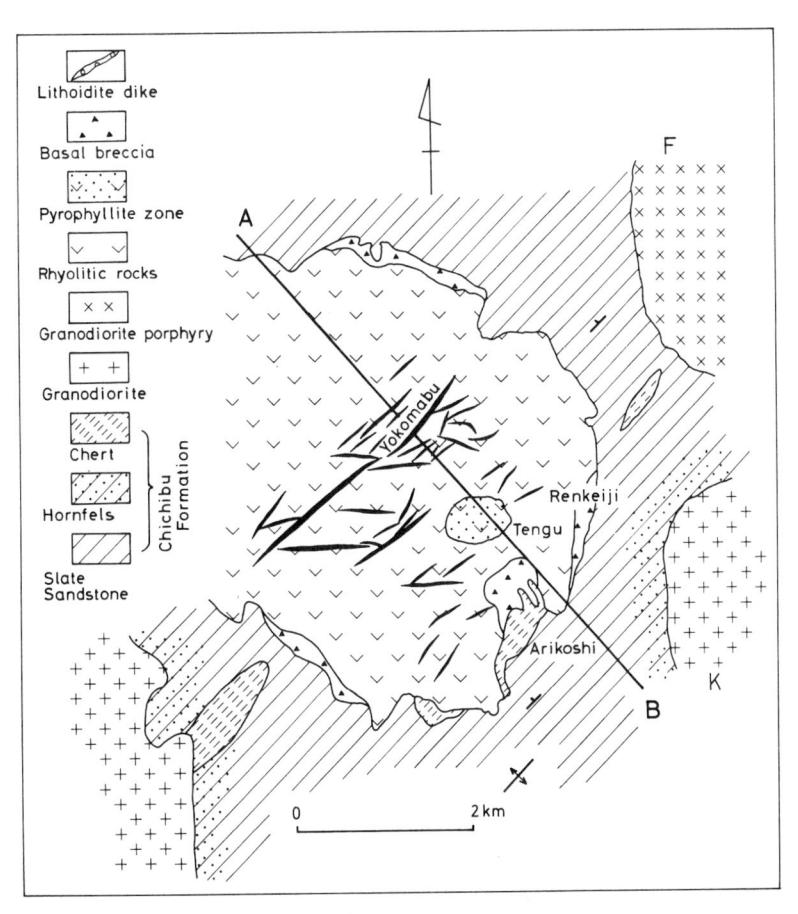

(a) Geologic map of the Ashio mine.
K: Kobugahara granodiorite, F: Fukazawa granodiorite porphyry.

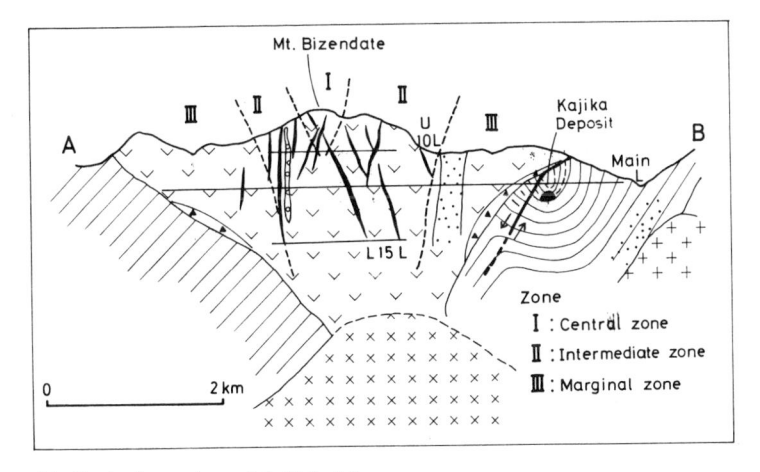

(b) Geologic section of A–B in (a).

FIG. II–15–21.

112   D. Xenothermal Veins

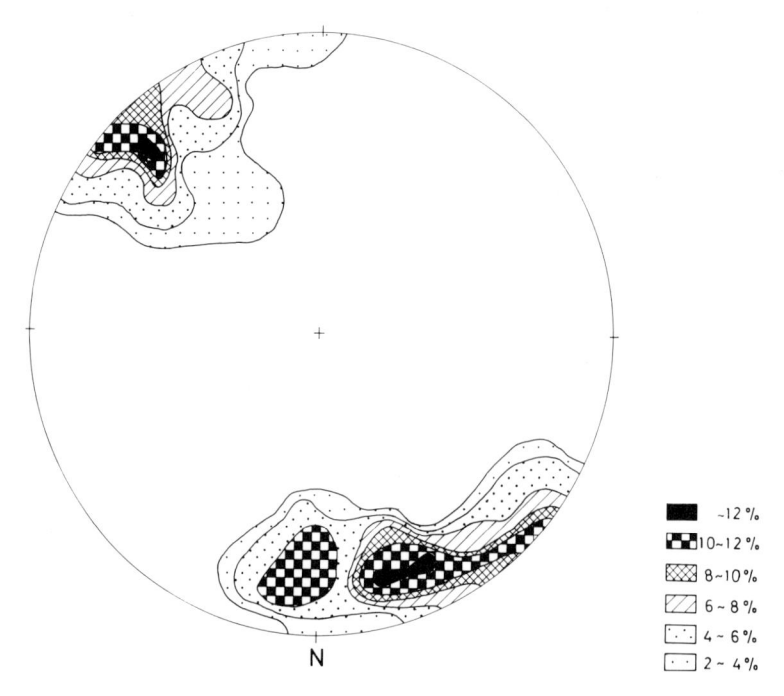

FIG. II-15-22. Stereographic projection of the poles of the veins in the Ashio Rhyolite Mass. Total: 84 veins.

Ore Deposits of the Ashio Mine, Tochigi Prefecture

It is believed that this deposit was discovered in 1610 and since then it has been the largest copper producer in Japan, having produced some 700,000 t of copper up to 1973 when the mine was closed.

This mine contains two kinds of deposit, a vein-type deposit and a Kajika-type deposit.

Vein-type deposit: The majority of producing veins occur in the Ashio Rhyolite Mass (Fig. II-15-21). About 1,400 potentially producing veins occur in the Ashio Rhyolite Mass, and some 20 veins occur in the surrounding Permo-Carboniferous sedimentary rocks. The main vein is the Yokomabu vein, which has a strike length of 2,100 m and extends for over 1,000 m down the dip (Fig. II-15-21). A pre-ore lithoidite dike of about 1 to 1.5 m in width runs nearly parallel to the Yokomabu vein in the Ashio Rhyolite Mass. Similar acidic dikes are observed in the areas of the Akenobe and Tada mines.

The fissure veins are classified into three groups (Kusanagi, 1957) according to their strike directions, as shown in Fig. II-15-22.

(1) The 45° vein group (e.g., Yokomabu vein and Kosei vein)—These veins strike N35°–55°E (average N45°E), dip steeply to NW or SE and are rich in clayey materials. The horizontal striation of the slickensides of these veins is often recognized on their walls, indicating that the original fissure is wrench fault. The displacement of the wall rocks along the fissures is recognized at some places in the adit.

(2) The 90° vein group (e.g., Shinsei vein and Tengu vein)—These veins strike N85°–110°E (average EW) and dip steeply to N or S. This group is similar to the 45° vein group in character, i.e., filling wrench faults.

TABLE II–15–6. Minerals of the different zones of the Ashio mine (mainly after Nakamura, 1961; Nakamura and Aikawa, 1973).

Hypogene mineral zone	Ore minerals	Gangue minerals
Central Zone	Cassiterite Ferberite Native bismuth Bismuthinite Native gold Wittichenite Bornite Arsenopyrite Stannite* Stannoidite* Mawsonite* Canfieldite* Chalcopyrite Pyrite	Quartz Sericite Chlorite
Intermediate Zone	Cassiterite* Scheelite* Arsenopyrite Chalcopyrite Pyrite Pyrrhotite* Marcasite Sphalerite Galena Stannite* Tetrahedrite* Tennantite* Bornite*	Chlorite Quartz Sericite Fluorite Apatite
Marginal Zone	Arsenopyrite Chalcopyrite Sphalerite Galena Stannite* Pyrite Stibnite*	Chlorite Quartz Sericite Calcite Siderite

* Rare.
Besides these minerals, ikunolite [Bi$_4$(S,Se)$_3$], cosalite Pb$_2$Bi$_2$S$_5$, vivianite Fe$_3$(PO$_4$)$_2$ · 8H$_2$O, ludlamite Fe$_3$(PO$_4$)$_2$ · 4H$_2$O, etc. are reported.

(3) The 68° group (e.g., Ebisu vein and Eisei vein)—These veins strike N65°–75°E (average N68°E) and dip steeply to SE or NW. They have a shorter strike length but a wider thickness than the former two groups and contain barren horses of country rock. The horizontal striations of the slickensides are never observed on the walls of the veins and no displacement of the wall rocks has been recognized at any place in the adits. These facts indicate that the fissures of this group are tension cracks.

From the fissure patterns of these three groups (Fig. II–15–22), it is concluded that the fissures were formed by lateral pressure exterted in a N68°E–S68°W direction. The foliation planes of the welded tuff of the Irohazaka Formation strike in a NNW-SSE direction with an E or W dip (Fig. II–15–20). The elongated direction of the granodiorite masses of Kobu-

D. Xenothermal Veins

gahara, Fukazawa, Matsuki and Sori in the Ashio-Nikko district is also NNW-SSE (Fig. II–15–20). This might suggest that the folding or undulations of the Irohazaka Formation occurred nearly at the same time as the granodiorite intrusion, probably during the Late Cretaceous or Early Tertiary. Consequently, a prevailing ENE-WSW pressure was exerted at that time, producing the above types of fissures.

Ore minerals of these veins are chalcopyrite accompanied by pyrite, sphalerite, galena, arsenopyrite and other minerals in minor amounts with quartz. The Ashio deposit exhibits three distinct zones, without no sharp boundary between them. The minerals characteristic of each zone are listed in Table II–15–6 (Nakamura, 1954, 1961; Nakamura and Aikawa, 1973). Following is a brief description of each zone (Fig. II–15–21(b)):

(I) The central tin-tungsten zone occurs around the Bizendate, which is 1,272.8 m above sea level, corresponding to the highest elevation in the Ashio mining area. It is characterized by the presence of cassiterite, wolframite, stannite and stannoidite. According to Nakamura (1954) and Kusanagi (1963), the grade of tin decreases remarkably with depth.

(II)(1) The intermediate copper zone is the main copper-producing zone and it surrounds the central zone vertically and horizontally.

(II)(2) The intermediate pyrite zone (the non-productive intermediate zone) is mainly composed of pyrite and lacks chalcopyrite and other minerals. The intermediate copper and pyrite zones grade into each other.

(III) The marginal zinc zone surrounds the intermediate zone and is characterized by the predominance of sphalerite and galena.

The type of mineral zonation exhibited in the Ashio mining area is a reverse zoning, as described by Kutina (1957). The pattern of zoning in the mine is the opposite of that described for the Akenobe and Ikuno mines. Examples of reverse zoning are reported in the cassiterite-wolframite deposits of Kiangsi, South China, by Tu Kwang-Chi and Liu Y-Mao (1965) and in the cassiterite-pyrite-pyrrhotite deposit of Lifudzin, Far East Siberia, by Kigai (1963). The former was explained as due to the continual uplift of the area during the whole period of mineralization and the latter as due to the development of fissures during tectonic movement and the constant change in the ore-forming fluids. The present writers explain the reverse zoning at the Ashio mine as due to the withdrawal or retreat of the source of the ore-forming fluid, i.e., the magma chamber (Fig. II–15–7). This is related to the cauldron subsidence at the central part of the rhyolitic eruption. The banded structure of the veins at the Ashio mine reveals its normal sequence of deposition. Nakamura (1954) observed that in the upper levels of the Ashio mine the mineralization of chalcopyrite is later than that of cassiterite. Through magmatic withdrawal (retreat) the banded structure revealed the normal (direct) sequence of deposition, as shown in Fig. II–15–7.

Wall-rock alterations of the veins include silicification, sericitization and chloritization, which are genetically related to the mineralization. The tin-tungsten zone is characterized by a silicified wall that ranges from several centimeters to 1 m in width. On the other hand, the copper and zinc zones are characterized by the development of sericite and chlorite, with the sericite occurring along the vein in about a 10 cm width which grades into a chlorite zone of about 10 m width. In the rhyolitic mass, however, pyrophyllite, topaz, diaspore, corundum, zunyite and andalusite occur in a pipe-like body near the Tengu Valley (Fig. II–15–21). This body has no obvious relationship to the ore mineralization (Kusanagi, 1955).

Kajika-type (massive replacement or impregnation) deposit: Kajika-type is a local name applied by miners to the massive impregnated or replacement deposits of copper and zinc in the Ashio mine. The main occurrence of this type of deposit is restricted to the chert surrounding the Ashio Rhyolite Mass.

The Kajika deposit occurs in the Arikoshi area, i.e., in the southeastern part of the Ashio

Part II: Vein-type Deposits

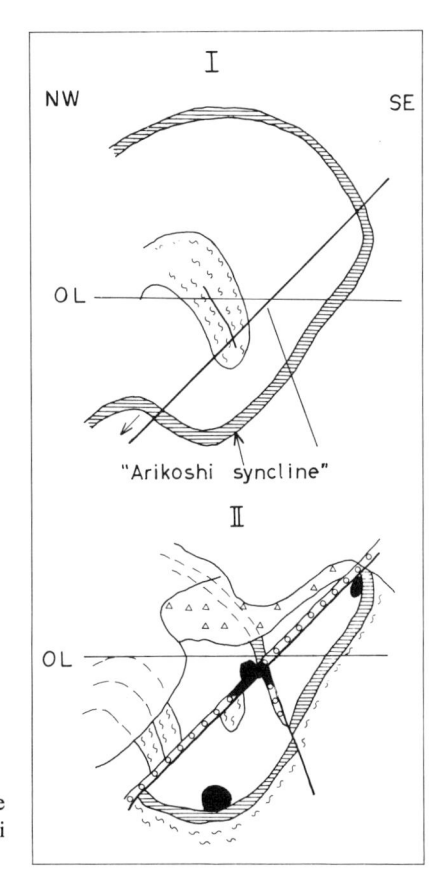

FIG. II–15–23. Diagrammatic formation mechanism of the nappe structure and cauldron subsidence in the Arikoshi area (cross section). Legend: refer to Fig. II–15–24.

Rhyolite Mass where the cherty member of the Permo-Carboniferous formation strikes to the NE-SW and dips steeply to NW or SE (Fig. II–15–21).

Figure II–15–23 shows a diagrammatic cross-section of the geologic structure of the Arikoshi area. It is postulated here that the main feature of the folded cherty member reveals a nappe structure with its basal part resembling a syncline (Arikoshi syncline). The axis of this apparent syncline plunges 35° in a N30°E direction (Fig. II–15–24(a), (b)). Along this axis the cherty rocks are replaced or impregnated by copper ore, which forms elongated massive ore bodies distributed in rows (Fig. II–15–24(c)). These are known as the Arikoshi Kajika group, which extend for more than 1,000 m in the plunge direction; the maximum section of the ore body in the horizontal plane is 50 m by 100 m.

The nappe structure is cut by a normal fault displacing the hanging wall block about 300 m down the dip (Asano, 1950) (Figs. 11–15–23, and –24). Along this normal fault a breccia dike is forcefully intruded, forming the Renkeiji Main Dike (RMD in Fig. II–15–24). The maximum width of this dike is 30 m. This normal fault has a horseshoe shape and dips inward (Fig. 11–15–24(a)). This means that the fault structure is a cauldron subsidence at the central part of the rhyolitic eruption. The process of the formation of the nappe structure, the cauldron subsidence and the breccia dike intrusion is shown in Fig. II–15–23. The same kinds of breccia dikes are abundant in the sedimentary rocks as well as in the rhyolitic rocks of the Arikoshi area, although they are less frequent in the latter. In some parts they run along the boundary between the sedimentary rocks and basal breccia (Fig. II–15–24(b)). These breccia dikes are mainly composed of fragments of chert, slate, sandstone and occasionally rhyolite in a rhyolitic matrix. They represent an important path

D. Xenothermal Veins

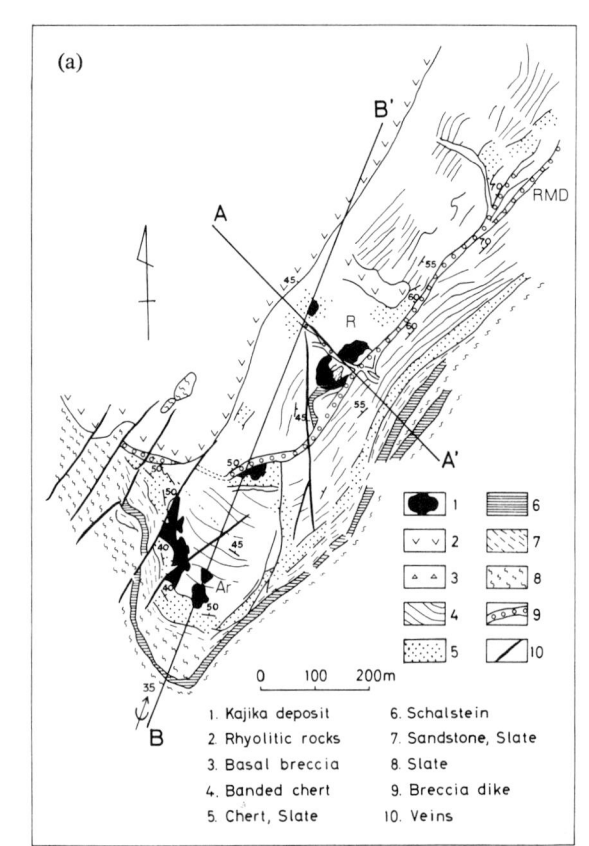

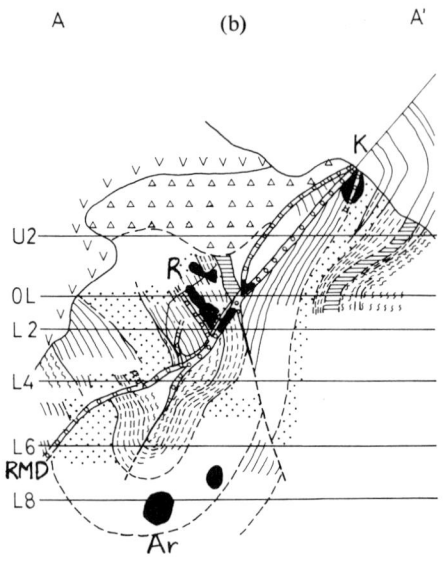

(a) Underground geologic map of the Kajika deposits in the Arikoshi and Renkeiji area. 0m level. Ar: Arikoshi Kajika, R: Renkeiji Kajika, RMD: Renkeiji Main Dike.
(b) Geologic section of A—A′ in (a). K: Kinryuzan Kajika.
(c) Geologic section of B—B′ in (a).

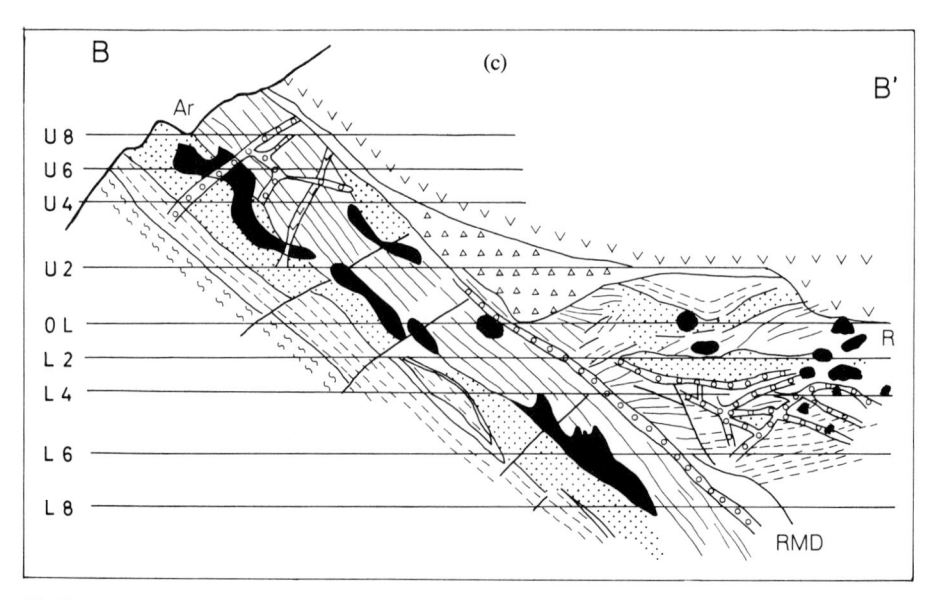

1. Kajika deposit
2. Rhyolitic rocks
3. Basal breccia
4. Banded chert
5. Chert, Slate
6. Schalstein
7. Sandstone, Slate
8. Slate
9. Breccia dike
10. Veins

FIG. II-15-24. Geologic map and sections of the Arikoshi area.

of the ore-forming fluids. In some parts of the Renkeiji Main Dike there are replacement and impregnated ore bodies of chalcopyrite (Fig. II–15–24).

Along the Renkeiji Main Dike some other replacement and impregnation ore bodies exist in the local folded parts of the cherty rocks (Fig. II–15–24(b)) as well as in areas overlain by basal breccia, which acts as a cap rock (Fig. II–15–24(b)). These ore bodies are known as the Renkeiji and Kinryuzan Kajika groups, respectively. The two groups were mineralized by ore-forming fluids ascending along the Renkeiji Main Dike.

The ore minerals of the Kajika-type deposit are chalcopyrite, pyrrhotite, arsenopyrite and sphalerite with small amounts of cassiterite, stannite, tetrahedrite, pyrite, galena, native bismuth, bismuthinite and native gold in gangues of quartz, sericite, chlorite, apatite and calcite (Nakamura, 1961). The main ore minerals are chalcopyrite and pyrrhotite, but in some ore shoots, sphalerite and pyrrhotite are predominant. This type of Kajika deposit corresponds to the mesothermal replacement-type and is rich in pyrrhotite, but lacks bornite, stannoidite and mawsonite which are characteristic of xenothermal vein-type deposits.

Fluid inclusion studies of quartz from vein-type deposits of the Ashio mine: Fluid and solid inclusions in quartz from mineralized veins and adjacent rocks were studied. Several types of inclusions were recognized under the microscope: (1) glass inclusions, (2) solid crystal inclusions, (3) liquid-rich inclusions, (4) gas-rich inclusions, and (5) monophase liquid inclusions. The inclusions of (1) and (2) are unrelated to mineralization.

(1) Glass inclusions were found solely in quartz from the surrounding rhyolitic rocks. They were mainly composed of glass and a gas bubble but sometimes contain solid crystals.

(2) Solid crystal inclusions were also observed in quartz from the surrounding rhyolitic rocks. They occur as rod-shaped short, prismatic or radial aggregates of needle-like crystals.

(3) Liquid-rich inclusions are the most common type in the veins and are composed of liquid and gas phases. They are characterized by a high degree of liquid filling compared with gas. All these inclusions homogenize into liquid with heating. No daughter mineral was recognized in this type of inclusion.

(4) Gas-rich inclusions are also composed of liquid and gas phases but are characterized by a larger gas bubble. On heating they homogenize into a gas phase at comparatively high temperatures.

Gas-rich and liquid-rich types of inclusions sporadically occur together in the same quartz crystal (Fig. II–15–25). Such an occurrence was observed in some parts of the mine, which suggests that both types of inclusions were formed from a heterogeneous mixture of

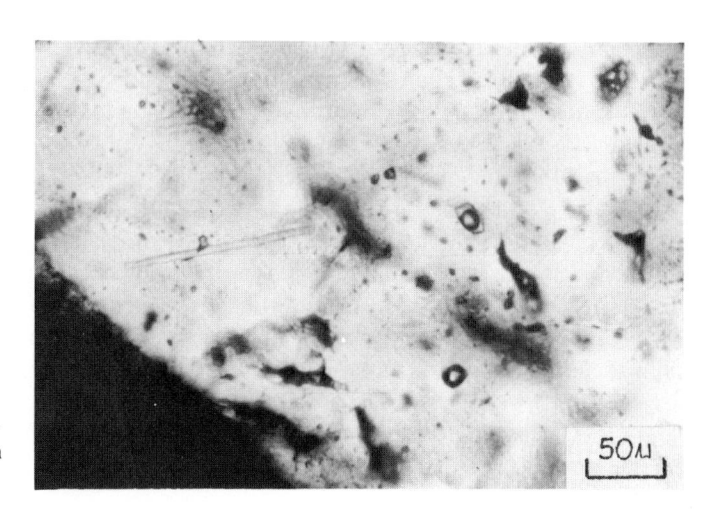

Fig. II–15–25. Gas-rich inclusions and liquid-rich inclusions in quartz.

D. Xenothermal Veins

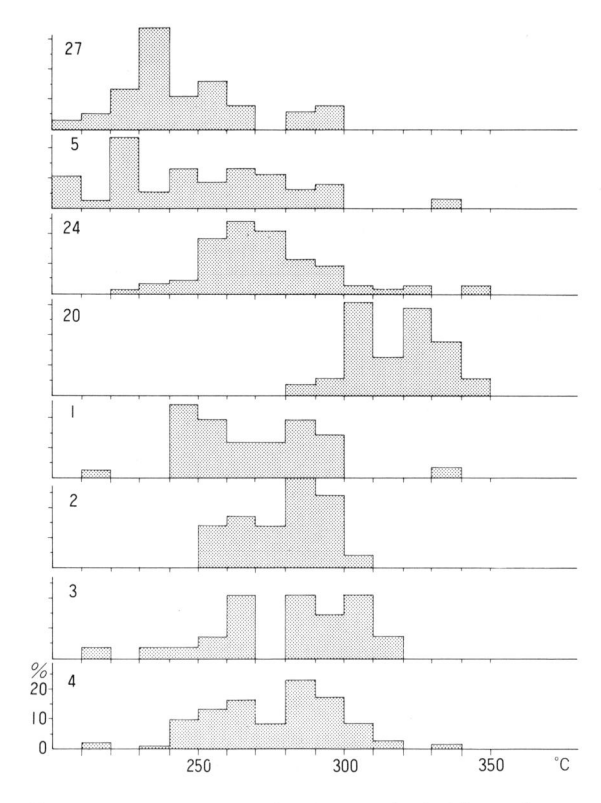

FIG. II-15-26. Histograms of the filling temperatures of fluid inclusions in quartz.
Nos. 1-4: tin-tungsten zone. No. 20: intermediate copper zone. Yokomabu No. 1 vein. Lower 15th level. No. 24: intermediate copper zone. Nanjin Uwaban vein. Lower 15th level. No. 5: intermediate copper zone. Yabushiki Tateire vein. Upper 10th level. No. 27: marginal zinc zone. Ginsei vein. Lower 2nd level. Refer to Fig. II-15-21.

liquid and gas. It may be due to the boiling phenomenon caused by a steep pressure drop due to the fissure opening at a comparatively shallow depth.

(5) Monophase fluid inclusions have no recognizable gas phase. They were probably produced by necking down of the liquid-gas inclusions at lower temperatures.

An inclusion which contains some daughter minerals of salt or liquid CO_2 phase has not been observed in this deposit. However, quartz belonging to the later stage of mineralization from the hypothermal or pegmatitic tin-tungsten veins contains inclusions rich in liquid CO_2 (Takenouchi and Imai, 1971; Kim et al., 1972).

Liquid-rich and gas-rich inclusions in quartz crystals from the veins were studied by heating-stage and freezing-stage microscopes. Filling temperatures for 951 fluid inclusions in 31 samples from the vein-type deposit were determined. Some of the histograms of the filling temperatures and their frequencies according to the zonal distribution are shown in Fig. II-15-26. No correction for the pressure has been made.

(I) Central tin-tungsten zone: All inclusions studied from this zone are from quartz veins. Inclusions in cassiterite and wolframite are difficult to measure because of their small size. The filling temperatures of the inclusions in quartz crystals are comparatively low. About 95 % of them are within the range 310° to 240°C (for example, Fig. II-15-26, Nos. 1-4). Kim et al. (1972)* determined the filling temperatures of fluid inclusions in eight cassiterite crystals from the Ohtani mine and found that temperatures range between 345°C and 297°C. However, filling temperatures of inclusions in quartz from the same mine were found to be predominantly below 300°C with some exceptions. The same writers also studied fluid inclusions in quartz and wolframite from the Kaneuchi mine. The filling temperatures of wolframite were between 337°C and 286°C, while those of quartz were between 308°C and 231°C. They interpret these differences as being due to the differences in the stages of

---

* Refer to Chapter II-4.

mineralization, as the quartz used for measurements generally belongs to the later stage of mineralization. The quartz of the earlier stage, however, which was contemporaneous in deposition with cassiterite and wolframite, contains very small inclusions which are inadequate for measurement. Kelly and Turneaure (1970) measured the filling temperatures of 390 primary and pseudosecondary fluid inclusions in the cassiterite from the ore deposits of the eastern Andes of Bolivia and determined deposition temperatures of 510°C to 410°C. Tugarinov and Naumov (1972) measured the filling temperatures and the decrepitation temperatures of the fluid inclusions in cassiterite from 75 ore deposits around the world and stated that 78 % of their measurements were in the range of 450°C to 300°C. As to wolframite, Tugarinov and Naumov found that 89 % of their measurements were in the range of 450°C to 250°C. From these facts, the present writers estimate that the formation temperatures of cassiterite and wolframite in the Ashio mine would be between 350°C and 250°C.

(II) Intermediate copper zone: Measurements of 579 fluid inclusions in 19 samples from this zone were carried out. In this zone, remarkable changes are recognized. A typical horizontal change in the filling temperatures is recognized in the Lower 15th level of the Yokomabu area (Fig. II–15–21). The Yokomabu No. 1 vein shows high filling temperatures, where 80 % of the measurements are in the range 340°C to 300°C (for example, Fig. II–15–26, No. 20). On the other hand, the filling temperatures of quartz from the Nanjin Uwaban vein, which is located some 350 m north of the Yokomabu vein, are generally lower. About 86 % of the measurements are in the range of 300°C to 250°C (for example, Fig. II–15–26, No. 24), and the majority are in the range of 270°C to 260°C. Generally speaking, a temperature drop of 40°C is observed in a horizontal distance of 350 m from center to margin. The samples studied covered a vertical distance between the Upper 10th level and the Lower 15th level of the mine, corresponding to 750 m (Fig. II–15–21). The filling temperatures in the area of Yabushiki Tateire of the Upper 10th level are even lower. It is estimated that 95 % of the measurements are concentrated in the range of 300°C and 200°C (for example, Fig. II–15–26, No. 5). To compare these values with the Yokomabu No. 1 vein in the lower 15th level, a temperature drop of 50°C to 60°C is estimated for a vertical distance of 750 m between the Lower 15th and Upper 10th levels.

(III) Marginal zinc zone: In this zone, quartz is rather rare in the veins. Only two samples were collected from the Ginsei vein. The filling temperatures of these samples range from 270°C to 200°C (for example, Fig. II–15–26, No. 27). These are also comparable to the marginal part of the copper zone.

Some 49 measurements of freezing temperatures in two samples, collected from the copper zone, were carried out with a freezing-stage microscope. These measurements showed

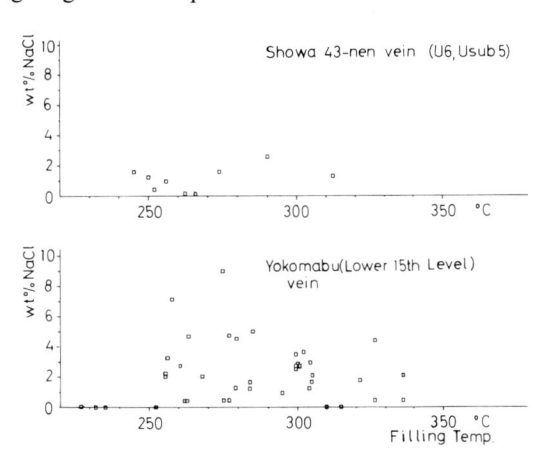

FIG. II–15–27. NaCl equivalent concentrations and filling temperatures of two-phase fluid inclusions.

that the NaCl equivalent concentration is generally low and ranges from 0.0 to 9.0 wt. %. About 96 % of the measurements were in the range of 0.0 to 5.0 %.

In some other mines, such as the Taishu (Imai *et al.*, 1971) and the Ohtani and Kaneuchi mines (Kim *et al.*, 1972), the writers observed that the filling temperatures drop with the decrease of NaCl equivalent concentration. This decrease has been explained as being due to the dilution of the ascending ore-forming fluid by underground water. In the Ashio mine, however, this tendency has not yet been observed (Fig. II–15–27).

Discussion and Conclusions

From the above descriptions and discussion, the writers can generalize on and characterize Sn-W-Cu-Zn-Pb-Ag xenothermal deposits in Japan:

(1) They occur in districts where rhyolitic rocks and equivalent pyroclastics coexist with intrusive granitic rocks. Both extrusive and intrusive rocks belong to the same period of igneous activity, the Late Cretaceous or Early Tertiary. In the German literature (for example, Schneiderhöhn, 1955), such deposits are known as the subvolcanic (subvulkanische) type. But subvolcanic deposits do not seem to be related genetically to plutonic rocks. As far as the xenothermal deposits in Japan are concerned, therefore, they cannot be exactly classified as the subvolcanic type of Schneiderhöhn. Buddington (1935, 1959) noted that in the Cascade Mountain Range the granite coexists with volcanic rocks which are presumably the products of the same igneous activity as the granite, and that xenothermal veins occur in the area.

In Japan, Miocene volcanic activities are represented by a succession of effusive rocks and are accompanied by many kinds of ore deposits. Most of them are classified as epithermal deposits. According to Kato (1928, 1937), the effusive rocks and the associated ore deposits are evidently consanguineous or hold a relation of sisterhood, and the ore-feeder was concealed far beneath. Kato called effusive rocks of this kind "sister rocks." Epithermal deposits in Japan are always accompanied by effusive rocks and include, in some cases, higher-temperature mineral assemblages, such as magnetite, pyrrhotite, bismuthinite, etc., showing some xenothermal-type characteristics. Takahashi (1963) discussed this problem from the viewpoint of minor elements such as In, Sn, Bi and Co.

Xenothermal veins bearing Sn-W-Cu-Zn-Pb-Ag ores discussed in this chapter would be related to Late Cretaceous or Early Tertiary igneous activities, different from the above type.

(2) Underneath a mining area with xenothermal-type deposits, cryptobatholiths of granitic composition are thought to occur. In the Akenobe mining area the cryptobatholith runs in a NE-SW direction, and in the Ikuno mining area the row of granitic stocks extends to the west. Both join at the Ariga mining area (Fig. II–15–2). Several granitic stocks occur in the Ashio mining area (Fig. II–15–20). It is probable, then, that a cryptobatholith of granodiorite or granodiorite porphyry exists underneath the Ashio mine. It is the same case in the Tada mining area. It is noticeable that dikes of rhyolite, andesite, basalt and granophyre occur frequently in these mining areas. Some of them are post-ore, while others are epigenized by the ores. These dikes often run along the veins.

(3) The mineralized fissures are wrench faults and tension cracks formed by the lateral pressure which caused the folded structure in each region, as well as normal faults due to the upheaval or withdrawal (including cauldron subsidence) of the granitic magma.

The strike directions of the tension cracks and wrench faults depend on the direction of the lateral pressure, but the attitude of the dip would be related to stress from underneath caused by upheaval pressure due to doming-up of the cryptobatholith. Doming-up of the cryptobatholith exerts pressure at right angles to the boundary plane of the magma reservoir with the surrounding rock. These pressures correspond to two principal stresses, i.e., maximum principal stress (σ_1) and intermediate principal stress (σ_2). This relation was verified in the pegmatitic or hypothermal tungsten-tin veins (Fig. II-4-7). However, in xenothermal veins the igneous mass exists as cryptobatholith; hence the above relation cannot be verified.

Through upheaval of the magma, normal faults might be produced that correspond to the diapiric structure. This kind of normal fault is observed at the Akenobe and Tada mines. These fissures are similar to those accompanied by an intrusive mass of the bysmalith type in the mechanism of fissure formation discussed by Imai (1966b).

Normal faults caused by cauldron subsidence, and probably due to the withdrawal of magma, are reported at the Ashio and Ikuno mines. At the Ikuno mine, the conjugated wrench faults formed by the lateral pressure in the initial stage would be displaced later as normal faults due to the cauldron subsidence.

(4) In the areas composed of Permo-Carboniferous sedimentary rocks intruded by granitic rocks of Late Cretaceous or Early Tertiary period adjacent to regions with predominating volcanics and xenothermal deposits, there occur some pegmatitic or hypothermal tungsten-tin veins and mesothermal chalcopyrite-and pyrrhotite-bearing veins or replacement deposits. They exist in the basement complex. But the Akenobe deposit, which is a typical xenothermal type, occurs in the basement complex near the border of the overlying volcanic sedimentary formations.

(5) At the Akenobe and Ikuno mines, both vertical and horizontal zones are recognized: tin-tungsten, tin-copper, copper-zinc, zinc-lead and silver, arranged from bottom upward and from center to margin. The distribution of these zones generally follows the type of zoning described by Emmons (1924) and Hosking (1964). At the Akenobe mine the mineralizing fluid may have ascended through faults formed by magmatic upheaval and flowed into tension cracks formed by lateral pressure, producing the zonal distribution of the minerals. In general, zoning in the above mines might be called monoascendant zoning (Park, 1963). However, upheaval (advance) or withdrawal (retreat) of the magmatic body could cause disordering of the zonal distribution. At the same time, by upward or downward motion (pulsation) of the magmatic body, a banded structure in the vein is developed. At the Ashio mine, reverse zoning is recognized and is interpreted as due to withdrawal of the granitic magma.

Graton (1933, 1940) advocated that the channelway of the mineralizing fluid was subjected to rising temperature through continued flow of the solution and that hypogene mineralization ordinarily proceeds on an ascending temperature scale rather than a descending one. Turneaure and Gibson (1945) followed this interpretation, based on the fact that cassiterite is late in the sequence and that with galena it forms a fine intergrowth replacing teallite. This corresponds to "rejuvenation" of mineralization (Berg, 1928; Kutina, 1957). Another possible explanation of reverse zoning could follow the suggestion of Kelly and Turneaure (1970) as due to the boiling of mineralizing fluid which would favor the deposition of quartz and cassiterite. In the Ashio mine, however, the boiling, as indicated from our study of fluid inclusions, is sporadically manifested in the three different zones and not confined to the central tin-tungsten zone. The present writers recognize the normal zoning and the prevailing normal (direct) banding of the vein in the typical xenothermal deposits as recognized at the Ikuno mine. The disorder of the zoning or reverse banding in the vein would be due to the upheaval or withdrawal of the magmatic chamber.

(6) Sn-W-Cu-Zn-Pb-Ag-type veins in Japan are characterized by the presence of primary bornite, primary chalcocite, stannoidite and mawsonite, which are not recognized in pegmatitic or hypothermal tungsten-tin veins, and mesothermal chalcopyrite- and pyrrhotite-bearing veins or replacement deposits in the adjacent areas. On the other hand, hypothermal tungsten-tin veins and mesothermal chalcopyrite- and pyrrhotite-bearing deposits are

characterized by the presence of pyrrhotite (cubanite and mackinawite), which are not recognized in the xenothermal veins. Tourmaline, beryl and lithium muscovite are not recognized in the xenothermal veins described above. These minerals together with topaz are generally found in the hypothermal tungsten-tin veins* (Imai et al., 1972; Takenouchi and Imai, 1971). Topaz was recognized as an accessory mineral by Kato (1920) in the Akenobe mine and by Yamaguchi (1939) in the Ikuno mine. Tourmaline occurs in the xenothermal veins at the Obira mine, described in the next chapter, and in those of Potosi, Bolivia.

(7) The paragenetic sequence of mineralization at the Tada mine is as follows:

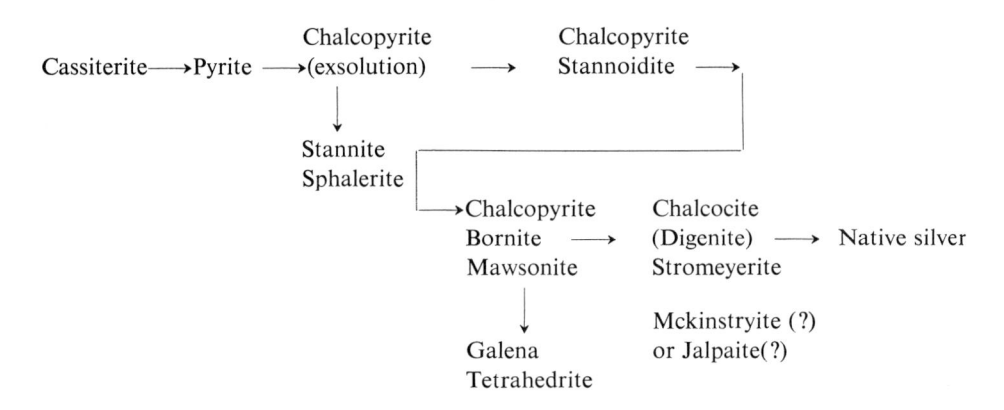

```
                   Chalcopyrite              Chalcopyrite
Cassiterite→Pyrite →(exsolution)      →     Stannoidite      →
                        │
                        ▼
                    Stannite
                    Sphalerite
                              └─→Chalcopyrite      Chalcocite
                                 Bornite      →    (Digenite)   →  Native silver
                                 Mawsonite         Stromeyerite
                                     │
                                     │                Mckinstryite (?)
                                     ▼                or Jalpaite(?)
                                 Galena
                                 Tetrahedrite
```

A similar paragenesis is observed in the Akenobe and Ikuno mines.

According to the hydrothermal syntheses (Lee et al., 1975), fugacity of sulfur in the univariant assemblage mawsonite + bornite + stannoidite + chalcopyrite is located between that of bornite + pyrite + chalcopyrite and that of pyrite + pyrrhotite, declining with a decrease in temperature. Mawsonite is stable under conditions of higher fugacity and/or lower temperature than the assemblage bornite + stannoidite + chalcopyrite. Similarly, stannoidite is stable under higher sulfur fugacity and/or lower temperature conditions than the assemblage stannite + bornite + chalcopyrite. It is thought that mineralization at the Tada mine proceeded toward the lower sulfur fugacity and lower temperature, crossing the boundary curves of the respective stability areas of stannite, stannoidite and mawsonite in the f_{S_2}-T (1/T) diagram.

(8) In the Western Kinki Metallogenetic Province, native silver occurs in the silver zone of the xenothermal deposits. This occurrence is related to the decrease of the sulfur fugacity toward the later stage of mineralization (Fig. II–15–13). In deposits in Cornwall, England, and Potosi, Bolivia, primary native silver is not found in the silver zone (Hosking, 1964; Turneaure, 1971). Nor has native silver been recognized in the silver zone of the Ashio mining area. However, in Ag-Co-Ni-U-As deposits in Echo Bay, Canada, native silver and mckinstryite have been reported (Robinson and Morton, 1971). In the Tada mine, an exsolution relation between chalcocite and stromeyerite is recognized.

(9) In the Sn-W-Cu-Zn-Pb-Ag-type xenothermal deposits in Japan, indium and selenium are concentrated in the veins, as verified from spectroscopic and chemical analyses of the minerals from these deposits. The existence of ikunolite and sakuraiite in the Ikuno mine, and of roquesite in the Akenobe mine, indicates a rather high content of indium and selenium. On the other hand, tantalum and niobium contents are lower in the cassiterite of xenothermal deposits than in that of pegmatitic or hypothermal deposits.

(10) From fluid inclusion studies of the quartz from the Ashio mine, it is recognized that

---

* Refer to Chapters II-3 and -4.

boiling occurred during mineralization. Fluid inclusions containing either CO_2 gas or daughter minerals have not been observed. However, CO_2-rich fluid inclusions are recognized in quartz from the hypothermal tungsten-tin deposits. Filling temperatures of liquid-rich and gas-rich inclusions from the Ashio mine range from 350°C to 200°C; decrepitation temperatures of minerals from the Akenobe mine are nearly coincident and range between 355°C and 155°C.

II-16. Obira Mine, Ohita Prefecture (Nishio et al., 1953)

H. Imai

This mine is situated in central Kyushu, geologically lying in the Outer Zone of Southwest Japan.

This district is composed of Chichibu Paleozoic sedimentary rocks intruded by acidic rocks of the Miocene epoch (for example, Okueyama granite, 21 m.y.) The Chichibu Formation consists of sandstone, clayslate, limestone, quartzite and schalstein.

Acidic rocks include lithoidite, granite porphyry and granite. They are affiliated with one another, and lithoidite was intruded by the other two.

The ore deposits are genetically related to granite or granite porphyry. Miyahisa (1953–1954, 1961) studied the mineral assemblage of these deposits. The ore deposits are composed of the following two types of vein.

The tourmaline-quartz-tin vein running along a dike of granite porphyry is composed of quartz, tourmaline, fluorite and cassiterite, while the copper-arsenic vein is composed of pyrite, arsenopyrite, chalcopyrite, stannite, cassiterite, molybdenite, wolframite, pyrrhotite, cubanite, mackinawite, native bismuth, bismuthinite, sphalerite, galena, tetrahedrite, berthierite, stibnite, jamesonite, boulangerite, marcasite, etc. (Miyahisa, 1953–1954). It is noticeable that the marcasite in this mine contains 0.05 % of germanium (Kinoshita, 1953).

In the northern part of the mine, both types of veins run in parallel and strike N45°–50°E, dipping 70°–80° to E and W. In the southern part they join to become one vein, also striking N45°–50°E. From observations in the adit, the former is of earlier deposition than the latter.

The writers (Nishio, et al., 1953) studied the geothermometry of the minerals from this mine by the decrepitation method. Decrepigraphs indicate that the decrepitation temperatures of this deposit are 350°–220°C. Those of the two types of vein are nearly the same, though those of the tourmaline-quartz-tin vein are a little higher than those of the copper-arsenic vein. The decrepitation temperature of a quartz crystal found in a vug in the cassiterite-tourmaline-quartz vein was determined to be 225°C. This quartz would represent the latest stage of deposition. The decrepitation temperature of marcasite belonging to the later stage of the copper-arsenic vein is 250°C. Acidity rather than temperature would be the essential condition for deposition of marcasite.

From the geologic setting, this deposit may be classified as xenothermal. But the paragenesis of minerals in this mine is different from that of the Akenobe, Tada, Ikuno and Ashio mines. It rather resembles to those of the Ohtani, Kaneuchi, Takatori and Ilkwang mines, which were described as hypothermal in other chapters.

Also, pyrometasomatic deposits occur in this area. Adjoining this mining area, the same type deposit as at Mitate and Hoei mines exists (refer to Chapters III–8 and VII–3).

E. GENETICAL PROBLEMS OF SULFIDE MINERALS

II-17. Geology and Ore Deposit of the Ilkwang (Nikko*) Mine, Korea, with Special Reference to the Genesis of Gudmundite (Imai, 1942, 1949a)

H. Imai and M. S. Lee

This mine is located about 25 km northeast of Busan (Fig. I–3), belonging to hypothermal type.

Geology and Ore Deposit of the Mine

From a geological standpoint, this area is included in "Tsushima Basin"** which comprises the west end of Honshu, the northern Kyushu and south Korea. South Korea is composed of the Gyeongsang (Keisho*) Formation including the Naktong (Rakuto*) and Silla (Shiragi*) Formations, whose geological age is Jurassic and Cretaceous, intruded by igneous rocks such as granites, porphyries, porphyrites, etc. They are connected with the Late Cretaceous igneous activities of the Pulguksa (Bukkokuji*) Epoch.

In the area of the Ilkwang mine, a composite stock of hornblende biotite granodiorite and quartz monzonite is intruded into the Gyeongsang Formation, which consists of sandstone, shale and conglomerate. The diameter of this stock is about 1 km. The ore deposit is located in this stock. It consists of two breccia pipes composed of stock-work veinlets. The shape of each pipe is cylindrical. They stand north and south and are respectively called "North Ore Body" and "South Ore Body." On the surface they are separated by a distance of about 10 m, but they join at a lower level (Fig. II–17–1). The ore bodies plunge 70°–80° S.

According to Kang *et al.*, (1976) and Fletcher (1977) the origin of the breccia pipe is interpreted as of collapse type, formed by the post magmatic fluid rising along a cone-shaped fracture in roof rocks resulting from shock.

In the quartz monzonite surrounding ore bodies there are many patches of tourmaline crystal aggregates, whose maximum size is 4–5 cm in diameter. There have also been sericitization, chloritization and silicification of the country rock. The country rock in some parts has completely lost its original texture and in other parts preserves it more or less.

According to Fletcher (1977), the K-Ar dating indicated that the age of intrusion of the stock is 81 m.y. and the age of mineralization is 69 m.y.

The sericitization, chloritization, and silicification are limited to the close periphery of the ore bodies, whereas tourmalinization is a little more widely developed. Also in the country rock close to ore bodies there are occasionally aggregates of reddish-brown garnet, whose sizes range up to 3 cm.

The writers identified the following vein-forming minerals:

(1) Metallic minerals: wolframite, scheelite, reinite*** (Fig. II–17–2), arsenopyrite, pyrrhotite, chalcopyrite, sphalerite (Fig. II–17–3), galena, native bismuth, tetrahedrite, bournonite, boulangerite, gudmundite (FeSbS) and meneghinite (?) (Fig. II–17–4).

---

*   Pronunciation in Japanese.
**   Geologic unit in Japan.
*** Wolframite after scheelite.

Part II: Vein-type Deposits

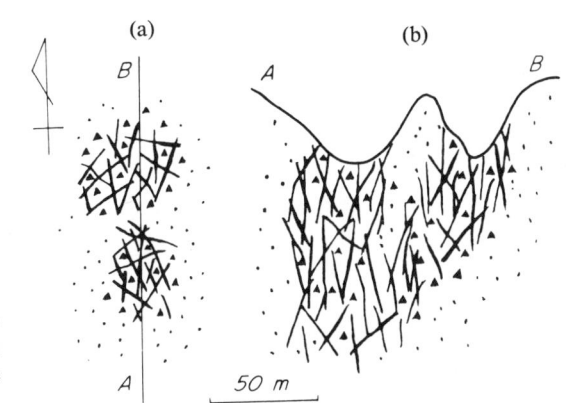

FIG. II–17–1. Diagrammatic geologic map and section.
 (a) Plan (b) Section
Groups of stock-work veinlets are surrounded by tourmalinized granite (dotted), and sericitized, chloritized, or silicified granite (triangular).

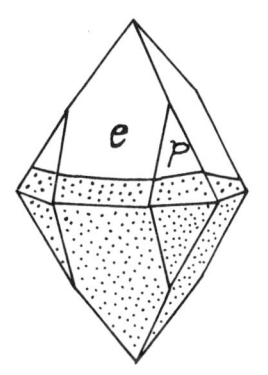

FIG. II–17–2. Scheelite and wolframite. e: (101), p: (111). Dotted: wolframite. Undotted: scheelite.

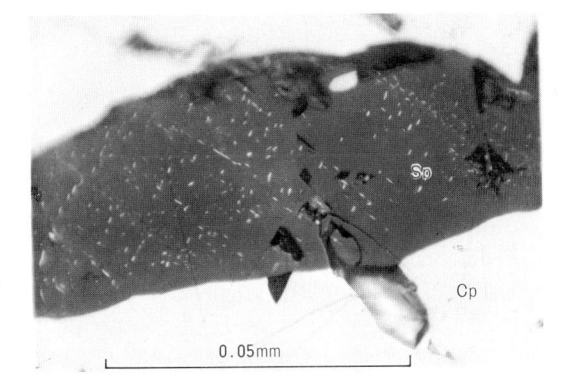

FIG. II–17–3. Photomicrograph. Polished section. One nicol. Oil immersion. Very fine dots of chalcopyrite are included in sphalerite (Sp). Cp: chalcopyrite.

(2) Gangue minerals: tourmaline (Fig. II–17–5), manganous garnet*, apatite, zircon, quartz, sericite, chlorite, calcite and others.

Among the metallic minerals, chalcopyrite and pyrrhotite are predominant. This mine was worked as a copper mine producing a small amount of tungsten ore.

The deposit is genetically connected with the igneous activity of the Pulguksa Epoch responsible for the granitic rocks and other igneous rocks.

In the "Tsushima Basin" there are similar ore deposits, such as the Yakuoji mine in the western region of Honshu (Japan proper) and the Hanan (Kanan) mine in south Korea. They were described by Kato (1912), and Kato and Oyama (1923), respectively.

---

* Pyralspite group (Fletcher, 1977).

126   E. Genetical Problems of Sulfide Minerals

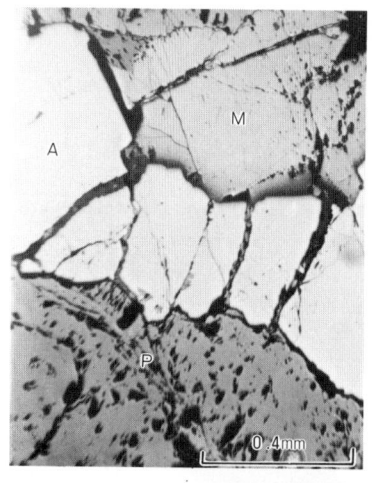

FIG. II-17-4. Photomicrograph. Polished setion. One nicol. A: arsenopyrite, P: pyrrhotite, M: meneghinite (?).

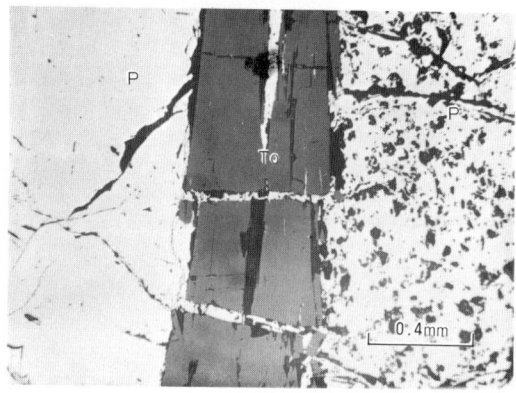

FIG. II-17-5. Photomicrograph. Polished section. One nicol. To: tourmaline, P: pyrrhotite.

Some Problems Regarding Paragenetical Relations

Succession of mineralization: The deposit is composed of groups of stock-work veinlets, and no vein reveals the appearance of a composite vein.

From microscopic observations, it is recognized that tourmaline, garnet, wolframite, scheelite and arsenopyrite are penetrated or corroded by chalcopyrite, pyrrhotite, sphalerite, galena, calcite, etc. (Fig. II-17-5). Consequently, it may be stated that mineralization is divided into at least two stages, and that minerals which have hitherto been said to be formed at high temperatures were deposited at the earlier stage and that the sulfide minerals, except arsenopyrite, belong to the later stage.

Paragenesis of sulfantimonide minerals, with special reference to the genesis of gudmundite: As described in the previous section, sulfantimonide minerals, namely tetrahedrite, bournonite, boulangerite and gudmundite, together with galena and native bismuth, are present within chalcopyrite and pyrrhotite (Fig. II-17-6). Figures II-17-7, -8 and -9, are photomicrographs of some parts of the aggregate of these minerals which were observed by high-power objective lens. This texture is the same as the relation between galena, native bismuth and bournonite in the Bayerland mine, Oberpfalz, which was described by Maucher (1940). He named this texture "Eutektischer Verwachsung." But in the Ilkwang mine the assemblages of sulfantimonide differ according to the mineral which surrounds them. When the surrounding mineral is chalcopyrite, the assemblage is galena, native bismuth, tetrahedrite, bournonite and boulangerite (Figs. II-17-7, and-8). When

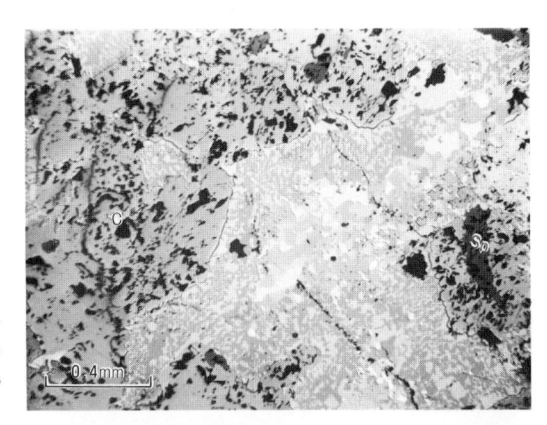

Fig. II-17-6. Photomicrograph. Polished section. One nicol. Aggregates of galena, native bismuth and sulfantimonides. C: chalcopyrite, Sp: sphalerite.

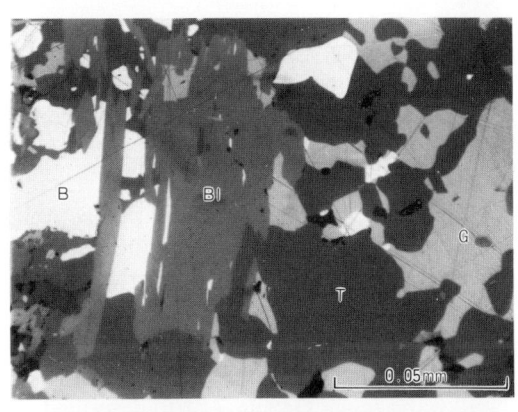

Fig. II-17-7. Photomicrograph. Polished section. One nicol. Oil immersion. G: galena, B: native bismuth, T: tetrahedrite, Bl: boulangerite.

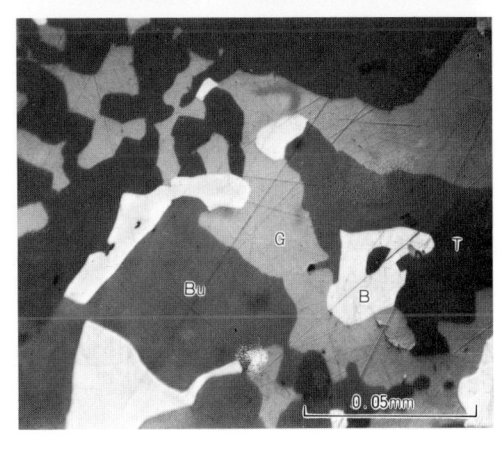

Fig. II-17-8. Photomicrograph. Polished section. One nicol. Oil immersion. B: native bismuth, G: galena, T: tetrahedrite, Bu: bournonite.

the surrounding mineral is pyrrhotite, the assemblage is galena, native bismuth, gudmundite and chalcopyrite (Fig. II-17-9). In the latter case, tetrahedrite, bournonite and boulangerite are sometimes present together with the above minerals. In some cases, though the surrounding mineral is pyrrhotite, the combination is the same as in the case of chalcopyrite. In a word, gudmundite does not exist in chalcopyrite, but exists only in pyrrhotite. In gudmundite, dots of pyrrhotite are found in the pseudoeutectic texture (Fig. II-17-9).

The writers interpret these observations as follows: When the solution which contained elements such as Bi, Sb, Cu, Pb, and S was included in the chalcopyrite (or in the solution

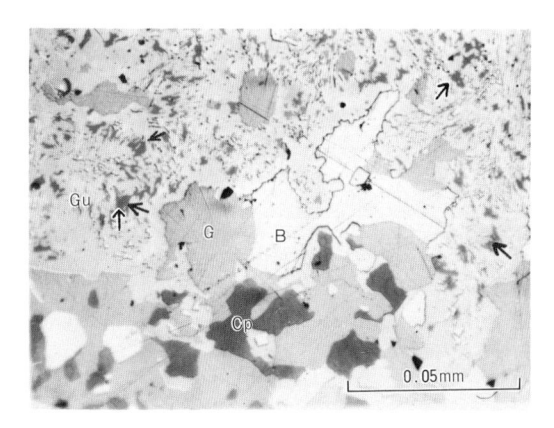

FIG. II-17-9. Photomicrograph. Polished section. One nicol. Oil immersion. B: native bismuth, G: galena, Cp: chalcopyrite, Gu: gudmundite. Pyrrhotite (dark dot) in gudmundite is shown by arrows.

which deposited the chalcopyrite), it deposited galena, native bismuth, tetrahedrite ($5Cu_2S \cdot 2(Cu, Fe)S \cdot 2Sb_2S_3$), bournonite ($Cu_2S \cdot 2PbS \cdot Sb_2S_3$) and boulangerite ($5PbS \cdot 2Sb_2S_3$).

But when the above solution was included in pyrrhotite (or in the solution which deposited pyrrhotite), antimony reacted with iron which was supplied from the surrounding pyrrhotite. Thus gudmundite (FeSbS) was formed. In that case copper was deposited as chalcopyrite instead of tetrahedrite or bournonite. Therefore, the combination of minerals is galena, native bismuth, gudmundite and chalcopyrite. If this reaction was incomplete, a small amounts of tetrahedrite, bournonite and boulangerite would be deposited. Gudmundite was not always formed, even if a solution containing Bi, Sb, Cu and Pb was included in pyrrhotite. In that case, the assemblage of minerals is the same as in the case of chalcopyrite.

There are some localities of gudmundite which have a genetical relation similar to that of the Ilkwang mine. Gavelin (1936) described gudmundite in the Malånäs district which was deposited by the reaction of Sb-bearing solution on pyrrhotite. Also, gudmundite which was formed by "Zerfall" of tetrahedrite or as a reaction rim was studied by Maucher (1940) at the Bayerland mine and by Ramdohr (1938b) at Jakobsbacken.

Sampson (1941) generalized the previous studies as follows:

1. Gudmundite favors heavy sulfide deposits rich in iron.
2. In these deposits it belongs to a group of minor minerals later than the main sulfides.
3. Arsenic may be present as arsenopyrite in early main mineralization.
4. Lead sulfantimonides and fahlerz (usually tetrahedrite) accompany gudmundite.
5. Gudmundite occuring in massive sulfantimonides commonly shows extremely complex intergrowth with a number of other minerals.
6. Although the early and minor minerals of the deposit may be from intermediate- to high-temperature episodes of deposition, gudmundite and associated minerals are from much lower temperatures, ending with the deposition of ruby silvers.

These characteristics generally coincide with the paragenetical relations of gudmundite in the Ilkwang mine.

Ramdohr (1950) described the myrmekitic intergrowth of gudmundite with pyrrhotite in Broken Hill, which may be similar to the Ilkwang mine.

Silver content: The silver content in the crude ore of this mine is 50–80 g/t. Of the minerals which the writers studied, silver carriers are galena, tetrahedrite, chalcopyrite, bournonite and boulangerite.

The rich ore for which analytical data are shown in Table II-17-1 consists of pyrrhotite, chalcopyrite, sphalerite, galena, native bismuth, tetrahedrite, bournonite, boulangerite, gudmundite, etc. According to Warren (1935), Lasky (1935), Watanabe (1940) and Tatsumi (1942), the maximum content of silver in galena is 10–16 g/t versus 1% of lead. The

TABLE II–17–1. Chemical analyses of ores from the Ilkwang mine.

	(A)	(B)
Au	0.1g/t	0.7g/t
Ag	494g/t	584g/t
Cu	8.70%	13.02%
Pb	1.17	1.26
Sb	0.34	tr.
As	0.005	2.86
Bi	0.45	tr.
Ni	tr	0.08
Co	nil.	n.d
Fe	50.43	37.88
S	36.65	31.46
Zn	n.d.	1.19
WO_3	n.d.	nil.
SiO_2	n.d.	5.26
Al_2O_3	n.d.	6.21
MgO	n.d.	0.17
CaO	n.d.	0.18
MnO	n.d.	nil.
Total	97.794	99.928

(A) Analysis of rich ore in this mine. Collector: Imai. Locality: open pit of the "North Ore Body." Sample no. Nikko 11. Analyzed by an analyst in the Beashi mine. (By courtesy of the late T. Matsuoka).

(B) Analysis of the copper concentrate from the mill. Data from the mine (1942).

percentage of the lead content in this ore is 1.17 % as shown in the Table. From these results, it may be said that the silver attributed to galena is present in small amounts. But Ramdohr (1938a) stated that in a high-temperature ore deposit galena might often contain $AgBiS_2$ molecules, to which a high silver content would be attributable. This is probable in this mine, but has not yet been confirmed.

Tetrahedrite is an important silver carrier in many deposits. It was reported that some tetrahedrite from Mount Isa was charged with about 42.5% silver (Riley, 1974). Even if all the silver content in Table II–17–1 were contained in tetrahedrite, it is by far below the content of Mount Isa tetrahedrite.

In this ore, silver content due to chalcopyrite is at most 50 g/t, according to the data of Warren (1935) and Lasky (1935). Also, bournonite and boulangerite are not important silver carriers. Thus, silver content in this ore is mainly attributed to galena and tetrahedrite, though its proportion in the two minerals is unknown.

Genetical Considerations

Judging from the above-described geologic setting and paragenesis of minerals, this deposit seems to be the hypothermal type. Of the minerals deposited at the earlier stage, tourmaline is the representative pneumatolytic mineral. It has been reported that pyralspite rich in manganese is deposited by pneumatolysis (Kozu *et al.*, 1940, 1941).

In this case, it cannot be determined whether the fluid was above or below its critical temperature. But, the writer uses the term "pneumatolysis" in the sense that it puts the emphasis on gas phase action, regardless of whether it is above or below the critical temperature, as Fenner (1933) and Graton (1940) stated.

Wolframite, scheelite and arsenopyrite are generally the minerals found in a high-temperature deposit. But there have been silicification, sericitization and chloritization of hornblende biotite granite, which is the country rock of the ore bodies. That is, there have been hydrothermal alterations. And there exist bournonite, boulangerite and gudmundite, which are comparatively low-temperature minerals. So it is probable that the main mineralization—that is, the deposition of chalcopyrite, pyrrhotite and other sulphide minerals—

belongs to the hydrothermal stage. These minerals might be deposited from hot alkaline solution.

It is noticeable that, notwithstanding the silicification, sericitization and chloritization of country rock which might be responsible for the main mineralization the tourmaline and garnet which had been formed by pneumatolysis did not alter. It is not clear whether these minerals have strong resistance against the later hydrothermal solution or whether they existed in equilibrium with it.

Conclusions

(1) The Ilkwang mine consists of stock-work veins which occur in the Upper Cretaceous hornblende biotite granite intruded in the Upper Jurassic and Lower Cretaceous formations. They are tourmaline copper veins. There have been tourmalinization, silicification, sericitization and chloritization of the country rock. These veins are genetically connected with the late Cretacous igneous activity responsible for the above granite.

(2) These veins belong to the hypothermal deposit. The vein-forming minerals are tourmaline, garnet, apatite, zircon, quartz, sericite, chlorite, calcite, wolframite, scheelite, arsenopyrite, pyrrhotite, chalcopyrite, sphalerite, galena, native bismuth, tetrahedrite, bournonite, boulangerite, gudmundite, meneghinite (?), etc.

(3) As for the succession of minerals, it might be said that tourmaline, garnet, wolframite, scheelite and arsenopyrite were deposited at an earlier stage and that the main mineralization belonged to the succeeding hydrothermal stage.

(4) The paragenetical relation of galena, native bismuth and sulfantimonides (tetrahedrite, bournonite, boulangerite and gudmundite) which were observed under an ore microscope is characteristic. From microscopic observations, the writers interpret the assemblage as follows: When the solution which contained elements such as Bi, Sb, Cu, Pb and S was included in chalcopyrite (or in the solution which deposited chalcopyrite), it deposited galena, native bismuth, tetrahedrite, bournonite and boulangerite. But when the above solution was included in pyrrhotite (or in the solution which deposited pyrrhotite), antimony reacted with iron which was supplied from pyrrhotite. Thus gudmundite was formed. In that case, copper was deposited as chalcopyrite instead of tetrahedrite or bournonite.

(5) The crude ore in this mine is comprised 50–80 g/t of silver. The silver-bearing minerals are galena, chalcopyrite, tetrahedrite, bournonite and boulangerite. Of these, galena and tetrahedrite may be the main carriers.

II–18. Paragenesis of Cu-Fe-S Minerals in the Komori Mine (Imai and Fujiki, 1963; Fujiki, 1963)

Y. Fujiki, S. Takenouchi and H. Imai

Introduction

The Komori mine is composed of pyrrhotite-chalcopyrite veins occurring in ultrabasic rocks. The paragenetic relations in the ore minerals from this mine were reported by Fujiki (1963, 1964a,b). Cu-Fe-S minerals in the ores are composed of chalcopyrite, pyrrhotite, cubanite and mackinawite. Cubanite occurs sporadically in large quantities and is accompanied by above minerals. They often reveal interesting exsolution textures formed by cooling from a solid solution at high temperatures. Pentlandite also accompanies them in exsolution textures. Imai and Fujiki (1963) analyzed mackinawite and pentlandite existing in the crystallographic intergrowth by use of electron microprobe. The results are as follows: mackinawite; Fe 52.1 wt. %, Ni 5.6 wt. %, Co 1.0 ~ 2.0 wt. %, Cu 4.0 wt. %, and pentlandite; Fe 28.7 wt. %, Ni 25.9 wt. %, Co 15.8 wt. %, Cu 0.1 wt. %. Mackinawite from the Komori mine is usually rich in nickel and cobalt, and pentlandite is also rich in cobalt (refer to Chapter II–5).

Typical Exsolution Textures

The exsolution textures observed under the reflection microscope are classified into five types (Fujiki, 1963). The formation process of the textures of the five types is tabulated below.

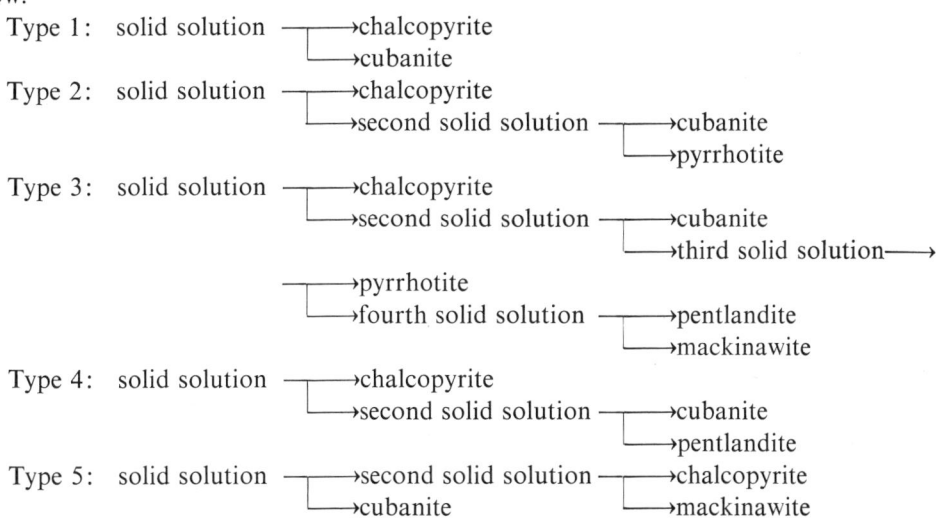

```
Type 1:  solid solution ──┬─→chalcopyrite
                          └─→cubanite
Type 2:  solid solution ──┬─→chalcopyrite
                          └─→second solid solution ──┬─→cubanite
                                                     └─→pyrrhotite
Type 3:  solid solution ──┬─→chalcopyrite
                          └─→second solid solution ──┬─→cubanite
                                                     └─→third solid solution──→
                          ┬─→pyrrhotite
                          └─→fourth solid solution ──┬─→pentlandite
                                                     └─→mackinawite
Type 4:  solid solution ──┬─→chalcopyrite
                          └─→second solid solution ──┬─→cubanite
                                                     └─→pentlandite
Type 5:  solid solution ──┬─→second solid solution ──┬─→chalcopyrite
                          └─→cubanite                └─→mackinawite
```

The Type 1 texture is most predominant (Fig. II–18–1). In Type 2, the second solid solution which exsolved cubanite and pyrrhotite is noticeable, because the chemical composition of the solid solution is assumed to be richer in iron than that of cubanite (Fig. II–18–2). The exsolution process of Type 3 is most complicated. The second and third solid solutions in this type are also assumed to be rich in cobalt and nickel, because Co-bearing pentlandite and Ni-bearing mackinawite are exsolved as final products from this solid solution (Fig. II–18–3(a), (b)). Type 4 is similar to Type 2, but the exsolved minerals from the second solid solution consist of cubanite and pentlandite (Fig. II–18–4). Type 5 is apparently different from the other types, because mackinawite exsolved from the second solid solution exists in chalcopyrite lamellae (Fig. II–18–5). The chemical composition of this second solid solution is considered to be richer in iron and nickel than that of chalcopyrite.

Another mode for exsolution textures of Cu-Fe-S minerals in Komori ores has been described by Fujiki (1963). They are called non-lamellar textures because of the granular form of chalcopyrite and cubanite. In these textures, mackinawite also occurs as lamellae in granular chalcopyrite, or along the boundary zones between chalcopyrite and pyrrhotite. Pentlandite lamellae are found in granular cubanite, or along the grain boundaries of cubanite.

Similar exsolution textures have been reported from many copper-ore deposits in Japan (Takeuchi and Nambu, 1953, 1958; Soeda, 1960).

Formation of Exsolution Textures

Phase diagrams of the system Cu-Fe-S, especially the phase relations between chalcopyrite and pyrrhotite, are important in the consideration of the formation of exsolution

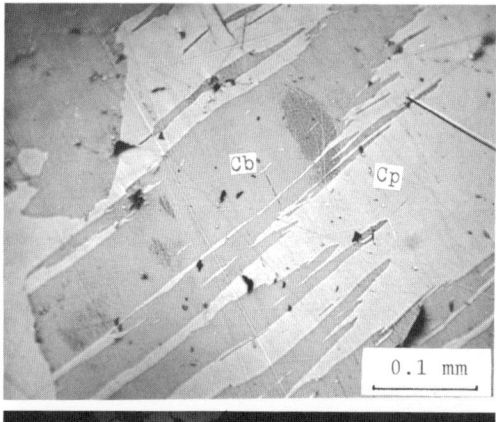

FIG. II–18–1. Photomicrograph of Type 1. One nicol. Exsolution texture consisting of chalcopyrite (Cp) and cubanite lamellae (Cb).

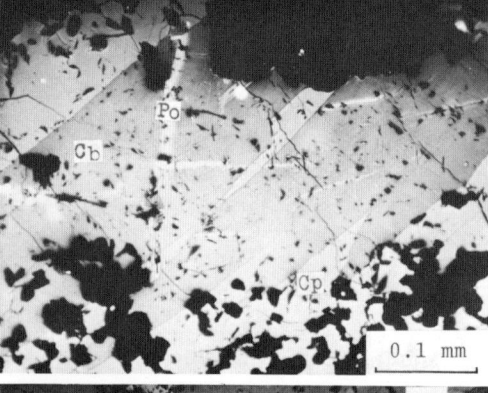

FIG. II–18–2. Photomicrograph of Type 2. Crossed nicols. Exsolution texture consisting of chalcopyrite (Cp) and cubanite lamellae (Cb), in which pyrrhotite (Po) lamellae are exsolved in the crystallographic intergrowth.

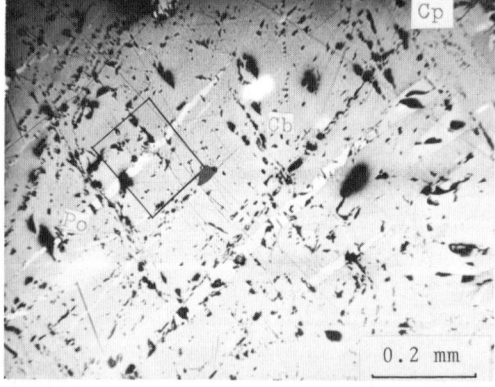

FIG. II–18–3(a). Photomicrograph of Type 3. Crossed nicols. Exsolution texture consisting of chalcopyrite (Cp) and cubanite lamellae (Cb), in which pyrrhotite (Po) lamellae including pentlandite (Pt) and mackinawite (Ma) are exsolved in the crystallographic intergrowth.

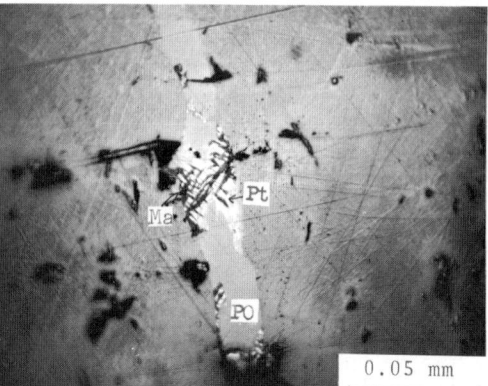

FIG. II–18–3(b). Photomicrograph. An enlarged view of the quadrated area in Fig. 3(a). One nicol. Exsolution texture consisting of pentlandite (Pt) and mackinawite (Ma) lamellae in pyrrhotite (Po).

FIG. II–18–4. Photomicrograph of Type 4. Crossed nicols. Exsolution texture comprised of chalcopyrite (Cp) and cubanite lamellae (Cb) in which pentlandite (Pt) lamellae are exsolved in the crystallographic intergrowth.

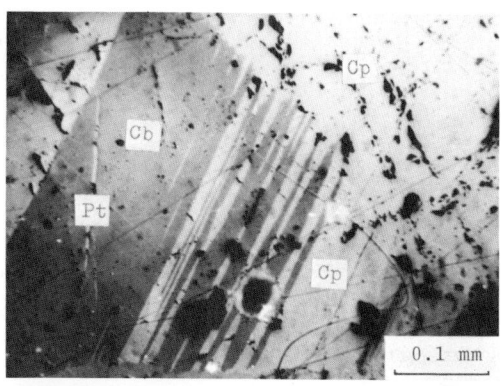

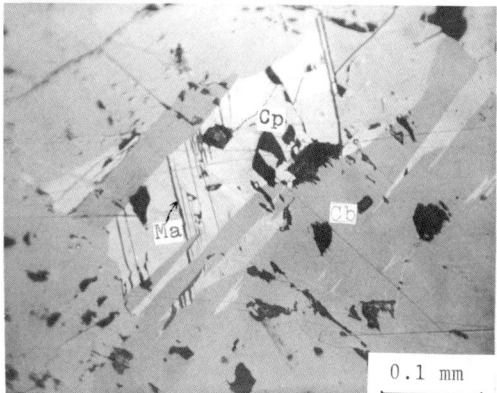

FIG. II–18–5. Photomicrograph of Type 5. Crossed nicols. Exsolution texture comprised of cubanite (Cb) and chalcopyrite lamellae (Cp) in which mackinawite (Ma) lamellae are exsolved in the crystallographic intergrowth.

textures in ores from the Komori mine. Borchert (1934) first tried to explain the formation of exsolution textures using a phase diagram. He reported the existence of three solid solutions—i.e., chalcopyrite solid solution, chalcopyrrhotite solid solution and pyrrhotite solid solution—at temperatures above 255°C. Merwin and Lombard (1937) experimentally showed two solid solutions—i.e., intermediate (chalcopyrite-cubanite) solid solution and pyrrhotite solid solution, but they did not recognize the chalcopyrrhotite solid solution. Yund and Kullerud (1961, 1966) experimentally established the phase relations of the system of Cu-Fe-S at various temperatures. According to them, the cubic chalcopyrite-cubanite solid solution is separated into a tetragonal chalcopyrite and cubic cubanite at temperatures below 550°C. However, the field of cubic cubanite solid solution does not extend toward the pyrrhotite side beyond the composition of cubanite. They could not synthesize the chalcopyrrhotite solid solution, and suggested that this solid solution might be a high-temperature modification of cubanite.

On the other hand, Ramdohr (1960, 1963) reported the occurrence of chalcopyrrhotite from several localities. It is found in chilled inclusions of cupriferous sulfides in basalts and other volcanic rocks, and in some stony meteorites. The writers consider that the mode of occurrence of chalcopyrrhotite is related to a rapid cooling process.

However, the inferred chemical compositions of the second solid solutions of Type 2 and Type 3 are richer in FeS than in cubanite as described above. This composition is similar to a chemical composition of the chalcopyrrhotite solid solution in Borchert's concept. A phase diagram similar to Borchert's is shown in Fig. III-10–8. He wrote that a chalcopyrite-chalcopyrrhotite solid solution at high temperatures is separated into chalcopyrite solid solution and chalcopyrrhotite solid solution below 450°C, and this chalcopyrrhotite solid solution exsolves cubanite and pyrrhotite at temperatures below 255°C (which is an

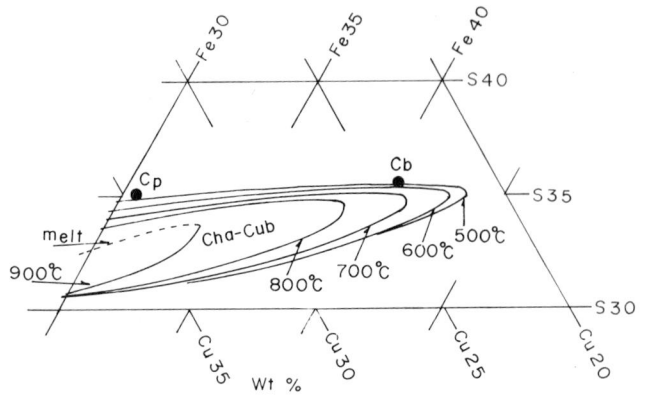

FIG. II-18-6. Fields of chalcopyrite-cubanite solid solution at temperatures of 500°C, 600°C, 700°C, 800°C and 900°C. Cp: chalcopyrite, Cb: cubanite, Cha-Cub: chalcopyrite-cubanite solid solution.

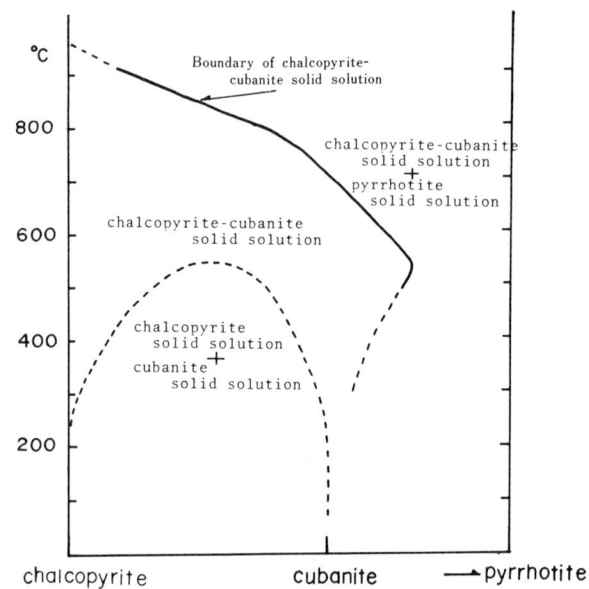

FIG. II-18-7. Projection of the field of chalcopyrite-cubanite solid solution on the chalcopyrite-pyrrhotite section.

eutectic temperature in the chalcopyrite-pyrrhotite pseudobinary system). He thought that cubanite was formed by the exsolution from this chalcopyrrhotite solid solution. The cubanite-pyrrhotite exsolution textures as shown in Type 2 can be interpreted well with this phase diagram, but the chalcopyrite-cubanite exsolution textures as in Type 1 cannot be explained.

Recently, Kaneda et al. (1975) have shown that natural orthorhombic cubanite heated up to 235°C separates into tetragonal chalcopyrite and an iron-rich cubic phase, and this iron-rich cubic phase is similar to chalcopyrrhotite in optical properties and chemical composition (refer to Chapter III-10).

Field of the Chalcopyrite-Cubanite Solid Solution

The existence of the chalcopyrrhotite solid solution is an important problem with regard to the formation of the exsolution texture consisting of cubanite and pyrrhotite. The writers assumed that this solid solution is intimately related to the cubic cubanite solid solution. Therefore, the field of the chalcopyrite-cubanite solid solution in a temperature range from 500°C to 900°C has been studied in detail with the rigid silica glass tube technique (Takenouchi and Fujiki, 1968a,b). The results are shown in Fig. II-18-6. The field of this solid solution at temperatures below 500°C is separated into the solid solutions of tetragonal chalcopyrite and cubic cubanite, although this is not shown in Fig. II-18-6. It was found that the field of the chalcopyrite-cubanite solid solution extends toward the pyrrhotite side beyond the composition of cubanite at temperatures below 600°C and reaches the maximum at about 550°C. Figure II-18-7 represents the projection of the area of the chalcopyrite-cubanite solid solution on the chalcopyrite-pyrrhotite section. In this figure, a field more iron-rich than cubanite composition, i.e., iron-rich cubanite solid solution, may correspond to the concept of the chalcopyrrhotite solid solution. This iron-rich cubanite solid solution can exsolve cubanite and pyrrhotite in the textures of Type 2.

The existence of the iron-rich cubanite solid solution was recognized at temperatures of 525°C and 350°C in the investigation of the solid solution by Mukaiyama and Izawa (1970), and at 600°C by Cabri (1973). Recently, the phase relations at low temperatures have been studied under hydrothermal conditions by Sugaki et al. (1975), and the iron-rich cubanite solid solution (called intermediate solid solution by these investigators) is certainly shown at 350°C.

In Fig. II-18-7, the iron-richest composition of the cubanite solid solution at 550°C was determined to be CuS·2.3FeS (Cu 21%, Fe 44% and S 35% by weight). This is much less in FeS content than in the composition of chalcopyrrhotite (CuS·4FeS) expressed by Borchert (1934). If pyrrhotite lamellae in cubanite exsolves from the solid solution having a composition of CuS·2.3FeS (or $CuFe_2S_3$·0.3FeS), the maximum volume of the unmixed pyrrhotite should be about 12%. This value seems to be in good agreement with those observed in Figs. II-18-2 and -3.

The exsolution texture consisting of cubanite and pyrrhotite observed in the Komori ores can be explained by the exsolution from the iron-rich cubanite solid solution. On the other hand, the exsolution textures consisting of chalcopyrite and cubanite are well interpreted by the unmixing process of the chalcopyrite-cubanite solid solution. However, the formation of the exsolution textures comprising chalcopyrite and mackinawite can not be explained by the present phase diagram between chalcopyrite and pyrrhotite.

II-19. "Wurtzite" from the Hosokura Mine (Imai, 1941, 1947)

H. Imai

The fibrous or prismatic zinc sulfides in Japan have frequently been studied by Japanese mineralogists and chemists (Ogawa, 1935; Iwasaki et al., 1940; Imai, 1941, 1947; Watanabe, 1941; Akizuki, 1969). They concluded that most fibrous or prismatic zinc sulfides in Japan are sphalerites revealing a special form.

Ogawa (1935) morphologically studied prismatic sphalerite from the Ashio mine, and concluded that it is a special form of sphalerite repeating the twin on (111).

Iwasaki et al. studied the "wurtzite" from the Hosokura mine in detail by X-ray. They concluded that "wurtzite" from the Hosokura mine is a sphalerite, elongating to [111].

Watanabe (1941) studied the prismatic zinc sulfide mineral from the Ohmori mine, Fukushima Prefecture, Northeast Japan, and stated that it was wurtzite at the time of deposition and was later transformed into sphalerite. In other words, according to him, it is sphalerite after wurtzite.

The writer (Imai, 1941, 1947) studied "wurtzite" from the Hosokura mine and states his conclusions in the following sections (refer to Chapter II-13).

136   E. Genetical Problems of Sulfide Minerals

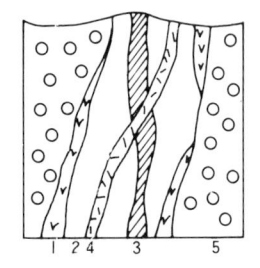

FIG. II-19-1. Schematic section of the vein of the Hosokura mine. 1: pyrite (first stage). 2: aggregates of quartz, chlorite, chalcopyrite, galena and sphalerite, etc. (second stage). 3: vein of "wurtzite" (third stage). 4: vein of fluorite and quartz (fourth stage). 5: country rocks.

Henmi (1941) recognized wurtzite from the Okoppe mine in morphological and X-ray studies.

Akizuki (1969) observed the polysynthetic twin on (111) in "wurtzite" from the Hosokura mine by electron microscope.

The Occurrence of "Wurtzite" in the Hosokura Mine

As stated before, in the Hosokura mining area many epithermal or subvolcanic veins are developed. "Wurtzite" occurs frequently over the whole area.

From observations in adits, the writer determined four stages of vein formation, as shown in Fig. II-19-1. The order of deposition is as follows:

First stage: Deposition of pyrite.
Second stage: Deposition of the main vein-forming minerals, such as sphalerite, galena, chalcopyrite, chlorite, quartz, etc.
Third stage: Deposition of "wurtzite" (accompanied by marcasite, pyrite, galena and chalcopyrite).
Fourth stage: Formation of veinlets composed of fluorite and/or quartz accompanied by sphalerite.

The second stage is the most predominant, and most of the ore minerals were deposited during this stage. In the course of the deposition of sulfides, "wurtzite" belongs to a later stage. In addition, hematite, tetrahedrite, stibnite, calcite, barite, etc., are present in very small amounts.

Macroscopic Appearances of "Wurtzite" and Microscopic Properties of its Thin Sections

There are two types of "wurtzite" in this mine, and they are different in some properties and mineral assemblages:

Type 1:   Macroscopically black, fibrous (Fig. II-19-2). It is barely translucent in extremely thin section, containing about 8 wt. % of iron (Imai, 1941).

FIG. II-19-2. Crystal aggregates of Type 1 "wurtzite".

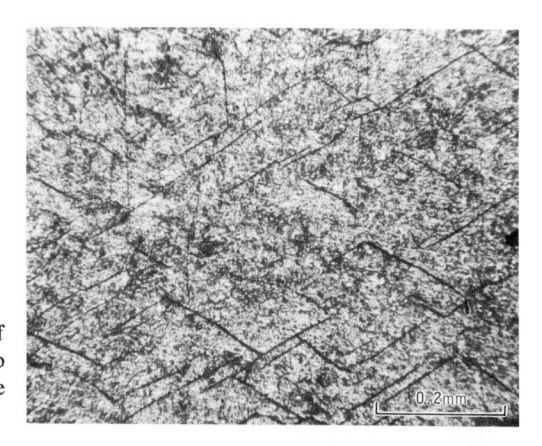

Fig. II-19-3. Photomicrograph. Thin section of Type 2 "wurtzite". One nicol. Perpendicular to fibre, showing three directions of cleavage trace which intersect at 120° with one another.

Fig. II-19-4. Photomicrograph. Thin section of Type 2 "wurtzite". One nicol. W: Type 2 "wurtzite". Sp: sphalerite, Q: quartz. Parallel to fibre; the direction of fibre is up and down. The trace of (11$\bar{2}$0) cleavage stretches up and down, and that of (0001) cleavage is indistinctly recognized perpendicular to the former. Sphalerite (Sp) and quartz (Q) belong to a later stage than "wurtzite". They penetrate into it.

Type 2: Macroscopically black, but with a tint of brownish color, and fibrous. Thin sections with appreciable thickness are translucent and reddish-brown in color. It contains about 5 wt. % of iron (Imai, 1941).

These two types have cleavages of the same kind. In sections perpendicular to the fibre, the traces of cleavages are recognized in three directions intersecting one another at 120°,

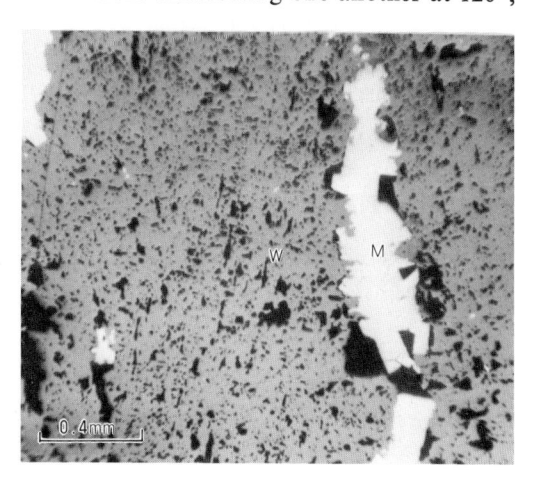

Fig. II-19-5. Photomicrograph. Polished section of Type 1 "wurtzite". One nicol. W: Type 1 "wurtzite". M: marcasite (+ pyrite). Parallel to fibre; the direction of fibre is up and down. Showing the relation between Type 1 "wurtzite" and marcasite (+ pyrite).

138   E. Genetical Problems of Sulfide Minerals

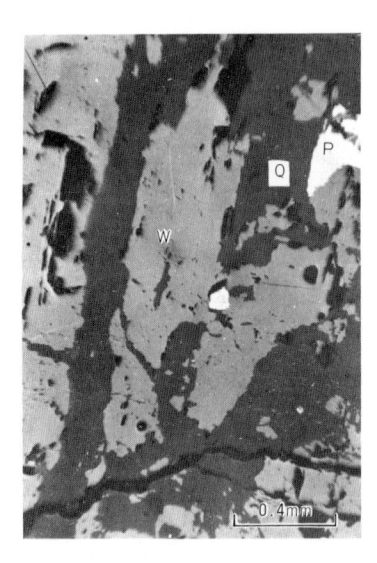

FIG. II–19–6. Photomicrograph. Polished section of Type 2 "wurtzite". One nicol. W: Type 2 "wurtzite", P: pyrite, Q: quartz. Parallel to fibre. Type 2 "wurtzite" is penetrated by quartz veinlets accompanied by pyrite.

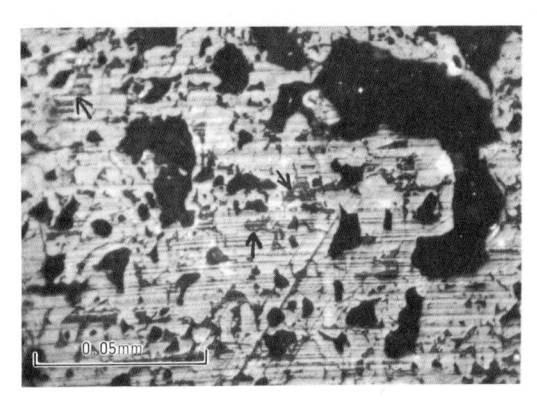

FIG. II–19–7. Photomicrograph. Polished section of Type 1 "wurtzite". One nicol. Oil immersion. Etched by saturated HI solution. Parallel to fibre. The direction of fibre is up and down. Polysynthetic twinning by the spinel law is shown, and the composition plane is perpendicular to fibre. In the host, fine grains of the same material exist, as if in poikilitic relation (shown by the arrows).

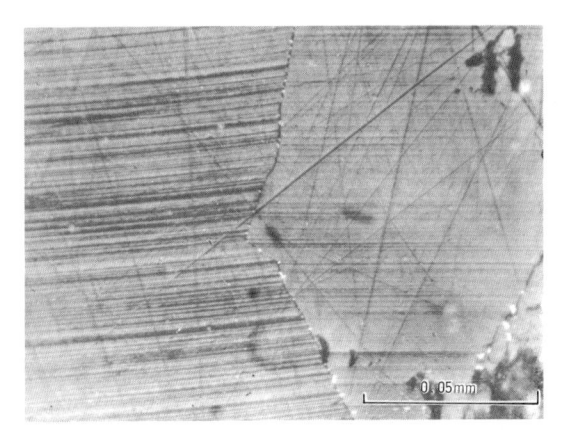

FIG. II–19–8. Photomicrograph. Polished section of Type 2 "wurtzite". One nicol. Oil immersion. Etched by saturated HI solution. Parallel to fibre. The direction of fibre is up and down. By the boundary line which runs above and below in the middle part of the Figure the two parts are different in the state of twinning and degree of etching. The boundary line represents that of original wurtzite individuals. Along this line, fine globules of chalcopyrite (white color in the photograph) are observed.

as shown in Fig. II–19–3. In sections parallel to the fibre, traces of these cleavages are observed as striations along the direction of the fibre (Fig. II–19–4). These traces are those of $(11\bar{2}0)$ cleavage which is characteristic of wurtzite. In sections parallel to the fibre, the trace of (0001) which is perpendicular to the fibre is also recognized (Fig. II–19–4), and any cleavage traces oblique to the fibre do not exist.

FIG. II-19-9. Photomicrograph. Polished section of Type 2 "wurtzite." One nicol. Oil immersion. Etched by saturated HI solution. W: Type 2 "wurtzite," P: pyrite, Sp: sphalerite. Parallel to fibre. The direction of fibre is right and left. Pyrite and sphalerite came with the quartz veinlet into Type 2 "wurtzite". In Type 2 "wurtzite," fine globules of chalcopyrite are observed, while in sphalerite they do not exist.

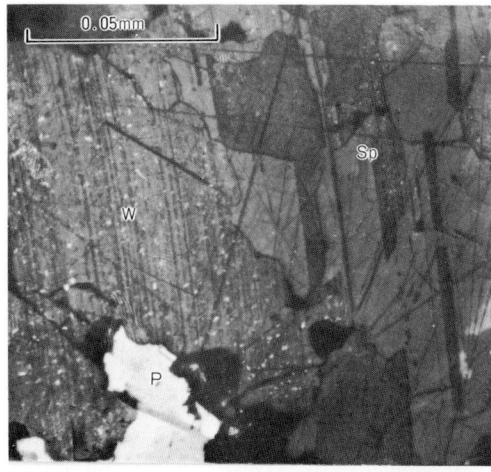

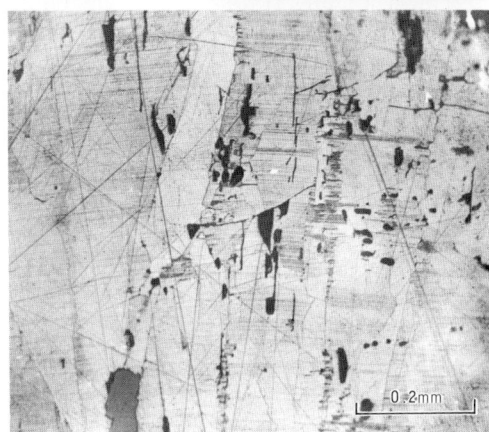

FIG. II-19-10. Photomicrograph. Polished section of Type 2 "wurtzite." One nicol. Etched by saturated HI solution. Parallel to fibre. The direction of fibre is up and down.

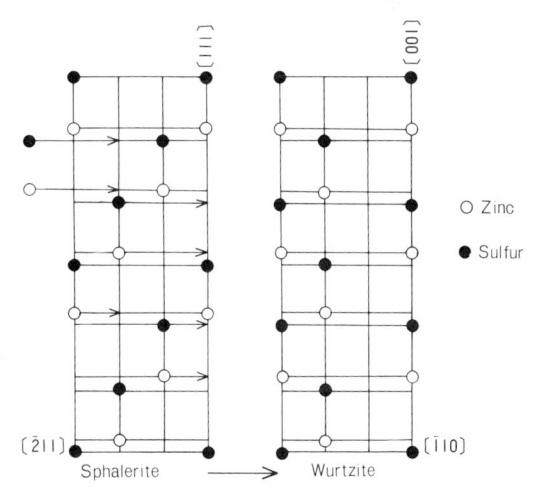

FIG. II-19-11. The transition of sphalerite into wurtzite (after Aminoff, 1923). The plane of this paper represents (0$\bar{1}$1) or (01$\bar{1}$) in sphalerite and (11$\bar{2}$0) in wurtzite. The zones [001] and [$\bar{1}$10] in wurtzite means directions perpendicular to (0001) and to (1$\bar{1}$00), respectively.

Pleochroism is not recognized in thin sections. Double refraction cannot be observed in Type 1, but is partly recognized in sections of Type 2, which are parallel to the fibre. In the latter it exhibits straight extinction. However, in sections of Type 2 perpendicular to the fibre, double refraction does not exist and in conoscopic observation a figure characteristic of uniaxial material is not seen.

Properties of the Polished Sections of "Wurtzite" under the Reflection Microscope

Properties of polished surface: Both types can be well polished. As Type 1 has abundant voids, there are many pits in the polished section (Figs. II–19–5 and –7). Type 2 has few voids and reveals a smooth surface (Figs. II–19–6 and –8).

The polished sections of both types are dark grey in color and have low reflecting power. Anisotropism and reflection pleochroism are lacking in both types. Internal reflection is observed in both types, but is stronger in Type 2 than in Type 1. With the oil immersion lens, reflectivity decreases very much, and anisotropism and reflection pleochroism cannot be observed. The properties of these minerals under the reflection microscope closely coincide with those of sphalerite or wurtzite, so they cannot be distinguished by these optical properties.

Structure etching: The polished sections were etched by a saturated solution of hydrogen iodide which was used by Van der Veen (1925). The zinc sulfides reveal the inner structures. Figures II–19–7 — –10 are photomicrographs of the etched zinc sulfides. All these are sections parallel to the fibre. The polysynthetic twins perpendicular to the fibre are characteristic of "wurtzite." The sections perpendicular to the fibre do not reveal the twin with etching.

There is a difference between Type 1 and Type 2. In Type 1, small and irregularly formed grains of the same material are included in the host, as if in poikilitic texture (Fig. II–19–7). The same kind of twin is recognized in these small grains as in the host, and the orientation of its composition plane is identical with that of the host.

According to Ramdohr (1969), the zinc sulfide which exhibits the twin through structure etching is sphalerite and not wurtzite. From this point of view, "wurtzite" in the Hosokura mine is substantially sphalerite.

Interpretations of the Microscopic Observations

Ehrenberg (1931) stated that discrimination between wurtzite and sphalerite requires the combination of three methods—that is, observations of thin sections, studies of polished sections and X-ray methods.

Iwasaki *et al.* (1940), studied "wurtzite" from the Hosokura mine by X-ray methods. They said that "wurtzite" from the Hosokura mine is substantially sphalerite and that the direction of fibre is [111] of sphalerite. This coincides with the writer's results with the etched structure of polished sections. That is, the multiple twinning revealed by etching with a saturated solution of hydrogen iodide seems to be that of spinel, and the composition plane (111) is perpendicular to the direction of the fibre.

Nevertheless, as described above, the cleavages recognized by observation of thin sections are those of wurtzite and not of sphalerite.

From these results, the writer concludes, as Watanabe (1941) did, that zinc sulfide was deposited originally as a crystal having the structure of wurtzite and later it changed into a sphalerite-type crystal. It is an example of a paramorph. The observed twin is that of spinel. So the plane (111) is perpendicular to the direction of the fibre.

Further, it was observed in the etched polished sections that aggregates of twinned sphalerite after wurtzite are partly different by domain in the degree of etching and the state of polysynthetic twinning (Figs. II–19–8, and–10). Namely, aggregates of sphalerite which exhibit polysynthetic twinning are divided into domains. The form of each domain is elongated along the direction of the fibre (Fig. II–19–10). In Type 2 "wurtzite," small globules of chalcopyrite frequently exist along the boundary between two adjacent domains (Fig. II–19–8).

The writer states that one domain of twinned sphalerite would have been one individual wurtzite crystal at the time of deposition. The plane (0001) of the latter would be converted into the plane (111) of the former. The one individual of the latter would change into an aggregate of the former which is subject to polysynthetic twinning by the spinel law.

Aminoff (1923) studied the transition of sphalerite into wurtzite. He stated that by the gliding of some zinc and sulfur atoms in sphalerite to the direction of [$\bar{2}$11] by $a/\sqrt{6}$ or $2a/\sqrt{6}$ (a: length of unit cell), sphalerite changes into wurtzite, and that (111) of the former is converted into (0001) of the latter (Fig. II–19–11).

Also, Shoji (1933) recognized experimentally by X-ray studies that (111) of sphalerite changes into (0001) of wurtzite. Müller (1952) discussed the thermodynamic problems of sphalerite-wurtzite transition, and found experimentally that hexagonal zinc sulfide at high temperature transforms completely to the cubic modification when tempered below 870°C.

Mitchell and Corey (1954) studied the structural similarity of hexagonal and cubic zinc sulfides, and observed a Vycor U-tube the oriented overgrowth (epitaxy) of both sulfides.

As to the formation of spinel twinning of sphalerite, there are two interpretations. The first is that the individual wurtzite crystal was converted into a sphalerite a described above, and later this sphalerite was subjected to polysynthetic twinning by deformation. Buerger (1928) determined experimentally that sphalerite becomes subject to twinning by the spinel law through deformation. From the point of view of crystallographic structure, it can be considered that, by the gliding of some of zinc and sulfur atoms in sphalerite to the direction of [$\bar{2}$11] by $a/\sqrt{6}$ or $2a/\sqrt{6}$, the twin in the spinel law is formed. This process is shown in Fig. II-19-12. The change to the right in the Figure indicated by the arrow shows this case. Consequently, the twin thus formed is that of rotation, not of reflection. The change to the left in Fig. II-19-12, namely the formation of a reflection twin, is impossible insofar as the atoms glide to one direction.

FIG. II–19–12. Gliding of atoms in the formation of twinning from one individual crystal of sphalerite by deformation. The planes ($\bar{2}$11) have a distance of $a/\sqrt{6}$ (a: length of unit cell) from each other. The plane of the paper represents (01$\bar{1}$) or (0$\bar{1}$1).

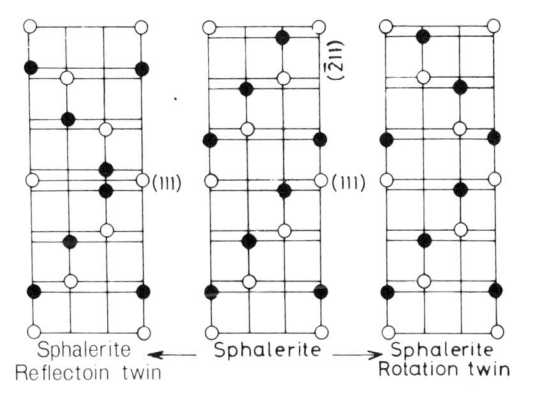

Sphalerite ← Sphalerite → Sphalerite
Reflectoin twin   Rotation twin

Another interpretation is that the individual wurtzite crystal changed directly into an aggregate of polysynthetically twinned sphalerite. In this case, some zinc and sulfur atoms in wurtzite glided to the direction perpendicular to (1$\bar{1}$00) by $a/\sqrt{6}$ or $2a/\sqrt{6}$; thus the individual wurtzite changed into an aggregate of sphalerite showing polysynthetic twinning. This process is shown in Fig. II–19–13. The change to the right in this figure shows this process. Also in this case the formation of a reflection twin (the change to the left) is impossible, insofar as the atoms glide to one direction.

The writer concludes, as did Iwasaki *et al.* (1940) and Watanabe (1941), that the double refraction observed in thin section of Type 2 "wurtzite" is a result of stress.

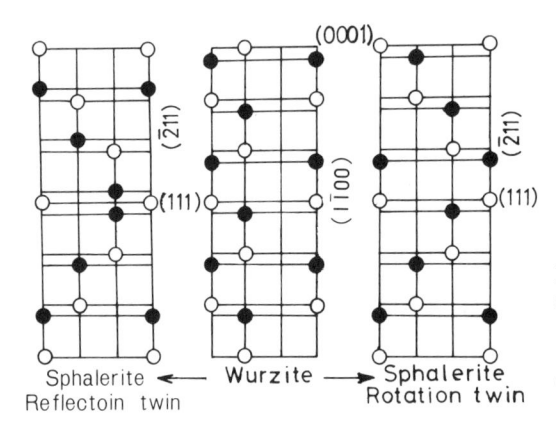

FIG. II-19-13. Gliding of atoms in the formation of twinned individuals of sphalerite from one individual of wurtzite.

The plane of this paper represents (11$\bar{2}$0) in wurtzite and (0$\bar{1}$1) or (01$\bar{1}$) in sphalerite.

Mineral Assemblage of "Wurtzite"

The two types are different in their mineral assemblage. Type 1 is accompanied by marcasite and galena. They are arranged parallel to the fibre of Type 1 "wurtzite" (Fig. II-19-5). Grains of marcasite are smaller than 0.5 mm. Marcasite is anisotropic and has reflection pleochroism. With an objective lens of high magnification, it was observed that the marcasite is accompanied by pyrite. The pyrite has no anisotropism and no reflection pleochroism, and has a yellow tint compared with marcasite. In some cases pyrite grains are present among marcasite grains, and in other cases they exist in a relation of parallel intergrowth (oriented overgrowth), i.e., in epitaxy (Fig. II-19-14).

Type 2 has chalcopyrite globules of very small size, detectable only with high magnification, in emulsion texture (Figs. II-19-8, and-9). The chalcopyrite globules are generally scattered at random, but in some cases they are arranged along boundary lines of the "wurtzite" individuals (Fig. II-19-8).

This texture is generally considered to be due to exsolution (Ramdohr, 1969). Experiments by Roberts (1965) indicated that the chalcopyrite emulsion in sphalerite might be formed at low temperatures. Imai (1941, 1947) suspected that chalcopyrite globules in sphalerite might be due to the emulsion of the former in the latter at the time of deposition. Robert (1965) also stated that this kind of texture might be explained by emulsion. Fujii (1970) noticed that exsolution is a phenomenon of phase separation in solid states which does not involve open spaces, and that the solid solute phase partly emplaces the solid solvent phase, which complicates the criteria for replacement or unmixing of textures.

In some cases, grains of chalcopyrite of comparatively large size are present (about 0.3 mm in diameter).

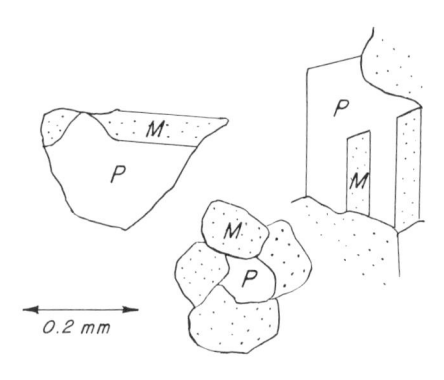

FIG. II-19-14. The relation of marcasite and pyrite observed in polished section of Type 1 "wurtzite." M: marcasite, P: pyrite.

Type 2 is also accompanied by galena bordering in mutual relation. Also in Type 2, a tetrahedrite group mineral exists with the comparatively large-sized chalcopyrite described above. The maximum size of the tetrahedrite group mineral is 0.3 mm in diameter.

It is characteristic of Type 2 that it is irregularly penetrated by quartz veinlets, as shown in Fig. II–19–6. Veinlets consist of quartz, pyrite, marcasite and sphalerite. Marcasite is present in far smaller amounts than pyrite. Sometimes both occur in parallel intergrowth as described above. It is remarkable that this veinlet is accompanied by sphalerite, not wurtzite. The discrimination between the host "wurtzite" and this sphalerite in unetched polished sections is sometimes difficult for unfamiliar eyes. Through structure etching, the two are quite different in the twinning state (Fig. II–19–9). In Fig. II–19–9, it is noticed that this sphalerite does not include globules of chalcopyrite. The two can also be discriminated by thin section. The thin section of "wurtzite" parallel to the fibre exhibits cleavage traces (11$\bar{2}$0) and (0001), as already described. Anisotropism is partly observed. But the sphalerite is very complicated in the state of cleavage and does not reveal anisotropism.

It has generally been said that the precipitation of wurtzite and marcasite in nature would require acid solution, after the experiments of Allen *et al.* (1914). But, according to Scott and Barnes (1972), wurtzite would be formed at lower f_{S_2}, condition than sphalerite. As discussed in the next chapter, Buerger (1934) suggested, and Scott and Barnes (1972) recognized, that wurtzite compositions lie predominantly to the S-deficient side of stoichiometric ZnS.

Summary

(1) The fibrous zinc sulfide ("wurtzite") from the Hosokura mine which was deposited during a later stage in the course of sulfide deposition is wurtzite having a hexagonal system, judging from the observation of the thin sections. But from the states of the etched structure in the polished section, it is sphalerite subject to polysynthetic twinning (spinel twin), whose composition plane is perpendicular to the fibre. From these observations, the writer concludes as follows: The fibrous zinc sulfide might have been deposited as wurtzite at the time of mineralization and later converted into sphalerite. Plane (0001) of wurtzite might change into the plane (111) of sphalerite, and an individual crystal of the former might change into the aggregates of the latter, subject to polysynthetic twinning on (111).

(2) There are two types of "wurtzite" in the Hosokura mine. The writer has described the differences between macroscopic and microscopic characters and the mineral assemblages of the two types.

II–20. Problems of Enargite, Luzonite and Famatinite (Imai, 1943, 1949b)

H. Imai

In the Circum-Pacific region, there are some arsenical copper (gold) deposits. Butte, Montana, U.S.A.; Cerro de Pasco, Peru; Mankayan, Philippines (Fig. II–20–1); and Chinkuashih, Taiwan, are examples of this type. The same type of deposit exists in Japan. They are characteristic in geologic occurrence, paragenesis and genesis. The important copper minerals of these deposits are enargite and luzonite. The problems of enargite, luzonite and famatinite have been studied and discussed by many mineralogists, as described in the following.

Previous Work on Enargite, Luzonite and Famatinite

The mineral luzonite which occurred in the Mankayan mine as the aggregate of crystals was described and named by Weisbach (1874). He stated that it has the same composition* as enargite (Cu₃AsS₄). This mineral was studied also by Klockmann (1891), Moses (1905), Frebold (1927), Jong (1928), Waldo (1935), Harcourt (1937) and others. Their conclusions were not necessarily coincident.

Schneiderhöhn and Ramdohr (1931) recognized that luzonite has a crystal structure distinct from that of enargite and concluded that two solid solution series are present:

Enargite (Cu₃AsS₄) ———Enargite structure——— Famatinite (Cu₃SbS₄)

Luzonite (Cu₃AsS₄) ———Luzonite structure——— Stibioluzonite (Cu₃SbS₄).

In the first series they did not believe that complete isomorphism exists; in the second series they considered that almost complete miscibility prevails. They said that the first series belongs to the rhombic system and the second series perhaps belongs to the monoclinic system.

Harcourt (1937) studied these minerals by X-ray powder photographs and spectrochemical analyses. He excavated minute samples from very small areas in a polished section by means of an improved micro-drill. He learned that the pinkish mineral which has the crystal structure of luzonite is rich in antimony. Mainly from this, he concluded that luzonite and famatinite refer to the same mineral series and are synonymous. Since famatinite has date priority and refers to an antimony-rich mineral, he asserted that this name should be retained for the series. In that case, there are two series; the proportion of antimony in Cu₃AsS₄ may increase only up to a certain amount and still have the crystal structure of enargite. Similarly, the proportion of arsenic in Cu₃SbS₄ may increase only to a certain amount and still retain the structure of famatinite:

Cu₃AsS₄————Cu₃(As, Sb)S₄,   Cu₃(Sb As,)S₄————Cu₃SbS₄.
Enargite   Famatinite

Harcourt (1937) observed under the microscope that the two minerals occur together. In the polished section they are identical in hardness, etch tests, anisotropism and constituent elements. Luzonite differs from enargite only in its color, its characteristic lamellar twinning and its lack of cleavage.

Hiller (1940) followed Schneiderhöhn and Ramdohr (1931) in his paper on the classification of sulfide minerals, but, referring to Harcourt's paper, he added that the question was still undecided.

Sawada (1944) studied the crystallography of small-sized luzonite from the Hokuetsu mine in Niigata Prefecture, Japan, by goniometry and X-ray diffraction. He concluded that the crystal is twinned and belongs to the pseudoisotropic tetragonal system. Its lattice constants are $a_1 = a_2 = 5.28_0$ Å, $c = 10.44_4$ Å, $a:c = 1.978$. Its space group belongs to D_{2d}^{11}-I $\bar{4}$2m. He also determined its crystal structure.

Imai (1943, 1949b) studied the microscopic properties of luzonite and enargite from the Mankayan mine in the Philippines and the Chinkuashih (Kinkaseki) mine in Taiwan. His results are nearly the same as those described in this chapter. Opposed to Harcourt's opinion, he maintained that the name luzonite should be retained.

M. Watanabe (1943, 1951) said that luzonite and enargite from the Hokuetsu mine would be the products of solfataric action in the crater wall of the recent volcano, Mt. Sumon.

T. Watanabe (1943a) reported luzonite and enargite in the Teine mine, as described in the next section. Minato et al. (1954) recognized that the luzonite in the Teine mine is rich in antimony.

Gains (1957) studied the mineralogical characters of luzonite, famatinite and some related

---

* Analyzed by Cl. Winkler, Fe: 0.93%, Cu: 47.51%, Sb: 2.15%, As: 16.52%, S: 33.14% (Weisbach, 1874; Doelter and Leitmeier, 1926).

minerals. He synthesized artificial luzonite and famatinite. According to him, a complete solid solution series extends between the tetragonal minerals, luzonite Cu_3AsS_4 and famatinite Cu_3SbS_4; luzonite is dimorphous with the orthorhombic mineral enargite, while the solubility of Sb in enargite is restricted, and the orthorhombic dimorph of famatinite is not known in nature. His study on the crystal structure of luzonite yielded nearly the same results as Sawada's study.

Skinner (1960) stated that enargite is the high-temperature polymorph of luzonite, and that luzonite and famatinite belong to an isomorphous series, after a study of the syntheses of Cu-As-S phases. He proposed that a two-phase assemblage enargite plus a luzonite-famatinite series may be useful as a geologic thermometer for determining the chemical compositions of these minerals.

Moh and Ottemann (1962) found natural and artificial tin-bearing enargite and luzonite.

McKinstry (1963) made a comprehensive attempt to deduce the possible stable phase assemblages from the Cu-Fe-As-S mineral paragenesis in nature. Gustafson (1963) developed McKinstry's opinion and discussed the assemblage of Cu-Fe-As-S minerals from the view point of the fugacity or activity of these elements.

Strunz (1966, 1970), Ramdohr (1969) and Uytenbogaardt (1971) classified these minerals as the following two series: enargite——stibioenargite (orthorhombic), and luzonite——stibioluzonite (tetragonal).

Lévy (1967) studied in detail the X-ray powder diffraction patterns, reflectivities and DTA curves of these minerals.

Barton and Skinner (1967) described the Cu_3AsS_4-Cu_3SbS_4 minerals from the viewpoint of phase stability. They determined that the inversion point from low-temperature phase luzonite to high-temperature phase enargite is 320°C. According to them, the fact that luzonite is always richer in Sb than the coexisting enargite indicates that the two-phase field of enargite-luzonite must slope upward from the luzonite-enargite inversion temperature (320°C).

Springer (1969) recognized that enargite could contain as much as 6 wt % antimony, and a complete solid solution was observed between the tetragonal phases of luzonite and famatinite. According to him, the compositions of enargite and luzonite-famatinite solid solutions were found to be very variable even within one polished section.

Maske and Skinner (1971) studied the phases and phase relations of the minerals in the system Cu-As-S mainly between 300°–665°C by experiments in sealed, evacuated silica glass capsules. Also, Skinner *et al.* (1972) conducted similar studies on the phases and phase relations in the system Cu-Sb-S.

Feiss (1974) studied the phase relation of tetrahedrite-tennantite coexisting with enargite-famatinite by means of dry syntheses.

Geologic Settings

Mankayan mine* (Fig. II–20–1): The Mankayan mine is situated in the Mountain Province, Luzon Island, about 50 km to the NNE of Baguio City. The production of this mine in 1973 was about 25,000 t of copper, 3 t of gold and 14 t of silver. The deposit of this mine was studied by Arsenio Gonzalez (1956). According to him, this district is near the eastern limb of a broad anticline with a N-S axis. It is composed of metavolcanics perhaps of the Cretaceous-Paleogene epoch and dacitic rocks of the Late Miocene or Pliocene epoch. Around the mining area, the dacitic rocks are mainly pyroclastic beds overlying unconformably the irregularly eroded surface of the metavolcanics.

---

* The writer thanks Dr. A. Gonzales, Dr. M. H. Tupas and Mr. R. A. Conception of the Lepanto Consolidated Mining Co. for their help in surveying the Mankayan mine, which is managed under this company.

FIG. II–20–1. Index map of northern Luzon.

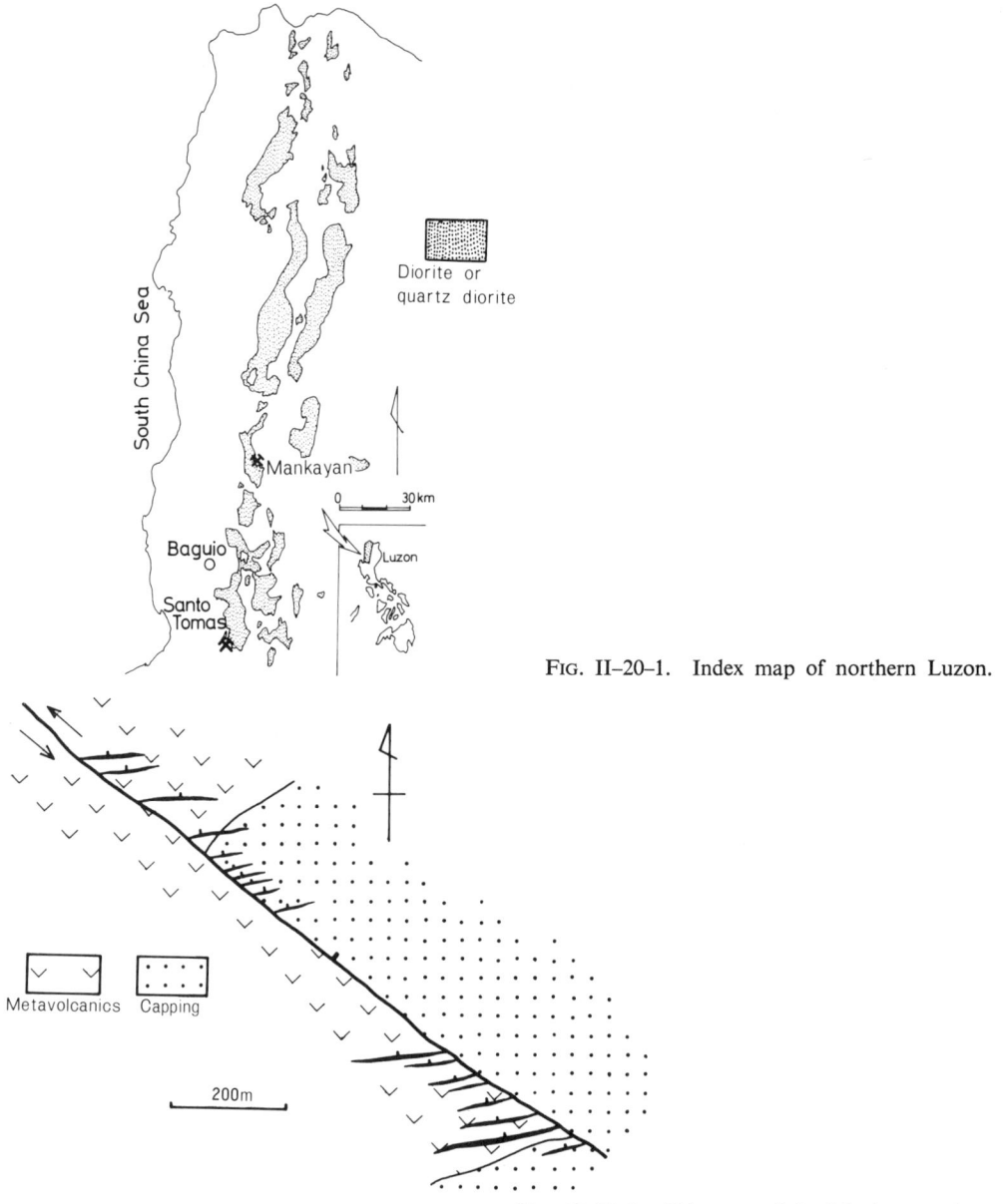

FIG. II–20–2. Vein map of the Mankayan mine.

At the core of the anticline there is an intrusive mass of trondhjemitic composition. This intrusive is also overlain unconformably by dacitic rocks.

The deposit exists in the metavolcanics and dacitic rocks. Copper mineralization occurs along the fault fissure, which extends at least 1.5km to N55°W, dipping steeply to NE (Fig. II–20–2). At both sides of the fissure, there are branches or splits which resemble the horsetails observed in the Butte mine. They strike nearly E-W, dipping N steeply.

The writer suggests that this master fault fissure is a left-handed wrench fault, as Bryner (1969) stated, and that the branches or splits are secondary tension cracks or shears of the second order. These fissures would be formed by the lateral pressure from E-W which caused the fold structure in this district. Along these fissures, the mineralizing fluid ascended, filling them and replacing the surrounding metavolcanics and dacitic rocks. Especially along the unconformity zone between the above two, ore shoots were formed. So the dacitic rock flow is called "capping" in this mine.

It is probable that the master fault displaced vertically as the normal fault after the mineralization, as Bryner (1969) stated. Thus, it corresponds to the fissure revealing multiple movement or displacement, as discussed in Chapter II-1.

The ore minerals are pyrite, enargite, luzonite, chalcopyrite, chalcocite, tennantite, sphalerite, galena and native gold with small amounts of gold-silver tellurides such as hessite, calaverite (?), krennerite (?), etc. The gangue minerals are quartz, kaolinite, barite, alunite, etc. The decrepitation temperature of a enargite sample from the Mankayan mine was determined to be 250°C by K. Nagano in the writer's laboratory.

Three different, but probably related, types of alteration have been recognized in the surrounding rocks of the deposits (Gonzalez, 1956). They are silicification, argillization and chloritization. Silicification affects both the metavolcanics and dacitic rocks (capping), and forms a halo around the ore body. Alteration to kaolinite forms a halo around the silicified zone. Alteration to montmorillonite-type clay and hydromica occurs principally around the zone of kaolinization in the capping, but chloritization forms a corresponding zone in the metavolcanics.

The exact age of mineralization is unknown. The mineralization of the "capping" indicated that the ore deposition was definitely younger than the "capping"—i.e., late Miocene or Pliocene. According to Gonzalez (1956), the trondhjemite intrusive is overlain unconformably by the "capping." So he stated that the only likely source of ore-forming fluid was the volcanic center from which the "capping" rocks were extruded. This would date the mineralization as either late Miocene or early Pliocene. According to Tupas (1960), the mineralization at Mankayan is perhaps Quaternary in age.

In northwestern Luzon, the quartz diorite complex is essentially a batholithic mass disposed in a N-S belt which intruded into the Cretaceous-Paleogene and Early- to Middle-Miocene rocks (Fernandez and Pulanco, 1967) (Fig. II-20-1). The quartz diorite corresponds to trondhjemite in the Mankayan mining area. Many porphyry copper deposits were formed in this district genetically related to this rock. The data on the absolute age of this complex are not sufficient, but two dates (K-Ar method) are available, i.e., 14.8 m.y. and 9.7 m.y. (Wolfe, 1972). Also, the absolute age (K-Ar method) of some diorite in Mindanao is 6.7 m.y. (Wolfe, 1972) extending to Latest Miocene or Pliocene epoch.

It is most probable that the mineralization in the Mankayan mine occurred later than the mineralization of the porphyry copper deposits in northwestern Luzon. Still, it is also possible that mineralization in the Mankayan mine was genetically related to the underneath acidic igneous activity of the same Alpine orogenesis, which would have continued after the dacitic flow eruption.

Chinkuashih (Kinkaseki*) mine, Taiwan (Fig. I-3): This mine is situated near the northern end of Taiwan Island (Formosa). The production attained its maximum in 1935–1941, with an annual output of 2–2.5 t of gold and 6,000–7,000 t of copper.

Recently many papers on the geology and mineralogy of this mine have been published by Huang** (1955, 1963, 1964, 1965, 1972, 1973, 1974), Wang (1955, 1973), Folinsbee *et al.* (1972) and others.

According to Huang (1955), the Chinkuashih area is composed mainly of Miocene sandstone and shale intruded by a chonolith of dacite. The diameter of the chonolith is about 1 km. The extrusion of this rock is said to have taken place in the Pleistocene, though there

---

* Pronunciation in Japanese.
** Prof. Huang had presented to the writer a sample of the famatinite of this mine, which the writer studied.

are no data on absolute age dating. The deposit occurs in dacite and in the surrounding Miocene sediments. The main factor controlling the location of the ore deposit is the presence of N-S normal faults (Wang, 1959; Huang, 1963; Yen, 1974), which would have been formed by the doming-up of the dacite chonolith. The faults were mineralized as veins, along which there are some pipe-like ore bodies or irregular ore masses formed by replacement or impregnation.

The ore minerals are pyrite, enargite, luzonite, famatinite, tennantite, chalcopyrite, sphalerite and native gold, accompanied by small amounts of bournonite, cinnabar, wurtzite, marcasite, etc. (Huang, 1955, 1965, 1972, 1973, 1974). The gangue minerals are quartz, barite, alunite, kaolinite, diaspore and native sulfur. According to Huang (1974), enargite in this mine ranges from 0.30 wt. % to 4.11 wt. % of Sb, while luzonite-famatinite ranges from 1.80% to 20% of Sb, supporting the complete solid solution of the system. Enargite and luzonite show a tendency to increase in Sb content with depth. There are three lateral zones in this deposit: (1) copper and gold zone, (2) copper-bearing gold zone and (3) gold zone, in order from the center towards the outside.

According to Wang (1973), the country rocks of this mine are consistently characterized by a well-developed argillic assemblage of quartz (cristobalite in some places)-alunite in the center and successively enveloped by kaolinite, sericite, montmorillonite and chlorite zones. It is mineralogically noteworthy that interstratified sericite-montmorillonite and chlorite-montmorillonite occur in the altered country rocks. Wang studied the filling temperatures of fluid inclusions of quartz mainly from the silicified zone of the deposit. They ranged between 290°C and 210°C. He concluded that the physicochemical conditions of the hydrothermal solution responsible for advanced argillic facies and the concurrent ore deposition might be as follows: pH 3.5–4.5, fo_2 10^{-30}–10^{-34} atm., fs_2 10^{-6}–10^{-8} atm., with sufficiently high activities of SiO_2, K^+ and SO_4^{2-}. According to Huang (1955) and Wang (1973), the early stage of mineralization was characterized by quartz and pyrite; the middle stage by enargite, luzonite and native gold; and the late stage by alunite, wurtzite, kaolinite and nacrite, and these mineralogic indications tell that the ore-forming fluid changed from alkaline in the early stage to acidic in the later.

Folinsbee et al. (1972) studied the fluid inclusion thermometry of barite and the sulfur isotope thermometry of pyrite and barite in this mine. The result of his fluid inclusion thermometry nearly coincides with Wang's results. Sulfur isotopes for sulfide minerals in this mine yield values close to meteoritic, and typical of magmatic hydrothermal deposits, but the equilibrium sulfate species of barite is 2.5‰ heavier in ^{34}S. Data on the barite-pyrite pairs suggest formation temperatures in the range of 255°-305°C, a range compatible with fluid inclusion temperatures in barite (for the most part between 228° and 305°C).

Teine mine, Hokkaido (Fig. I–3): The Teine mine is situated in a suburb of Sapporo City in Hokkaido. It was abandoned in 1972. Production attained its maximum in 1940–1942, with an annual output of 1.8t of gold, 25t of silver and 800t of copper.

This area is geologically composed of Late Miocene andesitic tuff breccia and tuff, mudstone and propylite. The strike of tuffaceous rocks and mudstone is N 25°–45°W, dipping 35°–50° to NE (Fig. II–20–3). The propylite extrudes into the tuffaceous rocks and mudstone.

The veins occur mainly in the propylite. They are in some cases accompanied by shale dikes. They are grouped into three areas: (1) Bannozawa, (2) Mitsuyama and (3) Koganezawa. The veins of the former two groups reveal a radiated pattern (Fig. II–20–3). In the Koganezawa area, the veins occur in a trapezoid pattern. The radiated pattern would be due to the pushing up of the propylite magma. In this case the fissures revealing the radiated pattern should have the properties of tension cracks. Though they have in many cases slickensides on the walls of the country rocks, they opened at the same time as the shearing move-

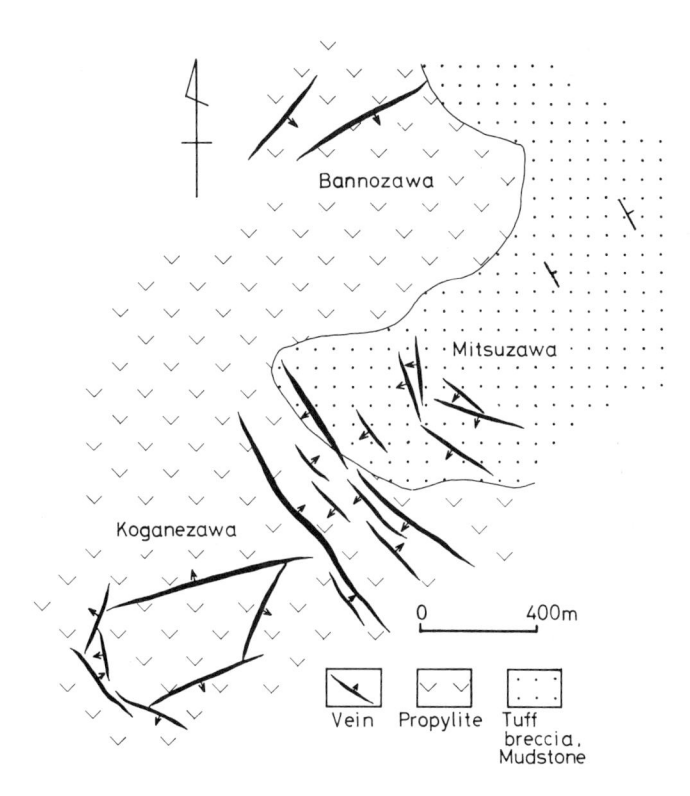

FIG. II–20–3. Geologic map of the Teine mine.

ment. A similar phenomenon is discussed in Chapter II–2 on the Ohya mine. The trapezoid-like pattern at the Koganezawa area would be also due to pushing-up of propylite. The striations observed in the vein wall of the Ohtoyo vein in this group are of dip direction. This vein pattern corresponds to a bysmalith in the magmatic emplacement, as discussed in Chapter I–1. The problem of the vein pattern in the Teine mine has also been discussed by Sugimoto (1952).

This deposit was worked as a gold-silver mine in the initial stage of development. Afterwards, copper ores, mainly enargite-luzonite ore, were found in some parts of the veins, especially in the Toriyabe vein, Mitsuyama area (Fig. II–20–3).

As ore minerals, there exist native gold, sylvanite, petzite, ricardite, pyrargyrite, tetrahedrite, enargite, luzonite, chalcopyrite, chalcocite, bornite, bismuthinite, emplectite, klaprothite, sphalerite, galena, stibnite, orpiment, realgar, pyrite, marcasite, etc. The gangue minerals are quartz, barite, calcite, rhodochrosite, etc. As supergene secondary minerals, tellurite and teineite* were found. Watanabe (1943a) discussed the reaction rim between enargite-luzonite and bismuthinite. He recognized in some specimens copper-bismuth minerals such as emplectite ($Cu_2S \cdot Bi_2S_3$), klaprothite ($3Cu_2S \cdot 2Bi_2S_3$) and tetrahedrite (bismuth-bearing?) along the boundary between enargite-luzonite and bismuthinite under the microscope. According to him, a mineral resembling goldfieldite also exists.

Gold ore is apt to occur in the upper levels, and copper ore in the lower levels. Luzonite, enargite and bismuth minerals are mainly found in the Mitsuyama area. Calcite, rhodochrosite, realgar and orpiment are rich in the Koganezawa area. Gold and copper ores are poor in the latter group. The Koganezawa area represents the marginal part of the mining area.

---

* Yoshimura, T. (1939) Teineite, a new tellurate mineral from the Teine mine, Hokkaido, Japan: *Jour. Fac. Sci. Hokkaido Imp. Univ.* Series 4, **4**, 465–470.

The country rocks are characterized by silicification and kaolinization.

This deposit is similar to the deposit of the Goldfield mine, studied by Tolman and Ambrose (1934).

Suttsu mine, Hokkaido (Fig. I–3): This mine is situated about 90 km southwest of the Teine mine. It was abandoned in 1962.

Veins occur in Miocene andesitic tuff, tuff breccia, mudstone and propylite. The veins contain mainly pyrite, sphalerite and galena. In some parts of the vein, ore is composed of tetrahedrite, chalcopyrite, enargite, luzonite, famatinite, cassiterite, stannite, tetradymite, bismuthinite, stibnite, etc. Moh and Ottemann (1962) described "stannoenargite" in this mine.

This deposit is a type of xenothermal vein type.

Hokuetsu mine, Niigata Prefecture (Fig. I–3): The small copper deposit of the Hokuetsu mine is situated in Niigata Prefecture, central Japan. It occurs in andesite or andesitic tuff. According to M. Watanabe (1943, 1951), the deposits are situated in dissected walls of caldera of the Quaternary volcano, Mt. Sumon.

The ore is composed of quartz, pyrite, marcasite, enargite, luzonite, sphalerite, galena and sporadic native sulfur. The country rocks underwent silicification and kaolinization.

Watanabe (1943, 1951) stated that the deposit was of solfataric origin.

Akeshi and Kasuga mines, Kagoshima Prefecture (Fig. I–3): The gold-quartz deposits of the Akeshi and Kasuga mines are situated in Kagoshima Prefecture, southern Kyushu. They occur in propylite and tuff breccia of the Miocene epoch. The network type of gold-quartz vein occurs in the silicified country rocks. The ore in the Akeshi mine is composed of quartz, native gold, luzonite and pyrite, with very small amounts of enargite and native sulfur. The ore in the Kasuga mine is composed of quartz, native gold, enargite and pyrite, with small amounts of luzonite-famatinite and berthierite (Tokunaga, 1954; Miyahisa, 1974).

The silicified country rocks are surrounded by kaolinized or alunitized parts. Alunite is accompanied by small amounts of diaspore and corundum. Rutile is found in these altered parts.

Kuroko deposit (Black Ore deposit): In some Kuroko deposits (for example, the Kamikita mine), enargite and luzonite are recognized in an intimate association with chalcocite, digenite, bornite, idaite, germanite and silver-bearing minerals in Kuroko Ore (Black Ore) (Matsukuma and Horikoshi, 1970). Refer to the Tsuchihata mine (Chapter II–11).

Kaise, Nagano Prefecture, and Masutomi, Yamanashi Prefecture (Fig. I–3): In the central part of Honshu Island, quartz diorite of perhaps Miocene epoch intruded into the Chichibu Permo-Carboniferous Formation (refer to Chapter III–6).

At the Kaise and Masutomi areas in this district, there occur enargite-pyrite quartz veins, genetically related to quartz diorite activity. But no luzonite is found. These veins are of mesothermal type.

Microscopic, X-Ray and Electron Microprobe Studies of Enargite, Luzonite and Famatinite from the Mankayan and Chinkuashih Mines

Microscopic study and X-ray diffraction photographs: (1) The writer observed enargite, luzonite and famatinite from the Mankayan and Chinkuashih mines under a reflection microscope. Enargite is greyish with a pale pinkish color under the microscope, while luzonite and famatinite are pink with a brownish tint. They reveal the same properties in etching tests (Short, 1940). Enargite has no twinning and no zonal structure, while the other two show twinning and a sporadically zonal structure (Figs. II–20–4, –5 and –8).

(2) The writer excavated under the microscope a small amount of enargite, and took a X-ray powder photograph. The rings in the photograph coincide with those of enargite.

(3) In the same way, the writer gouged out a small amount of a part recognized as luzonite or famatinite, which appears pink with a brownish tint and reveals polysynthetic

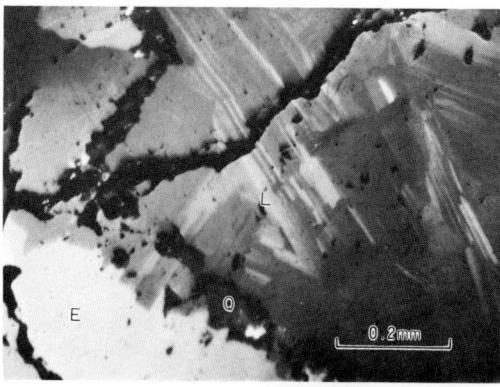

Fig. II–20–4. Photomicrograph. Polished section. Enargite and luzonite. Crossed nicols. E: enargite (white or grey part), L: luzonite (twinning prevails), Q: quartz. Black parts are cracks.

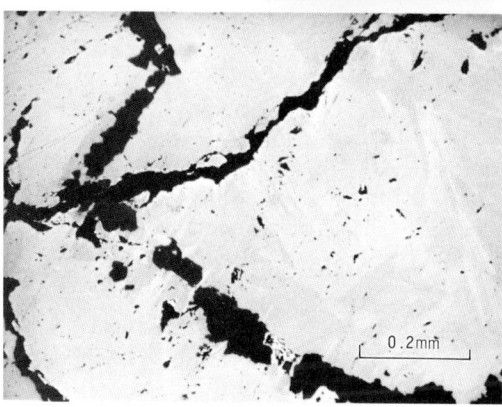

Fig. II–20–5. Photomicrograph. Same part as Fig. II–20–4. One nicol. On luzonite, twinning is revealed by reflection pleochroism.

twinning, and took an X-ray powder photograph. Its rings coincide with those recognized by Harcourt (1937) as having luzonite or famatinite structure.

(4) Next, the writer scratched a part which revealed no twinning but appeared pinkish with a brownish tint and closely resembled luzonite except for the absence of twinning. The rings of its X-ray powder photograph conicide perfectly with those of (3).

(5) Under the microscope, it is recognized that in the twinned luzonite the enargite-like mineral occurs in parallel growth with luzonite (Fig. II–20–6). This is verified by an X-ray powder photograph of the enargite-like mineral which was gouged out under the microscope. The rings of the X-ray powder photograph of the sample coincide with those of enargite which were recognized in study (1). This state is apt to be mistaken for the twinning of luzonite under the microscope. It is a kind of epitaxy or oriented overgrowth. The composition plane of polysynthetic twinning of luzonite (perhaps (112)) coincides with the plane of oriented overgrowth of this mineral with enargite, while the plane of oriented overgrowth of enargite is (001).

(6) In the twinned and untwinned famatinite under the microscope, an epitaxial enargite-like mineral does not exist.

X-ray powder diffraction pattern: The writer selected samples of macroscopically pure enargite, luzonite and famatinite from the ores of Mankayan and Chinkuashih, and studied them by X-ray powder diffractometer. Exact powder data from the diffraction charts of these samples are given in Table II–20–1 and Fig. II–20–7. At the same time, polished sections of these samples were re-examined under the microscope. In the polished section of luzonite, enargite in epitaxial relation with luzonite and quartz in a small veinlet were observed.

Some of the peaks of enargite in the chart of the luzonite sample would be due to the

152   E. Genetical Problems of Sulfide Minerals

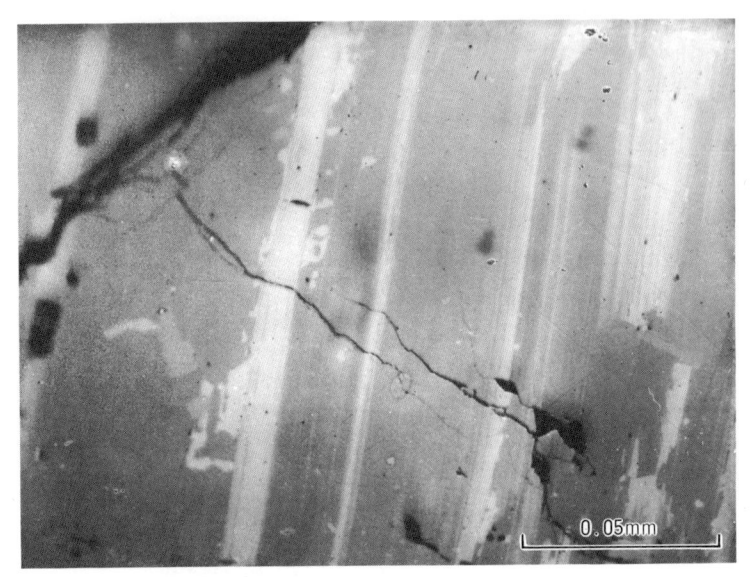

Fig. II–20–6. Photomicrograph. Polished section. Epitaxial relation of luzonite and enargite. One nicol. Oil immersion. White: enargite, Grey: luzonite. Black part: cracks.

(a) Enargite   (b) Luzonite   (c) Famatinite

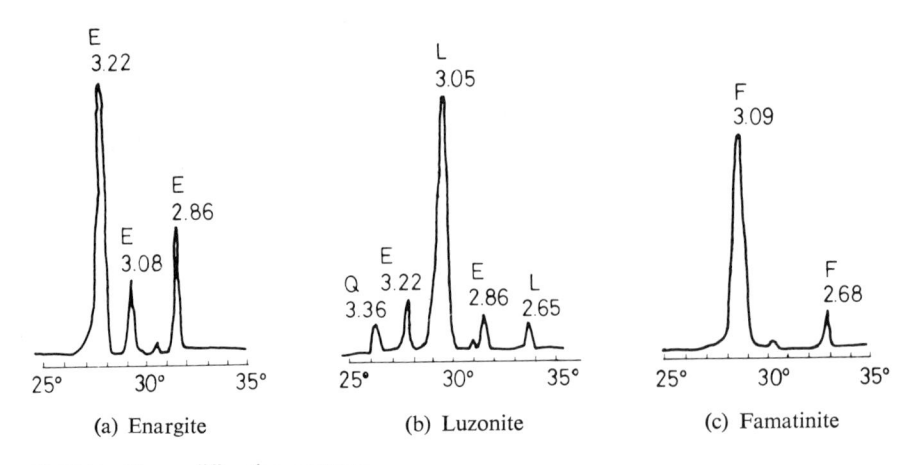

Fig. II–20–7. X-ray diffraction patterns.
E: enargite, L: luzonite, F: famatinite.

enargite being in epitaxial relation with luzonite. In the data for famatinite, no peak of enargite type is observed (Fig. II–20–7 and Table II–20–1).

The d values for the index planes in luzonite are smaller than corresponding ones in famatinite (Table II–20–1).

Electron microprobe studies: The writers studied the contents of Cu, As, Sb, Fe and S in enargite, luzonite and famatinite from Mankayan and Chinkuashih mines by means of electron microprobe.

It is generally said that luzonite has a higher content of antimony than coexisting enargite (Barton and Skinner, 1967). But in some polished sections (Table II–20–2), enargite has a higher antimony content than the coexisting luzonite. In luzonite and famatinite, arsenic and antimony contents vary part by part in the same polished section. A difference in the contents of arsenic and antimony is not recognized between luzonite and enargite in epitaxial relation.

TABLE II–20–1. X-ray powder data for enargite, luzonite and famatinite from the Mankayan mine, Philippines and the Chinkuashih mine, Taiwan. Cu Kα radiation, Ni filtered, 30 kV, 20mA.

Enargite, Mankayan			Luzonite, Mankayan			Famatinite, Chinkuashih		
d(Å)	I/I₀	(h k l)	d(Å)	I/I₀	(h k l)	d(Å)	I/I₀	(h k l)
6.46	2	(100)				5.34	3	(002)
4.87	7	(110)	4.75	3	(101)	4.80	10	(101)
			4.28	3	Quartz			
			3.36	2	Quartz	3.80	3	(110)
3.22	100	(200),(120)	3.22	20	Enargite			
3.08	40	(002)	3.05	100	(112)	3.09	100	(112)
2.95	1	(210)	2.93	1	(103)	2.97	3	(103)
2.86	50	(201),(121)	2.86	8	Enargite			
			2.65	5	(200)	2.68	13	(004), (200)
			2.62	1	(004)	2.34	2	(211)
2.22	18	(122), 202)	2.46	1	Quartz	2.21	2	(114)
2.06	2	(310)	2.23	1	Quartz	1.99	2	(213)
1.91	3	(222)	2.13	1	Quartz			
1.86	60	(040), (320)	1.86	40	(220)	1.89	50	(220)
1.73	20	(123), (203)	1.73	4	Enargite	1.79	2	(006)
1.64	1		1.60	21	(312)	1.61	27	(312)
1.62	3		1.58	9	(116)			
1.61	5	(240), (400)	1.52	2	(224)	1.55	2	(224)
1.59	15	(042), (322)	1.32	3	(400), (008)	1.34	2	(008), (400)
1.56	17	(241), (401)	1.31					
1.42	3	(242), (024) (402)	1.21		(332)	1.23	4	(332)
			1.08	5	(424), (228)			
1.35	1	(250), (430)	1.04	5	(512)			
1.26	3	(243), (403) (510)						

TABLE II–20–2. Chemical analyses of enargite, luzonite and famatinite by means of electron microprobe.

Mineral	Locality	Chemical composition (wt. %)				
		Cu	As	Sb	S	Total
Enargite	Mankayan 3	48.66	17.11	1.66	33.06	100.49
Luzonite	Mankayan 3	49.99	17.36	0.24	32.48	100.07
Famatinite	Chinkuashih 10–1	44.91	3.26	20.06	29.87	98.10
Famatinite	Chinkuashih 10–2	45.14	3.59	20.64	29.89	99.26
Famatinite	Chinkuashih 10–3	45.89	5.38	17.97	30.08	99.32

(Analyzed by T. Shoji)

Zoning is recognized in some luzonite, and it is generally unrelated to the twinning structure (Fig. II–20–8), but, in some parts, it is controlled by the twinning plane (Fig. II–20–8). Iron is not detected by electron microprobe though it is detected by a microchemical test under the microscope.

Mineralogical Problems of Enargite, Luzonite and Famatinite

On the basis of the results given in previous sections, the writer disagrees with Harcourt's opinion (1937) that the name luzonite should be abandoned, and believes that this name must be maintained. The luzonite from the Mankayan mine, where it was originally found, contains a large amount of arsenic and a small amount of antimony, and must therefore be distinguished from famatinite Cu_3SbS_4. In 1943 and 1949, the writer stated that the following three cases would be possible. They are reinvestigated in the present studies.

(a) Enargite and luzonite are in dimorphic relation, as Schneiderhöhn and Ramdohr

154   E. Genetical Problems of Sulfide Minerals

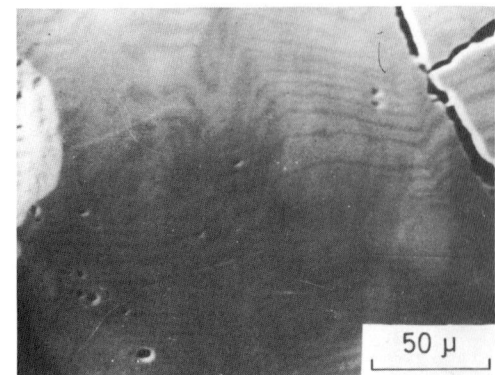

(a) Photomicrograph. Crossed nicols.

(b) Secondary electron image of electron microprobe.

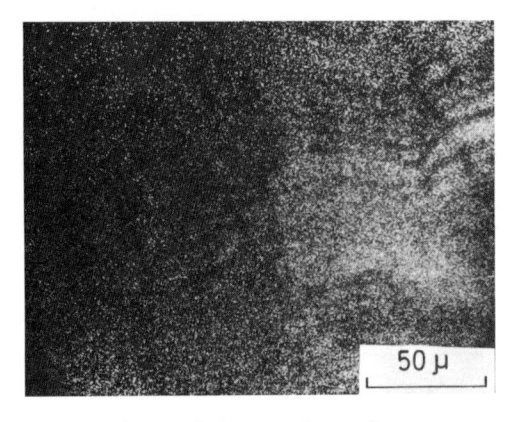

(c) Sb L_α image of electron microprobe.

FIG. II–20–8. Photomicrograph and electron microprobe photographs of luzonite. (a), (b) and (c) show the same part. (Photo by T. Shoji).

(1931), Springer (1969) and others have suggested. According to Sawada (1944) and others (Gains, 1957; Marumo and Nowacki, 1967), luzonite belongs to the tetragonal system.

(b) Luzonite belongs to a compound intermediate between Cu_3AsS_4 and Cu_3SbS_4. In the series of Cu_3AsS_4-Cu_3SbS_4 the mineral enargite corresponds to Cu_3AsS_4, while luzonite is an intermediate compound which has a little more antimony than enargite. This conclusion is somewhat similar to the opinion of Harcourt (1937). Also, this case includes the solvus relation between enargite and luzonite of Skinner's diagram (1960) as well as of Feiss's (1974).

(c) Both enargite and luzonite are composed of the same elements, such as Cu, As, Sb, S, etc., but they differ in chemical formula and they belong to different crystal systems.

As recognized by Harcourt (1937), Pauling and Weinbaum (1934), Sawada (1944) and others, the relation of the crystal structures of enargite and luzonite is like that of sphalerite and wurtzite. As stated above, enargite and luzonite occur in some cases in epitaxy. Mitchell and Corey (1954) found the epitaxy between sphalerite and wurtzite referred to in the preceding chapter. The present writer describes the epitaxial relation between pyrite and marcasite in the foregoing chapter. It was generally said that the relation between pyrite and marcasite and between sphalerite and wurtzite is dimorphic—that is, the relation corresponds to case (a) in the above three. It is supposed that the relation between enargite and luzonite is the same as these.

Buerger (1934) maintained that pyrite and marcasite are not dimorphic in the true sense and that they differ in chemical composition: He stated that the chemical formula of pyrite is FeS_2 and that of marcasite is $Fe|_{S_{2-x}}^{Fe_x}|$, $x \fallingdotseq 0.004$. Scott and Barnes (1972) said that the fs_2 dependence of the inversion requires wurtzite to be sulfur-deficient ZnS_{1-x} relative to sphalerite $Zn_{1-y}S$ at a given temperature. These results correspond to (b) or (c). If the relation is the same, that of enargite and luzonite corresponds to (b) or to (c).

But the present results indicate that the antimony content in enargite is not necessarily lower than that in luzonite. So the relation might not correspond to (b). Also, it is not known which form is in a sulfur-deficient condition at a given temperature.

This problem is correlated with that of dimorphism in mineralogy, especially in the sulfide minerals. But if the definition of dimorphism is widely interpreted to include the relationship between pyrite and marcasite or between sphalerite and wurtzite, then a similar relationship would hold between enargite and luzonite.

Scott and Barnes (1972) discussed the non-stoichiometry of sphalerite and wurtzite, and stated that these phases are not polymorphs within the rigid definition of the term. They suggested that the definition of polymorphism should be broadened to include phases with a maximum difference in composition on the order of 1 or 2%.

The writer recognizes that a complete solid solution series exists between the tetragonal minerals luzonite $Cu_3As\,S_4$ and famatinite $Cu_3Sb\,S_4$, and that luzonite is dimorphous with the orthorhombic mineral enargite in the broader sense. The solubility of Sb in enargite is restricted, and the orthorhombic dimorph of famatinite has not yet been observed in nature. So the writer proposes that the names stibioenargite and stibioluzonite should be abandoned.

The exact identification of the minerals belonging to the luzonite and enargite groups must collectively depend on microscope, X-ray and electron microprobe studies. But a rapid distinction between luzonite and enargite can be made only by the microscope. The characteristic differences under the microscope are color and reflection pleochroism: Enargite is pinkish-grey, while luzonite is pink with a brownish tint. This difference is more clearly observed by means of the oil immersion lens. It has been said that the existence of lamellar twinning is characteristic of luzonite. But, as stated above, luzonite showed no twin in some parts of the section. The other properties of luzonite and enargite closely coincide with each other.

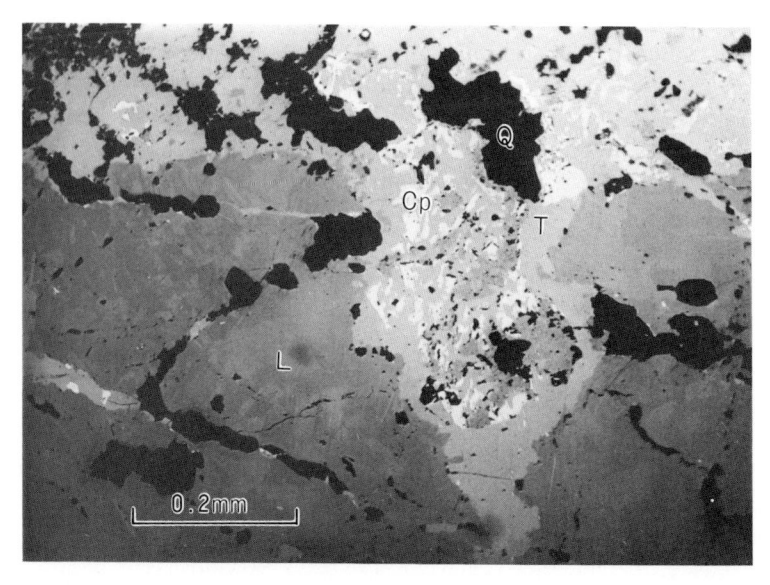

FIG. II–20–9. Photomicrograph. Polished section. One nicol. Aggregate of enargite and luzonite (L) is penetrated or replaced by veinlets consisting of quartz (Q), tennantite (T), and chalcopyrite (Cp).

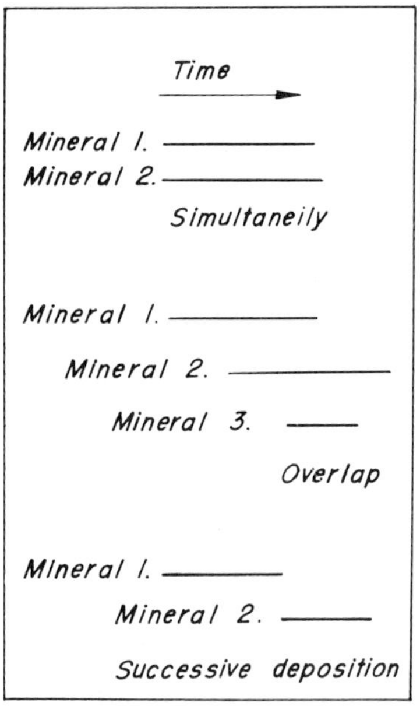

```
                Time
                ─────▶
Mineral 1. ─────────
Mineral 2. ──────────
         Simultaneily


Mineral 1. ─────────
   Mineral 2.  ─────────
      Mineral 3.   ──────
               Overlap


Mineral 1. ────────
    Mineral 2.  ──────
      Successive deposition
```

FIG. II–20–10. Diagram illustrating uses of the terms simultaneity, overlap and successive deposition (after Bastin et al., 1931).

As stated in the previous section, the epitaxial relation with enargite is recognized in luzonite but not in famatinite. This would be due to the fact that the antimony equivalent of enargite does not exist in nature, as Springer (1969) and Sugaki et al. (1976) recognized. This is an important point in discriminating between luzonite and famatinite under the microscope.

Luzonite is associated with enargite and pyrite, and they are penetrated by veinlets composed of tennantite, chalcopyrite and quartz (Fig. II–20–9). Examples of enargite penetrated by tennantite have been reported from the Butte mining district, Montana, the Cerro de Pasco mine, Peru, and other sites (Lindgren, 1927; McKinstry, 1963).

Under the microscope, in a sample from the Mankayan mine, the enargite and luzonite border is a mutual boundary, as has been described by Graton and Bowditch (1936) in samples from Cerro de Pasco and other sites, so the succession of mineralization is unknown in the exact sense. But the stage of mineralization would be the same for the two minerals, as was recognized by Frebold (1927). This does not necessarily mean that they were deposited simultaneously. It includes simultaneity, overlap and successive deposition (Fig. II–20–10). If the deposition is simultaneous, it may be conceivable that the nature of the ore-forming fluid is definite. If the deposition is overlapping or successive, a gradual change is conceivable.

Ramdohr (1969) maintained that luzonite was converted into enargite, for he observed that a part of the luzonite was transformed into enargite along the twinning planes, cracks and grain boundaries. But in the ore from the Mankayan mine, the writer did not observe any luzonite transformed along the twinning planes.

Some Genetical Problems of Enargite, Luzonite and Famatinite

Luzonite is found in epithermal deposits, accompanied by volcanic or pyroclastic rocks, and generally coexists with enargite. The deposits of the Mankayan, Teine and Chinkuashih mines are representative. In the Cerro de Pasco mine where luzonite and enargite coexist, cassiterite and wolframite are accompanied by low-temperature minerals such as aramayoite, polybasite, realgar, etc. (Petersen, 1965). A volcanic vent (complex explosion vent) which is closely related to the genesis of the ore deposit pierces the surrounding Meso-

TABLE II-20-3. Emission spectrochemical analyses of luzonite, famatinite, enargite and other sulfide minerals. (%)

	As	Sb	Bi	Ge	Ga	In	Sn	Ni	Co	Mo	Te	Ag	Mn	Cd	Fe	W
Famatinite, Chinkuashih	x~	x~	0.00x	0.00x	—	—	0.x~	0.00x	0.00x	0.00x	0.0x	0.0x~	0.000x	—	x~	—
Luzonite, Chinkuashih	x~x0	0.x~x	0.0x	0.00x	—	—	0.x	0.000x	0.000x	—	0.0x	0.00x	0.000x	—	0.x~x	—
Enargite, Chinkuashih	x~x0	0.x~x	0.000x	0.000x	—	—	0.00x	0.000x	—	—	0.0x	0.0x~	0.000x	0.00x	x~	—
Luzonite, Mankayan	x~x0	0.x~x	0.0x	0.0x	0.000x	—	0.0x	0.000x	0.000x	0.00x	0.0x	0.0x~	0.00x	—	0.x~x	—
Luzonite, Mankayan	x~x0	0.x~x	0.0x	0.0x	—	0.000x	0.0x	—	—	0.00x	~0.x	0.0x~	0.0x	—	x~	—
Enargite, Mankayan	x~x0	0x~x	0.00x	0.00x	—	—	0.00x	—	—	0.00x	0.00x	0.00x~	0.00x	0.00x	x~	—
Luzonite, Cerro de Pasco	x~x0	x~	0.0x	0.00x	0.00x	0.00x	0.x	—	—	0.000x	0.0x	0.0x~	0.0x	0.0x~	x~	0.0x
Famatinite, Teine	x~	x~x0	0.0x	0.00x	0.00x	—	0.x~x	—	—	—	0.00x	0.0x~	0.0x	0.00x	0.x~	—
Luzonite, Suttsu	x~x0	x~	0.00x	0.00x	—	—	0.x~	—	—	—	0.0x	0.00x	0.000x	—	0.0x	0.0x
Chalcopyrite, Osarizawa*	0.0x	0.00x~	0.000x	—	—	0.00x	0.000x	0.000x	—	0.00x	—	0.0x	0.0x	—	x-	—
Sphalerite, Hosokura*	0.0x	0.00x	—	—	0.00x	0.00x	0.00x	—	—	0.00x	—	0.00x	0.0x	0.x~x	x~x0	—
"Wurtzite," Hosokura*	0.00x	—	—	—	0.00x	0.00x	0.00x	—	—	0.00x	—	0.00x	0.0x	0.0x	x~x0	—
Galena, Hosokura*	0.00x	0.0x	0.00x	—	—	0.000x	—	—	—	0.000x	—	0.x	0.000x	0.00x	0.x~	—

* Refer to Chapters II-9 and -19. —: not detected.
Analyst: T. Kawashima, Central Research Laboratory, Mitsui Metal Mining and Smelting Co.

TABLE II-20-4. Se content in enargite, luzonite, famatinite and some other sulfides.

Mineral	Locality	Type of deposit	S content (%)	Se content (%)	Se/S (× 10⁻⁵)
Luzonite	Mankayan, Philippine	Subvolcanic (Epithermal, Xenothermal)	32	0.040	125
Enargite	ditto	ditto	32	0.050	156
Luzonite	Chinkuashih, Taiwan	ditto	32	0.014	43
Enargite	ditto	ditto	32	<0.001	3
Famatinite	ditto	ditto	30	0.013	43
Chalcopyrite	Osarizawa, Japan	Epithermal (Subvolcanic)	34	<0.001	2.9
Sphalerite	Hosokura, Japan	ditto	33	<0.001	3
"Wurtzite"	ditto	ditto	33	<0.001	3
Galena	ditto	ditto	13	0.007	53
Sphalerite, mill concentrate	Kamioka, Japan	Pyrometasomatic	33	0.004	12
Galena, mill concentrate	ditto	ditto	21	0.032	150

Analyst: M. Yoshida, Central Research Laboratory, Mitsui Metal Mining and Smelting Co.

zoic formation. It is composed of tuff breccia (agglomerate) and quartz monzonite porphyry. The deposit of the Tsumeb mine is very peculiar in its mineral assemblage. The deposit, which is genetically related to breccia pipe, is of meosthermal type. The ore contains galena, sphalerite, tennantite, chalcocite, bornite, digenite, enargite, plus subordinate pyrite, chalcopyrite, germanite, reniérite, molybdenite, wurtzite, luzonite, greenockite, stromeyerite, etc. (Park and MacDiarmid, 1964).

In the Teine mine, bismuth minerals exist as described above. Also, cassiterite, stannite and bismuthinite occur in the Suttsu mine accompanied by enargite and luzonite. Spectrochemical analyses of luzonite-group minerals show that the luzonite and famatinite are apt to contain comparatively large amounts of tin, bismuth, tellurium and germanium (Table II-20-3). In luzonite and enargite, the selenium content is comparatively high, as in the polymetallic xenothermal deposits (Tables II-20-4 and II-15-1). Uetani et al. (1966) described the selenian famatinite in gold ore from the Iriki mine, Kagoshima Prefecture. As described in the foregoing section, it is characteristic that the tellurium minerals are sporadically found in deposits producing enargite and luzonite. However, no tellurium minerals have been reported in polymetallic xenothermal deposits in Japan. Germanium is rich in enargite and luzonite, especially in luzonite. But indium is not so rich in these minerals, and no indium minerals have been reported in deposits containing enargite and luzonite. In the polymetallic xenothermal vein deposits, on the other hand, indium is concentrated in the vein materials.

These facts indicate that the deposits accompanying both enargite and luzonite have characteristics partly similar to polymetallic xenothermal deposits and partly different from them in some constituent elements, mineral assemblages and geological occurrences.*

The country rocks just adjacent to the ore deposit containing luzonite and enargite underwent kaolinization and silicification, as observed in Cerro de Pasco, Mankayan, Chinkuashih, Teine and other mines. According to Graton and Bowditch (1936), the ore-forming fluids at the start were alkaline at the Cerro de Pasco, but late in the period of mineralization in the upper reaches of the channelways, sulfuric acid was generated in the fluid by an

---

* According to Gains (1957), representatives of these minerals are found in low- to medium-intensity copper deposits, particularly in the former or in lower-intensity phases of higher-intensity deposits.

oxidizing reaction of water on some form of sulfur which the alkaline solutions had carried abundantly.

Enargite occurs in epithermal, mesothermal and even in pyrometasomatic deposits. The country rocks just adjacent to the ore deposits containing enargite without luzonite underwent sericitization and silicification, as observed in Butte, Montana (Sales, 1948) and Huanzala, Peru (Imai, unpublished).

As described above, in the Mankayan and Chinkuashih mines, the enargite-luzonite accompanied by pyrite is penetrated by tennantite accompanied by chalcopyrite (Fig. II-20-9). McKinstry (1963) recognized that replacement of enargite by tennantite would indicate a decrease in the activity of sulfur at any given temperature. In some deposits, such as those of the Chinkuashih, Hokuetsu and Akeshi mines, luzonite and enargite are accompanied by native sulfur. The writer concludes that the assemblage of enargite-luzonite-pyrite would be formed in a higher sulfur fugacity condition than that of tennantite-chalcopyrite, as shown in the following equation:

$$4 Cu_3As S_4 + 4 FeS_2 = Cu_{10}Fe_2As_4S_{13} + 2 CuFeS_2 + \tfrac{7}{2}S_2$$

enargite-luzonite + pyrite = tennantite + chalcopyrite + sulfur

Kozu and Takane (1938) observed that enargite is changed into tennantite and sulfur at 525°C. The writer also recognized a similar condition in luzonite (unpublished). So it is certain that the assemblage of tennantite-chalcopyrite is more stable at higher temperatures than that of enargite-luzonite-pyrite.

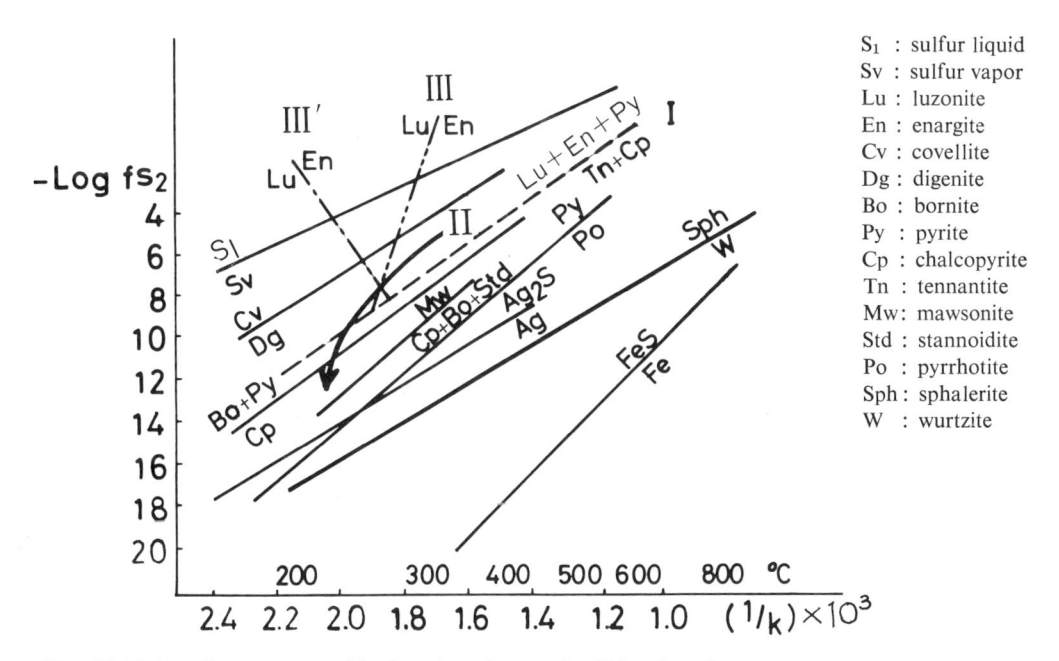

FIG. II-20-11. Temperature-sulfur fugacity relation of sulfide minerals.

From these facts, the relation between enargite (-luzonite)-pyrite and tennantite-chalcopyrite might be represented schematically by I in Fig. II-20-11. The mineralization in the Mankayan and Chinkuashih deposits would have proceeded along arrow line II. The f_{S_2}-T relation of sphalerite-wurtzite was studied by Scott and Barnes (1972), as shown in Fig. II-20-11. The relations of pyrite-marcasite and luzonite-enargite are undetermined. Marcasite is transformed to pyrite above 450°C. From the above discussion on the deficiency of sulfur in marcasite, it is supposed that pyrite would be stable under a high f_{S_2} condition: the univariant pyrite-marcasite boundary in the f_{S_2}-T diagram would decline with an increase in temperature.

The inversion temperature of luzonite to enargite was given as 275°C by Gains (1951)* and as 320°C by Barton and Skinner (1967). But in the case of luzonite-enargite, the exact chemical composition and thermochemical data (ΔH, etc.) are lacking. So the inclination of the boundary of two minerals in the f_{S_2}-T diagram might be considered in either of two cases III or III' as shown in Fig. II-20-11.

It is also likely that acidity would have a strong influence on the formation of luzonite.

Summary

In studies of enargite, luzonite and famatinite, the following three possibilities were investigated:

(a) Luzonite and enargite are in dimorphic relationship. The chemical formula is Cu_3AsS_4.

(b) Luzonite belongs to the compound intermediate between Cu_3AsS_4 and Cu_3SbS_4, containing a large amount of arsenic and a small amount of antimony. This luzonite differs from enargite (Cu_3AsS_4) in its crystal system.

(c) Luzonite and enargite are composed of the same elements, such as Cu, As, S, Sb, etc., but they differ in chemical formula and in crystal system.

It is probable that they have the same relation as that between pyrite and marcasite and between sphalerite and wurtzite. It has been generally accepted that the relation between pyrite and marcasite, or between sphalerite and wurtzite, is dimorphic, such as in case (a). The same would be true in the relation between enargite and luzonite.

But, this has already been questioned by Buerger (1934). He stated that the chemical compositions of pyrite and marcasite are not the same in the exact sense, suggesting the same relation in sphalerite and wurtzite. In this case, the relation between the above three pairs are in (b) or (c). It involves the definition of dimorphism in mineralogy, especially in sulfide mineralogy. If the definition is widely interpreted, as it generally is, to include pyrite and marcasite or sphalerite and wurtzite, enargite and luzonite would also be dimorphic. It is probable that the sulfur content is somewhat different in enargite and luzonite, resulting in (c) of the above three cases.

Scott and Barnes (1972) suggested that the definitions of polymorphism and polytypism should be broadened to include phases with a maximum difference in composition on the order of 1 or 2%.

The exact identification of the minerals belonging to the luzonite and enargite groups must be determined collectively by microscope, X-ray diffraction and electron microprobe (or microchemical) studies. But rapid discrimination between luzonite and enargite can be made only under the microscope. The characteristic differences are color and reflection pleochroism: Enargite is pinkish-grey, while luzonite is pink with a brownish tint. This color contrast is made more conspicuous with an oil immersion lens. It is said that the existence of lamellar twinning is characteristic of luzonite, but in limited parts of the polished section luzonite is not twinned. The other properties are quite similar to each other.

In luzonite, an epitaxial relation with enargite is sporadically recognized, as observed in pyrite and marcasite, but in famatinite it is not recognized. This is the important difference between the two minerals. So luzonite is discriminated from famatinite by the presence of epitaxy with enargite.

A complete solid solution series extends between the tetragonal minerals luzonite $Cu_3 As S_4$ and famatinite $Cu_3 Sb S_4$. The orthorhombic dimorph of famatinite is not yet known in nature. It is the reason why the epitaxy is not recognized in famatinite. In the writer's opinion, the names stibioenargite and stibioluzonite should not be used.

Enargite occurs in epithermal, mesothermal and even pyrometasomatic deposits; luzonite

---

* Unpublished datum. Refer to Maske and Skinner (1971).

is found mainly either in epithermal deposits or in high temperature-low pressure deposits due to shallow depth or to brecciation of country rocks in a moderately deep zone, and is generally accompanied by enargite. So the deposit containing both enargite and luzonite has some characteristics of xenothermal deposit.

In the Mankayan and Chinkuashih mines, it is observed under the microscope that the enargite-luzonite-pyrite assemblage is replaced by that of tennantite-chalcopyrite. The former would be deposited in higher sulfur fugacity than the latter. The mineralization sequence in these deposits would have proceeded towards lower sulfur fugacity and lower temperature.

II-21. Syntheses of the Cu-Fe-Sn-S Minerals (Lee *et al.*, 1974, 1975)

M. S. Lee, S. Takenouchi and H. Imai

Introduction

Stannite is a common Cu-Fe-Sn-S mineral in various ore deposits, but mawsonite and stannoidite characteristically occur in some xenothermal ore deposits such as the Ikuno, Akenobe and Tada, and Ashio deposits in Japan (refer to Chapter II–15). Occurrences of these two minerals are also reported in some pegmatitic and pyrometasomatic ore deposits in the world, for instance, in Australia by Markham and Lawrence (1965), in Portugal by Oen (1970) and in Canada by Petruk (1973).

Mawsonite and stannoidite are genetically associated with bornite and chalcopyrite, and mawsonite shows a replacement texture along the margins of stannoidite or the grain boundaries between stannoidite and bornite. The phase relations in the system Cu-Fe-Sn-S, especially at lower temperatures, are important for the elucidation of the genesis of the minerals, but studies on the syntheses of these two minerals are very limited. Therefore, this work on synthesis was carried out with particular attention to the assemblages among

TABLE II-21-1. Cell dimensions of stannite, stannoidite, and mawsonite (after Yamanaka and Kato, 1976).

Stannite	Stannoidite	Mawsonite
a = 5.461 ± 0.002 Å	a = 10.789 ± 0.009 Å	a = 10.745 ± 0.001 Å
	b = 5.413 ± 0.004	
c = 10.726 ± 0.007	c = 16.155 ± 0.009	c = 10.711 ± 0.006
I$\bar{4}$2m	I222	I$\bar{4}$2m, I4mm, I422, I4$_1$22

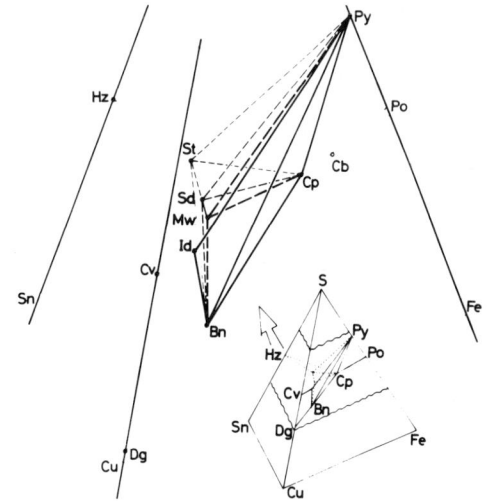

FIG. II–21–1. Schematic diagram of a portion of the tetrahedron Cu-Fe-Sn-S, showing mawsonite, stannoidite and stannite.
St: stannite, Sd: stannoidite, Mw: mawsonite, Id: idaite, Cv: covellite, Bn: bornite, Cp: chalcopyrite, Cb: cubanite, Po: pyrrhotite, Py: pyrite, Hz: herzenbergite, Dg: digenite.

bornite, chalcopyrite, stannite, stannoidite and mawsonite. The natural occurrences of the two minerals are described in Chapter II-15.

The composition of stannoidite has been reported as $Cu_5Fe_2SnS_8$ by Kato (1969) and that of mawsonite as $Cu_7Fe_2SnS_{10}$ by Markham and Lawrence (1965), but it is considered at present that their compositions are actually $Cu_8Fe_3Sn_2S_{12}$ and $Cu_6Fe_2SnS_8$, respectively (Springer, 1968; Petruk, 1973). Recently, Yamanaka and Kato (1976) have stated that the formulae for stannite, stannoidite and mawsonite might be expressed as $Cu_2{}^+(Fe^{2+}, Zn^{2+})Sn^{4+}S_4^{2-}$, $Cu_8{}^+Fe_2^{3+}(Fe^{2+},Zn^{2+})Sn_2^{4+}S_{12}^{2-}$ and $Cu_6{}^+Fe_2^{3+}Sn^{4+}S_8^{2-}$, respectively, and that Zn substitutes only for the ferrous ion. They also have studied the cell dimensions of these minerals, which are shown in Table II-21-1.

The schematic projection of the Cu-Fe-Sn-S tetrahedron shown in Fig. II-21-1 represents a general view of the compositional relations between mineral phases of the system.

Synthesis

Synthetic studies on the system Cu-Fe-Sn-S are few. Moh (1960) studied CuS-FeS-SnS by the flux method using sodium chloride, and reported fourteen synthetic phases including stannite, and later he carried out synthetic work on the binary system Sn-S (Moh, 1969). Moh and Otteman (1962) also investigated natural tetragonal stannite and experimentally found hexagonal, cubic and orthorhombic modifications. Franz (1971) worked on cubic stannite, and Lee (1972) studied the solid solution between stannite and sphalerite, while Bernhardt (1972) investigated the join stannite-chalcopyrite. Springer (1972) studied the pseudobinary system Cu_2FeSnS_4-Cu_2ZnSnS_4 at temperatures between 300°C and 1,000°C and found that stannite and kesterite phases are separated by an immiscibility region for temperatures below 680°C. Lee et al. (1975) studied the phase relations of stannite, stannoidite and mawsonite at 700°C, 500°C and 300°C, and discussed the genesis of these Cu-Fe-Sn-S minerals.

Synthetic methods: In the dry method, an ordinary rigid silica glass tube method was used. Synthesized materials were investigated by ore microscopy, X-ray powder diffraction and electron microprobe analysis. The flux method was used especially to synthesize the stannoidite and mawsonite phases at 350°C and 400°C. A mixture of 44 wt. % NH_4Cl and 56 wt. % LiCl was used as a flux. The weight ratio of flux to sulfide mixture was 2. A mixture of flux and sulfides of the desired composition was put in an evacuated silica glass capsule and kept at the desired temperature for one or two months.

The stannoidite and stannite phases were synthesized at 350°C and 400°C, but synthesis of mawsonite at 400°C was not successful. The product of synthesis of the mawsonite composition at 350°C contained a small amount of bornite, stannoidite and chalcopyrite, besides the mawsonite phase. The synthesized assemblage of stannoidite, mawsonite, bornite and chalcopyrite is very similar to the natural occurrence, and the synthesized mawsonite shows a well-developed twinning texture which is occasionally recognized in natural mawsonite.

Hydrothermal syntheses were carried out in ordinary test-tube-type pressure vessels

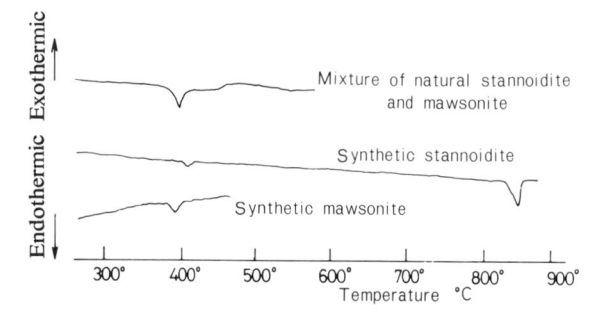

FIG. II-21-2. Differential thermal analyses of mawsonite and stannoidite.

with sealed collapsible gold capsules as sample containers, and in autoclaves with sealed pyrex glass capsules.

In the case of collapsible gold capsules, mixtures of copper, iron and tin sulfides which have compositions such as those shown below were generally used as the starting materials with 10 wt. % NaHCO$_3$ aqueous solution:

mawsonite, $Cu_6Fe_2SnS_8 = Cu_2S + 4CuS + 2FeS + SnS$
stannoidite, $Cu_8Fe_3Sn_2S_{12} = Cu_2S + 6CuS + 3FeS + 2SnS$
stannite, $Cu_2FeSnS_4 = 2CuS + FeS + SnS$

The internal pressure of the vessels was kept at about 600 bars by filling the vessels with appropriate volumes of water.

In the case of pyrex glass capsules, co-precipitated sulfide gel was used as the starting material. Cupric sulfate ($CuSO_4 \cdot 5H_2O$), ferrous ammonium sulfate ($Fe(NH_4)_2(SO_4)_2 \cdot 6H_2O$) and stannous chloride ($SnCl \cdot 2H_2O$) were used as copper, iron and tin sources, respectively. After mixing the three aqueous solutions, an excess amount of ammonium sulfide (($NH_4)_2S$) or sodium sulfide ($Na_2S \cdot 9H_2O$) was added to precipitate the sulfide gel. The sealed pyrex glass capsules were kept in an autoclave at water vapor pressure. The synthesized phases were checked only by means of the X-ray powder diffraction method.

Differential thermal analysis: A sample of 30 mg, sealed in a small evacuated silica glass capsule, was used for differential thermal analysis. The heating rate of the analysis was maintained at 2°C or 5°C per minute.

TABLE II–21–2. Electron microprobe analyses of heating products of natural stannoidite and mawsonite.

Heated miner.	Product	Composition	Cu	Fe	Sn
Stannoidite	Stannite	wt. %	31.83	9.16	25.64
		Atomic ratio	2.3	0.8	1.0
	Bornite	wt. %	56.84	13.81	0.51
		Atomic ratio	3.6	1.0	—
Mawsonite	Stannoidite	wt. %	39.37	12.33	15.09
		Atomic ratio	9.7	3.5	2.0
	Bornite	wt. %	55.93	12.67	0.69
		Atomic ratio	3.9	1.0	—

Analyzer: Shimazu ARL-SM; take-off angle 52.5°, specimen current 0.02μA, accelerating voltage 15 kV. Sulfur content was not determined. Synthetic stannoidite, stannite and bornite were used as standards. The stoichiometric compositions of stannoidite, stannite and bornite are (in weight percent): stannoidite: Cu = 39.17, Fe = 12.91, Sn = 18.29, S = 29.64; stannite: Cu = 29.57, Fe = 12.99, Sn = 27.61, S = 29.33; bornite: Cu = 63.22, Fe = 11.13, S = 25.55.

Endothermic peaks were recorded at 385°C for synthetic mawsonite, 400°C and 830°C for synthetic stannoidite, 915°C for synthetic stannite and 390°C for the mixture of natural mawsonite and stannoidite (Fig. II–21–2). The endothermic peak of mawsonite at 385°C probably represents decomposition to bornite and stannoidite, and those of stannoidite at 400°C and 830°C represent the phase transition to the high form and the decomposition to bornite and stannite, respectively. The peak of stannite at 915°C presumably represents melting. The endothermic peak of the mixture of natural stannoidite and mawsonite at 390°C approximately agrees with the peaks of synthetic stannoidite and mawsonite.

Heating experiments: Small samples having polished surfaces of about 9 mm^2 were sealed in evacuated silica glass capsules with powders of the same specimens. The powder reduces the loss of sulfur from the polished sections during the experiment. The polished sections were quenched after heating and checked under the microscope. The same polished section was heated repeatedly until some textural change was observed.

Natural mawsonite decomposed to bornite, stannoidite and chalcopyrite at 390°C. The

164   E. Genetical Problems of Sulfide Minerals

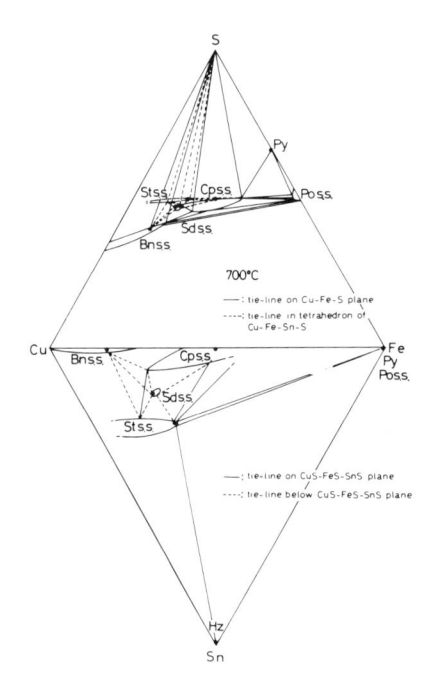

Fig. II–21–3. Phase relations in the system Cu-Fe-Sn-S at 700°C. (Abbreviations are the same as those in Fig. II-21-1.)

decomposed product of mawsonite was analyzed by electron microprobe, with the results shown in Table II–21–2. Chalcopyrite, however, was so small in amount that no analysis was practicable. As will be discussed later, the phase diagram indicates that the product from mawsonite at 500°C should be an assemblage of bornite, stannoidite, idaite and pyrite. However, idaite and pyrite were not observed in the decomposed product.

Synthetic zinc-free stannoidite decomposed to bornite, stannite and chalcopyrite at 830°C, while natural stannoidite containing 3 wt. % of zinc was stable below 500°C. It appears that the thermal stability of stannoidite is greatly affected by the zinc content. The compositions of bornite and stannite in the decomposed materials of stannoidite are shown in Table II–21–2. According to the phase diagram, the decomposed product of natural stannoidite at 500°C should have an assemblage of bornite, stannite, pyrite and idaite, but, as with mawsonite decomposition, pyrite and idaite were not recognized microscopically.

Experimental Results

Phase relations at 700°C: Solid solution phases of chalcopyrite, bornite, stannite, stannoidite and pyrrhotite, and herzenbergite exist stably at 700°C, as shown in Fig. II–21 3. On the projection of the Cu-Fe-Sn plane, the stannite solid solution extends from Cu_2FeSnS_4 to $Cu_{2.3}Fe_{0.6}Sn_{1.1}S_4$ and further to $Cu_{2.7}Fe_{0.2}Sn_{1.1}S_{4-x}$. Only the stannite solid solution crosses the CuS-FeS-SnS plane in the region between Cu_2FeSnS_4 and $Cu_{2.3}Fe_{0.6}Sn_{1.1}S_4$. The region of stannoidite solid solution is tentatively shown on the projection. Concerning the relation between solid solutions of chalcopyrite and stannoidite, it is supposed that the stannoidite solid solution extends further than that of chalcopyrite, because an exsolution texture of chalcopyrite in stannoidite is often recognized in synthetic products, indicating a wider range of stannoidite solid solution at high temperatures. The sulfur-rich side of each solid solution is connected with liquid sulfur at this temperature.

Phase relations at 500°C: The fields of solid solution of chalcopyrite, bornite, stannoidite and stannite are greatly reduced at 500°C with the first appearance of idaite, as shown in Fig. II–21–4. The assemblage stannite-pyrrhotite, which is stable at 700°C, be-

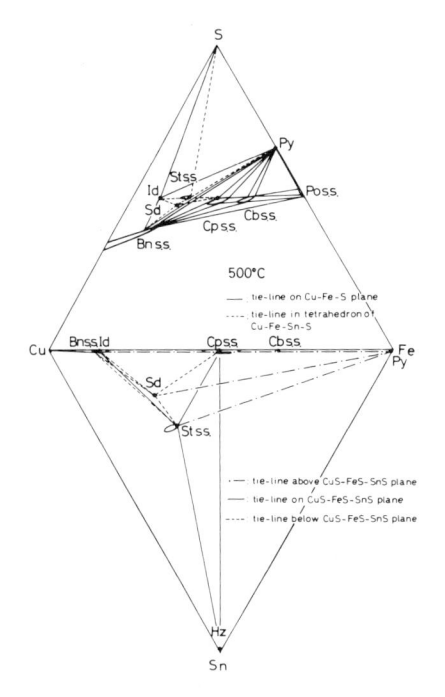

FIG. II-21-4. Phase relations in the system Cu-Fe-Sn-S at 500°C.

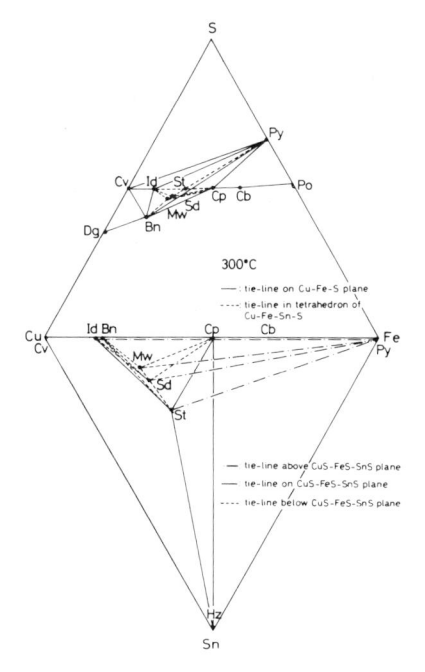

FIG. II-21-5. Phase relations in the system Cu-Fe-Sn-S at 300°C.

comes unstable at 500°C because of the appearance of the assemblage chalcopyrite-herzenbergite.

The principal invariant assemblages of the system at 500°C are:
stannoidite-stannite-chalcopyrite-pyrite
stannoidite-stannite-bornite-chalcopyrite
stannoidite-bornite-chalcopyrite-pyrite
stannite-chalcopyrite-pyrite-herzenbergite
stannoidite-idaite-bornite-pyrite
stannoidite-stannite-idaite-pyrite
stannoidite-stannite-idaite-bornite

Phase relations at 300°C: Phase relations in the system at 300°C are shown in Fig. II-21-5, which is drawn from the results of dry, flux and hydrothermal runs. Experiments were carried out for the ternary systems bornite-mawsonite-stannoidite, bornite-stannoidite-chalcopyrite, bornite-stannoidite-stannite and stannoidite-stannite-chalcopyrite at various temperatures. The remarkable difference between the phase diagrams at 500°C and those at 300°C is the appearance of mawsonite below 390°C. In general, stannoidite contains fine exsolved lamellae or specks of chalcopyrite, and chalcopyrite occasionally contains exsolved lamella of stannite. However, an exsolved texture of chalcopyrite in stannite has never been observed. Mawsonite can be synthesized at temperatures lower than 390°C, but it is always accompanied by a small amount of bornite and/or stannoidite and often occurs as fine networks in stannoidite.

The principal quaternary assemblages at 300°C, which are very similar to naturally-occurring associations in xenothermal ore deposits, are bornite-chalcopyrite-stannoidite-mawsonite and bornite-chalcopyrite-stannite-stannoidite.

It is also presumed that the following assemblages are stable: bornite-chalcopyrite-mawsonite-pyrite, bornite-digenite-stannite-stannoidite, bornite-digenite-stannoidite-mawsonite, chalcopyrite-stannite-stannoidite-pyrite and chalcopyrite-stannoidite-mawsonite-pyrite; while the assemblages bornite-chalcopyrite-stannoidite-pyrite and bornite-chalcopyrite-stannite-pyrite are unstable. It is known from phase diagrams that the tie-line between mawsonite and chalcopyrite pierces the plane bornite-stannoidite-pyrite, and that of stannoidite and chalcopyrite runs through the plane bornite-stannite-pyrite (Figs. II–21–1, and –5). Therefore, the above-mentioned two assemblages cannot exist at 300°C because the assemblages of chalcopyrite-mawsonite and chalcopyrite-stannoidite are stable.

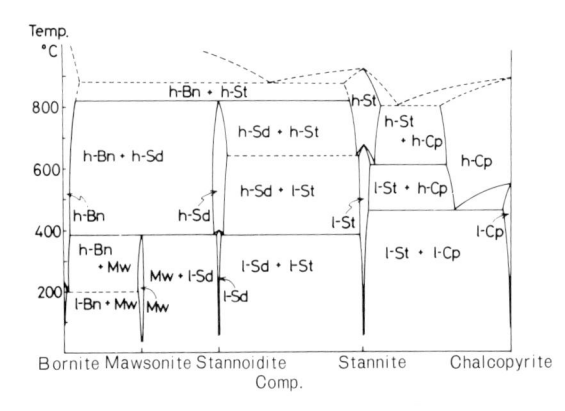

FIG. II–21–6. Pseudobinary system bornite-mawsonite-stannoidite-stannite-chalcopyrite join. (Data on the stannite-chalcopyrite join are taken from Bernhardt, 1972). Abbreviations are the same as those in Fig. II–21–1. Prefixes h- and l- represent the high-temperature form and low-temperature form, respectively.

Pseudobinary system bornite-mawsonite-stannoidite-stannite-chalcopyrite: The pseudobinary system along the join bornite-mawsonite-stannoidite-stannite-chalcopyrite is shown in Fig. II–21–6. The phase relation between stannite and chalcopyrite is plotted from data given by Bernhardt (1972).

From the pseudobinary phase diagram, it is presumed that mawsonite decomposes to bornite, stannoidite and chalcopyrite (or idaite and pyrite) at 390°C. In Fig. II–21–6, only the major phases of bornite and stannoidite are shown as the products of decomposition. It is also presumed that zinc-free stannoidite would have a phase transition at about 400°C, though the crystal system of the high form is not yet determined. The decomposition of zinc-free stannoidite to bornite, stannite and chalcopyrite occurs at 830°C, although only stannite and bornite are shown in the pseudobinary diagram. Heating experiments, however, showed that the zincian stannoidite with 3 wt. % zinc decomposed at about 500°C. Accordingly, the decomposition temperature of zincian stannoidite depends largely on the zinc content.

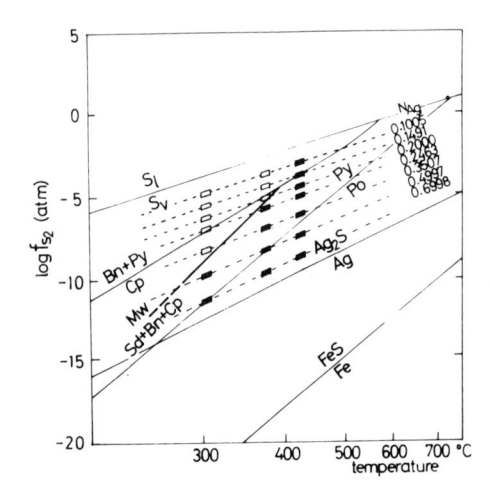

FIG. II–21–7. Sulfur fugacity-temperature diagram showing the stability curve of mawsonite.

Heating experiments with natural stannoidite and mawsonite were carried out by Oen (1970). He reported that mawsonite decomposed to bornite and stannite at 500°C, while stannoidite (called hexastannite in his paper) exsolved some bornite at 400°C and decomposed to isotropic stannite and bornite at 500°C.

Sulfur fugacity of the assemblage of bornite, chalcopyrite, stannoidite and mawsonite: Determination of the sulfur fugacity (f_{S_2}) of the univariant assemblage of bornite, chalcopyrite, stannoidite and mawsonite was carried out at 430°C, 370°C and 300°C by the electrum tarnish method developed by Barton and Toulmin (1964). Seven compositions of electrum were prepared having 0.1005, 0.1491, 0.2000, 0.2463, 0.3507, 0.4997 and 0.6998 mole fractions of silver.

All types of electrum were completely tarnished at 430°C. Electrum with a mole fraction higher than 0.2463 and that higher than 0.4997 were tarnished at 370°C and 300°C, respectively. Consequently, the stability line of mawsonite was established on the $\log f_{S_2} - T$ (1/T) diagram, as shown in Fig. II–21–7. The assemblage bornite-chalcopyrite-stannoidite is stable below the line while mawsonite is stable above it. The line is located between the stability lines of chalcopyrite and pyrite, and has a steeper slope than these two lines.

Discussion

The compositional point of mawsonite in the tetrahedron Cu-Fe-Sn-S is located in the sulfur-rich side of the plane bornite-stannoidite-pyrite and in chalcopyrite on the sulfur-poor side. Since the tie-line between mawsonite and chalcopyrite is stable below 390°C, the assemblage bornite-stannoidite-pyrite can be stable only above this temperature. The tie-line between pyrite and bornite on the Cu-Fe-S ternary diagram is stable at 500°C. Idaite coexists with pyrite and bornite on the sulfur-rich side of the tie-line, while chalcopyrite exists stably with bornite and pyrite on the sulfur-poor side (Yund and Kullerud, 1966). Therefore, the mawsonite point is included in the tetrahedron bornite-stannoidite-pyrite-idaite rather than in bornite-stannoidite-chalcopyrite-pyrite in the phase diagram at 500°C.

Furthermore, the mawsonite point in the tetrahedron bornite-stannoidite-pyrite-idaite is situated very close to the bornite-stannoidite-pyrite and bornite-stannoidite-idaite planes. This means that the major constituent minerals in the decomposed product of mawsonite are mainly bornite and stannoidite. The following equation is considered to be reasonable for the decomposition of mawsonite:

$$20(Cu_6Fe_2SnS_8) = 7(Cu_5FeS_4) + 10(Cu_8Fe_3Sn_2S_{12}) + 2(FeS_2)$$
$$\text{mawsonite} \qquad \text{bornite} \qquad \text{stannoidite} \qquad \text{pyrite}$$
$$+ (Cu_{5.5}FeS_{6.5}) + S_2$$
$$\text{idaite}$$

It was found experimentally, however, that mawsonite decomposes to stannoidite, bornite and chalcopyrite, for which the equation may be:

8 mawsonite = 3 bornite + 4 stannoidite + chalcopyrite + S_2.

In the case of stannoidite, synthetic zinc-free stannoidite exists as a stable phase up to 830°C but decomposes to stannite, bornite and chalcopyrite, releasing sulfur above this temperature. Natural stannoidite containing 3 wt. % zinc, however, decomposes at about 500°C. The tie-lines between chalcopyrite solid solution and liquid sulfur on the Cu-Fe-S system are stable at 700°C. Accordingly, the products of decomposition of zinc-free stannoidite would be bornite, stannite and chalcopyrite, according to the following equation:

$$4(Cu_8Fe_3Sn_2S_{12}) = 3(Cu_5FeS_4) + 8(Cu_2FeSnS_4) + (CuFeS_2) + S_2$$
$$\text{stannoidite} \qquad \text{bornite} \qquad \text{stannite} \qquad \text{chalcopyrite}$$

As the tie-line bornite-pyrite is stable at 500°C, chalcopyrite cannot coexist with liquid sulfur. Idaite, however, appears on the sulfur-rich side of the bornite-pyrite tie-line. The point corresponding to stannoidite composition is located on the sulfur-rich side of the bornite-stannite-pyrite ternary plane, and thus is situated in the tetrahedron bornite-stannite-pyrite-idaite, as in the case of mawsonite.

However, in the decomposition products of mawsonite and stannoidite, idaite and pyrite were not observed microscopically, and instead a small amount of chalcopyrite was identified. This discrepancy would be caused partly by temperature fluctuation in the experiments and partly by deviation from the stoichiometric composition of constituent phases. Idaite is stable below 501°C and decomposes to bornite, pyrite and sulfur above this temperature (Yund and Kullerud, 1966). This, however, does not explain the appearance of chalcopyrite in the products.

Deviation from the stoichiometric composition of the constituent phases may change the phase assemblage. This is especially significant in such cases, as the amount of some constituent phases is far larger than others. Since the decomposition products of mawsonite are mainly bornite and stannoidite, compositional deviations in the solid solutions of these two phases could have influenced the appearance of minor phases. As the mawsonite point is located very close to the bornite-stannoidite line in the tetrahedron bornite-stannoidite-idaite-pyrite, the loss of a small amount of sulfur from the assemblage, or an increase of sulfur content in the solid solutions of bornite and stannoidite, would form a new assemblage of bornite-stannoidite-chalcopyrite-pyrite, or even of bornite-stannoidite-chalcopyrite.

Naturally-occurring mawsonite often has a texture replacing stannoidite (Markham and Lawrence, 1965; Imai et al., 1975*). A similar texture was found in some synthetic products (Fig. II–21–8). The ratio of metal to sulfur of stannite, stannoidite and mawsonite is 1, 13/12 and 9/8, respectively, showing a successive decrease in sulfur. However, mawsonite decomposes to bornite, stannoidite and chalcopyrite, and it is stable under conditions of higher sulfur fugacity than that of the bornite-stannoidite-chalcopyrite assemblage. These facts suggest that natural mawsonite would have been formed by an increase of sulfur fugacity and/or lowering of temperature by replacing stannoidite at temperatures lower than 390°C.

Similarly, it is considered that stannoidite is stable under higher sulfur fugacities and/or at lower temperatures than stannite. Accordingly, it can be said that some ore deposits containing stannoidite and/or mawsonite have formed under higher sulfur fugacities than those having stannite, providing that temperature conditions were similar.

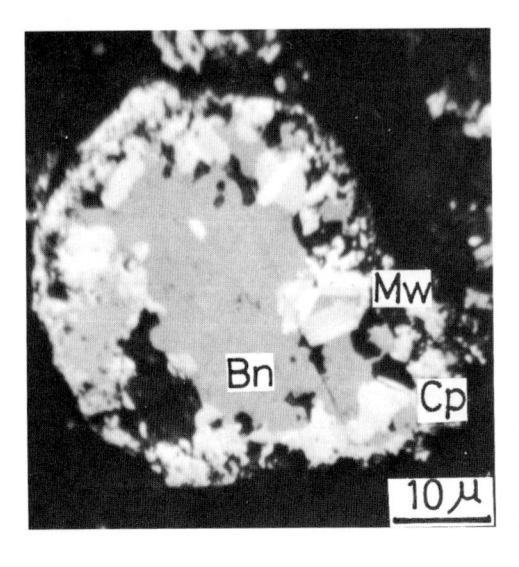

Fig. II–21–8. Photomicrograph. Polished section. One nicol. Synthetic bornite and twinned mawsonite. Synthesized at 350°C by flux method. Mw: mawsonite, Bn: bornite, Cp: chalcopyrite.

---

* Refer to Chapter II–15.

At present, it is unclear whether replacement of stannoidite by mawsonite is caused by a solid reaction between stannoidite and bornite or by reaction between stannoidite and hydrothermal solutions. Natural stannoidite generally contains zinc up to several percent, while the zinc content in natural mawsonite is very low. If mawsonite had replaced zincian stannoidite, it should be expected that sphalerite would be formed from the expelled zinc; otherwise the zinc content of stannoidite partly replaced by mawsonite is increased. However, neither of these processes has been observed in natural assemblages. It is probable that hydrothermal solutions played an important role in the formation of mawsonite and the transportation of sulfur and zinc.

PART III
PYROMETASOMATIC DEPOSITS

A. PYROMETASOMATIC DEPOSITS IN JAPAN AND KOREA

III-1. Introduction

H. Imai

In Japan and on the Korean Peninsula, there are many pyrometasomatic deposits. Most of them are genetically related to Mesozoic igneous activities. Country rocks of this type of ore deposit type in Japan are mostly limestone, mainly of the Permo-Carboniferous Chichibu Formation. The deposits discussed in the following chapters are all in limestone.

However, Shimazaki (1968b) recognized that there are skarn ore bodies in dolomitic marble in the Tsumo mine near the west end of Honshu Island. He found magesium skarn minerals, such as phlogopite, chondrodite, tremolite, etc.

Also, in the Chichibu Formation, bed-formed rhodochrosite deposits occur sporadically. In the Noda-Tamagawa district near the north end of Honshu, northeastern Japan, granitic rock occurs in the vicinity of a rhodochrosite deposit. Yoshimura (1952) and Watanabe *et al.* (1970b) recognized many manganese minerals such as hausmanite, braunite, tephroite, alabandite, etc. in this mine. This deposit may be a kind of pyrometasomatic type.*

On the Korean Peninsula, pyrometasomatic deposits occur in dolomitic limestone of Upper Proterozoic era. The Hol Kol mine, Suan (Suian) district, north Korea, studied by Watanabe (1943b, 1958), is characteristic; he described many magnesian skarn minerals such as forsterite, humite, kotoite ($Mg_3(BO_3)_2$), suanite ($Mg_2 B_2O_5$), warwickite (($Mg, Fe)_3Ti B_2O_8$), etc.

The occurrence of ore bodies is controlled by the shape of the limestone and the fissure pattern. The limestone occasionally yields a plastic deformation showing similar folding or diapirism at the time of folding, as discussed in Chapters III-2, -3 and -4. Fissure control is discussed in Chapters III-4, -5 and -6.

III-2. Kamioka Mine, Gifu Prefecture (Imai, 1963b)

H. Imai

This mine is situated in the Hida mountain area, central Japan (Fig. I-1). It is the largest

---
* Refer to Chapter VII-4.

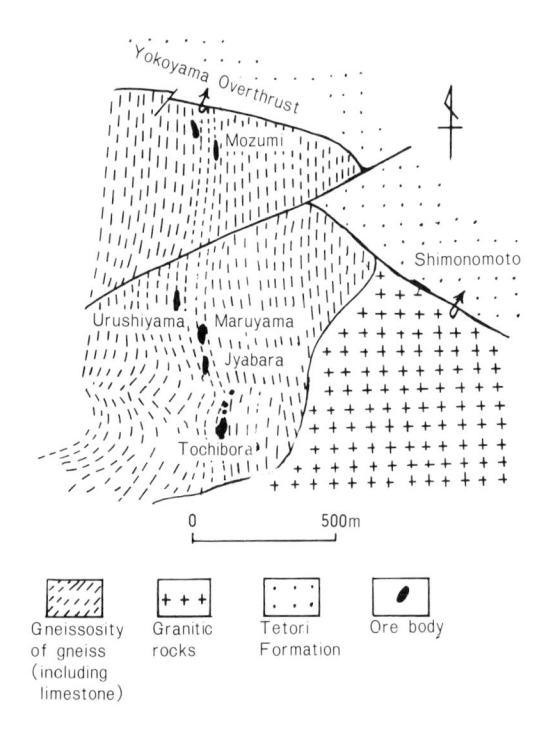

FIG. III-2-1. Geologic map of the Kamioka mine.

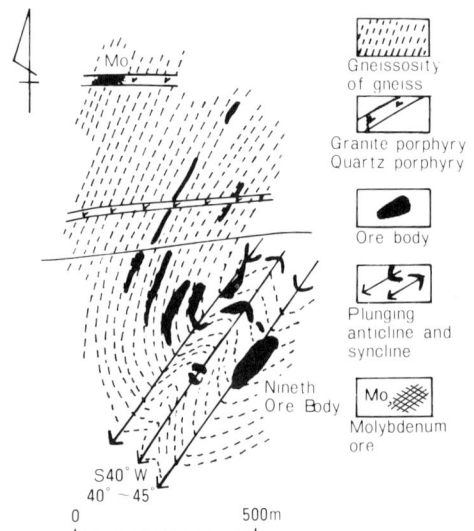

FIG. III-2-2. Ore bodies as related to the folded structures in the Tochibora area.

zinc and lead producer in Japan. This mine has produced 44 million tons of crude ore, containing Zn: 5.37%, Pb: 0.98%, Cu: 0.02% and Ag: 41 g/t.

Geologically, this district is composed of injection gneiss (Hida Gneiss), igneous rocks (granites), the Tetori Group (Jurassic system) and later acidic igneous rocks (including quartz porphyry). The injection gneiss in the Kamioka mining district is composed mainly of hornblende gneiss and is accompanied commonly by hornblende-biotite gneiss. In some places, small masses of amphibolite and metagabbro are found in the metamorphics. It is characteristic that Hida gneissic rocks are intercalated with a large number of limestone beds, as stated in Part I. Gneissic rocks are intruded by Funatsu granitic rocks, and both are unconformably overlain by the Tetori Group. In the northern or eastern parts of this area, gneisses and granitic rocks border the Tetori Group along an overthrust (Yokoyama overthrust) (Fig. III–2–1). The gneiss and granite have been thrust from south to north or from southwest to northeast over the Jurassic Tetori sediments. The overthrust trends roughly WNW-ESE and dips 30° to 60° S or SW.

Gneissic rocks, granites and the Tetori Group are penetrated by later igneous rock (quartz porphyry) which intruded after deposition of the Tetori Group and may be genetically related to the ore deposits of this district. Seki (1972) studied the geochronological age of porphyries in this district, using the Rb-Sr method, on whole rock, potash feldspar and plagioclase, and concluded that the absolute age of the quartz porphyry which accompanied the ore deposition is about 90 m.y.

The Kamioka mine includes several important groups of deposits (Fig. III–2–1), i.e., Tochibora, Jyabara, Maruyama, Urushiyama, Mozumi and Shimonomoto. These are, with the exception of the Shimonomoto belonging to a simple fissure vein in the overthrust, pyrometasomatic deposits which replaced limestone intercalated in gneissic rocks.

The typical pyrometasomatic deposit consists of masses of aggregates of irregularly oriented manganiferous hedenbergite crystals scattered with equigranular, black-colored grains of sphalerite and occasional grains of galena. In some cases, the aggregates of hedenbergite are accompanied by garnet, epidote, wollastonite, actinolite, ilvaite, etc.

One of the centers of mineralization is located in the area between the Todhibora and Maruyama areas (Figs. III–2–1 and –2), where dikes of granite porphyry and quartz porphyry are intruded. Molybdenite is observed in some of these dikes.

Other noticeable ore minerals are pyrrhotite, which is very abundant and characteristic of the Jyabara deposit; scheelite, which also occurs in the Jyabara deposit with quartz and epidote in hedenbergite skarn; chalcopyrite, which occurs sparingly but is widely distributed, mostly with galena or in some cases as stars or drops in sphalerite; hematite; tetrahedrite; matildite, which is the important silver-carrier of this deposit; argentite; pyrargyrite; jamesonite; native bismuth; electrum and finely crystalline graphite, which occurs in the crystalline limestone as well as in the skarn.

Recently Sasaki *et al.* (1975) recognized a new mineral, kamiokite ($Fe_2Mo_3O_8$), in the gneiss surrounding the porphyry dikes described above.

Magnetite which is not martitized occurs in the Maruyama deposit. The latest mineralization in the deposit is characterized by the common occurrence of druse, containing beautiful tetrahedral crystals of sphalerite as well as crystals of galena with etched surfaces, flat or lamellar calcite, apophyllite, quartz and hairy nests of native silver.

Iwafune (1952) recognized that the ore bodies occur in the folded limestones, and the large ones are concentrated along anticlinal or synclinal axes. They are elongated in the direction of plunges of the fold axes (Fig. III–2–2). In the Tochibora area, gneissic rocks and intercalated limestones are folded with axes plunging S40°W40°–45°. Limestones are thick at the axial parts of the folds and thin at the limbs. In other words, the fold belongs to similar fold. Drag folds are developed in limestone, and are especially concentrated in the axial parts of the folds. In the Tochibora area, nearly all the limestones are skarnized and mineralized. The Ninth Ore Body is the largest in the Tochibora group, being 260 m × 80 m in the maximum horizontal section and extending more than 600 m in the direction of plunge. It occurs at the anticlinal axis of similar folding, where the limestone which was thickened by the folding was almost entirely skarnized (Fig. III–2–2).

In the Mozumi area, many fissures extending in various directions are mineralized. Ore shoots exist at the junctions of limestones with the mineralized fissures.

The writers (Nishio *et al.*, 1953) studied fluid inclusion geothermometry by the decrepitation method. A two-phase fluid inclusion in hedenbergite is shown in Fig. III–2–3.

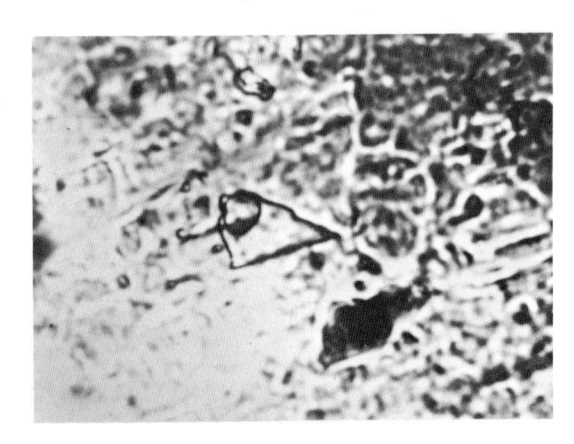

FIG. III-2-3. Photomicrograph. Fluid inclusion in hedenbergite (Nishio et al., 1953). × 660.

The decrepitation temperatures of the minerals in this deposit are 325°–200°C. In some parts the decrepitation temperatures do not reflect the order of deposition, but in general they decline progressively in the course of deposition.

Shiobara (1961) studied the decrepitation temperatures of the minerals from this mine. He stated that the decrepitation temperature of hedenbergite ranges from 455°C to 320°C. The area between the Tochibora and Jyabara deposits shows higher temperatures, where porphyry dikes run and molybdenum and tungsten minerals are found (Fig. III-2-1).

In the Shimonomoto area, granite and gneiss were thrust over the Tetori Group along the Yokoyama fault (Fig. III-2-1). The Tetori Group strikes nearly E-W and dips 20°–70° to the north. Dikes and sills of quartz porphyry and lithoidite trending E-W have intruded the granite and Mesozoic formation. Veins occur along the Yokoyama fault and paralel fissures in the Tetori Group. The veins trend N70°W and dip 50°–60°SW. The deposits are banded veins, consisting of quartz, sphalerite, galena, pyrite, arsenopyrite, jamesonite, pyrargyrite, tetrahedrite, native gold, etc.

In the outcrops of the skarn-type deposits, brownish or reddish residual soil is developed, which contains the metals shown in Table III-2-1. It is oxidized ore of zinc, lead and silver. The writer studied this soil by microscope, X-ray powder diffraction, electron microprobe and other methods, and recognized the minerals as shown in Table III-2-2. It is noteworthy that anglesite and smithonite have not yet been found.

TABLE III-2-1. Chemical composition of oxidized ore (rich ore).

Zn: 10–14 %	Ag: 300–400 g/t
Pb: 1– 4 %	SiO_2: 25–30 %
Cu: 1– 2 %	Fe_2O_3: 40 %

TABLE III-2-2. Minerals in the oxidized ore in the Kamioka mine.

Zinc minerals:	hemimorphite $Zn_4(OH)_2Si_2O_7 \cdot H_2O$
	sauconite $(Zn, Al, Mg, Fe)_3 (Al, Si)_4 O_{10}(OH)_2 \cdot Ca/2 \cdot K \cdot Na \cdot 4H_2O$
	chalcophanite $(Mn, Zn)O \cdot 2MnO_2 \cdot 2H_2O$
	hydrozincite $Zn_5(OH)_6(CO_3)_2$, aurichalcite $(Zn, Cu)_5(OH)_6(CO_3)_2$,
	veszelyite $7(Cu, Zn)O \cdot P_2O_5 \cdot 9H_2O$
Silver minerals:	argento-plumbojarosite $(Ag_2,Pb)Fe_6(SO_4)_4(OH)_{12}$, native silver(?)
Lead minerals:	cerussite $PbCO_3$
Copper minerals:	native copper (tin-bearing), malachite, azurite, chrysocolla, brochantite
Cadmium mineral:	greenockite (hawleyite)
Other minerals:	amorphous Zn-Pb-Cu-hydrous silica mineral, hisingerite or canbyite $(Fe_2O_3 \cdot 2SiO_2 \cdot n\ H_2O)$, hematite, limonite, gypsum, quartz, etc.

TABLE III-2-3. Formation of secondary minerals from primary minerals.

Primary minerals	Decomposed ion	Pyrite Fe^{3+}	Sphalerite Zn^{2+}	Galena Pb^{2+}	Ag^+	SO_4^{2-}	Chalcopyrite Cu^{2+}
Hedenbergite	Fe^{3+}	Hematite, Limonite		Zn-Pb-Cu-SiO$_2$-H$_2$O-	Native silver(?)	Argento-	Native copper
	SiO_3^{2-}	Hisingerite	Hemimorphite	amorphous mineral		plumbo jarosite	Chrysocolla
	Mg^{2+}	Quartz					
Garnet	Ca^{2+}		Sauconite				
Epidote	Al^{3+}						
	Mn^{4+}		Chalcophanite				
Calcite	CO_3^{2-}		Hydrozincite	Cerussite			Azurite
	HCO_3^-		Auricalcite				Malachite, etc.

The minerals in Table III–2–2 were formed by the decomposition of mainly skarn minerals and ore minerals, and by recombination of the elements.

In Table III–2–3, the first and second vertical columns on the left side show the kinds of leading skarn minerals and their decomposed ions, and the first and second horizontal lines in the upper part represent the kinds of ore minerals and their decomposed ions, respectively. By the respective combination of these ions in the vertical columns and horizontal lines of Table III–2–3, the supergene secondary minerals were formed.

From the results of experimental studies by Takahashi (1960) and Garrels and Christ (1965), the formation circumstances of the supergene minerals would be pH 6 ∼ 7.5, Eh + 0.6V ∼ − 0.05V and ΣCO_2 in water below 10^{-2} mole/l, at 25°C and where ΣS in solution is 0.1 mole/l (Fig. III–2–4).

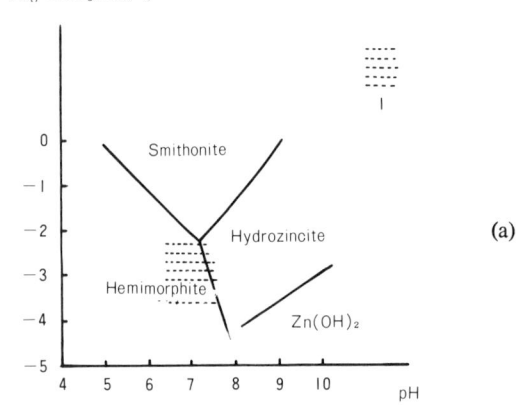

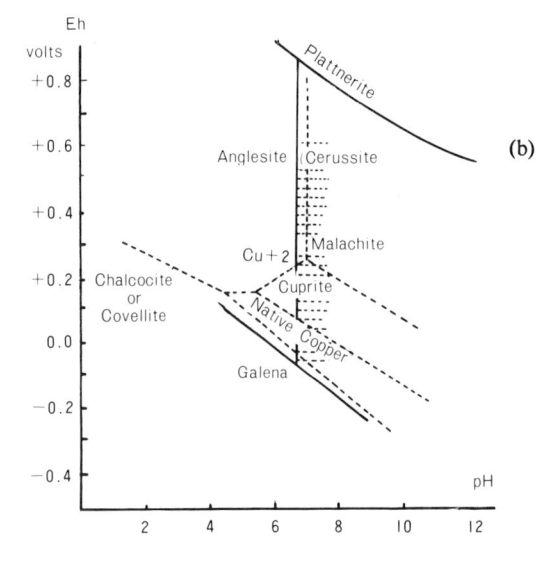

Fig. III–2–4. (a) Possible field of log ΣCO_2 mole/l-pH diagram at the outcrop in the Kamioka mine. (b) Possible field of Eh-pH diagram at the outcrop in the Kamioka mine. Copper: in broken line, Lead: in solid line.

1: possible field. 25°C at (a) and (b), $P_{CO_2} = 10^{-3.5}$ atm. and $\Sigma S = 10^{-1}$ mole at (b).

III–3. Nakatatsu Mine, Fukui Prefecture

H. Imai

This mine is situated in the district adjoining the western part of the Hida mountain area (Fig. I–3).

Part III: Pyrometasomatic Deposits

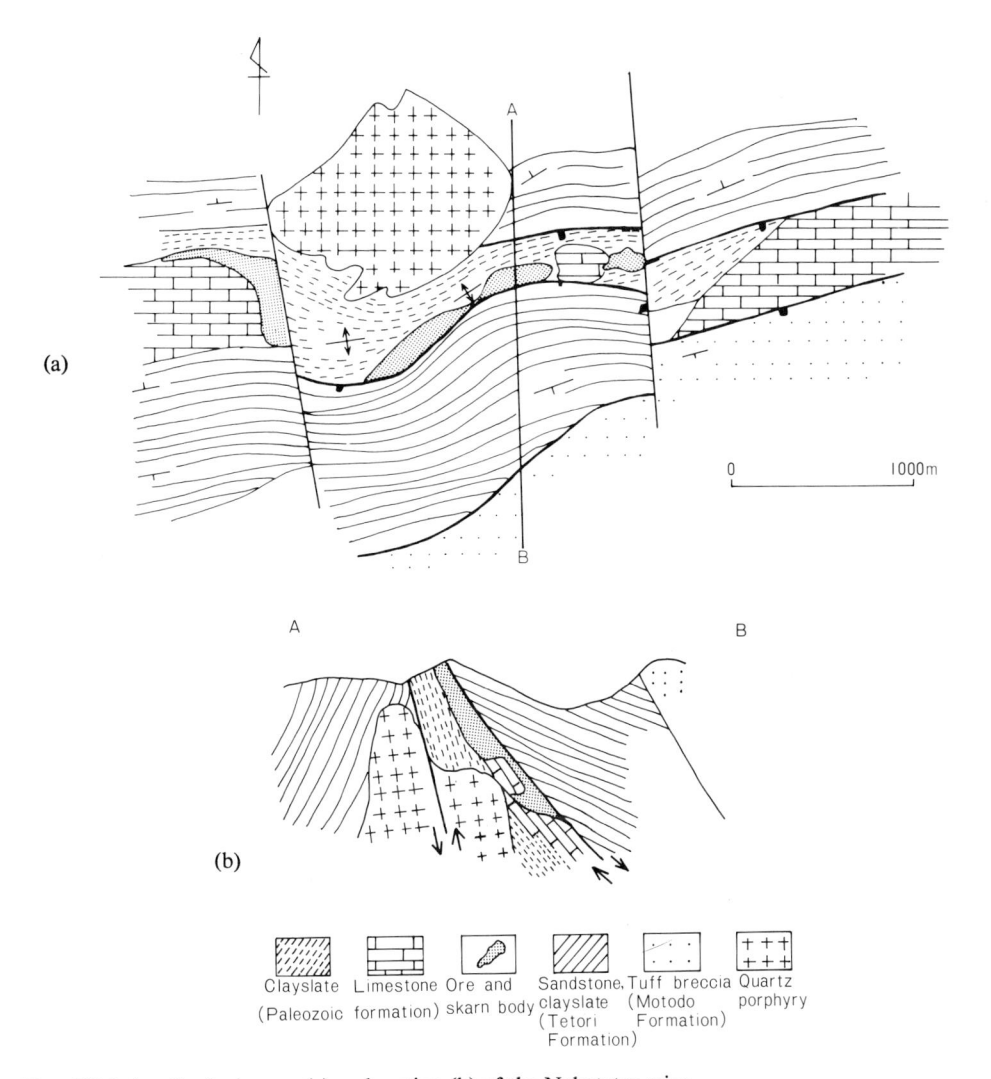

FIG. III-3-1. Geologic map (a) and section (b) of the Nakatatsu mine.

Clayslate (Paleozoic formation) | Limestone | Ore and skarn body | Sandstone, clayslate (Tetori Formation) | Tuff breccia (Motodo Formation) | Quartz porphyry

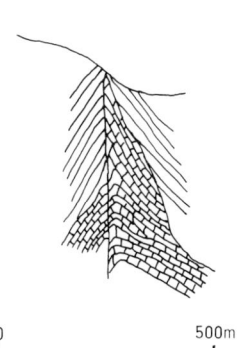

FIG. III-3-2. Diapiric limestone at the Nakatatsu mine. Rectangle: limestone of the Paleozoic formation, Hatch: shale of the Mesozoic formation.

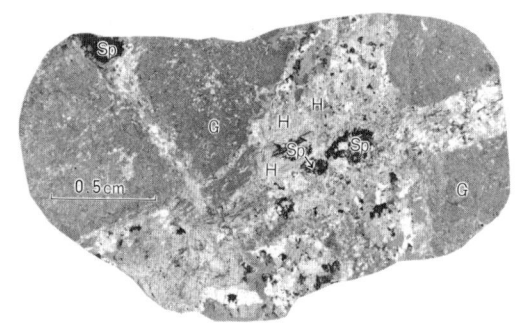

Fig. III-3-3. Close-up of thin section. Crossed polaroids. Andradite skarn (G) is penetrated by the aggregate of hedenbergite (H), calcite and sphalerite (Sp).

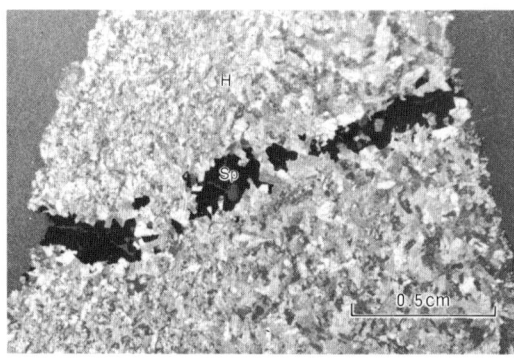

Fig. III-3-4. Close-up of thin section. Crossed polaroids. Hedenbergite (H) is penetrated by sphalerite (Sp).

It is worked for zinc and lead with a small amount of silver. Monthly production of crude ore is 40,000 t, containing 5.7 % of zinc, 0.45% of lead, 0.06% of copper, and 29 g/t of silver.

The mining region is geologically composed of Paleozoic (Permo-Carboniferous) sedimentary rocks, Tetori Jurassic Formation and Motodo undifferentiated Mesozoic Formation, which were intruded by quartz porphyry of the late Mesozoic or early Tertiary period. The sedimentary rocks generally extend E-W. They are folded with the axis stretching nearly E-W (Fig. III-3-1). The Paleozoic formation is composed of sandstone, clayslate, limestone, quartzite and schalstein, while the Jurassic Tetori Formation consists of conglomerate, clayslate and sandstone. Limestone is lacking in the Tetori and Motodo Formations. But in the mining area, limestone of Paleozoic origin was pushed into the Tetori Formation as a plug due to diapirism at the anticlinal part, just as observed in salt domes (Fig. III-3-2). To the south of the mining area, undifferentiated Motodo Formation occurs, bordering on the Tetori Formation with a fault. It is composed mainly of conglomerate and breccia.

An ore deposit of the pyrometasomatic skarn-type occurs in the Paleozoic limestone at the anticlinal part of the folding (Fig. III-3-1(b)).

Limestone was partly injected upwards, cutting through the anticline of the Paleozoic and Tetori formations. This is due to diapirism at the time of folding (Fig. III-3-1). This limestone was skarnized and mineralized.

A stock of quartz porphyry occurs in the midst of the ore deposit, stretching E-W.

The ore minerals are sphalerite, galena, chalcopyrite, pyrrhotite, pyrite, magnetite, molybdenite, scheelite, bismuthinite, etc. The skarn minerals are clinopyroxene (mainly manganoan hedenbergite), garnet (mainly andradite and a small amount of grossular), wollastonite with subordinate epidote, prehnite, fluorite, vesuvianite, quartz, calcite, etc. Tokunaga (1965) recognized small amounts of bustamite and ferro-johannsenite.

The hedenbergite skarn is widely developed, but in the lower levels of the adits andradite skarn becomes abundant. Andradite was replaced by hedenbergite (Fig. III-3-3).

Allen and Fahey (1957) stated that garnet was the first mineral to crystallize and was closely followed, or overlapped in time of deposition, by manganous hedenbergite at the Princess mine, New Mexico. But Loughlin and Koschmann (1942) considered that garnet formed later than hedenbergite at Linchburg Tunnel, New Mexico. Shimazaki (1969) recognized the replacement of clinopyroxene skarn by andradite skarn in the Yaguki mine. He discussed oxygen and sulfur fugacity in the course of the deposition of skarn and ore minerals in this mine. Morgan (1975) stated that magnesian clinopyroxene is followed by garnet coexisting with a more iron-rich clinopyroxene in the Mount Morrison Pendant, Calif. He studied oxygen fugacity in the formation of skarns in correlation with CO_2 gas dissociated from carbonate rock.

Concerning the Nakatatsu mine, zinc rich ore bodies exist in the hedenbergite skarn or in the transition zone between the hedenbergite and andradite skarns. The sphalerite was deposited at the same time as or a little later than hedenbergite (Fig. III–3–4), and was closely affiliated with this mineral (Fig. III–3–3), occasionally accompanied by calcite and fluorite. The andradite skarn is not favored with zinc ore (Figs. III–3–4). Zinc ore in andradite skarn is in most cases accompanied by a small veinlet of hedenbergite. Unlike in this mine, andradite is poor in the Kamioka mine. Graphite is widely found in the skarn in the Kamioka mine, while in the Nakatatsu mine it has not been observed. According to Burt (1972), a common association in many skarn zinc deposits in the southwestern United States, Mexico and other areas is manganiferous hedenbergite with sphalerite. He has interpreted that the retrogressive break-down of hedenbergite might tend to reduce through-going solutions and this would lower sphalerite solubility (for example, by converting SO_4^{2-} ions to S^{2-} ions), resulting in the preferential association of sphalerite with hedenbergite.

From the above facts, it might be inferred that skarnization began in an oxidizing condition depositing andradite, and transferred to a reducing condition depositing hedenbergite. The deposition of sphalerite would be closely related to the reducing condition.

Magnetite and pyrrhotite occur in the andradite skarn.

Molybdenite and scheelite are found in the skarn, especially near the intrusive body of quartz porphyry (Fig. III–3–1).

According to Shibata and Ishihara (1974), sericite from a later veinlet cutting the zinc-and lead-bearing skarn body in this mine is 60.1 m.y old by the K-Ar method.

III–4. Kuga, Fujigatani and Kiwada Mines, Yamaguchi Prefecture (Imai and Ito, 1959; Takenouchi and Imai, 1975)

H. Imai, S. Takenouchi and K. Ito

Kuga Mine

The Kuga tungsten and copper mine is located about 50 km southwest of Hiroshima City, southwest Japan (Figs. I–3 and III–4–1). It includes tungsten and copper deposits of two types: pyrometasomatic skarn-type and scheelite-quartz vein-type.

Geologically, this district is mainly composed of the Carboniferous formation consisting of clayslate, sandstone, quartzite and limestone. The formation displays a folded structure with the axis plunging N60°–70°W30° (Figs. III–4–1 and–2). The lineation (wrinkling) in clayslate is parallel to this direction.

Ore deposits occur along the anticlinal part (Fig. III–4–2). The limestone beds were deformed into a pipe-like mass elongated parallel to the anticlinal axis, and were concentrated at the anticline by the compressive force which caused similar folding (Figs. III–4–3 and –4).

Pyrometasomatic skarn-type deposits were formed by the replacement of limestone beds

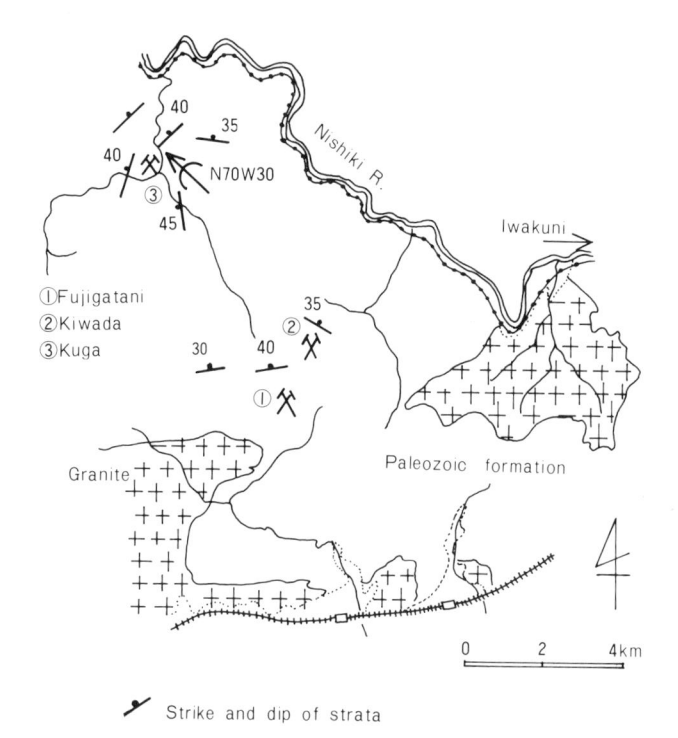

FIG. III-4-1. Index map of the Kuga mining area.

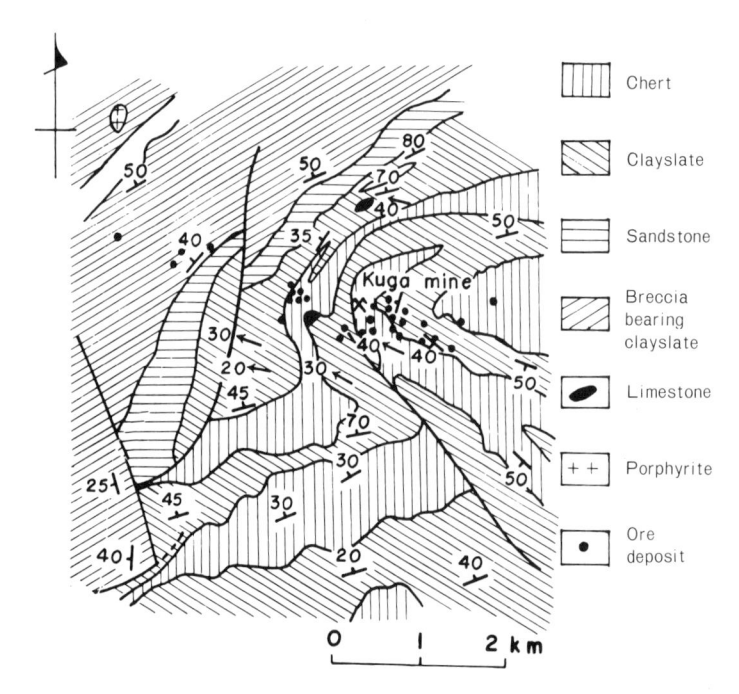

FIG. III-4-2. Geologic map of the Kuga mining area (after Kashiwagi, 1953).

with the mineralizing fluid which ascended through tension cracks. The tension cracks strike N20°–30°E, dipping 60°SE—that is, they are perpendicular to the folding axis of the Carboniferous formation (Figs. III-4-3 and-4).

These tension cracks would be formed by lateral pressure causing the folding and upheaval pressure due to the doming-up of magma, as discussed in Chapter II-4. Deposits of

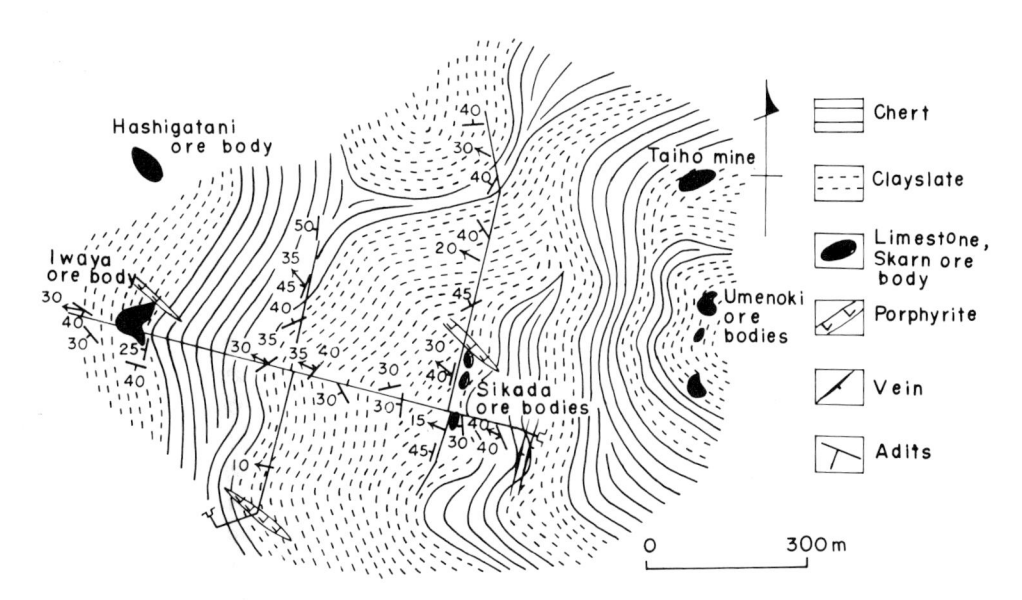

FIG. III–4–3. Geologic map of the Kuga mine.

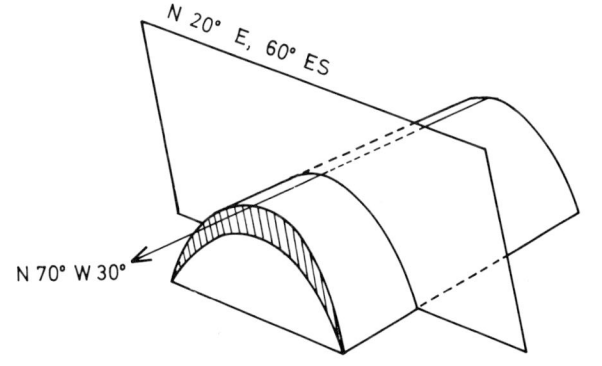

FIG. III–4–4. Folded structure and tension crack in the Kuga mine. The limestone (hatched part) becomes thick at the anticline, showing similar folding.

the scheelite-quartz vein-type filled these tension cracks. The cracks were also the paths of ore-forming fluid in the deposition of skarn-type deposits.

The locations of the ore deposits are controlled by the folded structure belonging to similar fold, the loci of the tension cracks and the shape of the limestone bed.

Granite occurs 7 ~ 8 km south and east of the Kuga mine (Fig. III–4–1). But, judging from the hornfelsic metamorphism of the Carboniferous formation near the ore deposits, a cryptobatholith exists underneath the anticlinal part of the formation.

Minerals of the pyrometasomatic skarn type are as follows:

Skarn minerals: grossular (Shimazaki, 1977), hedenbergite (Shimazaki, 1977), wollastonite, epidote, actinolite.

Ore minerals: scheelite, cassiterite, chalcopyrite, stannite, bismuthinite, native bismuth, sphalerite, galena, marcasite.

Gangue minerals: biotite, lithia mica*, albite, quartz, apatite, fluorite, calcite, chlorite.

Minerals of the scheelite-quartz vein type are as follows:

---

* LiO_2 content is 0.24%, analyzed by flame photometer in Geological Survey of Japan. $\gamma = 1.598$, $2V(-) = 5°–8°$.

Ore minerals: scheelite, chalcopyrite, pyrrhotite, arsenopyrite, pyrite.
Gangue minerals: quartz, biotite, muscovite, lithia mica, albite, apatite, beryl, calcite.

In most parts, the quartz veins include no albite, but in some parts they do. The latter may be regarded as pegmatitic veins. Both grade into each other.

The decrease of the filling and decrepitation temperatures of the minerals in paragenetic relation generally reflects the order of ore deposition, i.e., the highest is 308°C for hedenbergite by the decrepitation method, and the lowest is 160°C for quartz filling the skarn minerals by the heating-stage method.

Fujigatani and Kiwada Mines

These are situated about 6 km southeast of the Kuga mine. They are tungsten mines with almost no copper minerals.

The geologic structure is nearly the same as in the Kuga mine. The anticlinal structure with the same direction of the axis as in the Kuga area is rather flat and undulating. The Carboniferous sedimentary rocks underwent thermal effect and changed into hornfels. Limestone bodies which were deformed into pipe-like shapes generally plunge to N60°W, 30°, i.e., the direction of the axis of anticline. The ore-bearing skarn occurs at the periphery of the limestone body. According to Ito (1962), the arrangement of the zoned skarn in limestone is as follows, from the hornfels towards the limestone: (1) cordierite-biotite hornfels, (2) epidote-hornblende rock, (3) epidote-clinopyroxene (salite) rock, (4) garnet (grossular) rock, (5) hedenbergite rock, (6) garnet (grossular)-vesuvianite rock, (7) wollastonite rock, (8) limestone. Ito suggested that these zoned skarns represent a type of reaction zone between limestone and hornfels at elevated temperatures: Si, Al, Mg and Fe have been transferred towards the limestone, and Ca towards hornfels from limestone.

Scheelite-quartz veins occur in both mines, as in the Kuga mine. The strikes of scheelite-quartz veins are N20°–60°E, mostly N25°E, dipping E and W, but some of them are nearly flat-lying. They are all tension cracks.

The vein fissures dipping E are the same as in the Kuga mine, but some of those dipping W and lying flat would correspond to tension cracks due to the subsidence accompanying the solidification of magma, as discussed in Chapters II-1, II-3 and II-4.

In recent drilling it has been found that the depth of the granitic rock is about 300 m below the surface of the Fujigatani mine.

According to Shibata and Ishihara (1974), pegmatitic muscovite occurring in the scheelite-bearing skarn deposits yields ages of 92.1 m.y. at the Fujigatani mine and 95.8 m.y. at the Kiwada mine using the K-Ar method.

Fluid inclusions in quartz from veins and granite of the above drilling cores in the Fujigatani mine were studied microscopically. The fluid inclusions observed were all two-phase liquid inclusions. No cubic salt crystal was found in them. A few inclusions in quartz veins showed well-shaped negative crystals, while many of the inclusions found in quartz of granite were of irregular shape. In general, fluid inclusions in granite are very small, less than 10 microns, and are distributed along cracks or as clusters. From the shape and distribution of inclusions, it is considered that fluid inclusions in granite are secondary in origin.

The filling temperatures of inclusions in granite lie mostly in the range between 180°C and 280°C. No remarkable temperature difference was observed between samples from the top and the bottom of drilling cores in granite. Owing to the small size, the inclusions in granite are generally inadequate for the measurement of freezing temperatures. However, some data on salinity were gathered from larger inclusions. Most values of salinity lie in the range between 4 and 8 wt. % in NaCl equivalent concentration.

The filling temperature and salinity of fluid inclusions in quartz veins of the Fujigatani and

Kiwada mines are shown in Chapter VII–3. No remarkable difference is found in temperature and salinity between the inclusions in quartz veins and those in granite. The distribution of measured values is nearly parallel or slightly inclined to the salinity axis. This suggests that the temperature of hydrothermal fluids was kept fairly constant, but that salinity varied during the hydrothermal activity.

The similarity in the temperature and salinity of fluid inclusions in quartz veins and granite suggests that the fluids which permeated the granite body and those which formed quartz veins were probably the same, though the relation between quartz veins and granite has not yet been determined in the field.

III–5. Kamaishi Mine, Iwate Prefecture

H. Kaneda, T. Shoji and H. Imai

The Kamaishi copper and iron mine is located 20 km west of Kamaishi harbor, north Japan (Fig. I–3). It is said that this mine was opened early in the seventeenth century. For the past several years, its annual production of iron-copper crude ore has been about 1,500,000t.

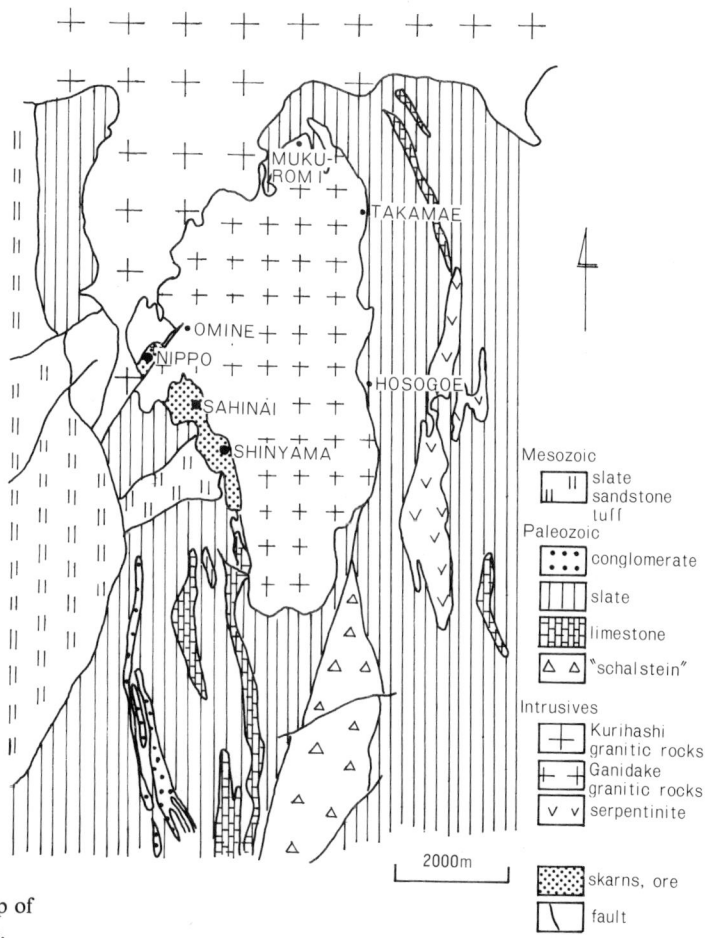

FIG. III–5–1. Geologic map of the Kamaishi mining district.

The grade ranges from 26 to 28% iron and about 0.6% copper. Besides, the mine produces 500,000 t of crude copper ore a year. The grade is from 1.5 to 2.5% copper.

Geologic Setting

This district is composed of non-metamorphosed and thermal-metamorphosed sediments of Paleozoic and Mesozoic age and intrusives of Late Cretaceous age (Fig. III–5–1).

The sediments of this district form an anticlinorium structure trending in a N-S direction. The Paleozoic sediments, ranging from Carboniferous to Permian age, consist chiefly of slate and "schalstein". Sandy slate and limestone are intercalated in slate, and some thin conglomerate beds are also found. The Paleozoic sediments generally strike from N-S to N30°W and dip about 60°–80°E or W. The Mesozoic sediments corresponding to Cretaceous age are distributed at the western limb of the anticlinorium in this district. They border on the Paleozoic formation with a fault which trends N-S and dips steeply W. The Mesozoic sediments strike between N-S and N20°W, dipping about 60°W. They consist mainly of fine-grained sandstone including sandy slate, slate, rhyolitic or andestic tuff and "schalstein".

Intrusive rocks in this district are prominently granitic rocks including granodiorite, diorite and diorite-porphyrite, and subsidiarily gabbro, quartz-monzonite and dikes of lamprophyre, pegmatite, porphyrite and aplite. They intruded after the prelusive activity of serpentinite. Two distinct intrusive bodies of granitic rocks are observed in this mining district. One is Ganidake granitic rocks, which form the core of the anticlinorium, and the other is the Kurihashi granitic rocks. The strikes of dikes generally range from N50°E to EW and their dips range from 60°SE-S to almost vertical. All the intrusions of these rocks may have occurred in Late Cretaceous age. The absolute ages of Ganidake and Kurihashi granitic rocks are 109 m.y. and 120 m.y., respectively (Takeuchi, 1967).

Ore Deposits

General remarks: The Kamaishi ore deposits are located along the boundaries between the sediments (especially limestone) of Carboniferous age and the Ganidake granitic rocks of Cretaceous age. Two ore zones are known at the west and east sides of the Ganidake granitic rocks, i.e., the West Ore Zone and the East Ore Zone. The ore deposits in the West Ore Zone, which consist of the Shinyama iron and copper, Sahinai iron, Omine iron and Nippo copper ore deposits have far larger amounts of the proved and probable ores than those of the East Ore Zone including the Hosogoe iron, Takamae iron and Mukuromi copper and iron ore deposits (Fig. III–5–1).

The Ganidake granitic rocks are genetically related to the formation of ore deposits. Though the Kurihashi granodiorite is older than the Ganidake granitic rocks in the absolute age as described above, from various field evidences the Kurihashi granodiorite is the post-mineralization intrusive. Two kinds of dikes are observed in the Shinyama ore deposit. One is a dike very closely related to sulfide mineralization and is assumed to have intruded at the later stage of mineralization of the Shinyama deposit, cutting early garnet and epidote skarns, and the other is post-mineralization. The post-mineralization intrusives, except for Kurihashi granodiorite, are lamprophyre, aplite, pegmatite, porphyrite and gabbro. They have not suffered skarnization or alteration.

The Kamaishi deposits occur in large masses of skarn. The skarn consists mainly of garnet and clinopyroxene. The garnet skarn is distributed in or close to the Ganidake granitic rocks side, while the clinopyroxene skarn is on the limestone side. The copper ore bodies are closely associated with clinopyroxene skarn, while iron ore bodies are with garnet skarn.

The Shinyama ore deposit is the largest in this mine and is composed of iron and copper ore bodies in the *massive skarn*.

An underground geologic map of the 450m level is shown in Fig. III–5–2, and a sche-

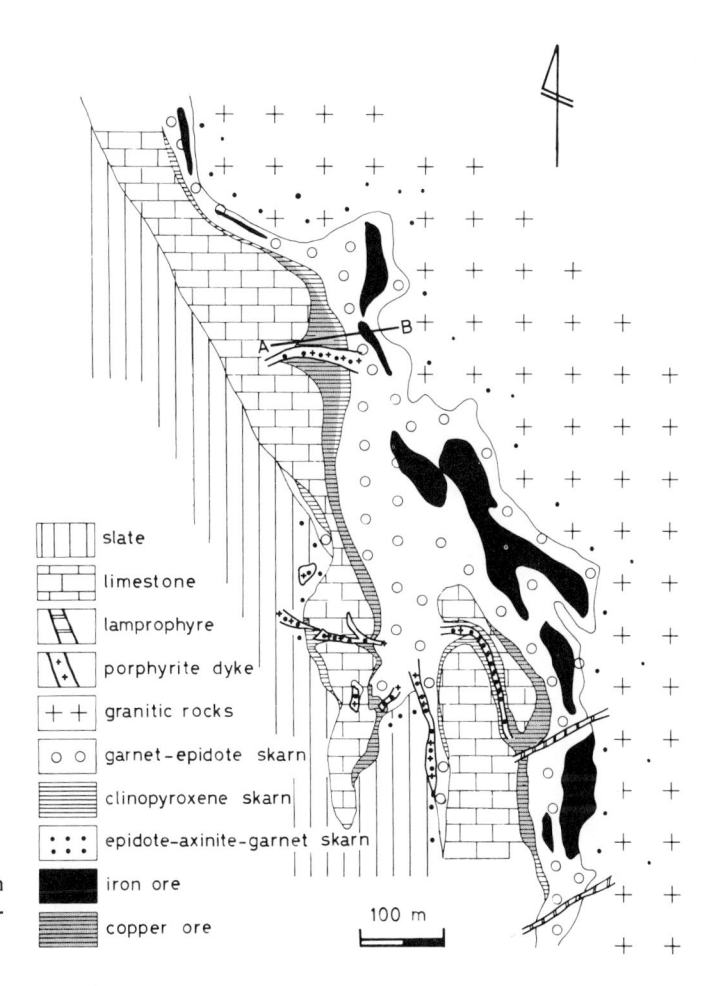

FIG. III-5-2. Geologic plan map of 450m level of the Shinyama ore deposit.

matic section of A–B in this map is exhibited in Fig. III–5–7(a). The iron ore bodies occur as irregular massive bodies extending vertically. The marginal part of the Ganidake granitic mass becomes diorite or porphyrite, bordering on ore bodies. It is skarnized to garnet, epidote and axinite, and sometimes altered to chlorite and biotite. The copper ore bodies occur as tabular masses adjoining to limestone, predominantly along dike rocks.

While, to the west of the Omine iron deposit (Fig. III–5–1), the Nippo copper deposit is composed of the pipe-like ore bodies in hornfels, and the flat ore bodies in limestone margin. An underground geologic map and a combined cross section along A–B and A′–B′ of the Nippo deposit are shown in Fig. III–5–3(a), (b). The Nippo deposit consists of five ore bodies. The ore bodies of D_1, D_2, and D_3 occur as skarnized breccia pipe in hornfels, not in limestone. While those of D_4 and D_5 occur in limestone in the same occurrence as in the Shinyama copper ore bodies, showing flat and lenticular shapes in contact with limestone. The skarnized breccia pipes consist of rounded or semiangular breccia fragments of hornfels rimmed by K-feldspar which is cemented by matrices consisting of clinopyroxene with small quantities of copper ore (chalcopyrite, pyrrhotite and cubanite), garnet, wollastonite, calcite and sphene (Fig. III–8–1 (b)). The breccia pipe ore body horizontally occupies 120 m × 60 m in area, and extends vertically 400 m. It occurs as a column or pipe cutting the hornfels. Though the breccia pipe forms an ore body, the productive ore is in the marginal parts, showing a ring-like shape in the horizontal section. This breccia-bearing skarn in breccia pipe is called breccia-skarn. The brecciated

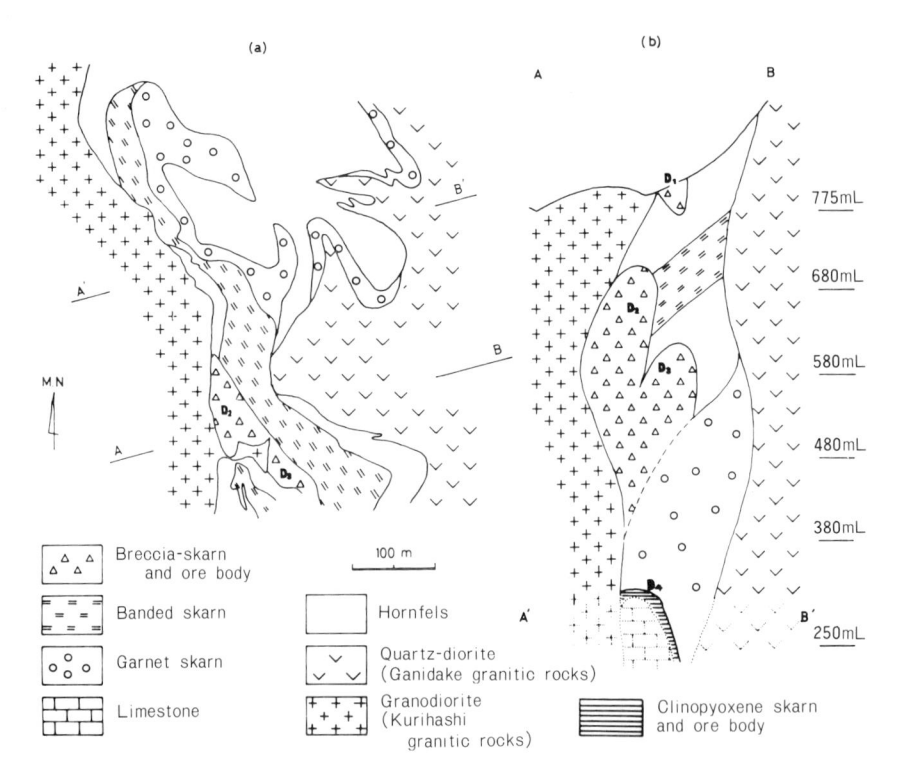

FIG. III–5–3. (a) A geologic map of the 680m level in the Nippo ore deposit. (b) A compiled section along A–B and A'–B' in (a) (after Shoji, 1972).

structure of the hornfels becomes indistinct at lower levels, and changes gradually to the *massive* garnet skarn. The genesis of the hornfels breccia has been discussed by several authors (Takeuchi and Yamaoka, 1965; Kano, 1965; John, 1968). According to Takeuchi and Yamaoka, and John, brecciation of hornfels would be accompanied by granodiorite intrusion. The present writers agree with this. In this case, lime in skarn surrounding the hornfels breccia might have been supplied from the underlying limestone.

Skarns: The *massive skarn* in the Kamaishi ore deposit (especially in the Shinyama ore deposit) consist of the following minerals: garnet, clinopyroxene, epidote, axinite, tourmaline, magnetite, and ludwigite.

The original rocks of *massive skarn* are considered to be limestone, slate, diorite, porphyrite and other rocks. The successive zones of skarns are found from the intrusive rock side towards the limestone in this district (especially in the Shinyama ore deposit), that is, 1) epidote-axinite-tourmaline-garnet zone, 2) garnet-epidote zone including iron ore (bearing small amount of ludwigite), 3) clinopyroxene zone including copper ore (Fig. III–5–2).

Garnet is widely developed in the marginal part of the Ganidake granitic intrusives and in the adjacent limestone, and clinopyroxene is mostly in the limestone (Fig. III–5–2). Epidote and axinite occur in the intrusives and slate. Tourmaline, apatite and sphene are abundant in the intrusives, while ludwigite is in magnetite ore bodies with calcite, tourmaline and small amounts of apatite, chalcopyrite, chlorite, diopside, pyrrhotite and sphalerite. Locally, ludwigite is seen to cut through garnet skarn along its joints (Tsusue, 1961). The garnet in the Shinyama ore deposit is classified into two types, i.e., earlier-stage garnet and later-stage garnet. Megascopically the later one is more reddish in color, but both are nearly the

Part III: Pyrometasomatic Deposits

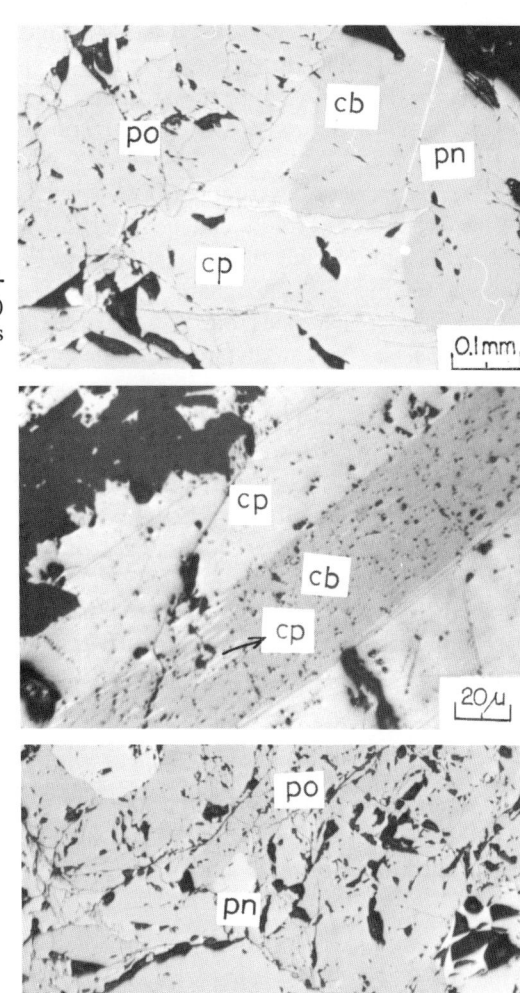

FIG. III-5-4. Photomicrograph. One nicol. Chalcopyrite (cp), pyrrhotite (po) and cubanite (cb) occur in mutual boundaries. Pentlandite (pn) is in the rim of pyrrhotite.

FIG. III-5-5. Photomicrograph. One nicol. Chalcopyrite contains cubanite exsolution lamellae with exsolved chalcopyrite.

FIG. III-5-6. Photomicrograph. One nicol. Exsolved pentlandite with bladed form occurs in cubanite matrix.

same in chemical composition. According to X-ray powder diffraction, compositions of both types of garnet range from 60% to 35% in andradite molecules.

On the other hand, in Fig. III-5-3, *banded skarn* is drawn. It is free from ore minerals. It may be called a kind of thermal metamorphosed rock showing a rhythmic structure; i.e., hornfels—feldspar band—diopside band—feldspar band—hornfels (Fig. III-8-1). Each band is 5-50 cm in width. The band coincides with the original bedding of the sedimentary rocks (refer to Chapter III-8).

Parageneses of Sulfides: In the Shinyama copper ore bodies, principal sulfides are chalcopyrite, pyrrhotite and cubanite. Sphalerite occurs in small amount associated with these three minerals. Pentlandite and mackinawite are sporadically found in the above-mentioned three principal minerals. The other sulfides in minor amounts are Ag-pentlandite (Mariko *et al.*, 1973, 1974), pyrite, siegenite, bornite, millerite, chalcocite, arsenopyrite, and molybdenite. The occurrence of molybdenite is restricted to aplitic dikes of post-mineralization.

Chalcopyrite and pyrrhotite are distributed widely throughout all the copper ore zones, while pyrite is restricted in the garnet side and cubanite in the limestone side where is the richest zone of sulfide ore bodies (Fig. III–5–7). Three principal minerals occur on mutual boundaries (Fig. III–5–4) or in exsolution relations (Fig. III–5–5).

The zonal distribution of Cu-Fe-S minerals is recognized in the copper ore body (①–④ in Fig. III–5–7).

The three phases of pyrrhotite (monoclinic, hexagonal and troilite) occur in the Shinyama copper ore bodies. Troilite can be discriminated from hexagonal pyrrhotite by etching with HCl solution and the distinction between hexagonal and monoclinic phases can be made by the method of magnetite colloid adherence and by X-ray powder diffraction.

In the copper-rich zone (zone ③ or ④ in Fig. III–5–7), a small amount of very fine-grained minerals such as pentlandite, mackinawite and Ag-pentlandite usually exists. Microscopically, pentlandite usually occurs as irregular coarse- and fine-grains in the margin of pyrrhotite grains (Fig. III–5–6). Unmixing intergrowths of pentlandite with graphic or needle form are also observed in cubanite (Fig. III–5–4), pyrrhotite and very often in chalcopyrite.

Ag-pentlandite is a very rare mineral and optically very similar to bornite. This mineral is easily tarnished in air. It occurs only in chalcopyrite grains and is less than 0.05 mm in diameter. Some grains are surrounded by mackinawite. Mackinawite always exists only in chalcopyrite grains. The hair-like curving texture of this mineral, as if it were derived from pyrrhotite, is very commonly observed in the margin of chalcopyrite grains bordering on pyrrhotite grains, and an oriented fine graphic texture is also found within chalcopyrite grains. The data from chemical analyses of mackinawite reported by John (1968) shows that it includes Ni and Co.

Genetical Remarks

The zonal distribution of copper ore and skarns including magnetite from the intrusives to the limestone side is observed as shown in Figs. III–5–2 and –7.

From the mode of occurrence of skarns and ores which reveals continuous changes, it

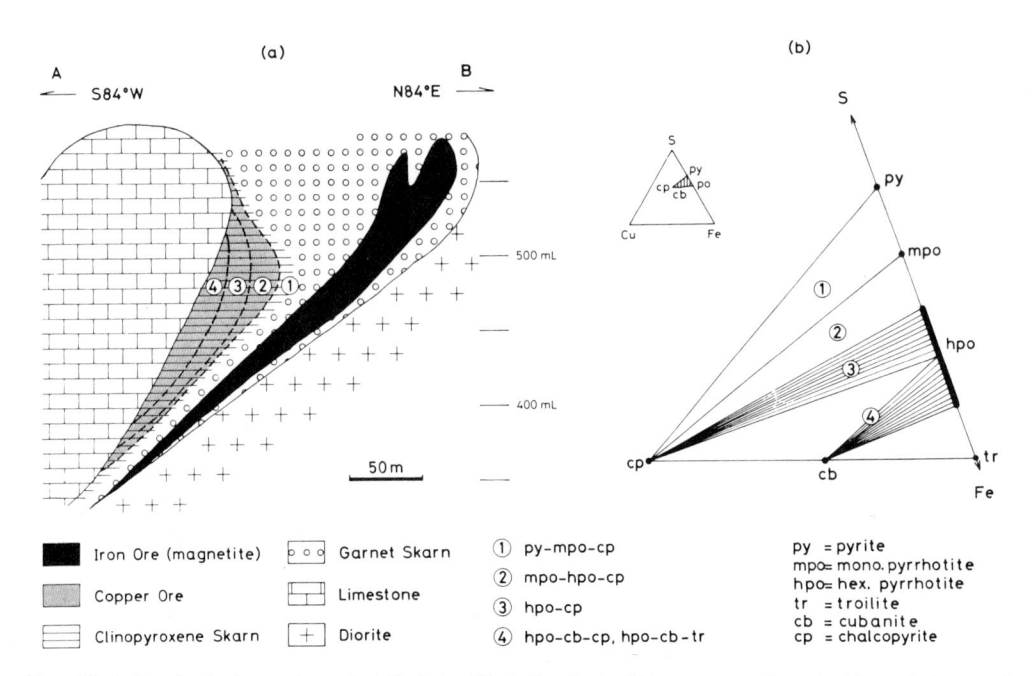

FIG. III–5–7. Geologic section of A-B (Fig. III–5–2) of the Shinyama ore deposit (a), and the zonal distribution of sulfides (b).

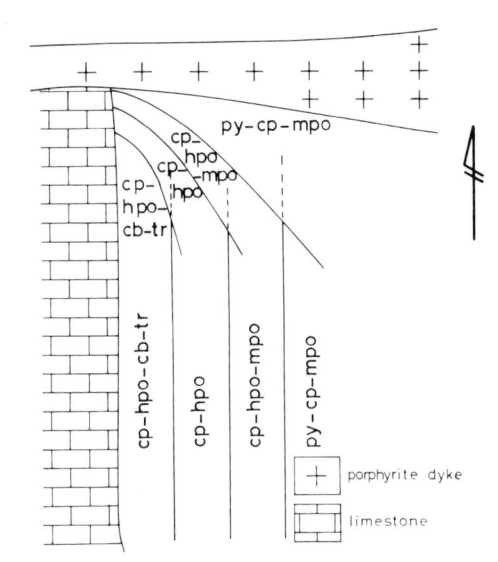

Fig. III-5-8. Schematic zonal arrangements of sulfides near dike rocks.

may be concluded that the interaction between the original rocks and the mineralizing fluid mainly took place through the diffusion process.

In the Shinyama deposit, two successive processes of mineralization are assumed from the field evidence shown in Figs. III-5-2, -7(a), and -8. The one occurrence was caused by the diffusion process regarded as the mineralization of early stage genetically related to the Ganidake granodiorite, and the other was due to deposition from the solution migrating along dikes, belonging to the later mineralization. The later stage mineralization overlaps the early formation in the parts near dikes (Fig. III-5-8). The early garnet near dikes is replaced by the later garnet. The similarity of the two kinds of garnet suggests a similar composition of the ore fluid.

Judging from the field evidence that garnet is distributed in the granitic rock side, and from the decrepitation temperatures of this mineral reported by Shoji (1970), it may be assumed that the formation temperature of skarns is relatively lower at the hedenbergite side including sulfides, than that of the garnet side including magnetite.

Gustafson (1974), Burt (1971, 1972) and Kurshakova (1970, 1971) reported that andradite, hedenbergite and magnetite occur in the paragenetic relation at a limited temperature and oxygen fugacity, that andradite is stable at higher fo_2 than hedenbergite, and moreover that the assemblage of andradite and magnetite is stable at higher fo_2 than hedenbergite.

Therefore, it is considered that the skarnization proceeded with the decrease of temperature and of oxygen fugacity from garnet to clinopyroxene skarns.

In the Shinyama ore deposit, large amounts of pyrrhotite occur vertically from the 350m level to the 550m level, horizontally from the intrusive rocks the to limestone side. As the results of d(102) measurement by the method of X-ray powder diffraction and with microscopic observation, three phases of pyrrhotite (troilite, hexagonal pyrrhotite, monoclinic pyrrhotite) are recognized. It has become clear that the sulfur content in pyrrhotite increases from the limestone side towards igneous rocks (Fig. III-5-7).

From a report by Natarajan and Garrel (1958), it may be concluded that among the minerals in the Cu-Fe-S-O system, the sulfides occupy relatively lower fields of oxygen fugacity as compared with oxides, and the chalcopyrite-pyrrhotite assemblage occupies relatively lower sulfur and oxygen fugacity than the chalcopyrite-pyrite-pyrrhotite assemblage. Burt (1972) stated that an assemblage such as chalcopyrite-cubanite-pyrrhotite occupies a relatively lower and more limited stability field of fo_2-fs_2 relation than andradite.

Judging from the above-mentioned considerations, the sequence of mineralization of the

Shinyama ore deposit can be assumed to have taken place from igneous rocks to the limestone side with the decrease of temperature, sulfur fugacity, and oxygen fugacity as follows: 1) garnet, 2) magnetite-garnet, 3) hedenbergite-garnet-pyrite-chalcopyrite-monoclinic pyrrhotite, 4) hedenbergite-monoclinic pyrrhotite-hexagonal pyrrhotite-chalcopyrite, 5) chalcopyrite-hexagonal pyrrhotite-hedenbergite, 6) chalcopyrite-hexagonal pyrrhotite-cubanite-troilite. In this deposit, nickel- and cobalt-bearing minerals such as pentlandite and mackinawite exist in zones 3), 4), 5) and 6).

Generally speaking, the occurrence of cubanite seems to be closely related to ultrabasic or basic rocks (Lawrence and Plimer, 1969; Lawrence and Golding, 1969). For example, in the Komori ore deposit large amount of cubanite can be observed in serpentinite*. It is probable that the cubanite in the Kamaishi mine is also due to the existence of basic and ultrabasic rocks.

III-6. Chichibu Mine, Saitama Prefecture

T. Shoji and H. Imai

The Chichibu mine is located in the Kanto mountain area, about 90 km W or NW of

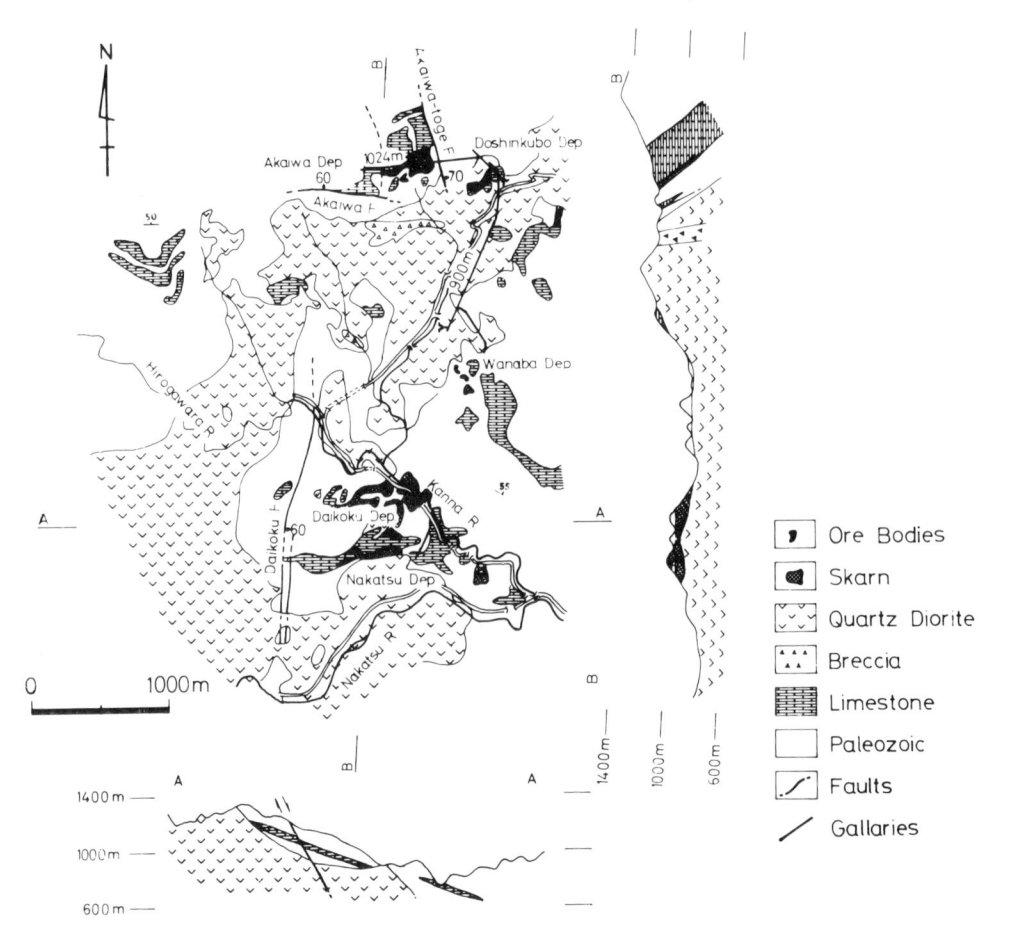

FIG. III-6-1. Geologic map of the Chichibu mine (after Kanada *et al.* 1961).

---
* Refer to Chapter II-5.

Tokyo (Fig. I–3). Before 1973, the mine produced 500,000 t of iron, copper, lead and zinc ores per year, and copper ore alone from 1973 to 1976. However, the mine has been abandoned.

General Geology

The Paleozoic sediments, which are widely developed in this district, consist of slate, sandstone, chert, limestone and "schalstein" of Upper Carboniferous or Lower Permian age. Generally speaking, the Paleozoic sedimentary rocks trend from WNW to ESE, and dip NNE. Limestone is generally divided into three layers. From the south to the north, they are called the first, second and third limestone layers.

Quartz diorite, the K-Ar and fission track ages of which are reported to be 7.9 and 8.2 m.y., respectively (Ueno *et al.*, in prep.), has intruded into the Paleozoic sediments (Fig. III–6–1) and metamorphosed them. The elongation of the quartz diorite body is approximately perpendicular to the general strike of the sedimentary rocks. The ore bodies exist in limestone at the peripheral part of quartz diorite.

Dikes of quartz porphyry and porphyrite occur in the vicinity of the margin of quartz diorite. The directions of these dikes are perpendicular to the margin of quartz diorite. It is considered that they are the latest facies of quartz diorite magma. A breccia dike trending W-E also occurs in Paleozoic sediments and quartz diorite (Fig. III–6–1). Breccia consists of recrystallized chert, hornfels, quartz diorite and crystalline limestone, and the matrix consists of andesitic tuff. It is distinguished from intrusive breccia of quartz diorite, which is sporadically found at the margin of quartz diorite.

Ore Deposits

Several ore bodies were worked in this mine (Fig. III–6–1). The Daikoku, Nakatsu, Takiue-Takishita and Yamadorikubo deposits lie in the first limestone layer; the Rokusuke, Wanaba and Saitozawa deposits occur in the second; and the Akaiwa and Doshinkubo deposits occur in the third. Figure III–6–2 shows the variation of ore minerals in each ore deposit (Kanada, 1967). Zoning of ore minerals is controlled by quartz diorite. From the

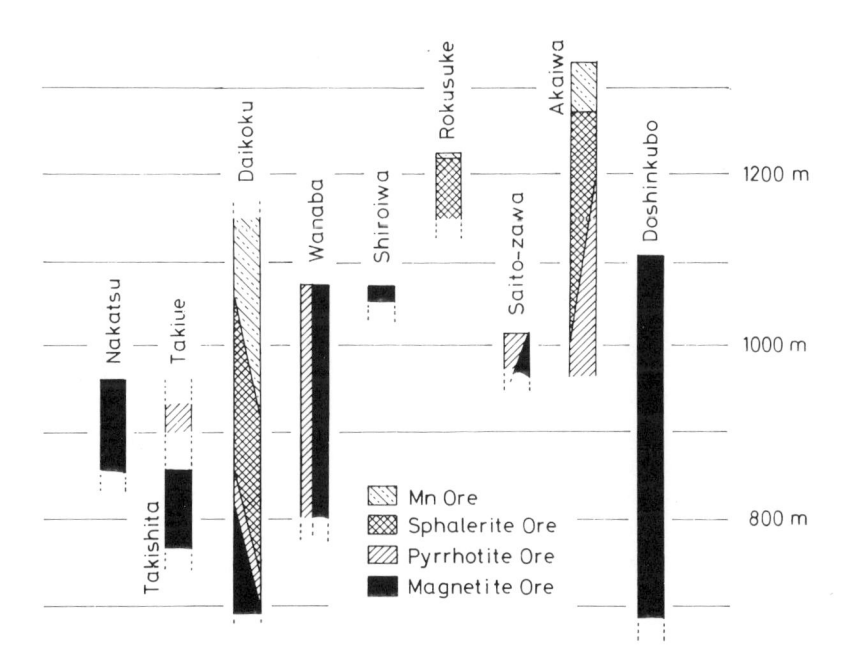

FIG. III–6–2. Vertical zoning of ore minerals in each deposit of the Chichibu mine (after Kanada, 1967).

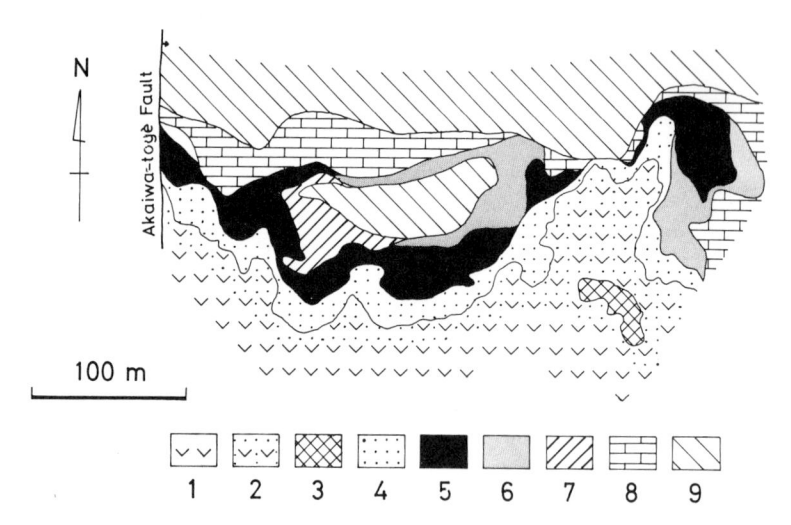

FIG. III-6-3. Underground geologic map (1024 m level) of the Doshinkubo deposit, the Chichibu mine (after Shoji, 1975).
1: quartz diorite, 2: mixed rock, 3: vesuvianite-xanthophyllite xenolith, 4: garnet-epidote skarn, 5: magnetite ore body, 6: pyrite-limonite ore body, 7: hedenbergite-ilvaite skarn, 8: marble, 9: chert. Fault: Akaiwa-toge fault.

vicinity of quartz diorite outwards, ore consists mainly of magnetite, pyrrhotite, sphalerite and manganese minerals.

Fracture Systems Caused by Intrusion of Quartz Diorite

The main deposits of the mine are the Daikoku, Doshinkubo and Akaiwa deposits, which are located, respectively, along the Daikoku, Akaiwa-toge and Akaiwa faults (Fig. III-6-1). Besides these major faults, there are many minor faults and joints in the mining area. An ore shoot of the skarn-type ore deposits is located occasionally on the intersecting position between limestone and fractures, which were a pathway for ore-forming fluids. For example, in the main ore body of the Doshinkubo deposit (Fig. III-6-3), the copper grade increases towards the west. This fact may show that the ore-forming fluid was supplied through the Akaiwa-toge fault. A small-scale zinc ore deposit is located on the west contact of the fault.

The formation of the Daikoku and Akaiwa-toge faults was related to the intrusion of quartz diorite. A normal fault formed by magmatic upheaval has the following three characteristics (Iddings, 1898; Imai, 1966a, b): (i) Its strike coincides with that of the front of the igneous body; (ii) its dip is steeper than that of the front; and (iii) the fault cuts across the boundary between the igneous body and the sedimentary rock (the front of the igneous body), and extends in both rocks.* The strike of the Daikoku fault is N10°E, dipping 60°E (Fig. III-6-1). The displacement of limestone on both sides of the fault shows that the hanging wall slid down about 35 m relative to the footwall (Shoji et al., 1969). In other words, the Daikoku fault is normal. This fault is considered to have been formed by magmatic upheaval, because it has the above-mentioned characteristics (i) and (ii). Fractures or joints at the eastern periphery of diorite intrusives in the area of the Wanaba deposit, striking N-S with a dip of 90°, were also formed by magmatic upheaval (Shoji et al., 1969).

The general trend of the boundary between quartz diorite and the Paleozoic sediments is varied on both sides of the Akaiwa-toge fault (Fig. III-6-1). The direction is N60°E, 65°

---
* Refer to Chapter II-1.

NW on the west side, and W-E, 85°N on the east side. When the orientation of the front of the magmatic invasion changes abruptly, the magmatic pressure would cause a kind of hinge fault which contains rotation and vertical glide. The Akaiwa-toge fault would have been produced in such a way (Shoji et al., 1969). According to this model, the relation between the apparent horizontal dislocation and the height was calculated geometrically as follows: $\Delta x = 0.7 \cdot \Delta h$, where Δx is the difference of apparent horizontal dislocation between the two levels, and Δh is the difference in height between the two levels (Shoji et al., 1969). The apparent horizontal dislocation of limestone is 35 m between the 1074 m and 1024 m levels, and 85 m between the 1024 m and 900 m levels. These data agree with the above equation. The igneous rock protruding into the sedimentary rocks near this fault scarcely shows dislocation on either side. This fault makes a wedge-shaped cavity, according to the above-mentioned mechanism. The protrusion of the igneous rock is explained by the interpretation that unconsolidated magma intruded later into the cavity (Shoji et al., 1969).

Joints formed by the solidification of magma are observed in quartz diorite as typical fractures. These joints are divided into two systems. The one is N25°W in strike and 80°SW in dip, while the other is N55°W and 20°NE. These two systems are perpendicular to each other. In some places, the former is observed to be a tourmaline vein and the latter to be a calcite vein. These fractures seem to be pathways for ore-forming fluids.

Zoned Skarn of the Doshinkubo Deposit

The skarn of the Doshinkubo deposit shows a typical zonal arrangement from quartz diorite to marble. Harada (1962) classified it as follows: mixed rock zone, vesuvianite-xanthophyllite zone, garnet-chlorite-epidote zone, magnetite ore body, hedenbergite-ilvaite zone. Shoji (1975), however, considered that the vesuvianite-xanthophyllite zone is a xenolithic part of the mixed-rock zone and not a definite zone by itself (Fig. III–6–3).

The quartz diorite consists essentially of quartz, plagioclase, hornblende and biotite, with small amounts of pyroxene, sphene, and apatite. The mixed rock is an endogenous skarn with a dioritic texture. This rock consists mainly of plagioclase, clinopyroxene, hornblende, and garnet with accessory sphene. The vesuvianite-xanthophyllite xenolith consists of garnet, vesuvianite, calcite, clinopyroxene, plagioclase, xanthophyllite and quartz with accessory epidote. Vesuvianite occurs as reaction rims along the boundary between garnet and calcite. The garnet-chlorite-epidote zone consisted originally of garnet, calcite, vesuvianite, epidote and clinopyroxene. Chlorite in the zone was formed subsequent to sulfide mineralization. The magnetite ore body consists predominantly of magnetite with chalcopyrite, pyrite and sphalerite in subordinate amounts, and garnet, calcite and ilvaite as gangue minerals. The grade of the ore is about 40% Fe and 0.4% Cu. It is probable that magnetite was formed simultaneously with the skarn minerals. Calcite in this zone was formed in a later stage. The hedenbergite-ilvaite zone consists of clinopyroxene, ilvaite and garnet with accessory magnetite and pyrite.

In the eastern part of the deposit, the pyrite ore body exists in contact with the magnetite ore body (Fig. III–6–3). A narrow band (less than 1 cm) of hematite is frequently found along the boundary between the magnetite and pyrite ore bodies. Many globes, consisting of magnetite in the core and pyrite in the crust, are observed in the marble near the pyrite ore body. The size of these globes varies between less than 10 cm and a few meters. A hematite band is also found between the core of magnetite and the crust of pyrite.

The Takiue-Takishita Deposit

The Takiue-Takishita copper deposit located at the eastern adjacent part of the Daikoku deposit has been mined for the past seven years (Fig. III–6–1).

The deposit lies along the boundary between limestone and chert, slate or quartz diorite

A. Pyrometasomatic Deposits in Japan and Korea

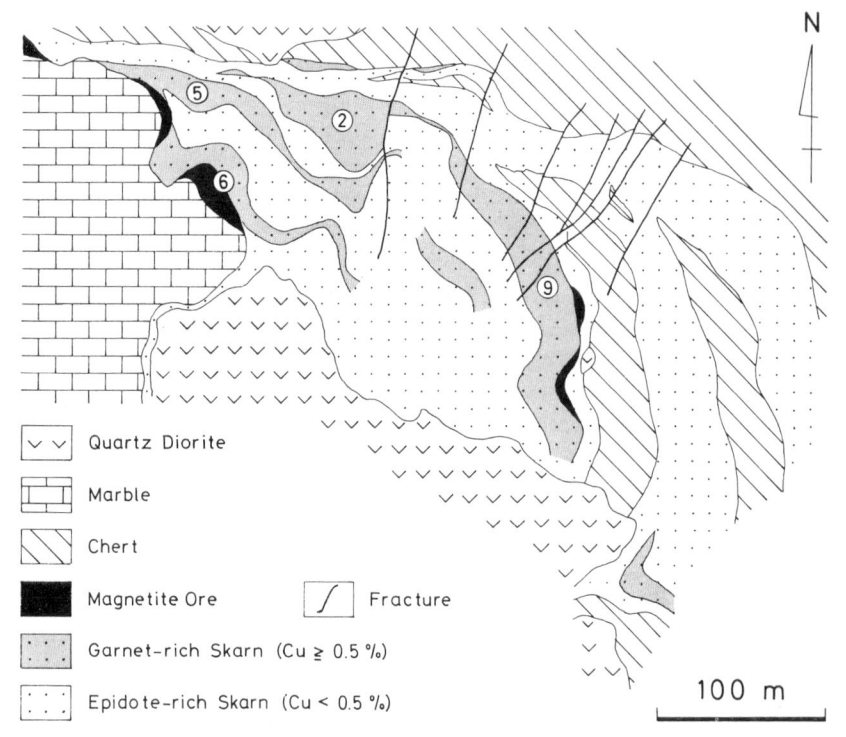

FIG. III-6-4. Underground geologic map (800 ml) of the Takishita deposit, the Chichibu mine. Numerals represent the numbers of the ore bodies.

(Fig. III-6-4). The skarn of this deposit consists of garnet, clinopyroxene, epidote, actinolite, calcite and quartz. The skarn mass does not show a zonal arrangement as a whole, but consists of several alternating layered epidote and/or garnet zones (Fig. III-6-4). The trend of each zone coincides approximately with the geologic structure of the Paleozoic sediments in this area. The skarn mass shows a partial zonal arrangement along the original boundary between limestone and slate. That is, the skarn in contact with limestone consists mainly of garnet, while the skarn contacting with the relict of hornfels consists predominantly of epidote. These occurrences suggest that the skarn is derived from argillaceous limestone, and that the garnet and epidote zones correspond to the calcareous and argillaceous beds

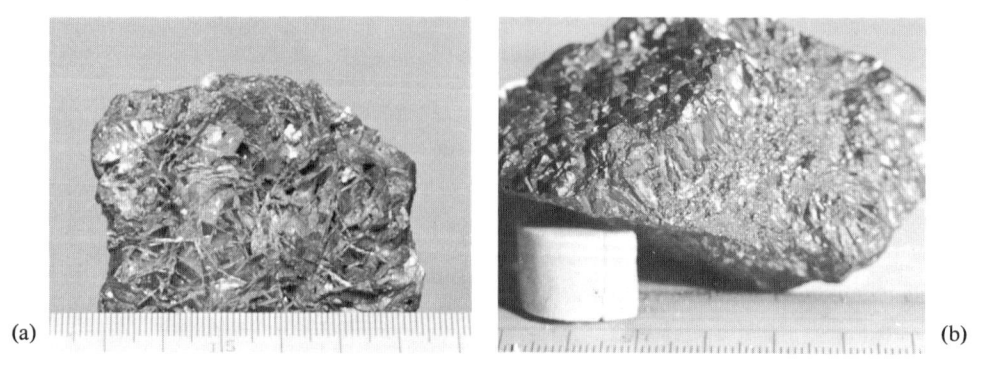

FIG. III-6-5. Photographs of anomalous crystal forms of magnetite from the Chichibu mine. (a) Foliate magnetite from the Daikoku deposit (after Shoji, 1969b). (b) Short-columnar magnetite from the Doshinkubo deposit.

of the original rock, respectively. Electron microprobe analyses indicate that garnet and epidote coexist over wide compositional ranges, and that compositional zoning of each mineral is caused by change in the partition coefficient due to temperature decreases (Kitamura, 1975).

The ore consists of chalcopyrite, sphalerite, pyrite and magnetite with the above-mentioned skarn minerals. The copper grade is about 0.45%. Copper ore bodies are restricted in the garnet zone of skarn. For this reason, the trend of the ore body is parallel to the geologic structure. Generally, ore minerals occur in the grain boundary of garnet. Chalcopyrite appears to be controlled by the bedded structure of the original rock (Fig. III–6–4), while sphalerite is controlled by the fracture in the skarn mass (Kanada, 1968). A small amount of scheelite is found along the quartz veins.

Besides copper ore bodies, magnetite ore bodies occur along the boundary between the skarn mass and limestone, or within the skarn mass (Fig. III–6–4). The zones of skarn consist of garnet (-chalcopyrite-pyrite), magnetite, hematite-pyrite, and limonite (Shoji, 1969a). The hematite-pyrite zone is accompanied by foliate magnetite which was formed from hematite.

Anomalous Crystal Form of Magnetite

Three crystal forms of magnetite are found in the mine: granular, foliate and short-columnar. Foliate magnetite is found in the Daikoku and Takishita deposits (Fig. III–6–5(a)). In the Daikoku deposit, foliate magnetite occurs along pyrite and/or siderite veins. The developed plane of foliate crystal is (111) (Shoji, 1969b). The reduction from hematite to magnetite is carried out by the sliding of oxygen and iron layers of the crystal, the migration of iron ions in each iron layer, and the addition of iron ions. Since the foliate magnetite has no measurable amount of vacant positions, it is formed by the addition of iron ions into hematite crystal (Shoji, 1969b).

Short-columnar magnetite is found in the Doshinkubo deposit (Fig. III–6–5(b)). It occurs along veins or cavities. The crystal elongates to the direction of [100]. The elongation is perpendicular to fractures or cavities. Short-columnar magnetite seems to be formed towards open spaces at the latest stage of the formation of magnetite, when nuclei are few.

Fluid Inclusions

Several authors have reported the filling temperatures of fluid inclusions and the decrepitation temperatures of skarn minerals, including calcite and quartz (Takahashi *et al.*, 1955; Park and Miyazawa, 1971a, b; Enjoji, 1972; Miyazawa and Enjoji, 1972). The filling temperatures are distributed between 200°C and 375°C (Table III–8–2), while the decrepitation temperatures are between 200°C and 360°C (Table III–8–1). Fluid inclusions are found not only in skarn minerals, such as garnet (Fig. III–6–6), hedenbergite and axinite, and quartz and calcite coexisting with them, but also in quartz constituting quartz diorite (Miyazawa, 1959a). Some of them consist of two phases of gas and liquid, and others contain one or

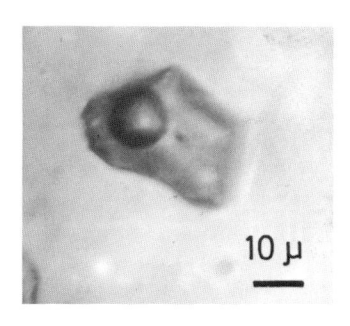

Fig. III–6–6. Fluid inclusion in garnet from the Daikoku deposit, the Chichibu mine (after Takahashi *et al.*, 1955).

more solids in addition. The salinity of two-phase inclusions was reported to be from 3.2 to 11.6 wt. % in NaCl equivalent concentration (Enjoji, 1972).

III-7. Sangdong Mine, Korea

Y. W. John

Introduction

Sangdong tungsten mine, located in the mountainous southeastern part of the Korean Peninsula (Fig. I-3), has been known as one of the largest tungsten producers in the world since 1951. Besides tungsten ore, it produces substantial amounts of bismuth and molybdenum ores as by-products.

Klepper (1947) is the first person who described and discussed the geology and genesis of the mine. Microscopic study of the minerals of the mine was conducted by the writer, and the geology was discussed, including mineral paragenesis, zonal arrangement and genesis (John, 1963). The writer, with Hong, conducted regional studies on the mineralization of the Sangdong mining district (Hong and John, 1965). They also studied the filling temperatures of the liquid inclusions in the quartz veins from Sangdong mine, using the heating-stage microscope, for the purpose of determining the formation temperature of the vein minerals and the relationships between the filling temperatures and mineral assemblages (Hong and John, 1966). Chung (1964) discussed the origin of zoning in the Sangdong ore deposit. Both Chung (1966) and Yun (1966) discussed the relationships of the geologic structures and mineralization of the Sangdong mining district.

General Geology

The stratified rocks in the vicinity of the Sangdong mine range in age from Precambrian to Ordovician. Rocks of the Precambrian Taebaeksan Series consist of interbedded biotite schist, sericite schist, quartzite, crystalline limestone, hornfels and hornblende schist (Fig. III-7-1).

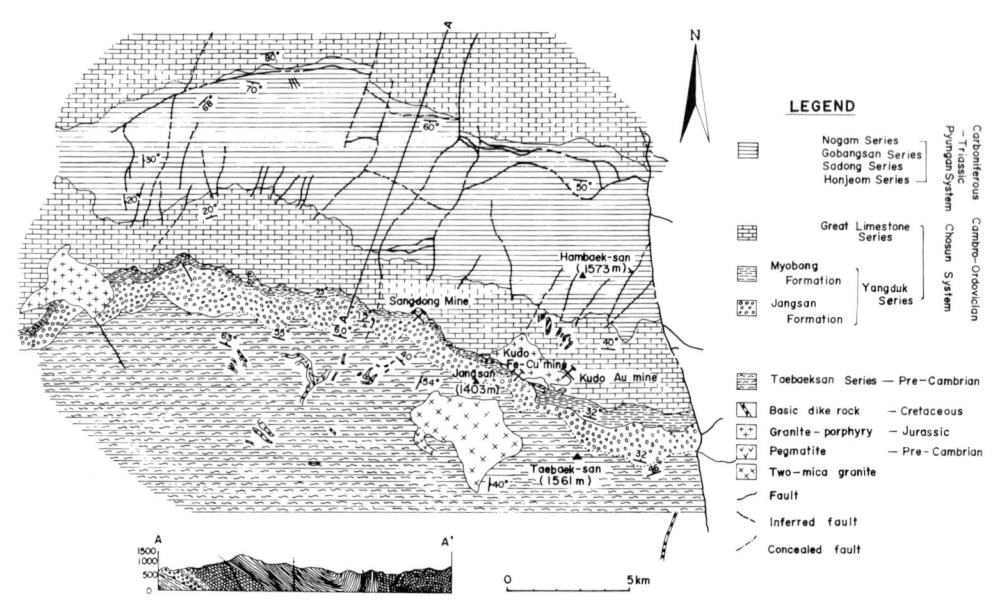

FIG. III-7-1. Geologic map of the Sangdong mine.

The Cambrian rocks unconformably overlying the Taebaeksan Series are of the Yangduk Series, which is subdivided into the Jangsan and Myobong Formations. The lower Cambrian Jangsan Formation is a hard, well-jointed, cliff-forming quartzite, and its thickness is about 200 m in the vicinity of the mine. It is overlain conformably by the Myobong Formation of middle and upper Cambrian age. The Myobong Formation consists of interbedded thin to thick bedded marl, sandstone, phyllite, shale, limestone, and locally hornfels, with a total thickness of approximately 200 m.

The Great Limestone Series of Ordovician age overlying the Myobong Formation consists mainly of thin- to thick-bedded white to dark gray limestones.

The stratified rocks of Cambrian and younger periods near the mine strike between N 75°W and N80°W, and form an asymmetrical syncline. The northern limb of the syncline has steeper dips, even overturned, whereas the dip of the southern limb is around 30°NE.

Intrusive rocks in the Sangdong mine area are granite, granite porphyry, pegmatite, lamprophyre, porphyrite and quartz porphyry.

A stock of two-mica granite located about 4 km SSE of the Sangdong mine has intruded the Precambrian Taebaeksan Series, occupying approximately 9 km² of the area. The essential minerals of the rock are quartz, microcline, albite, orthoclase, muscovite and biotite. The K-Ar age of the granite determined with muscovite is 1,530 m.y. (Ueda, 1968), which implies Precambrian age.

Another granitic intrusive exposed over an area of approximately 3 km² at the Kudo mine about 4 km east of the Sangdong mine, is hornblende-biotite granite and porphyry (Fig. III–7–1). The K-Ar age of the granite porphyry determined with biotite is 169 m.y., which implies Mid-Jurassic (Ueda, 1968). Most of the calcareous rocks bordering the granite are either replaced by skarn or mineralized.

Ore Deposits

The ore deposits in the Sangdong mine are localized in several calcareous beds within the Myobong Formation. They are more or less lenticular in shape, and are essentially parallel to the stratification of the host rocks.

The main ore bed, located approximately 40 m below the top of the Myobong Formation, strikes between N 75°W and N 80°W, and dips 15° ∼ 30° NE. The thickness of the bed fluctuates upward from 3 m, averaging 4.5 m. The vertical depth from the outcrop to the present deepest working level is about 500 m.

The ore minerals of this mine are scheelite, wolframite, molybdenite, bismuthinite and tetradymite. The other opaque minerals are, in the order of abundance, pyrite, pyrrhotite, chalcopyrite, sphalerite, arsenopyrite, magnetite, cassiterite and gold. Scheelite is the most important ore mineral of the mine, and molybdenum, bismuth and gold are recovered as byproducts.

Non-opaque gangue minerals are quartz, diopside, garnet, epidote, hornblende, biotite, chlorite, sericite, fluorite, calcite and some apatite.

Penetration of innumerable quartz veins subsequent to metasomatism of the ore bed has intensely altered the wall rocks. Significant wall rock alterations are silicification, sericitization, chloritization and biotitization.

Quartz Veins

The occurrence of quartz veins is one of the most interesting characteristics of the ore deposit; they are very important in that the ore body is substantially enriched by them. The thickness of the quartz veins ranges from a fraction of a centimeter to several centimeters. Some of them are as thick as 1 m, however, those more than 20 cm thick are not very common. In the upper part of the ore-shoot, innumerable quartz veins irregularly intersect one another, forming a network. Nearly 200 quartz veins were studied in an attempt to find their general trends. Although the result was not so satisfactory as was hoped, it has been possible to classify them into two groups in terms of their relationships with the ore bed: veins concordant to the ore bed, and discordant veins. Several relatively thick quartz veins near the footwall of the main ore bed are apparently parallel to the ore bed. They are fairly consistent throughout the ore bed in both horizontal and vertical dimensions. Since the concordant veins occur within 1 m from the footwall of the main ore bed, they serve as a useful guide to the ore bed in underground mine workings. Most of the quartz veins, however, are discordant to the ore bed. The veins cutting the ore bed strike to almost any direction with varying dips. The more common trends of the discordant veins, however, are N 10° $\sim$ 30°W, 50° $\sim$ 70°NE and N 65°E to E $\sim$ W, 40° $\sim$ 60°NW. The quartz veins are most abundant in the central ore shoot, and least in the marginal poor ore zone. Vertically, they are more abundant in the upper part of the ore body, becoming thinner and less abundant downwards.

Many quartz veins carry scheelite, molybdenite and bismuthinite with a small quantity of pyrite, sericite, calcite and fluorite. Although less common, massive sulfide minerals such as bismuthinite, chalcopyrite, sphalerite, arsenopyrite and pyrite occur in some quartz veins. The thin quartz veins contain little or no scheelite and molybdenite, and they are probably formed at relatively low temperatures. There are also some barren quartz veins with or without pyrite.

The filling temperatures of the liquid inclusions in the quartz veins have been determined with the heating-stage microscope. Specimens of vein quartz have been taken from 22 different localities. The inclusions have been recognized to consist of liquid phase and gas phase; no polyphase inclusion was observed. The size of the inclusions has been generally small, ranging from 2 to 20 microns.

As the result of the filling temperature determination it has been possible to classify them into three groups in terms of temperature range and the mineral association (Hong and John, 1966): (1) The high-temperature quartz veins with average filling temperatures of 257°C . . . barren or contain small quantities of pyrite and scheelite. (2) The medium-

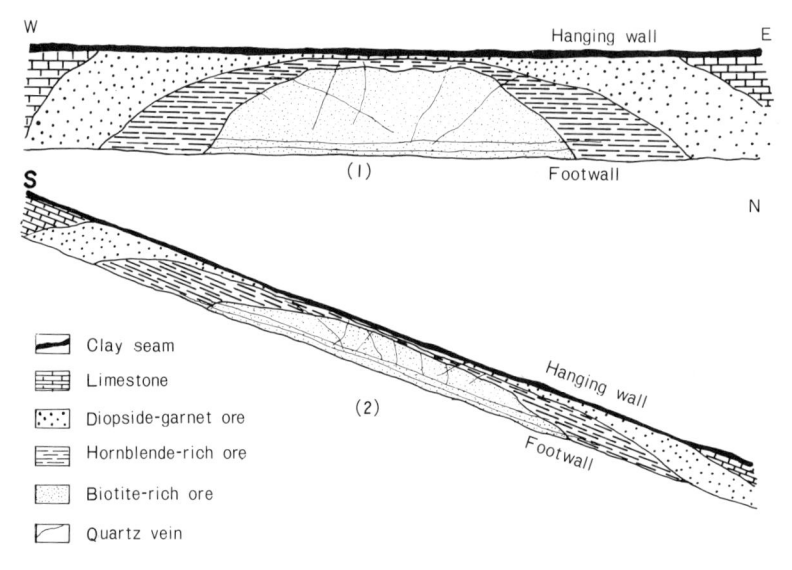

FIG. III-7-2. Schematic sections of the Sangdong mine. (a) Along the strike. (b) Along the dip.

temperature quartz veins with an average filling temperature of 217°C . . . carrying scheelite and some pyrite and chalcopyrite, but no or little molybdenite. (3) The low-temperature quartz veins with an average filling temperature of 170°C . . . high in molybdenite, but relatively low in scheelite and pyrite.

From the above classification, it is revealed that the quartz veins rich in molybdenite have been formed at relatively low temperatures, whereas those rich in scheelite have been of higher-temperature origin. These facts may be part of the explanation for the decrease in molybdenite downward in the ore bed.

Zoning of the Ore Deposits

The main ore bed of the Sangdong mine is divided into three distinct zones, in terms of the ore grade and the kind of gangue minerals, as follows: (1) the central high-grade zone (ore-shoot) with predominant quartz, biotite, sericite and chlorite; (2) the intermediate zone with hornblende and quartz as the predominant gangue minerals; and (3) the marginal poor ore zone which is characterized by diopside and garnet (Fig. III–7–2). The zoning is symmetrically distributed in the plan of the ore body. The ore shoot is a banded zone at the central part of the ore bed extending downward roughly in the direction of the dip. The ore shoot is sandwiched by the intermediate zones, which in turn are surrounded by the marginal poor ore zone.

Ore shoot: The ore-shoot in the central portion of the main ore bed extends down the dip of the bed plunging approximately N 10°W 20°. The gangue minerals of the ore-shoot are, roughly in order of abundance, quartz, biotite, sericite, hornblende and chlorite, with minor quantities of calcite, fluorite, apatite and zircon. Among them, quartz is the most abundant mineral of the zone, and the ore consists of almost 60% by volume of quartz. Quartz veins are most abundant in this zone.

The ore shoot locally contains irregular isolated masses and/or thin bands of diopside and garnet along the hanging-wall side. It grades into the adjoining intermediate zones, and one can't draw a sharp boundary between them.

Intermediate zone: The intermediate zone lies between the central ore-shoot and the marginal poor ore zone.

Microscopic study of the thin sections reveals a significant increase in hornblende and a decrease in biotite, sericite, quartz and chlorite. The proportion of quartz in the ore has decreased to approximately 40% by volume.

The band or masses of calc-silicate minerals along the hanging-wall becomes thicker than in the ore-shoot. The mineralized quartz veins become less abundant than in the ore-shoot, and the ore grade consequently becomes lower.

Poor ore zone: The poor ore zone lying in the outer margin of the ore body grades inwards into the intermediate zone, and outwards into the host rock.

The most abundant and characteristic gangue mineral of the zone is diopside, which occupies roughly 50% by volume of the ore; garnet is also a characteristic mineral of the zone, almost always associated with diopside. The number of quartz veins, again, decreases considerably, and the ore grade becomes too poor to be economical.

This zonal arrangement is also recognized in the vertical cross section of the ore bed. As is seen in Fig. III–7–2, there is a general tendency for the high ore-grade zone to be distributed on the footwall side, becoming leaner towards the hanging-wall side.

The characteristics of the zonal arrangement become gradually obliterated downwards. At −6 level or below, the zoning is so weak that careful observation is necessary to recognize the zonal distribution.

The zoning of the gangue minerals may best be explained by progressive hydrothermal alteration of the skarn, which consists predominantly of such anhydrosilicates as diopside and garnet. A hydrothermal solution rich in silica and metals is assumed to have been active for a substantial period after the formation of the skarn.

Genesis of the Ore Deposits

A granitic body intruding the rocks of the Chosun System about 14 km west of Sangdong mine shows signs of scheelite mineralization along its contact aureole. The Sangdong ore body shows evidence of local up-warping or anticlinal folding. From these facts, a granitic batholith is thought to extend throughout the area between the Kudo intrusive and the western intrusive; the possibility of the presence of a cryptobatholith underneath the Sangdong area becomes strong (Fig. III–7–1).

The result of K-Ar age determination of the Kudo intrusive and its association with scheelite and bismuthinite along the contact aureole are sufficient evidence that the granite porphyry is related to the mineralization at the Sangdong mine.

The bedded form of the ore body is primarily controlled by the stratigraphy. The zonal arrangement of the ore body and the mineral enrichment by quartz veins, on the other hand, are controlled by structural features. The problem of the structure which allowed the ascent of the mineralizing solution is of the utmost interest. Bedding-plane thrusts may have been formed at the time of formation of the Samchok Syncline, along both the footwall and hanging-wall of the limestone bed. On the hanging-wall side of the present ore body, there is a loose clay seam, which is presumably a fault clay.

In addition to the bedding-plane thrusts, more fractures originating at depth and cutting the limestone bed are developed. They may be thrust faults longitudinal to the synclinal structure, or subsequent transverse tension fractures, which usually cut the former. In the ore body, the thrust faults are represented by concordant quartz veins cutting the ore bed with an E-W trend; the tension fractures are represented by discordant quartz veins.

The mineralizing solution ascended through the above fractures, reacted with the limestone bed and formed the skarn rich in diopside and garnet. Tungsten mineralization, however, was not intense at this stage. After the formation of the skarn, innumerable irregularly trending fractures were formed within the ore bed. Subsequent hydrothermal solution relatively rich in tungsten permeated the fine fractures in the skarn and altered the diopside into uralitic hornblende, by which process enrichment of the tungsten ore was accomplished.

The hornblendization and scheelite enrichment were followed by biotitization of hornblende and another scheelite enrichment.

Although there seems to be a considerable difference in time between the stage of skarn formation and the formation of the quartz veins, the mineralizing solution in both cases may be derived from a single source as a series of continuous activities. Intense mineralization is accompanied by hornblendization of diopside, biotitization of hornblende and subsequent hydrothermal activities.

B. GENETICAL PROBLEMS

III-8. Skarn Formation

T. Shoji

Generally, most ore-bearing skarn is formed under a metasomatic process caused by a mixed fluid of water and carbon dioxide, which is generated through magmatic activity and the decomposition of carbonate minerals. For this reason, the physicochemical conditions for formation of ore-bearing skarn is controlled by the nature of the H_2O-CO_2 mixture supplied to carbonate rocks.

Temperatures

There are very few data which indicate the formation temperatures of mineralized skarn. Stabilities of most common minerals found in skarn are as follows: Grossular is stable up to 750°C or 850°C under water pressure of a few kilobars (Yoder, 1950; Shoji, 1974), and andradite up to more than 1,000°C (Huckenholz and Yoder, 1971; Suwa et al., 1976). Diopside is stable up to 1300°C (Yoder, 1965; Eggler, 1973), and hedenbergite up to more than 800°C (Gustafson, 1974). Since these minerals are stable up to high temperatures, their stabilities do not indicate formation temperatures.

In some pyrometasomatic ore deposits, potash feldspar and plagioclase occur with skarn minerals (Shoji, 1972). The modes of occurrence of these feldspars are classified into three types: (i) in banded skarn, (ii) in breccia-bearing skarn (refer to Chapter III-5), and (iii) along diopside veins (Fig. III-8-1). In general, feldspars exist between biotite hornfels and clinopyroxene zones. From macroscopic and microscopic observations, it is suggested that these feldspar-bearing skarns are formed by the reaction,

biotite + calcite + quartz
= K-feldspar + anorthite + clinopyroxene + CO_2 + H_2O.   (A)

The equilibrium p-T curve of Reaction (A) can be determined by that of the simplified reaction,

$KMg_3AlSi_3O_{10}(OH)_2 + 3CaCO_3 + 6SiO_2$
$= KAlSi_3O_8 + 3CaMgSi_2O_6 + 3CO_2 + H_2O$.   (B)

Figure III-8-2 shows the equilibrium X_{CO_2} − T curve of Reaction (B) as a parameter of total pressure. According to Fig. III-8-2, Reaction (B) occurs from left to right at about 500°–600°C under 500–2,000 atm. This condition corresponds to pyroxene-hornfels facies. If the fluid contains a smaller amount of carbon dioxide, the p-T curve passes through lower temperature regions.

Generally, vesuvianite does not coexist with ore minerals, especially sulfides. Vesuvianite occurs occasionally in skarns which appear to be formed at relatively high temperatures. For example, it is an important member of a xenolithic part of quartz diorite in the Chichibu mine (Fig. III-6-3), or of a narrow zoned skarn between granodiorite and marble in the Mitate mine (Fig. III-8-7). The water-free composition of vesuvianite is very similar to grossular. Therefore, the instability of vesuvianite seems to owe not to bulk compositions but to temperatures in the skarn formation. Vesuvianite is stable between 400°C and 600°C (Ito and Arem, 1970; Shoji, 1971). At lower temperatures, it decomposes to the assemblage grossular-diopside-dellaite. If the fluid contains a small amount of carbon dioxide, dellaite

202   B. Genetical Problems

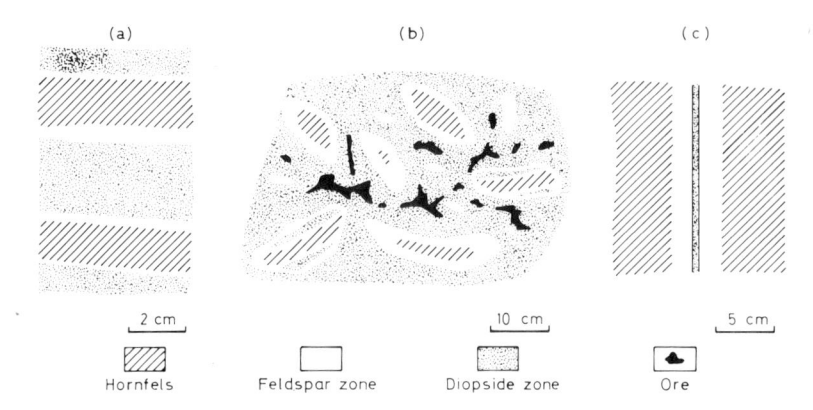

FIG. III–8–1. Sketches of feldspar-bearing skarn (after Shoji, 1972).
(a) Banded skarn from Nippo deposit, the Kamaishi mine.
(b) Breccia-bearing skarn (or breccia skarn) from the Nippo deposit.
(c) Diopside vein in the Chichibu mine.

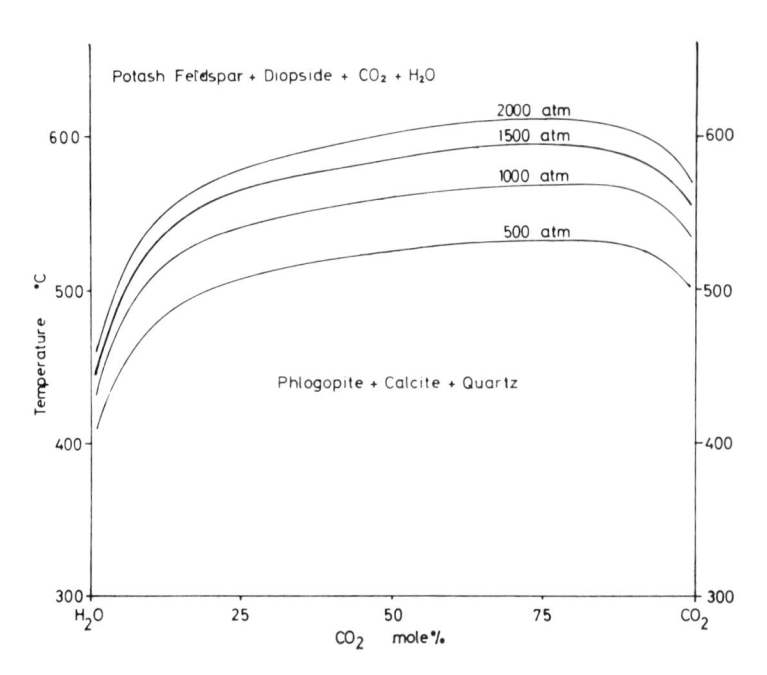

FIG. III–8–2. Equilibrium curves for Reaction (B) represented by compositions of H_2O–CO_2 mixtures and temperatures (after Shoji, 1972). Parameters represent fluid pressures.

seems to decompose to the assemblage calcite-quartz. The assemblage garnet-clinopyroxene-calcite-quartz is commonly found in ore-bearing skarns. That is, the lower temperature limit of vesuvianite stability indicates the upper limit of temperatures for the formation of ore-bearing skarn.

A few investigations have reported fluid inclusions found in skarn-type ore deposits. Table III–8–1 shows the filling or homogenization temperatures of fluid inclusions in sphalerite, quartz and others from some skarn-type ore deposits and their vicinities. Most of them show temperatures between 150°C and 350°C. If skarn minerals were formed at earlier

TABLE III–8–1. Filling or homogenization temperatures of fluid inclusions in the minerals from some skarn-type ore deposits and their vicinities.

Mine	(Deposit)	Temperature	Mineral*	Reference**
Kamaishi	(Sahinai)	300– °C	gt	E2
	do.	253–307	ep, ax, qz	E1, E2
Akagane	(Hozumi)	210–320	qz, ca	M-N
	(Sakae)	210–320	qz, ca	M-N
Yaguki		260–	sch	E2
		255–293	ep, qz	E2
Akatani		250–315	qz	E1, M-E
Igashima		126–200	fl	E1
Chichibu	(Akaiwa)	200–375	qz, gt, hd, ca	P-M1, P-M2, E1, M-E
	(Wanaba)	200–230	qz	P-M2
	do.	335–370	ax	P-M1
	(Takiue)	281–355	gt, hd, qz	E2
Kamioka	(Mozumi)	150–350	sp	M-M
	(Maruyama)	150–350	qz(druse)	S-K
	do.	251–310	hd, qz	E2
		250–510	qz(granite porphyry)	T-I
Tsumo	(Horai)	300–	gt	E2
	do.	271–312	hd, qz	E2
Kiwada		195–260	qz(vein)	T-I
Fujigatani		160–305	qz(granite)	T-I
Sangdong		161–305	qz	H-J
Tyrny-Auz		280–480	gt, fl, qz, sh	L
		505–510	fl	L

* ax = axinite, ca = calcite, ep = epidote, fl = fluorite, gt = garnet, hd = hedenbergite, qz = quartz, sh = scheelite, and sp = sphalerite.
** E1 = Enjoji (1972), E2 = Enjoji (1977), H-J = Hong and John (1966), L = Lesnyak (1957), M-E = Miyazawa and Enjoji (1972), M-M = Mukaiyama and Miyazaki (1973), M-N = Muramatsu and Nambu (1976), P-M1 = Park and Miyazawa (1971a), P-M2 = Park and Miyazawa (1971b), S-K = Shoji and Kihara (1970), and T-I = Takenouchi and Imai (1975).

TABLE III–8–2. Decrepitation temperatures of the minerals from some skarn-type ore deposits.

Mine	(Deposit)	Temperature	Mineral*	Reference**
Kamaishi	(Shinyama)	299–375°C	gt	Sh
	(?)	315–400	ca, gt	T
Chichibu	(Daikoku)	200–360	sp, po, mt, py, iv, gt, hd, ca	T
Kamioka	(Tochibora)	198–325	ca, qz, hd, ap, sp, gn	N
	(Mozumi)	255–315	ca	N
	(Tochibora / Maruyama / Mozumi)	200–475	ca, qz, gt, hd, ep, sp, iv	Si
	(Tochibora)	310	hd	T
Kuga		140–308	qz, hd, sh, po, sp, fl.	T, I
Obira	(Okura)	218	fl	N
	(Kurauchi)	279–397	ax, gt, iv, hd	N

* ap = apatite, ax = axinite, ca = calcite, ep = epidote, fl = fluorite, gt = garnet, gn = galena, hd = hedenbergite, iv = ilvaite, mt = magnetite, po = pyrrhotite, py = pyrite, qz = quartz, sp = sphalerite, and sh = scheelite.
** I = Imai and Ito (1959), N = Nishio et al. (1953), Sh = Shoji (1970), Si = Shiobara (1961), and T = Takahashi et al. (1955).

stages than sphalerite and quartz, the formation temperatures of skarn minerals may have been higher than those listed in Table III–8–1.

Decrepitation temperatures were measured for the minerals from some skarn-type ore

204   B. Genetical Problems

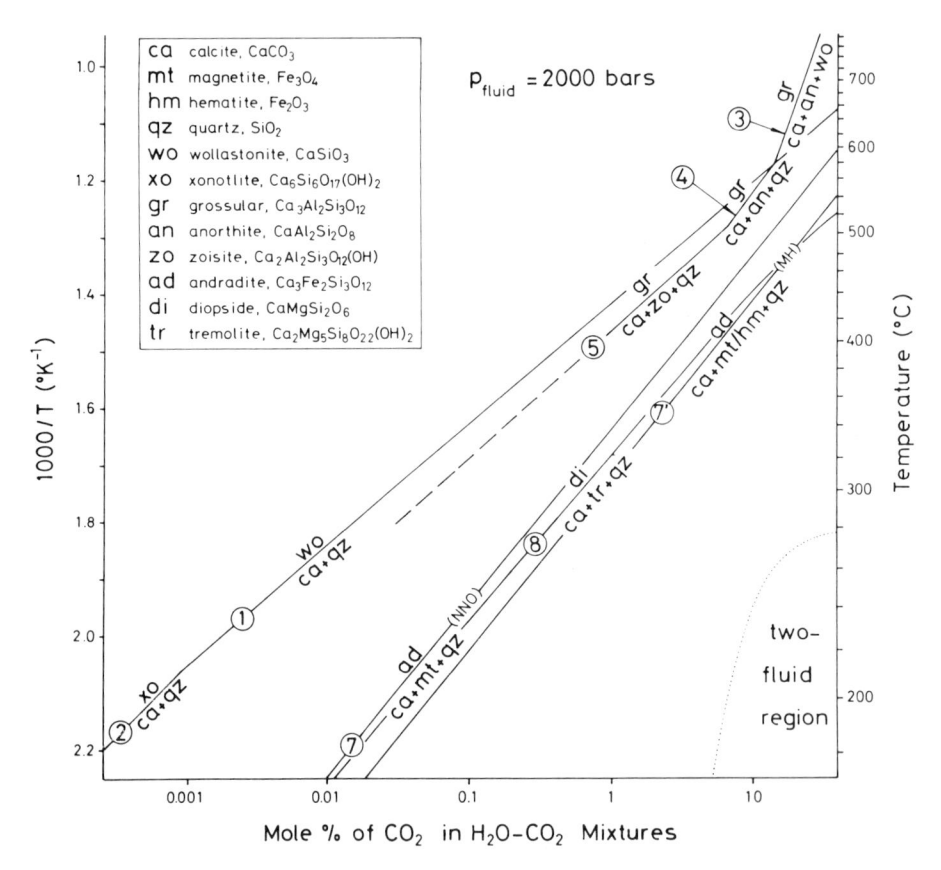

FIG. III-8-3. Equilibrium boundaries for some stable reactions in the systems CaO-SiO$_2$-H$_2$O-CO$_2$, CaO-Al$_2$O$_3$-SiO$_2$-H$_2$O-CO$_2$, CaO-Fe$_2$O$_3$-SiO$_2$-H$_2$O-CO$_2$, and CaO-MgO-SiO$_2$-H$_2$O-CO$_2$. Numerals represent the reaction numbers stated in the text. Buffer: MH: magnetite-hematite; NNO: Ni-NiO.
References: 1: Greenwood (1967), Shoji (1976b), 2: Shoji (1976b), 3, 4: Gordon and Greenwood (1971); 5: Shoji (1977a); 7, 7': Shoji (1977c); 8: Skippen (1974).

deposits (Table III-8-2). Most of them are between 200°C and 400°C. This range of temperatures is approximately 50°C higher than that of the filling temperatures. Generally, the decrepitation temperature correlates with the filling temperature, but overshoots it (Shoji and Kihara, 1970). Overshooting of decrepitation temperature depends on the heating rate and the size and amount of sample. However, independently of these factors, the logarithmic expression of the decrepitation frequency shows a linear relation to the temperature in the neighborhood of the knick point of the decrepigraph (Shoji, 1970). For this reason, if Shiobara (1961) had used a semilogarithmic graph, the overshoot of the decrepitation temperature would become still smaller. The difference between the filling and decrepitation temperatures can be interpreted by the fact that the latter overshoots the former, or that skarn minerals were formed at higher temperatures than calcite, sphalerite and quartz. It is impossible, however, to determine which of the two interpretations is correct.

Feldspar-bearing skarn is commonly found near skarn-type ore deposits. The equilibrium boundary for Reaction (B) at 2,000 bars passes approximately through the points: X_{CO_2} = 1 mole %, T = 430°C, and X_{CO_2} = 0.1 mole %, T = 370°C. Since a small amount of calcite always occurs in the country rock of a feldspar-bearing skarn, the fluid from which this skarn was formed seems to have been very poor in carbon dioxide. Therefore, Reaction (B) appears to occur at temperatures of about 400°C or lower. At the Doshinkubo deposit (Fig.

III–8–3), the zoned skarn with magnetite ore was formed at lower temperatures than the vesuvianite-bearing xenolith. Therefore, the formation temperature of the ore-bearing skarn is inferred to be below 400°C, as indicated by the stability relations of vesuvianite. These inferences coincide with most of the data on filling and decrepitation temperatures listed in Tables III–8–1 and –2, though they may not directly indicate the formation temperatures of minerals.

Carbon Dioxide

The skarn-forming fluid consists predominantly of water and carbon dioxide.

Greenwood (1967) determined experimentally the stability of wollastonite at temperatures above 500°C in supercritical mixtures of water and carbon dioxide at pressures of 1,000 bars and 2,000 bars. Wollastonite is relatively stable at high temperatures and in CO_2-poor fluids, while the assemblage calcite-quartz is stable at low temperatures and in CO_2-rich fluids. Shoji (1976b) calculated the stability of the assemblage calcite-quartz in H_2O-CO_2 mixtures at low temperatures on the basis of previous work in the systems CaO-SiO_2-CO_2 and $CaSiO_3$-H_2O. The stability field of the assemblage is restricted by the following reactions, from high to low temperatures:

$$CaCO_3 + SiO_2 = CaSiO_3 + CO_2 \tag{1}$$
$$6CaCO_3 + 6SiO_2 + H_2O = Ca_6Si_6O_{17}(OH)_2 + 6CO_2. \tag{2}$$

Although Buckner et al. (1960) and Gustafson (1974) gave two different results for the temperature at which xonotlite changes to wollastonite, the stability field of the assemblage calcite-quartz in H_2O-CO_2 mixtures is almost independent of both of the results. The equilibrium boundaries for Reactions (1) and (2), which were determined on the basis of Gustafson's (1974) results, are shown in Fig. III–8–3. The assemblage calcite-quartz is stable on the CO_2-rich side of the boundaries, while wollastonite or xonotlite is stable on the opposite side.

Gordon and Greenwood (1971) and Shoji (1977a) determined the stability of grossular in H_2O-CO_2 mixtures on the basis of previous work in the systems CaO-Al_2O_3-SiO_2-H_2O and CaO-SiO_2-H_2O-CO_2. The reactions limiting the grossular field are as follows, from high to low temperatures (Fig. III–8–3):

$$\text{grossular} + CO_2 = \text{calcite} + \text{anorthite} + \text{wollastonite} \tag{3}$$
$$\text{grossular} + CO_2 = \text{calcite} + \text{anorthite} + \text{quartz} \tag{4}$$
$$\text{grossular} + CO_2 + H_2O = \text{calcite} + \text{zoisite} + \text{quartz} \tag{5}$$
$$\text{grossular} + CO_2 + H_2O = \text{calcite} + \text{prehnite} \tag{6}$$

These reactions place strict limits on the formation of grossular believed to have grown in equilibrium with a fluid phase. Grossular is stable on the H_2O-rich (CO_2-poor) side of the

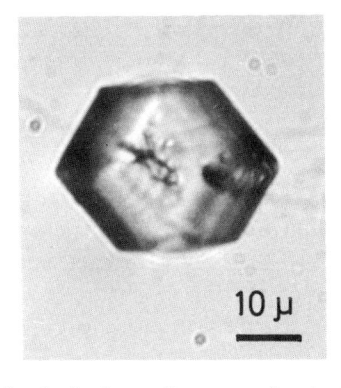

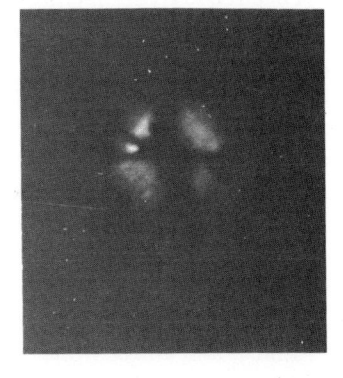

Fig. III–8–4. Synthesized grandite garnet showing birefringence. It was synthesized from the mixture of $CaCO_3$, Al_2O_3, Fe_2O_3 and SiO_2 at 504°C under a fluid pressure of 1,000 atm. (a) One nicol. (b) Crossed nicols.

boundaries for these reactions. The equilibrium boundaries for Reactions (3), (4) and (5) at 2,000 bars are shown in Fig. III–8–3. The decomposition curve of grossular approaches the H$_2$O-rich side in H$_2$O-CO$_2$ mixtures from high to low temperatures. This conclusion is supported by the observation that the amount of grossular synthesized from reagent mixtures of CaCO$_3$, Al$_2$O$_3$ and SiO$_2$ decreases from high to low temperatures (Christophe-Michel-Lévy, 1956; Shoji, 1974, 1975). For example, in the case where grossular is synthesized at the temperature at which Reaction (5) or (6) occurs, the growth of grossular and zoisite or prehnite ends on the equilibrium boundary for Reaction (5) or (6) due to the generation of carbon dioxide. In Fig. III–8–3, the equilibrium boundary for Reaction (6) is not shown, but that for Reaction (5) is shown to be extrapolated to the low-temperature regions, because in a system containing iron the assemblage prehnite-hematite is considered to be replaced by the assemblage grandite-epidote-quartz at a low temperature (Shoji, 1977a).

Shoji (1977c) calculated the stability of andradite in H$_2$O-CO$_2$ mixtures on the basis of previous work in the systems CaO-SiO$_2$-CO$_2$, CaO-Fe$_2$O$_3$-SiO$_2$ and CaO-FeO$_x$-SiO$_2$-O$_2$, and thermochemical data. The reactions,

$$\text{andradite} + CO_2 = \text{calcite} + \text{magnetite} + \text{quartz} + O_2 \tag{7}$$

and

$$\text{andradite} + CO_2 = \text{calcite} + \text{hematite} + \text{quartz} \tag{7'}$$

limit the andradite field under relatively high oxygen fugacities. The equilibrium boundaries for Reactions (7) and (7') at 2,000 bars is shown in Fig. III–8–3. Andradite is stable on the CO$_2$-poor side of the boundaries. The stability field of andradite in H$_2$O-CO$_2$ mixtures is wider than that of grossular, while the decomposition curve of andradite approaches the CO$_2$-poor side in H$_2$O-CO$_2$ mixtures from high to low temperatures, as does grossular.

Shoji (1975) investigated the effect of carbon dioxide on the stability of grandite garnet between 200°C and 700°C under a fluid pressure of 1,000 bars. The result suggests that grandite garnet is only stable under limited CO$_2$ pressures at low temperatures, as is grossular. The grandite garnet (except both end members) synthesized in CO$_2$-containing fluids shows birefringence (Fig. III–8–4), though the X-ray powder diffraction pattern indicates a cubic system. The result suggests that optically anisotropic garnet is formed only under conditions where the three components CO$_2$, Al$_2$O$_3$ and Fe$_2$O$_3$ are present. This suggestion is supported by observation of optically anisotropic garnet. Generally, optically anisotropic garnet shows compositional zoning between Al$_2$O$_3$ and Fe$_2$O$_3$, and occurs in an euhedral form surrounded by calcite. That is, compositional zoning indicates that the optical anisotropism should not be observable in either end-members of grandite garnet, and the presence of calcite indicates that carbon dioxide should have an important role in the formation of optical anisotropism.

Skippen (1974) derived stability relations among the minerals quartz, calcite, dolomite, talc, tremolite, diopside, forsterite and enstatite. The reaction limiting the diopside field is as follows:

$$\text{diopside} + CO_2 + H_2O = \text{calcite} + \text{tremolite} + \text{quartz}. \tag{8}$$

The equilibrium boundary for Reaction (8) at 2,000 bars is shown in Fig. III–8–3. Diopside is stable on the CO$_2$-poor side of the boundary.

Let us consider the amount of carbon dioxide in fluids which are favorable for the formation of ore-bearing skarns. An ore-bearing skarn, with a few exceptions, is characterized by the presence of calcite, quartz, clinopyroxene and/or grandite garnet, and by the lack of wollastonite and/or xonotlite. The condition of formation of an ore-bearing skarn is the same as that under which the assemblage calcite-quartz, grandite garnet and/or clinopyroxene are always formed, and where wollastonite and xonotlite are unstable. Figure III–8–3 shows the equilibrium boundaries for some stable reactions in the systems CaO-SiO$_2$-H$_2$O-CO$_2$, CaO-Al$_2$O$_3$-SiO$_2$-H$_2$O-CO$_2$, CaO-Fe$_2$O$_3$-SiO$_2$-H$_2$O-CO$_2$ and CaO-MgO-SiO$_2$-H$_2$O-

CO_2. Although the stability of grandite garnet in H_2O-CO_2 mixtures has not been determined yet, the limit of the grandite field probably passes through some places between or near the stability limits of grossular and andradite in accordance with its Fe_2O_3 content. Accordingly, the stability limit of andradite corresponds to the upper limit of stability of grandite garnet. On the other hand, the stability of hedenbergite in H_2O-CO_2 mixtures has not been determined at all. In this section, let us assume that the stability field of hedenbergite scarcely differs from that of diopside. The above-mentioned consideration and assumption indicate that the lower and upper limits of the amount of carbon dioxide in fluids under which an ore-bearing skarn can be formed is 0.5 and 6 mole % at 400°C, and 6×10^{-4} and 0.06 mole % at 200°C by the stability relations of the assemblage calcite-quartz and grandite garnet, respectively.

The equilibrium boundary for Reaction (7′) does not pass through the two fluid regions determined by Takenouchi and Kennedy (1964). However, it is highly possible that the boundary line intersects the two fluid regions, because the boundary line was determined with a considerable amount of error (Shoji, 1977c). Moreover, in general the boundary line passes through the more CO_2-rich side at lower pressures (Shoji, 1977a), while the two fluid region extends toward all directions in Fig. III–8–3 (Takenouchi and Kennedy, 1964). If any equilibrium boundary intersects the two-fluid region, many interesting modes of occurrence should be found. For example, hematite will be formed along the interface between liquid and vapor in accordance with the equilibrium boundary for Reaction (7′).

Oxygen Fugacity

Gustafson (1974) determined the hydrothermal phase relations for the bulk compositions $3CaO \cdot 2FeO_x \cdot 3SiO_2$ and $CaO \cdot FeO_x \cdot 2SiO_2$ with solid phase oxygen buffers to control f_{O_2}. At 2,000 bars, andradite is stable above an oxygen fugacity of 10^{-15} bar at 800°C and 10^{-32} bar at 400°C. At low oxygen fugacities, andradite gives way at successively lower temperatures to the assemblages magnetite-wollastonite, kirchsteinite($CaFe^{2+}SiO_4$)-wollastonite and kirchsteinite-xonotlite. On the other hand, hedenbergite is stable below an oxygen fugacity of 10^{-13} bar at 800°C and 10^{-28} bar at 400°C. Above these oxygen fugacities, hedenbergite breaks down to the assemblage andradite-magnetite-quartz. Figure III–8–5 shows an isobaric, 2000 bars fluid pressure, log f_{O_2}-1000/T diagram in which the projection of the end member of andradite and the end-member of hedenbergite stability fields have been plotted. Figure III–8–5 indicates that the assemblage andradite-hedenbergite is almost stable between Ni-NiO and iron-magnetite buffers. If the oxygen fugacity is controlled by carbon mixing into pure H_2O-CO_2 fluids (Yui, 1966), the amount of excess carbon is estimated to be 10^{-5} to 10^{-3} mole per H_2O-CO_2 of 1 mole.

Shimazaki (1974) discussed the genesis of a scheelite-bearing skarn. The main constituents of the skarn are exclusively ferric iron-poor minerals—that is, clinopyroxene and garnet close to hedenbergite and grossular, respectively. Opaque minerals associated with scheelite are usually chalcopyrite and pyrrhotite and sometimes magnetite, pyrrhotite and pyrite, but are always free from hematite. These mineral assemblages indicate that the scheelite-bearing skarn was formed under an environment not higher than the oxygen fugacity represented by the assemblage magnetite-pyrrhotite-pyrite. The isobaric log f_{O_2}-1/T relations for the reaction,

$$\text{magnetite} + \text{pyrite} = \text{troilite} + O_2,$$

in the system Fe-S-O can be determined using the thermochemical data compiled by Robie and Waldbaum (1968). The equilibrium boundary at 2,000 bars nearly coincides with the upper limit of the hedenbergite field determined by Gustafson (1974), and passes approximately through the vicinity of the buffer line of Ni-NiO (Fig. III–8–5). Therefore, the assemblage magnetite-pyrrhotite without pyrite indicates lower oxygen fugacities than the Ni-NiO buffer.

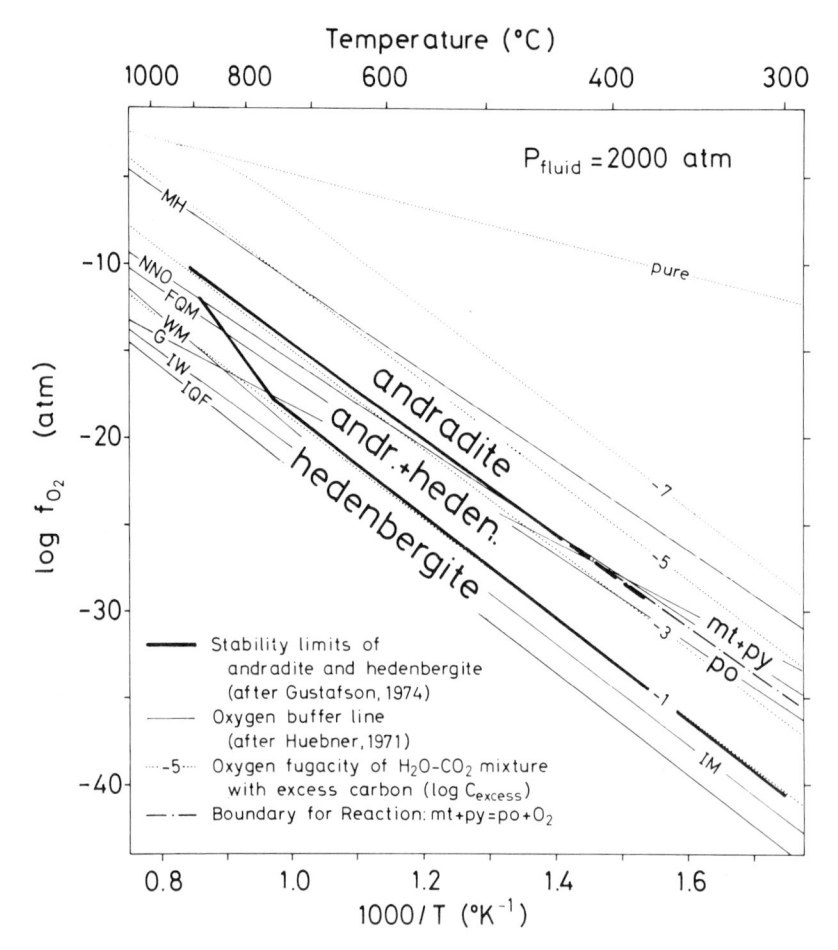

Fig. III–8–5. Projections of log f_{O_2}-1000/T stability range (thick lines) of pure andradite and hedenbergite at 2,000 bars fluid pressure (after Gustafson, 1974). Equilibrium boundary (a dash-dotted line) for the reaction, magnetite + pyrite = pyrrhotite + O_2, was calculated using the thermochemical data compiled by Robie and Waldbaum (1968). Thin lines represent buffer curves (after Huebner, 1971). Abbreviations of buffers: MH: magnetite-hematite, G: graphite, NNO: Ni-NiO, FMQ: fayalite-magnetite-quarts, WM: wüstite-magnetite, IM: iron-magnetite, IW: iron-wüstite, and IQF: iron-quartz-fayalite. Dotted lines represent oxygen fugacity of H_2O-CO_2 mixtures with excess carbon. Numerals indicate logarithms of mole fraction of excess carbon.

Iron-wollastonite of $Ca_5FeSi_6O_{18}$ composition is frequently found in some skarn-type ore deposits, such as the Sampo (Matsueda, 1973a, b), Kagata, Yamato, Ohta (Shimazaki and Yamanaka, 1973), Kamaishi (Matsueda, 1974), Tsumo (Matsuoka, 1976) and Kasugayama mines (Shimazaki and Bunno, 1976). Since iron-wollastonite is always associated with clinopyroxene of hedenbergite composition (Matsueda, 1973b; Shimazaki and Yamanaka, 1973), Shimazaki (1976) concluded on the basis of experimental results in the system $CaSiO_3$-$CaFeSi_2O_6$ by Rutstein (1971) and Liou (1974) that the assemblage of iron-wollastonite and hedenbergite seems to be stable under lower oxygen fugacities than the Ni-NiO buffer.

The writer considers that the lower limit of oxygen fugacity for the formation of ore-bearing skarns may be indicated by the mineral assemblage of scheelite-bearing and/or iron-wollastonite skarns. On the other hand, the upper limit may be represented by the zonal arrangement of magnetite, hematite and pyrite.

The mineralogical zoning in the system Fe-S-O is frequently observed in some parts of the

Chichibu mine (Shoji, 1969a; see Chapter III-6). In the Takishita deposit, the zones consist of magnetite, hematite or foliated magnetite, and pyrite from garnet skarn to marble. In the Doshinkubo deposit, a narrow band of hematite is frequently found between the magnetite and pyrite ore bodies (Fig. III-6-3). These zonal arrangements indicate that the oxygen and sulfur fugacities of the formation of these minerals are approximately represented by the magnetite-hematite-pyrite buffer, and that both fugacities increase from the magnetite to pyrite zones.

Zoning Found in Skarn-Type Ore Deposits

Skarn-type ore deposits frequently show mineralogical zonings. One of them corresponds to the district zoning found in vein-type ore deposits. The volume ratio of garnet skarn to clinopyroxene skarn decreases with increasing distance from the genetically related igneous rocks (Shoji, 1977b). The iron deposits consist mainly of garnet, while the zinc deposits consist mainly of clinopyroxene.

Miyazawa (1959b) classified the important skarn-type deposits of Japan into the following three types on the basis of their mineralogical assemblages and modes of occurrence: (i) Kamaishi type, (ii) Kamioka-Nakatatsu type and (iii) Chichibu type. Ore deposits of the Kamaishi type are located in the Kitakami and Abukuma mountain areas. The Kamaishi, Akagane and Yaguki deposits belong to this type. The ore deposits are typical pyrometasomatic type and occur close to the genetically related igneous rock. The ore consists predominantly of magnetite, chalcopyrite and pyrrhotite and is free from sphalerite and galena. Ore deposits of the Kamioka-Nakatatsu type are located in the Hida mountain area and its vicinity. The Kamioka and Nakatatsu deposits belong to this type. The ore deposits occur away from the related igneous rock. The ore consists essentially of sphalerite and galena, and a small amount of chalcopyrite and magnetite. Ore deposits of the Chichibu type are located in the Kanto mountain area. The ore deposits occur close to the related igneous rock, but seem to be formed in shallower positions than those of the Kamaishi type. The ore consists of magnetite, chalcopyrite, pyrrhotite, sphalerite and galena. The deposits of this type are also characterized by a large amount of pyrite. Ore deposits of the Kamaishi, Kamioka-Nakatatsu and Chichibu types proposed by Miyazawa (1959b) correspond to the hypothermal-mesothermal, mesothermal-epithermal and xenothermal ones of the vein-type, respectively.

Shimazaki (1975) showed the significance of Miyazawa's (1959b) classification by his statistical investigation. The ratios Cu/Zn-Pb and the chemical compositions and textures of genetically related igneous rocks indicate that Cu-rich deposits are associated with plutonic igneous activity, and Zn-Pb-rich ones with hypabyssal to effusive activity. Most of the total production of lead came from the deposits associated with hypabyssal to effusive rocks, and only 4.1 % of the production came from the deposits of the Chichibu and Bandojima mines, which are associated with plutonic quartz diorite.

The Cu-rich and Zn-Pb-rich belts are recognized as a regional zoning in the Inner Zone of southwest Japan (Shimazaki, 1975). Shimazaki (1975) suggested that this pattern was caused by a tilted erosion level through an acidic igneous activity of the Late Cretaceous to Early Paleogene.

In an ore-bearing unit, clinopyroxene skarn occurs closer to limestone than garnet skarn (Shoji, 1977b). This corresponds to the ore body zoning proposed by Riley (1936). The zoned skarns of the Doshinkubo deposit, the Chichibu mine (Fig. III-6-3) and the Yaguki deposits, which are developed along the contact between limestone and slate (Fig. III-8-5), are typical examples of this zoning. The former zoned skarn consists of mixed rock, garnet-epidote skarn, magnetite ore, and hedenbergite-ilvaite skarn, from quartz diorite to limestone (Fig. III-6-3). The latter skarn shows a zonal arrangement from slate to limestone: epidote skarn, andradite skarn, clinopyroxene skarn and quartz-scheelite rock (Shimazaki, 1969). In these examples, clinopyroxene occurs near limestone while garnet occurs away from limestone. On the other hand, the zoned skarn of the Mochikura mine consists of hedenbergite skarn and garnet skarn from granodiorite to limestone (Watanabe, 1932), and that of the Ofuku mine consists of pyroxene skarn, garnet skarn and wollastonite skarn from chert to limestone (Suzuki, 1932). That is, the zonal arrangement of garnet and clinopyroxene of these mines is reversed. As shown in Fig. III–8–5, andradite is stable at relatively high oxygen fugacities, while hedenbergite is stable at low oxygen fugacities (Gustafson, 1974). The dotted lines shown in Fig. III–8–5 indicate that the drop in oxygen fugacity caused by the mixing of carbon is on the same order for both water and carbon dioxide fluids. If the oxygen fugacity of CO_2 fluid is lower than that of the supplied aqueous fluid, because the CO_2 fluid is generated from graphite-bearing limestone, the oxygen fugacity of the mixed fluid decreases with increasing amount of carbon dioxide. In this case, hedenbergite is stable near limestone where the CO_2 pressure is high, while andradite is stable away from limestone where the CO_2 pressure is low. The zonal arrangement is expected to be normal. On the other hand, where the oxygen fugacity of aqueous fluid is lower than that of the CO_2 fluid, a reverse zonal arrangement is expected. This is one possible explanation for the genesis of the normal and reverse zonal arrangements of garnet and clinopyroxene (Shoji, 1976a).

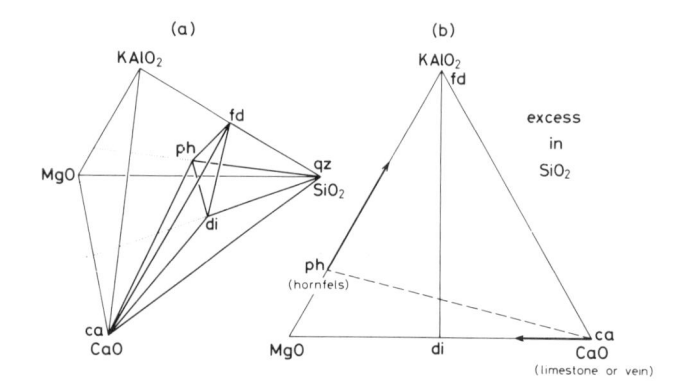

FIG. III–8–6. Phase relation among calcite, diopside, Kfeldspar, phlogopite and quartz (after Shoji, 1977b). (a) $KAlO_2$-CaO-MgO-SiO_2 tetrahedron. (b) Projection of the tetrahedron from the SiO_2 apex. Two arrows show the compositional change of hornfels and limestone caused by the migration of MgO.

Abbreviations: ca : calcite, di: diopside, fd: K-feldspar, ph: phlogopite, and qz: quartz.

The genesis of the zonal arrangement cannot be explained without the concept of ion migration, because the bulk composition of any zone is not the same. The zonal arrangement of feldspar-bearing skarn, which is shown in Fig. III–8–1, is interpreted as follows (Shoji, 1977b): The triangle calcite-phlogopite-quartz is cut by the tie-line between diopside and K-feldspar as shown in Fig. III–8–6(a). In case the system has an excess of SiO_2, the tie-line between calcite and phlogopite changes to the tie-line between diopside and K-feldspar, as

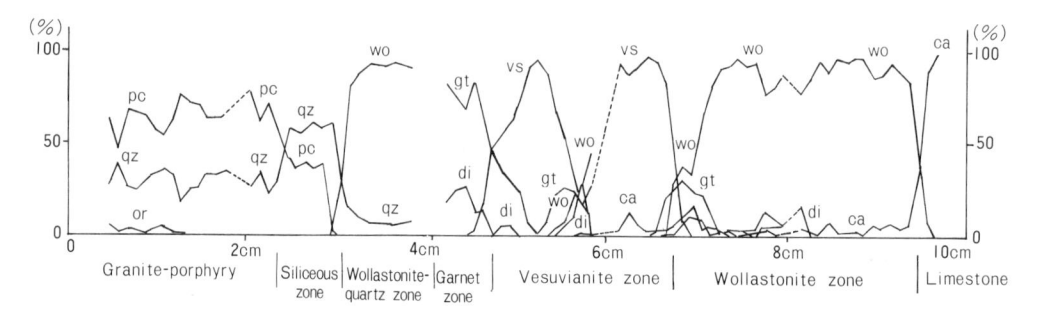

FIG. III–8–7. Modal analysis of the zoned skarn of the Obuki-nishi deposit, the Mitate mine (after Shoji, 1975). Abbreviations; or: orthoclase, pc: plagioclase, qz: quartz, wo: wollastonite, gt: garnet, di: diopside, vs: vesuvianite, and ca: calcite.

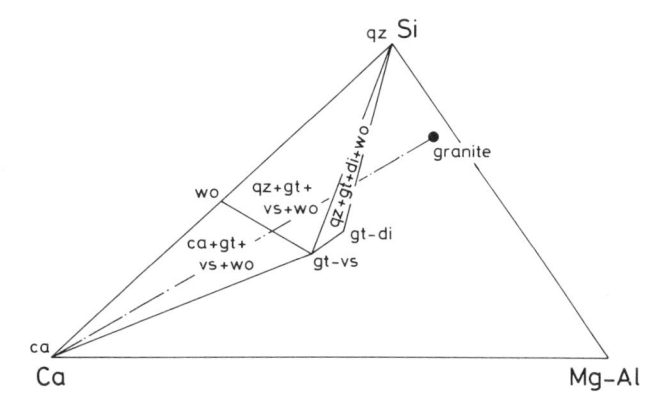

FIG. III–8–8. Profile of the Si-Ca-Mg-Al tetrahedron passing through the points: granite, Si and Ca. Dash-dotted line shows the compositional change in the reaction between granite and limestone.

shown in Fig. III–8–6(b). If MgO migrates faster than $KAlO_2$ and CaO, the bulk compositions of limestone and hornfels change toward the directions shown by the arrows in Fig. III–8–6(b). The difference in the migration rate of the components results in the formation of the zonal arrangement of feldspar-bearing skarn.

A similar interpretation can be applied to the formation of the zoned skarn between granite porphyry and marble of the Mitate mine (refer to Chapter III–9). From the contact with marble, the zoned skarn consists of wollastonite-quartz, vesuvianite, garnet and wollastonite (Fig. III–8–7) (Shoji, 1975). Granite porphyry consists essentially of plagioclase and quartz with orthoclase and biotite as accessory minerals. The amounts of quartz and sericite in the granite porphyry increase towards the contact. The wollastonite-quartz zone consists of a fine mixture of wollastonite and quartz. The garnet zone consists of grossular and diopside, while the other zones are almost mono-mineralic (Fig. III–8–7). Every zone contains an approximately constant amount of oxygen ion—that is, 0.08 mole/cm^3. For this reason, let us consider only the migration of cations in the following discussion. The main elements contained in the zoned skarn are calcium, magnesium, aluminum and silicon, except for oxygen, carbon and hyrogen. Figure III–8–8 shows a profile of the Si-Ca-Mg-Al tetrahedron, which passes through the points: Ca, Si, and the composition of granite. When a skarn is formed by the reaction between granite and marble, the expected zonal arrangement is as follows: quartz-wollastonite-garnet-diopside, quartz-garnet-vesuvianite-wollastonite and calcite-garnet-wollastonite, from the contact outwards. This arrangement is similar to that of the zoned skarn. However, the ratio of minerals in every zone differs from field observation. The above-mentioned arrangement is expected in a case where the migration rate of all ions is equal to one another. Accordingly, the fact that every ion has its own migration rate is one of the most likely reasons for the difference between the natural occurrence of a zoned skarn and the expected zonal arrangement.

Skarnization and Mineralization

A close relation between sulfide minerals and skarn minerals has been observed. Generally, sphalerite and galena are associated with pyroxene skarn, magnetite with garnet skarn, and chalcopyrite with both skarns.

Garnet coexisting with magnetite or chalcopyrite decrepitates frequently, while garnet without ore minerals does not (Shoji, 1970). The two types of garnet are similar from the viewpoint of composition. The differences in decrepitation pattern seem to be caused by the amounts of defect, such as fluid inclusions, formed during the crystal growth of garnet. Two reasonable interpretations are considered for the difference in the decrepitation pattern, on the assumption that ore minerals were formed later than garnet: (i) Ore minerals were deposited in skarn consisting of garnet with many defects. (ii) The defects in garnet crystals were increased by the ore mineralization. On the other hand, if ore minerals and skarn minerals were formed at the same time, ore bodies should be formed in the particular positions where garnets with many defects were formed. In any case, the difference in the decrepitation pattern suggests that garnet coexisting with magnetite and/or chalcopyrite was influenced by the ore mineralization, and that the common point is found between the formation of sulfide minerals and that of magnetite.

The relation between skarn and sulfide ore minerals in skarn-type ore deposits is similar to that between the intrusive body and ore minerals in porphyry copper deposits. Ore minerals of porphyry copper deposits occur in many veinlets as a network. Some authors state that ore minerals were derived directly from the intrusive magma. Others state that they were derived from the magma located on the lower levels. Similarly, two opposite opinions have been proposed for the relationship between skarn and sulfide minerals. Some authors suggest that skarn and ore minerals are deposited from the same fluid. Others suggest that ore minerals are formed later than skarn minerals. Occasionally, sulfide minerals occur on the grain boundary of skarn minerals or cut the skarn as veinlets. These modes of occurrence seem to suggest that sulfide minerals were formed after the formation of the skarn. Kitamura (1974) stated on the basis of thermochemical calculations that the previously formed hedenbergite skarn acts as an oxygen buffer for the formation of galena. On the other hand, Shimazaki (1967) concluded that the fluids from which an ore-bearing skarn was formed contained some amount of zinc. This conclusion is inferred from the fact that electron microprobe analyses suggested the presence of zinc in the skarn minerals associating with sphalerite, and its absence in those without sphalerite.

In the Shinyama deposit of the Kamaishi mine, pyrite is accompanied by garnet skarn, and pyrrhotite by clinopyroxene skarn (Fig. III–5–7). This field evidence is consistent with the phase relations shown in Fig. III–8–5. In a system with excess magnetite, the boundary between pyrite and pyrrhotite fields passes approximately through the upper stability limit of hedenbergite. It is suggested, therefore, that the formation of sulfide minerals was controlled by oxygen fugacity. For example, the assemblage, monoclinic pyrrhotite-hexagonal pyrrhotite-chalcopyrite was formed under the condition buffered by the reaction, hedenbergite = andradite + magnetite + quartz + O_2.

III–9. Malayaite (Takenouchi and Shoji, 1969; Takenouchi, 1971b)

S. Takenouchi

Malayaite, $CaSnO\ SiO_4$, was first found at Perak, Malay Peninsula (Ingham and Bradfood, 1960), and was named by Alexander and Flinter (1965). The crystal system of the mineral is monoclinic, with the lattice constants a = 6.66Å, b = 8.89Å, c = 7.15Å and β = 113°20′ (Ramdohr and Strunz, 1967), or a = 7.173 ± 0.009Å, b = 8.876 ± 0.005Å, c = 6.688 ± 0.008Å and β = 113.7 ± 0.1° (Takenouchi, 1971). The crystal structure has been refined recently by Higgins and Ribbe (1977).

Ramdohr (1935) reported the occurrence of tin-bearing sphene coexisting with ordinary sphene in aggregates of green garnet and wollastonite from Arandis, Southwest Africa. Although Ingham and Bradfood (1960) found that malayaite from Perak contains little titanium, it is reported that in some cases malayaite contains 10 ~ 20% titanium. From this fact, it is presumed that a solid solution exists between malayaite and sphene, and that malayaite is isomorphous with sphene. The position of titanium in the crystal structure of sphene is probably partially substituted for by tin (Ramdohr and Strunz, 1967). The occurrence of malayaite in a thin wollastonite and diopside bed has also been reported by

El Sharkawi and Dearman (1966) in Devonshire, England. John (1967) reported the optical character of malayaite from Pinyok, Thailand. In Pinyok, malayaite occurs with diopside, andradite, cassiterite, quartz and calcite.

In Japan, the occurrence of malayaite was first reported by N. Imai *et al.* (1967) in the Sampo mine, Okayama Prefecture, as a malayaite-like mineral. From the Tsumo mine, Shimane Prefecture, it was also reported by Shimazaki (1968a) as a tin-bearing sphene in a wollastonite-diopside-green garnet zone between limestone and massive skarn.

Malayaite was also found by Takenouchi and Shoji (1969) in tin-bearing skarn deposits of the Hoei mine, Ohita Prefecture, and the Mitate mine, Miyazaki Prefecture, and in tungsten-bearing skarn deposits of the Kuga mine, Yamaguchi Prefecture. They proposed the environments for formation of malayaite from field evidence and suggested the possibilities of finding malayaite in similar skarn deposits in southwestern Japan, such as the Toroku mine, Miyazaki Prefecture, and the Kiwada mine, Yamaguchi Prefecture. Later Miyahisa *et al.* (1975) reported the occurrence of malayaite in the Toroku mine, and Ogawa (1975) has reported data on malayaite from the Sampo mine.

Mitate Mine, Miyazaki Prefcture (Fig. I–3)

The ore deposit of the Mitate mine, Miyazaki Prefecture, is one of the cassiterite deposits which were formed related to the acidic igneous activity of the Miocene age (Miyahisa, 1961). The deposits in this district are the Obira, Toroku, Kiura, Hoei and Mitate mines.

Geologically, this area consists of limestone, sandstone, chert and slate of Permian age, and the so-called "Mitate Conglomerate," the age of which is controversial but is probably Tertiary (Miyahisa *et al.*, 1971). The ores consist mainly of pyrrhotite and cassiterite, and occur in two different modes. The first type, represented by the Mitate Hompi (Main Vein), develops in skarns formed along the boundary between limestone and conglomerate

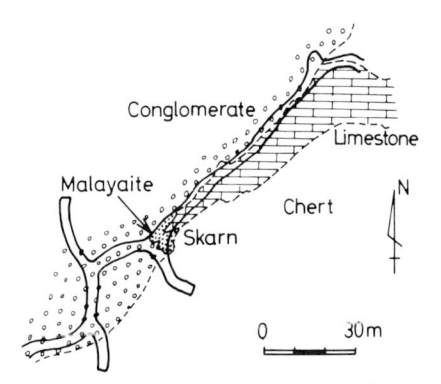

FIG. III–9–1. Underground geologic map of the Mitate Hompi (Main Vein).

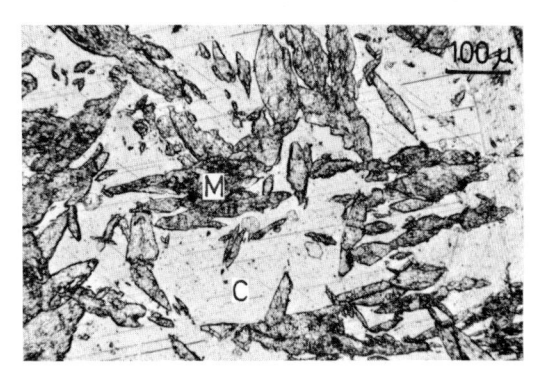

FIG. III–9–2. Photomicrograph of malayaite in calcite (Mitate). M : malayaite, C : calcite. One nicol.

or chert. The second type, represented by Ohbuki skarn ore bodies, occurs along the boundary between limestone and intruded granite porphyry.

Malayaite is found in a skarn formed along the boundary between "Mitate Conglomerate" and Paleozoic chert and limestone in the Mitate Hompi vein (Fig. III-9-1). The skarn, consisting mainly of clinopyroxene and calcite and a minor amount of cassiterite, green garnet and sphene, forms an irregular lens-shaped body, several meters in length.

Malayaite occurs in the aggregates of calcite which have contact with green skarn, forming slender streaks or small aggregates. When malayaite concentrates in calcite, parts of the calcite show a pale yellowish-brown color. Under the microscope, malayaite is very similar to sphene in its shape and optical character. It is easy, however, to distinguish malayaite from sphene because of the distinctive pale greenish-yellow fluorescent color emitted by short-wave ultraviolet ray. As shown in Fig. III-9-2, malayaite shows a slender spindle-like shape, the size of which is generally 50 microns × 20 microns but can be as large as 270 microns × 60 microns. It occurs in calcite but never in clinopyroxene or andradite.

Hoei Mine, Ohita Prefecture (Fig. I-3)

The ore deposit of the Hoei mine belongs to the same metallogenetic province of tin of Miocene age in Kyushu as the Mitate deposit. The Main ore body, which is formed in Paleozoic limestone, is a massive skarn at upper levels but platy at lower levels. The ore comprises sphalerite, pyrrhotite and some chalcopyrite, arsenopyrite and pyrite. Cassiterite occurs in dissemination in sulfide ores. The skarn consists mainly of andradite but also contains some clinopyroxene, wollastonite and fluorite.

Malayaite was found in several parts of the skarn body. In the ore body, malayaite occurs in calcite and quartz veinlets which cut massive skarns. Slender prismatic crystals of malayaite are found along the walls of the veinlets (Fig. III-9-3).

In the northeastern part of the ore body, malayaite occurs in the boundary zones of the network-like green skarn formed in the crystalline limestone. The boundary zones surround blocks of limestone and consist of wollastonite, clinopyroxene and garnet zones from the inner blocks to the outer. Malayaite is found in crystals of quartz and calcite of the wollastonite and clinopyroxene zones, forming slender prismatic crystals and, in some places, showing an incomplete radial aggregate.

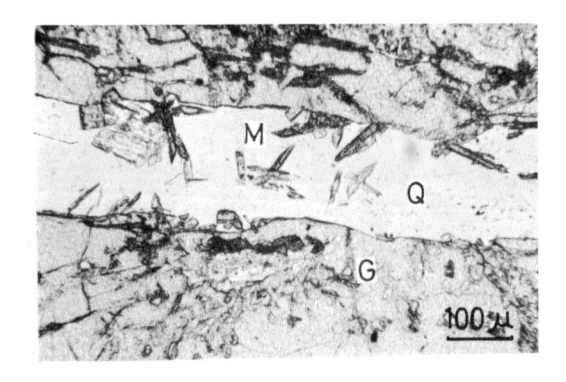

FIG. III-9-3. Photomicograph of malayaite in a quartz veinlet cutting garnet skarn (Hoei). Q : quartz, G : garnet. One nicol.

Toroku Mine, Miyazaki Prefecture (Fig. I-3)

According to Miyahisa et al. (1975), malayaite is generally found in wollastonite skarns with some arsenopyrite and löllingite, but is rarely found in hedenbergite and/or garnet skarn. Cassiterite was found not with malayaite but with diopside-hedenbergite. Some crystals of malayaite were as long as 3 cm. In general, the crystal shape was irregular, but sometimes it was platy or wedge-like. The chemical composition of malayaite from the Toroku is reported as

$Ca_{0.978}(Sn_{0.980}Ti_{0.002})Si_{1.002}\ O_{5.000}$.

Kuga Mine, Yamaguchi Prefecture (Fig. I–3)

The geology of the Kuga mine is described in Chapter III–4. The ore deposit consists of scheelite-bearing skarn bodies, such as the Shikada, De-ai, Hashigatani, Ideno-oku, Taiho, Umenoki and Han-ei deposits. Ore bodies in the northwestern part of this region contain a small amount of tin in the ores. For instance, the Shikada contains 0.03 ~ 0.08% Sn, Ideno-oku 0.6 ~ 0.8% Sn, and Hashigatani 0.2 ~ 0.4% Sn.

In the Shikada No. 3 ore body, malayaite was found in the lower siliceous part, with quartz, garnet, calcite and some clinopyroxene and wollastonite. The mode of occurrence was similar to that of the Mitate mine.

In the Ideno-oku No. 4 ore body, malayaite was found in the highly silicified peripheral zone of the skarn body, forming very fine striped patterns. Under the microscope, fine streaks consisting of minute crystals of malayaite (10 ~ 35 microns in size) penetrate the structure of aggregate of mozaic quartz and garnet.

Sampo Mine, Okayama Prefecture (Fig. I–3)

The ore deposit is the magnetite-bearing skarn type formed in Paleozoic limestone which is in contact with granitic rocks. N. Imai *et al.* (1967) first reported the occurrence of malayaite in this deposit, and Ogawa (1975) presented precise data on it. The writer has also found malayaite in the highly silicified hornfels which has contact with the massive skarn at the eastern end of the ore body (Fig. III–9–4). Cassiterite, sphene and tourmaline were observed with malayaite under the microscope.

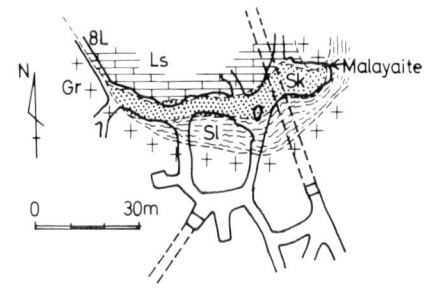

FIG. III–9–4. Underground geologic map of the No. 8 level of the Sampo mine.
 Ls: limestone, Gr : granitic rock, Sl: slate, Sk: skarn.

Environments for Formation of Malayaite

From field evidence, it appears that malayaite has been formed in an environment rich in silica and calcium, because it is generally found in quartz- or calcite-rich zones located between massive skarns and limestone or slate. In the other words, the mineral occurs in the peripheral zones of massive skarns. In any case, it occurs in the crystals of quartz or calcite but rarely in clinopyroxene or garnet, and when it is found in a massive skarn, it occurs in quartz or calcite veinlets which cut the skarn. In many cases, cassiterite accompanies malayaite, but this is not always the case. In the Mitate and Tsumo mines, tin-bearing green garnet accompanies malayaite.

From the results of macroscopic and microscopic studies of the genesis of malayaite, it is inferred that the mineral was formed by one of three processes:

1) When a small amount of tin existed at the skarnization, tin was concentrated at the front or the silicified zones of the skarn, and it reacted with calcium and silica to form malayaite.

2) Calcium- and silica-rich hydrothermal fluids reacted with tin-bearing minerals such as cassiterite and tin-bearing andradite in the skarn, and malayaite was formed using tin from these tin-bearing minerals.

3) Tin, calcium and silica were transported into veinlets by hydrothermal solutions, and malayaite was formed there with quartz and/or calcite.

Some preliminary hydrothermal experiments were carried out to synthesize malayaite from starting materials having the chemical composition of malayaite plus iron or magnesium. The products were mixtures of andradite or diopside, wollastonite and cassiterite. But when the starting material was deficient in iron or magnesium, malayaite mixed with andradite or diopside was the product. This fact agrees well with the natural occurrence of malayaite and suggests that malayaite would have been formed in an environment rich in calcium and silica but deficient in iron.

The complete solid solution between malayaite and sphene was experimentally determined, and it was found that the X-ray powder diffraction pattern changed successively from that of malayaite to that of sphene (ASTM Card 25–176). As shown in Fig. III–9–5, the top of the solvus is located at 615° ± 15°C, and the immiscibility region of the solid solution expands rapidly at lower temperatures, though the tin-rich side is somewhat wider than the titanium-rich side (Takenouchi, 1971).

From the shape of the immiscibility region of the malayaite-sphene system, it is considered that tin-rich sphene and titanium-rich malayaite seldom occur in nature, because the formation temperatures of these minerals in the skarn are probably lower than 500°C.

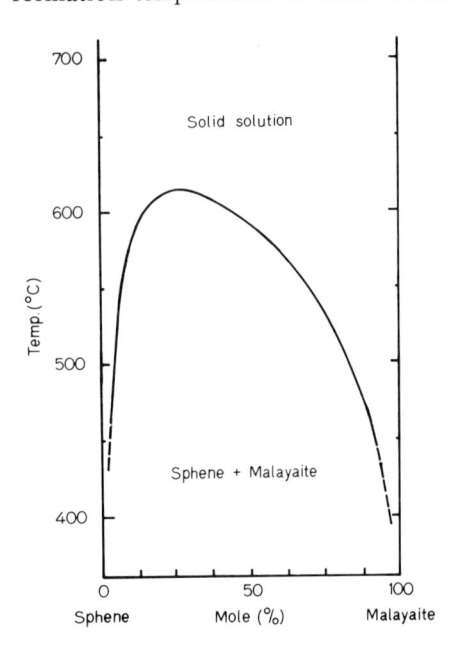

FIG. III–9–5. Phase diagram of the malayaite-sphene solid solution.

III–10. Cu-Fe-S Mineral Syntheses (Kaneda *et al.*, 1978)

H. Kaneda, T. Shoji, S. Takenouchi and H. Imai

The study of phase relations in the Cu-Fe-S system is necessary for a discussion of the genesis of copper sulfide minerals in ore deposits. The phase relations in the Cu-Fe-S system have been paid much attention since the contribution of Merwin and Lombard

(1937). Bartholomé (1958) and McKinstry (1959) showed a comparison of the assemblages in ore deposits with published experimental relations, and recognized many uncertainties in the phase relations, especially below 500°C. Yund and Kullerud (1966) determined the phase relations in the Cu-Fe-S system from 700°C to approximately 200°C. Barton and Skinner (1967) studied the 600°C isotherm in the Cu-Fe-S system. Mukaiyama and Izawa (1970) studied the phase relations in the copper-deficient part of the Cu-Fe-S system, and determined the phase transitions and thermal stability of pyrrhotite, chalcopyrite and cubanite. Cabri and Harris (1971), and Cabri and Hall (1972) discussed mineralogically the new minerals such as talnakhite, mooihoekite and haycockite. Cabri (1973) studied the phase relations in the central part of the Cu-Fe-S system at 100°C and 600°C. Sugaki *et al.* (1975) studied the Cu-Fe-S system at low temperatures under the hydrothermal conditions.

But, many questions still remain to be solved, especially below 300°C. Judging from the field evidence on the ore deposits, knowledge of the phase relations at low temperatures from 200°C to 300°C is a necessary basis for deciphering the genesis of the copper-sulfide deposits.

Description of Heating Experiments of Cubanite

The phase relations of cubanite can be discussed from the standpoint of polymorphism, and from the point of view of the decomposition of cubanite.

Up to the present, studies of cubanite from the standpoint of polymorphism have been generally accepted—that is, there have been many studies on the transformation of cubanite by heating. Phase relations based on the decomposition of cubanite were reported only by Borchert (1934); he stated that cubanite decomposes to two phases of chalcopyrite and chalcopyrrhotite above 235°C.

TABLE III-10-1. List of heating experiments with orthorhombic cubanite.

1)	Yund and Kullerud (1961)	ortho. < 200°C < tetra. < 270°C < cub.
2)	Sawada *et al.* (1962)	ortho. < 270°C < cub.
3)	Yund and Kullerud (1966)	ortho. < 213°C < tetra. < 252°C < cub.
4)	Genkin *et al.* (1966)	ortho. < 250°C < cub.
5)	Fleet (1970)	ortho. < 220°C < ortho.+ cub. < 230°C < cub.
6)	Mukaiyama and Izawa (1970)	ortho. < 200°C < ortho.+ cub. < 235°C < cub.
7)	Mizota (1971)	ortho. < 220°C < cub.
8)	Vassjoki (1971)	ortho. < 210°C < hex.
9)	Sugaki (1972)	ortho. < 230°C < ortho.+ cub. < 235°C < cub.
10)	Cabri *et al.* (1973)	ortho. < 200°C < ortho.+ cub. < 210°C < cub.
11)	Borchert (1934)	ortho. < 255°C < chalcopyrite + chalcopyrrhotite

ortho.: orthorhombic, tetra.: tetragonal, cub.: cubic, hex.: hexagonal.

Heating experiments on cubanite are listed in Table III-10-1. As shown in Table III-10-1, the transition temperature and the crystal systems of transition phases are not necessarily identical. Transformation to hexagonal phase has been suggested by Vaasjoki (1971), while the existence of a tetragonal polymorph between 213°C and 252°C ± 5°C was asserted by Yund and Kullerud (1966).

Cabri *et al.* (1973) reported that annealing of cubic cubanite below the inversion temperature results in chalcopyrite lamellae in cubic cubanite matrix. It does not coincide with the fact that orthorhombic cubanite is frequently found as exsolution lath in a chalcopyrite matrix.

Thus, though heating experiments on cubanite have been carried out with considerable attention since the contribution of Yund and Kullerud (1966), the phase relations of cubanite still remain in question.

Starting Materials and Techniques

Starting materials were as follows: (1) a small piece of polished section, (2) fine-grained cubanite powder from 100 mesh to 70 mesh separated by an isodynamic separator, (3) metallic iron, copper and elemental sulfur, (4) co-precipitated gel, (5) non-equilibrated sulfide compounds which were made by heating mixtures of copper, iron and sulfur at 300°C for several hours.

In the case of heating experiments on cubanite, the following two methods for the experiments were undertaken; (1) a small piece of polished section and fine-grained cubanite powder were sealed into evacuated rigid silica-glass tubes; (2) fine-grained cubanite powder was sealed in a collapsible gold tube prepared for a test-tube type pressure vessel.

In the case of mineral syntheses above 500°C, metallic iron, copper and elemental sulfur were used as starting materials.

In the case of co-precipitated gel, cupric sulfate ($CuSO_4 \cdot 5H_2O$) and ferrous ammonium sulfate ($Fe(NH_4)_2(SO_4)_2 \cdot 6H_2O$) were used for Cu and Fe sources. The equivalent sodium sulfide ($Na_2S \cdot 9H_2O$) solution as S source was added to the weighted salt compound solution with 10% Na_2CO_3 solution. In this case, a pyrex glass tube was used as a container, and the heating was carried out at only 180°C.

In the case of the non-equilibrated sulfide compounds, each starting material with equivalent bulk composition was ground in order to mix the samples homogeneously.

They were heated in test-tube type pressure vessels at temperatures from 200°C to 400°C. After heating, all products were quenched rapidly in cold water or in compressed air. The products were identified by the ore microscopic and X-ray powder diffraction methods, and in some cases by electron microprobe analyses.

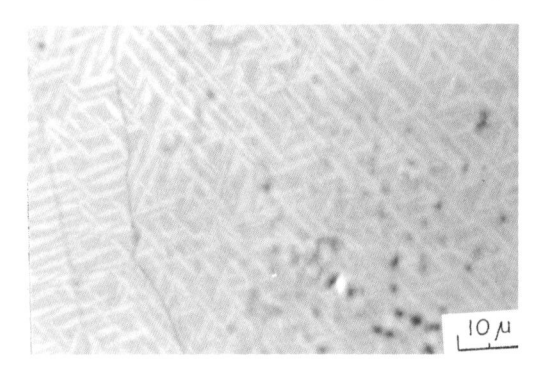

FIG. III–10–1. Photomicrograph. One nicol. Oil immersion. Cubanite heated at 400°C for 30 days. Darker area is chalcopyrrhotite, while lighter fine lamellae are chalcopyrite.

Results

CuFeS$_2$-FeS system

Heating experiments on cubanite: with microscopic studies, X-ray powder diffraction and electron microprobe analyses, it is recognized that cubanite decomposes into two phases.

A photograph under a high-powered microscope, X-ray powder diffraction patterns and the back-scattered electron image of cubanite heated at 400°C for 30 days are shown in Figs. III–10–1, –2 and –3.

The heated products of cubanite are composed of two phases of very fine texture, as Borchert (1934) stated. One phase, which forms the host of the texture, is optically very similar to chalcopyrrhotite, which was described by Ramdohr (1969). The other phase, which forms very fine lamellae in the host, corresponds optically to chalcopyrite.

Part III: Pyrometasomatic Deposits

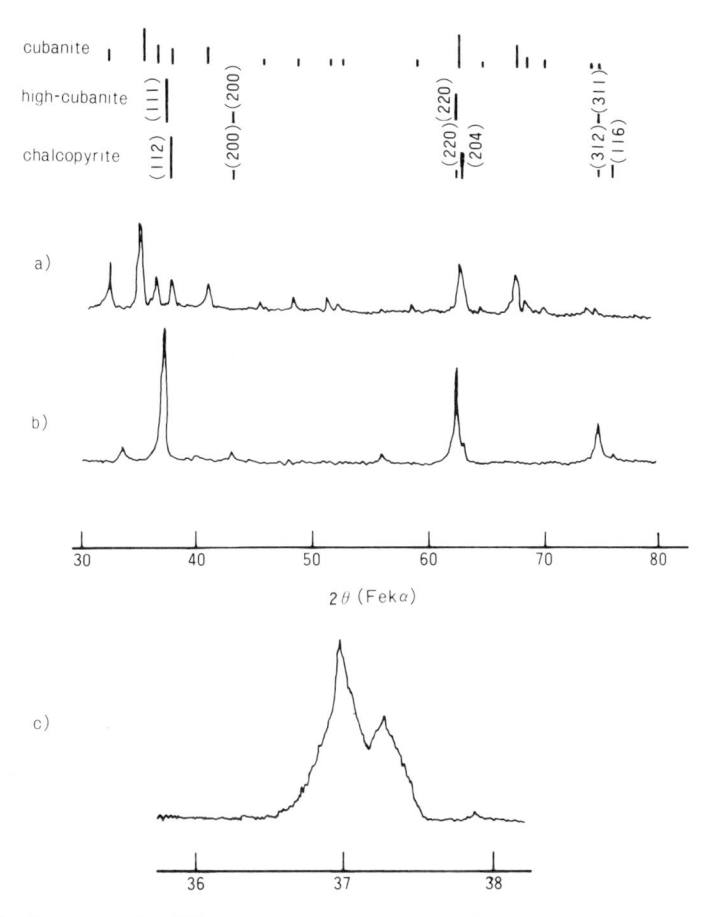

FIG. III–10–2. X-ray powder diffraction patterns. a) Pattern of the starting material before heating. The highest peak of chalcopyrite ($2\theta = 37.20°$) is not recognized. b) Diffraction pattern of heated cubanite at 400°C. c) Peaks of high-cubanite (chalcopyrrhotite) and chalcopyrite in the vicinity of 37° (2θ).
  a), b) FeKα, Mn-filter, slit: 1/2-1/2-0.15, scanning speed: 2°/min.
  c) FeKα, Mn-filter, slit: 1/6-1/6-0.15, scanning speed: 1/4°/min.

FIG. III–10–3. Back-scattered electron image of heated cubanite. Light areas and dark areas represent the Cu-rich and Fe-rich phases, respectively. Cross-cutting lines are images of scratch due to polishing.

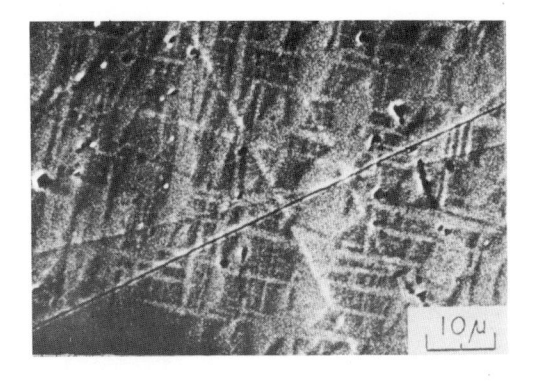

It was confirmed by high-temperature X-ray diffraction that these two phases of very fine texture were not formed by the process of rapid cooling.

As shown in Fig. III–10–2 (b), the peaks which are comformable with (204) and (116) of chalcopyrite and high cubanite are recognized. The peaks of chalcopyrite other than (204) and (116) correspond to those of high-cubanite except for the strongest peak. The strongest peaks of reported high-cubanite and chalcopyrite are recognized at very close sites.

Generally, in the case of diffraction at fast scanning speed (2°/min), two peaks are shown as one peak. The separated peaks at slow scanning speed are shown in Fig. III–10–2 (c). The one corresponds to the highest peak of high-cubanite, and the other to that of chalcopyrite.

Figure III–10–3 shows that the chemical compositions of the two phases are different; the host is richer in Fe-content and poorer in Cu-content than the guest.

As the texture of the two phases is very fine, exact quantitative analyses by electron microprobe are very difficult, but it is clear that the Fe content of the host is richer than that of cubanite, while the Cu-content of the guest is richer than that of cubanite.

As mentioned above, the host phase is optically very similar to chalcopyrrhotite as described by Ramdohr (1966), and coincident with the X-ray powder diffraction pattern of the so-called high-cubanite, but it is not sure whether the chemical composition is the same as that reported by Borchert (1934). The other phase, which forms very fine lamellae in the host, corresponds to chalcopyrite in chemical composition, optical properties under the microscope and X-ray powder diffraction pattern. In this paper, the former is called chalcopyrrhotite and the latter is presumed to be chalcopyrite.

The texture of the two phases becomes coarser with an increase in heating temperature or heating duration. In the case of lower heating temperature or shorter heating duration, the two phases can scarcely be distinguished because of the finer texture and may be regarded as only one phase (host) under the microscope and in the X-ray powder diffraction pattern.

The temperature of decomposition decreases with the increase of P_{H_2O} pressure under hydrothermal conditions, as shown in Table III–10–2; that is, under P_{H_2O} of 15 atm, 500 atm and 1500 atm, the temperature declines to 210°C, 200°C and 185°C, respectively.

In experiments using evacuated rigid silica-glass tubes, the decomposition temperature is 215°C.

Mineral syntheses in the $CuFeS_2$-FeS system: In order to determine the phase relations of chalcopyrrhotite, mineral syntheses in the $CuFeS_2$-FeS system have been carried out under

TABLE III–10–2. Results of heating experiments on cubanite.

Type of experiment	No.	Temp. (°C)	Duration (days)	Phases of products
Dry	Kamaishi*-3	210	30	cb
	Komori**-3	210	30	cb
	Kamaishi-11	215	30	cb, cpo, cp
	Komori-11	215	30	cb, cpo, cp
	Kamaishi-4	220	30	cpo, cp
	Komori-4	220	30	cpo, cp
	Kamaishi-7	550	30	cpo
Wet				
$P_{H_2O} = 15$ atm	Kamaishi-23	200	30	cb
$P_{H_2O} = 500$ atm	Kamaishi-44	190	30	cb
	Kamaishi-45	200	30	cpo, cp
	Kamaishi-46	210	30	cpo, cp
$P_{H_2O} = 1,500$ atm	Kamaishi-63	180	30	cb
	Kamaishi-65	185	30	cb, cpo, cp
	Kamaishi-64	190	30	cpo, cp

cb: orthorhombic cubanite, cpo: chalcopyrrhotite, cp: chalcopyrite.
* Refer to Chapter III–5.
** Refer to Chapters II–5 and–18.

hydrothermal and dry conditions. The methods and results of the experiments are shown in Table III–10–3.

i) Runs above 250°C

In all of the products above 250°C, chalcopyrrhotite, which is the same as the host phase of heated cubanite in optical properties and X-ray powder diffraction patterns, is obtained.

Bulk compositions of starting materials are on the $CuFeS_2$–FeS join. Synthetic products are grouped as follows: (1) chalcopyrrhotite-hexagonal pyrrhotite, (2) chalcopyrrhotite-chalcopyrite-hexagonal pyrrhotite, (3) chalcopyrrhotite-chalcopyrite, (4) chalocopyrrhotite.

The mineral assemblage of chalcopyrrhotite and hexagonal pyrrhotite is obtained in bulk compositions richer in Fe than $CuFe_4S_5$, which is the chemical composition of chalcopyrrhotite reported by Borchert (1934). The two phases are associated with each other very closely, that is, they occur as mutual boundaries in very fine granular aggregate ($< 10\mu m$), or pyrrhotite occurs as a network in fine grains of chalcopyrrhotite. In any case, chalcopyrrhotite grains are too small to analyze chemically by means of electron microprobe.

In the case of bulk compositions richer in Cu than $CuFe_4S_5$, products of mineral syntheses showed the following assemblages: (1) chalcopyrrhotite-chalcopyrite assemblage under dry conditions from 400°C to 500°C, (2) chalcopyrrhotite-chalcopyrite-pyrrhotite under wet conditions from 250°C to 350°C, (3) chalcopyrrhotite above 550°C.

That a three-phase assemblage is observed in wet conditions indicates that initial bulk compositions might shift into metal deficient one, in other words, a part of the metal would ionize to dissolve into hydrothermal solutions.

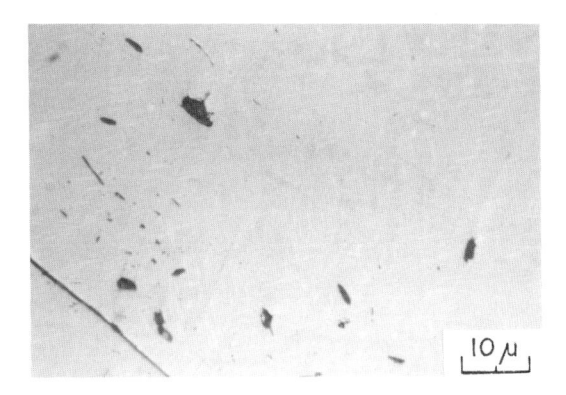

FIG. III–10–4. Photomicrograph. One nicol. Oil immersion. The hydrothermally synthesized product heated at 300°C for 60 days. Light parts of fine lamellae are chalcopyrite, and the darker areas are chalcopyrrhotite.

A photomicrograph of a synthetic product of the bulk compositions of $CuFe_2S_3$ at 300°C is shown in Fig. III–10–4. It is interesting that the texture is the same as that of heated cubanite.

At the bulk composition of $CuFe_2S_3$, the product at 500°C and 600°C showed only a single phase corresponding to chalcopyrrhotite in the X-ray diffraction pattern and the optical properties as well as the properties in heating experiments of cubanite. It is assumed that chalcopyrrhotite solid solution might be quenchable at the composition of cubanite.

ii) Runs below 240°C

As shown in Table III–10–3, in the case of P_{H_2O} of 500atm and 15atm, only the mineral assemblage of chalcopyrite and hexagonal pyrrhotite was obtained at any bulk compositions of starting materials, but in the case of P_{H_2O} of 1,500atm, and the bulk composition of $CuFe_2S_3$, the mineral assemblage of chalcopyrrhotite, chalcopyrite and a very small amount of hexagonal pyrrhotite was obtained at the heating temperatures of 185°C and 200°C, but at 160°C and 140°C, chalcopyrite and hexagonal pyrrhotite were obtained. As mentioned above, the formation temperatures of chalcopyrrhotite change with pressure, that is, the temperatures decrease with the increase of P_{H_2O}.

TABLE III-10-3. Results of syntheses in the CuFeS$_2$ – FeS system between 140°C and 600°C.

Bulk composition Reactants	Hydrothermal method											Dry method		
	P$_{H_2O}$ = 15 atm			P$_{H_2O}$ = 500 atm			P$_{H_2O}$ = 1,500 atm					
	Temp. (°C)	Duration (days)	Products	Temp. (°C)	Duration (days)	Products	Temp. (°C)	Duration (days)	Products	Temp. (°C)	Duration (days)	Products
CuFe$_9$S$_{10}$												
po, cp, cc	180	120	po, cp	200	90	po, cp				400	60	po, cpo
po, cp, cc				250	90	po, cpo				500	60	po, cpo
po, cp, cc				300	60	po, cpo						
CuFe$_4$S$_5$												
cp, po, cv	180	120	po, cp	200	90	po, cp				400	60	po, cpo
cp, po, cv				250	90	po, cpo				500	60	po, cpo
cp, po, cv				300	60	po, cpo						
CuFe$_3$S$_4$												
cv, cp, py	180	120	cp, po	200	90	cp, po						
cv, cp, py				250	90	cpo, cp, po						
Cu$_3$Fe$_7$S$_{10}$												
po, dig. cp, Fe	180	120	cp, po	200	90	cp, po				400	60	cpo, cp
po, dig. cp, Fe				250	90	cpo, cp, po				500	60	cpo, cp
po, dig. cp, Fe				300	60	cpo, cp, po						
CuFe$_2$S$_3$												
po, cv, bn, Fe	180	120	cp, po	200	90	cp, po	140	180	cp, po	400	60	cpo, cp
po, cv, bn, Fe				250	90	cpo, cp, po	160	180	cp, po	500	60	cpo, cp
po, cv, bn, Fe				300	60	cpo, cp, po	180	180	cp, po	550	60	cpo
							185	180	cpo, cp, po	600	60	cpo
cpo, cp*				190	30	cp, po	160	30	cp, po			
cpo, cp*				200	30	cpo, cp	180	30	cp, po			
cpo, cp*							185	30	cpo, cp			
Cu$_2$Fe$_3$S$_5$												
po, cv	180	120	cp, po	200	90	cp, po				400	60	cpo, cp
				250	90	cpo, cp				500	60	cpo, cp
				300	60	cpo, cp						

* Starting materials are cubanite heated at 300°C for 30 days. po: pyrrhotite, cp: chalcopyrite, cpo: chalcopyrrhotite, cc: chalcocite, cv: covelline, py: pyrite, dig: digenite, bn: bornite.

Part III: Pyrometasomatic Deposits

TABLE III-10-4. Results of syntheses in the CuFeS$_2$(Cu$_5$FeS$_4$)-FeS-FeS$_2$ system between 180°C and 300°C.

Bulk composition	Reactants	180°C Heating Duration (days)	180°C P$_{H_2O}$ (atm)	180°C Products	180°C Po d(102)	250°C Heating Duration (days)	250°C P$_{H_2O}$ (atm)	250°C Products	250°C Po d(102)	300°C Heating Duration (days)	300°C P$_{H_2O}$ (atm)	300°C Products	300°C Po d(102)
—	natural cb	30	15	cb									
—	natural cb	30	500	cb		30	500	cpo, cp		30	500	cpo, cp	
—	natural cb	30	500	cb									
—	cpo + cp*	30	15	cp, po		30	500	cpo, cp		30	500	cpo, cp	
—	cpo + cp*	300	500	cp, po									
—	cpo + cp*	30	1,500	cp, po									
CuFe$_2$S$_3$	CuSO$_4$·5H$_2$O Fe(NH$_4$)$_2$SO$_4$·6H$_2$O Na$_2$S·9H$_2$O	60	15	cp, po	2.070								
CuFe$_2$S$_3$	py, po, cv					30	500	cpo, cp, po	2.068	30	500	cpo, cp, po	2.068
CuFe$_2$S$_{2.9}$	cp, po, cv					30	500	cpo, po	2.069	30	500	cpo, cp, po	2.069
CuFe$_2$S$_{2.7}$	po, cv, cc, Fe	60	500	cp, po	2.079	50	500	cpo, cp					
CuFe$_2$S$_{2.5}$	cp, po, cv	72	15	cp, po	2.076	31	500	cpo, po, bn	2.089	32	500	cpo, po	2.071
CuFe$_2$S$_{2.25}$	cp, po, cc, Fe	72	15	cp, po	2.078	50	500	cpo, po, bn	2.090	50	500	cpo, po, bn	2.089
CuF$_2$S$_{1.75}$	CuSO$_4$·5H$_2$O Fe(NH$_4$)$_2$SO$_4$·6H$_2$O Na$_2$S·9H$_2$O	14	15	cp, py									
CuFeS$_3$	CuSO$_4$·5H$_2$O Fe(NH$_4$)$_2$SO$_4$·6H$_2$O Na$_2$S·9H$_2$O	16	15	cp, po	2.069								
CuFe$_6$S$_{13}$	py, po, cv	57	15	py, po, cp	2.056	41	500	cp, py, po	2.061				
CuFe$_7$S$_{12}$	po, py, cv	50	15	py, po, cp	2.057	50	500	cp, py, po, py	2.056	50	500	cp, py, po	2.057
CuFe$_3$S$_6$	py, cv, po	50	500	cp, py, po	2.061	41	500	cp, py					
Cu$_2$Fe$_7$S$_{11}$	po, py, cp, cv	57	15/500	cp, py, po	2.060/2.060	30	500	cp, po, py	2.062	30	500	cp, py, po	2.060
CuFe$_5$S$_{11}$	py, cv, po	60	15	cp, py, bn	2.072	60	500	cp, py, bn	2.076	60	500	cp, py, bn	2.072
CuFe$_4$S$_5$	cp, po, cv	60	15/500	cp, po	2.069	60	500	cpo, po		60	500	cpo, po	
CuFe$_9$S$_{10}$	po, cp, cc	60	15/500	cp, po	2.076/2.068	60	500	cpo, po	2.076	60	500	cpo, po	2.075
Cu$_7$Fe$_{13}$S$_{20}$	cp, dig, po	60	500	cp, po	2.072	60	500	cp, po	2.068	60	500	cpo, cp, po	2.067
Cu$_2$Fe$_3$S$_5$	po, cv, dig, cp					60	600	cp, po	2.067	60	500	cp, po	2.065
Cu$_4$Fe$_7$S$_9$	po, cp, bn	60	15	cp, po	2.076								
Cu$_8$Fe$_{13}$S$_{19}$	cp, cv, cc, po					60	500	cpo, cp		60	500	cpo, cp, po	2.068
Cu$_9$Fe$_{12}$S$_{19}$	cp, cv, po, dig									60	500	cpo, cp	

* Starting materials are cubanite heated at 300°C for 30 days. Abbreviations same as in Table III-10-3.

224　　　　　　　　　　　　　B. Genetical Problems

When the starting material was a chalcopyrrhotite-chalcopyrite mixture which was derived by heating cubanite, phases after heating under dry conditions at 180°C, 160°C and 140°C for 180 days were the same as the starting material (chalcopyrrhotite-chalcopyrite), but under hydrothermal conditions, the products were different from the heated cubanite and changed into an assemblage of chalcopyrite and hexagonal pyrrhotite at the following temperatures and pressures: (1) lower than 180°C at 1,500atm, (2) lower than 200°C at 500 atm, (3) lower than 210°C at 15atm.

All the experimental results in the CuFeS$_2$-FeS join are shown in the pressure-temperature diagram (Fig. III–10–5). In this diagram, it is recognized that the stable area of cubanite in the heating experiments with natural cubanite corresponds to that of the assemblage of chalcopyrite and hexagonal pyrrhotite in hydrothermal syntheses.

CuFeS$_2$ (Cu$_5$FeS$_4$)-FeS-FeS$_2$ system
Phase Relations at 180°C: The 180°C isothermal section is shown in Fig. III–10–6 (a),

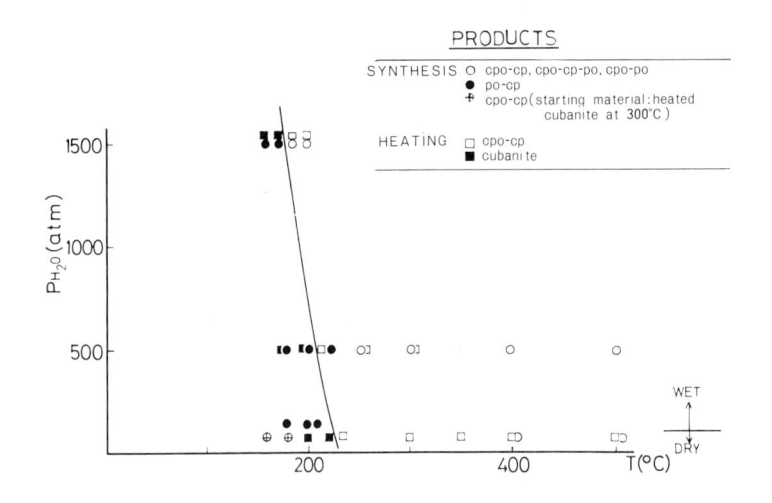

FIG. III–10–5. Temperature-pressure diagram of synthetic and heated products. cpo: chalcopyrrhotite, cp: chalcopyrite, po: pyrrhotite.

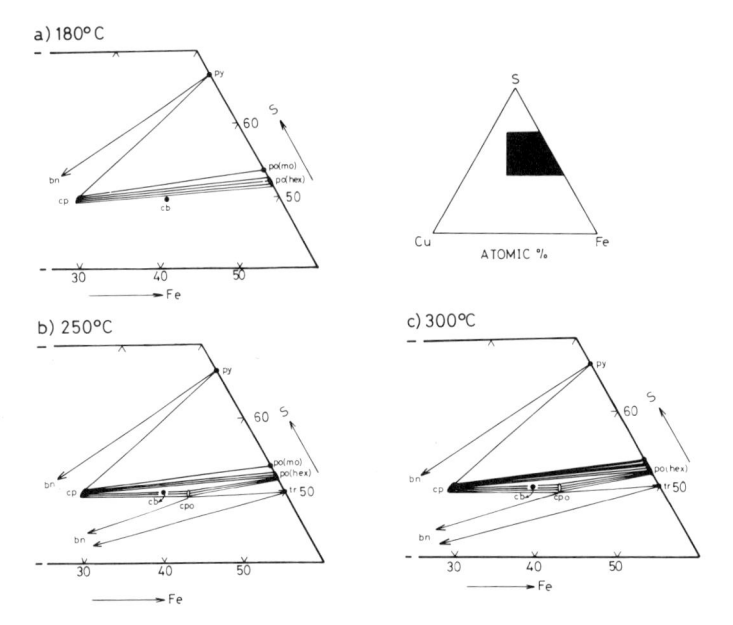

FIG. III–10–6. Isothermal sections in the CuFeS$_2$ (Cu$_5$FeS$_4$)-FeS-FeS$_2$ system at low temperatures.

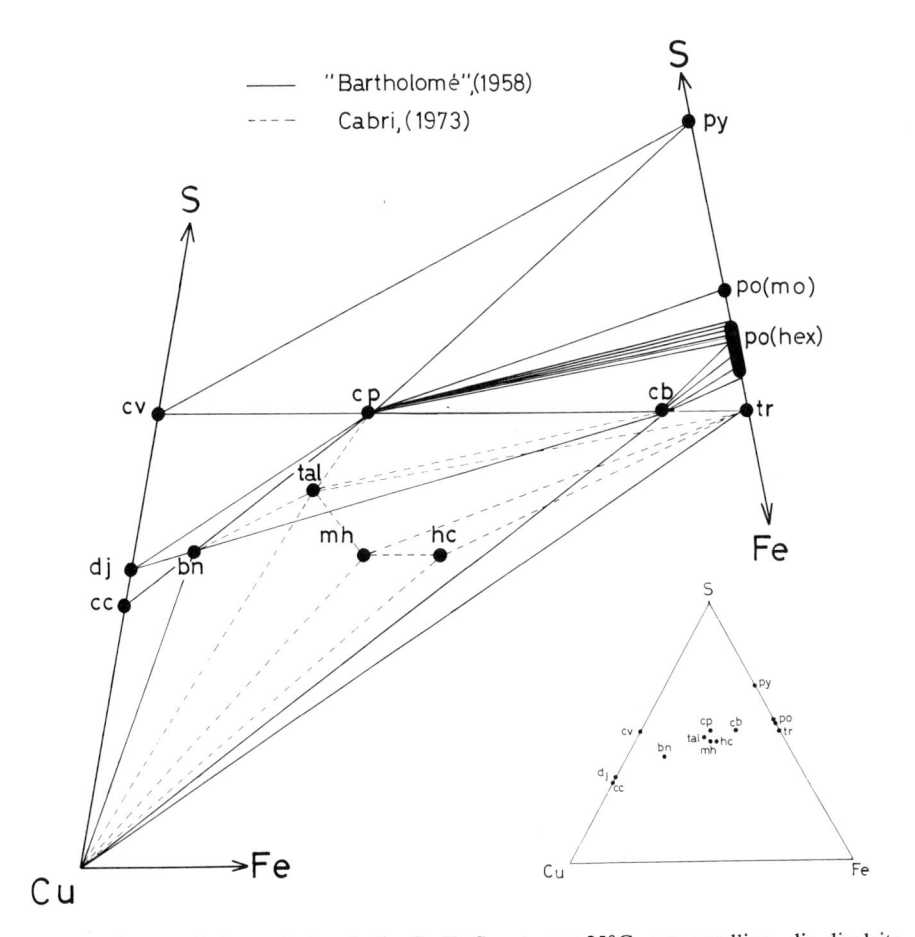

FIG. III-10-7. Presumed phase relations in the Cu-Fe-S system at 25°C. cv: covelline, dj: djurleite, cc: chalcocite, bn: bornite, cp: chalcopyrite, tal: talnakhite, mh: mooihoekite, hc: haycockite, cb: cubanite, tr: troilite, po(hex): hexagonal pyrrhotite, po(mo): monoclinic pyrrhotite, py: pyrite.

but at this temperature, cubanite was not obtained. The methods and products of experiments are shown in Table III-10-4.

The assemblages at 180°C are as follows: pyrite-chalcopyrite-bornite, pyrite-chalcopyrite-monoclinic pyrrhotite, chalcopyrite-monoclinic pyrrhotite-hexagonal pyrrhotite, chalcopyrite-hexagonal pyrrhotite.

The assemblage of cubanite, hexagonal pyrrhotite (troilite) and bornite was not obtained from the sulfur-deficient bulk compositions. It is assumed that the formation of this assemblage requires lower sulfur fugacity and longer heating duration.

Some tie-lines can be described between chalcopyrite and hexagonal pyrrhotite solid solution. The chemical composition of pyrrhotite was determined by the values of $d_{(102)}$ by means of X-ray powder diffraction as reported by Arnold (1966, 1967), while the chemical compositions of chalcopyrite solid solution were presumed on the grounds that the intensities of doublet peaks of $d_{(220)}$ and $d_{(204)}$ change with the decrease of sulfur content in chalcopyrite (Takenouchi and Fujiki, 1968a,b).

Phase relations at 250°C and 300°C: The 250°C isothermal section is very similar to the 300°C section as shown in Fig. III-10-6 (b), (c). But at 300°C, the crystal systems of obtained pyrrhotite are all hexagonal phase.

Chalcopyrrhotite, which agrees with a host of heated cubanite, is exhibited in the isothermal sections. As mentioned above, the chemical composition of chalcopyrrhotite is different from that of orthorhombic cubanite. The chemical compositions of chalcopyrrhotite solid solutions were inferred assuming that the obtained single phase corresponds to the bulk composition of the starting material.

The phases obtained at 250°C are as follows: pyrite-bornite-chalcopyrite, pyrite-chalcopyrite-monoclinic pyrrhotite, chalcopyrite-hexagonal pyrrhotite, chalcopyrite-chalcopyrrhotite-hexagonal pyrrhotite, chalcopyrrhotite-hexagonal pyrrhotite, chalcopyrrhotite-chalcopyrite, and troilite-chalcopyrrhotite-bornite. It is interesting that the assemblage of troilite, chalcopyrrhotite and bornite appears at 250°C as well as at 300°C and disappears at 180°C.

The products at 300°C are as follows: pyrite-bornite-chalcopyrite, chalcopyrite-pyrite-hexagonal pyrrhotite, chalcopyrite-hexagonal pyrrhotite, chalcopyrrhotite-hexagonal pyrrhotite, chalcopyrrhotite-hexagonal pyrrhotite, hexagonal pyrrhotite-bornite-chalcopyrrhotite.

Recently, in the central part of the Cu-Fe-S phase diagram, new phases such as mooihoekite, haycockite and talnakhite have been described. Cabri (1967, 1973) suggested new phase relations at 25°C in the Cu-Fe-S system from the natural paragenesis. The cubanite-bornite join proposed by Bartholomé (1958) is unreasonable because field observations suggest that the assemblage chalcopyrite-haycockite is stable at 250°C (Fig. III–10–7).

According to Yund and Kullerud (1966), cubanite is in a paragenetic relation with monoclinic pyrrhotite. But in the present results, chalcopyrrhotite is not in a paragenetic relation with monoclinic pyrrhotite. In nature, a cubanite-monoclinic pyrrhotite assemblage is not observed.

Discussion

As mentioned above, all studies on the heating experiments with cubanite except Borchert's (1934) have been carried out only from the standpoint of polymorphism. It is due to the following two reasons: (1) The fine texture of the two phases in heated cubanite cannot be identified without extremely high magnification ($\times$ 1,000); especially in the case of the shorter heating duration and lower heating temperatures near the decomposition temperature, the identification of the two phases is nearly impossible because of the very fine grain of chalcopyrite. (2) As the X-ray diffraction pattern of chalcopyrrhotite is very similar to that of chalcopyrite, when chalcopyrite exists in a small amount, the X-ray diffraction pattern of chalcopyrite may be concealed by that of chalcopyrrhotite.

As mentioned above, it is inferred from this study that cubanite is unstable above 200°C. Yund and Kullerud (1966) stated that the tie-line between high-cubanite (probably a mixture of chalcopyrrhotite) and pyrite changes into the tie-line from chalcopyrite to pyrrhotite at 328° $\pm$ 5°C. In the copper ore bodies of the Kamaishi mine, the mineral assemblage of chalcopyrite, hexagonal pyrrhotite and cubanite is commonly observed. It is presumed that cubanite is stable only at lower sulfur fugacity conditions than the assemblage of chalcopyrite and hexagonal pyrrhotite and at temperatures lower than 328 $\pm$ 5°C.

Cubanite has been thought of as a mineral that occurs in a deposits with high formation temperatures. The mines where cubanite occurs in Japan are the Kamaishi, Komori, Akagane, Shimokawa, Yaguki, Yagoshi, Tenryu, Okura, Nodatamagawa, Kawayama, Makimine, Iwai, Iwato Yoshihara, Kameyama and Mukuromi mines. But, all these are not necessarily deposits of high formation temperatures.

In the Kamaishi mine, cubanite occurs in the zone farthest away from the Ganidake granodiorite. That is, it is considered geologically that cubanite occurs at relatively low temperatures.

The above-mentioned mines are usually accompanied by basic or ultrabasic rocks. It is

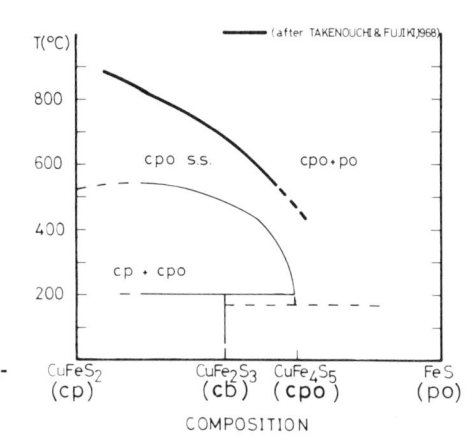

Fig. III-10-8. Schematic phase diagram in the CuFeS$_2$-FeS system (modified from Kaneda et al., 1978).

assumed that these rocks may be related to the genesis of cubanite, as discussed in Chapter II-5.

From the results of experiments in the CuFeS$_2$-FeS system, the schematic diagram between CuFeS$_2$ and FeS shown in Fig. III-10-8 was obtained. According to this figure, chalcopyrrhotite occurs stably above about 200°C, but the natural occurrence of this mineral has been reported by Geijer (1924) from the Kaveltorp mine and by Ramdohr (1969) from the Pirin mine, Kaveltorp mine, Ang mine, etc. It occurs very seldom, because (1) the phase diagram shown in Fig. III-10-8 is an unstable system, and (2) chalcopyrrhotite below 200°C changes into the the mineral assemblages of cubanite-hexagonal pyrrhotite, cubanite-chalcopyrite and chalcopyrite-pyrrhotite by retrograding effects.

Hereafter, the phase relations between CuFeS$_2$ and FeS must be studied in more detail. As compared with the reported isothermal section at 300°C and 200°C by Yund and Kullerud (1966), at 300°C as well as 250°C, the characteristic of the writer's experiments is the chemical composition of cubanite (cubic cubanite). In the phase diagrams of Yund and Kullerud (1966), the univariant assemblage of chalcopyrite and pyrrhotite was not described. But, considering its natural occurrence, cubanite is not in paragenetic relation with monoclinic pyrrhotite, and the mineral assemblage of chalcopyrite and pyrrhotite without cubanite can be very commonly observed.

The mineral assemblage of bornite, pyrrhotite and cubanite was synthesized not at 180°C but at 250°C and 300°C, seemingly because cubanite-bornite join is in cross-relation with mooihoekite-copper or talnakhite-copper join at 180°C. In nature, the assemblage of cubanite and bornite has not been observed.

Judging from these results, it may be suggested that new minerals such as talnakhite, mooihoekite and haycockite are formed at temperatures below 250°C.

III-11. Transition of Pyrite into Marcasite, Magnetite and Hematite by Chalcopyrite Mineralization at the Sasagatani Mine, Shimane Prefecture (Imai, 1951)

H. Imai

Introduction

The Sasagatani mine is situated near the western end of Honshu, and was an old copper mine which had been worked for more than six hundred years (Fig. I-3), but now it has been abandoned. The ore deposit is of the pyrometasomatic type.

The geologic structure of this region is comparatively simple (Fig. III-11-1). A series of

B. Genetical Problems

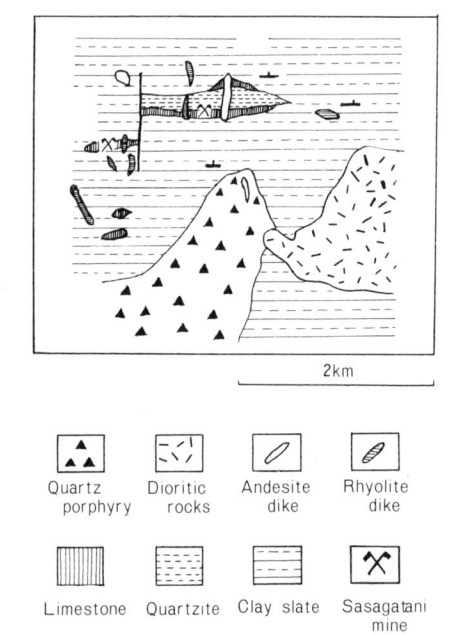

FIG. III-11-1. Geologic map of the Sasagatani mine and environs.

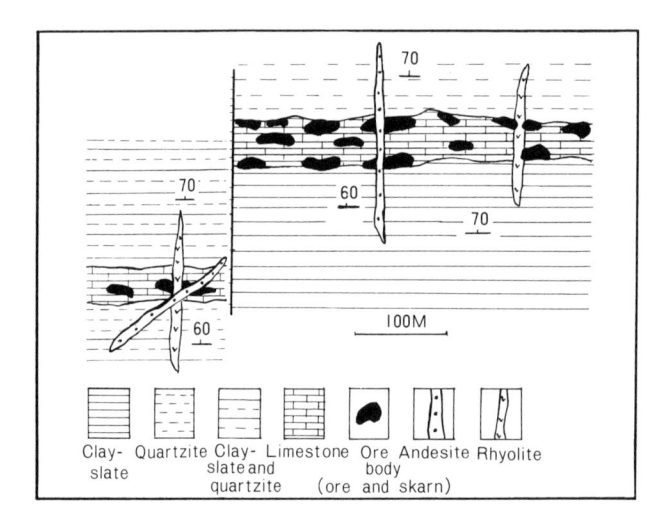

FIG. III-11-2. Geologic occurrence of ore bodies at the Sasagatani mine.

Carboniferous or Permian sedimentary rocks strikes E-W and dips 60°–70° N. They are composed of clayslate and quartzite intercalated by limestone lenses. These strata are intruded by masses of quartz porphyry and dioritic rock. Dioritic rock is assimilated by quartz porphyry at the boundary between them. Thus, it is concluded that the quartz porphyry belongs to a later intrusion than the dioritic rock. Dikes of rhyolitic rock, which would be offshoots of the quartz porphyry, occur in many places. These igneous rocks are generally considered to be Late Mesozoic in age.

Dikes of andesite, which are later than quartz porphyry or dioritic rock, are found here and there, and are perhaps of Tertiary age.

The ore deposit of the Sasagatani mine occurs as massive bodies in the lenticular limestone which is intercalated between beds of clayslate and quartzite. The dikes of rhyolitic rock adjoining the ore bodies are not or are only slightly skarnized (Fig. III-11-2); thus we know that the rhyolite belongs to an earlier intrusion than the skarn formation and that it was skarnized with much more difficulty than limestone. Andesite dikes clearly cut the ore bodies; they are later intrusives than the ore deposition. The ore deposit may be genetically related to the intrusion of quartz porphyry or rhyolite.

The skarn and gangue minerals are wollastonite, diopside-hedenbergite, garnet, epidote-clinozoisite, vesuvianite, tremolite, ilvaite chlorite, chalcedonic quartz and calcite; the ore minerals are chalcopyrite, arsenopyrite, pyrrhotite, magnetite, sphalerite, galena, native arsenic, hematite, marcasite and undetermined silver minerals.

Paragenesis of Marcasite, Magnetite and Hematite

Pyrite was deposited earlier than chalcopyrite. Under the microscope, it is observed that the pyrite is corroded by chalcopyrite. In some cases, the part of a pyrite grain which is surrounded by chalcopyrite is changed into marcasite at the margin (Fig. III-11-3). Also, in some cases there are tiny grains of magnetite or hematite in a pyrite grain surrounded by chalcopyrite (Fig. III-11-4). These two minerals are believed to have been formed by the oxidation of pyrite. The magnetite occurs as very small grains, but it is recognized by its color, its magnetic property and etching reaction by HCl (1:1).

These changes are found in the pyrite grains which are surrounded by chalcopyrite but not in those which are surrounded by sphalerite. From these facts, the writer infers that the changes are due to the mineralization of chalcopyrite.

At the Twin Butte mine, Arizona, Webber (1929) reported that marcasite exists in the pyrometasomatic deposit. He interpreted that it was deposited at a comparatively low temperature, from a neutral or acid solution, at a later stage than the deposition of pyrite and nearly at the same time as the chalcopyrite.

Here the writer considers the following facts:

(a) Allen *et al.* (1914) stated that marcasite can *only** be obtained from acid solution. According to Newhouse (1925), it is formed at room temperature from a neutral solution.

The marcasite was formed accompanying the chalcopyrite mineralization in the Sasagatani mine, and chalcopyrite would have been deposited from the acid solution.

(b) In the Sasagatani mine, marcasite was transformed from pyrite. According to Allen *et al.* (1912), pyrite and marcasite are in dimorphic relation; the latter possesses larger free energy than the former, and is in monotropic form. A paramorph of marcasite after pyrite does not exist.

However, Buerger (1934) stated that they are not dimorphic and that they are different in chemical formula. According to him, pyrite is FeS_2, while marcasite is $Fe \begin{vmatrix} Fe_x \\ S_{2-x} \end{vmatrix}$, where x is a small fraction in the neighborhood of 0.004. He proposed the following equation:

$$x H_2SO_4 + x FeSO_4 + FeS_2 = Fe \begin{vmatrix} Fe_x \\ S_{2-x} \end{vmatrix} + x H_2S + \underbrace{2x SO_4 + 4x \oplus}_{2x SO_2 + 2x O_2}$$

This equation shows that pyrite might change into marcasite under acid conditions. The above observations can be explained by his proposal.

(c) It is generally believed that marcasite is deposited from a low-temperature solution. But Allen *et al.* (1914) obtained it from an acid and comparatively high-temperature solution. Acidity rather than temperature is the essential condition for deposition of marcasite. The temperature conditions for the deposition of marcasite need to be further discussed, though it is probable that marcasite or chalcopyrite was deposited at a comparatively low temperature in the Sasagatani mine.

(d) The transition of pyrite to magnetite and hematite indicates that chalcopyrite was

---

* The italics are the present writer's.

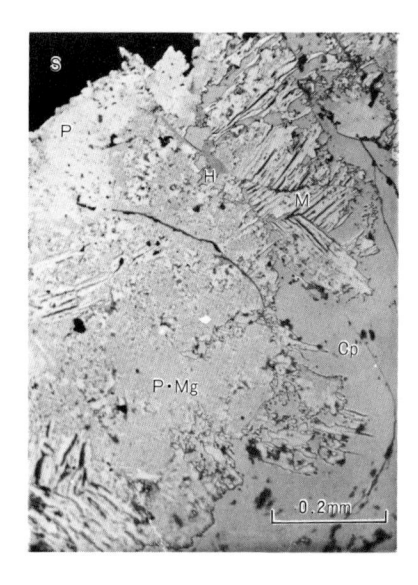

FIG. III-11-3. Photomicrograph. Polished section. One nicol.
P: pyrite, Cp: chalcopyrite, M: marcasite, P·Mg: aggregate of pyrite and magnetite, H: hematite, S: skarn. A part of the pyrite grain was transformed into marcasite, or an aggregate of pyrite, magnetite and hematite.

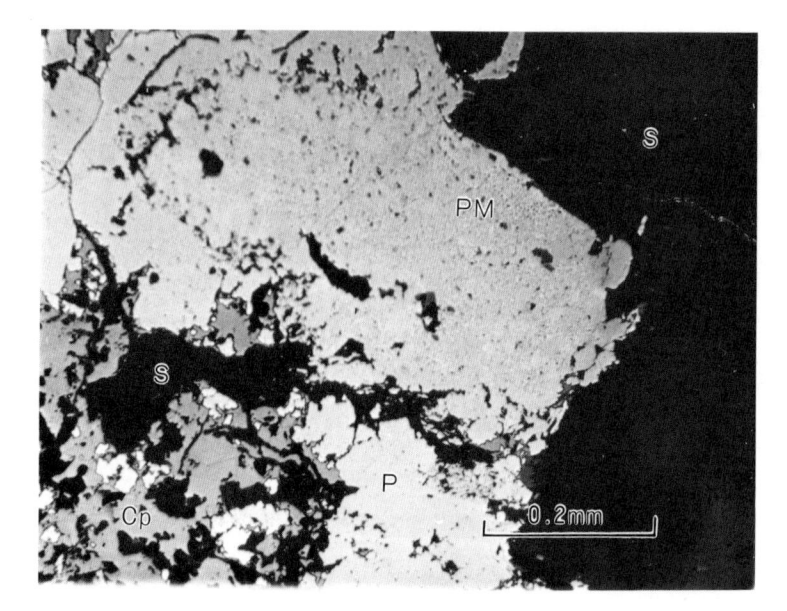

FIG. III-11-4. Photomicrograph. Polished section. One nicol.
P: pyrite, Cp: chalcopyrite, PM: aggregate of pyrite and marcasite, S: skarn. A part of the pyrite crystal was transformed into an aggregate of pyrite and marcasite.

formed under oxidizing conditions. According to Buerger (1934), the change of pyrite into marcasite accompanies the oxidizing agency, as is understood from the above equation.

(e) The problem of pyrite-marcasite formation must be reinvestigated from the points of f_{S_2}, f_{O_2}, pH and temperature conditions, as Scott and Barnes (1972) investigated sphalerite-wurtzite.

Genesis of Martite

Martitization of magnetite is observed under the microscope in ore from the Sasagatani mine. Veinlets of hematite irregularly penetrate into magnetite. It is generally said that this hematite was formed by hypogene oxidation of magnetite, but the process and the

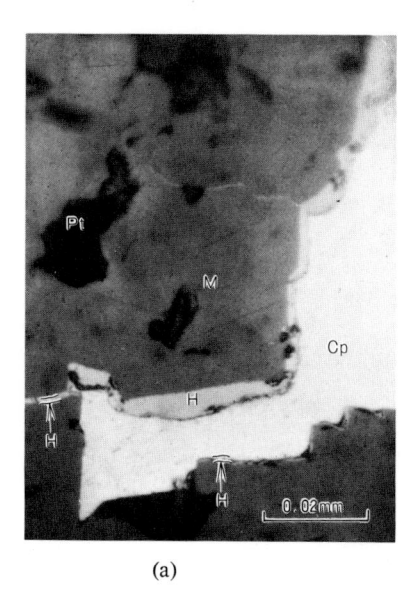

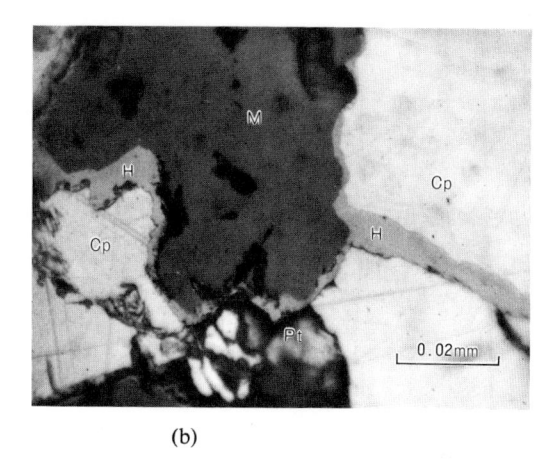

FIG. III-11-5. Photomicrograph. Polished section. One nicol. Oil immersion.
M: magnetite, Cp: chalcopyrite, H: hematite, Pt: pit.

time of this reaction are not yet certain. In the Sasagatani mine, when a magnetite grain exists among chalcopyrite grains which were deposited at a later stage than the former, the former is frequently coated by a narrow rim of hematite along the boundary with the latter (Fig. III-11-5). In some cases, there exist veinlets of hematite in magnetite, branching from the rimming hematite (Fig. III-11-5(a)).

The writer believes that martitization, if not all, was carried out at the time of chalcopyrite mineralization, as far as this deposit is concerned. Chalcopyrite was deposited under oxidizing conditions.

Also, in some cases, veinlets of hematite enter into chalcopyrite grains which are in contact with magnetite (Fig. III-11-5(b)). This indicates that hematite was deposited from a fluid; this hematite would not have been formed by the mere oxidation of magnetite in the solid state.

There is another form of ferric oxide called maghemite, which belongs to the cubic system; it is isomorphous with magnetite and has very strong magnetism. Newhouse and Glass (1936) studied lodestone and discussed its occurrence and genesis. According to them strong magnetism is due to maghemite which occurs as veinlets in magnetite. They said that perhaps its origin is supergene secondary. In Japan, maghemite is known in several

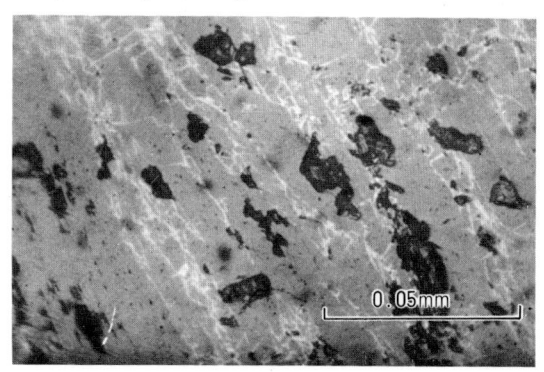

FIG. III-11-6. Photomicrograph. Polished section. One nicol. Oil immersion. Lodestone from the Taishi mine, Hiroshima Prefecture.
Grey: magnetite, White: maghemite.

localities*. It is found at the outcrops of magnetite deposits or of basic rocks containing magnetite. Under the microscope, tiny veinlets of maghemite irregularly penetrate into magnetite grains (Fig. III-11-6). The writer believes that it was formed by supergene oxidation, as did Newhouse and Glass. It is uncertain whether maghemite was formed by the mere oxidation of magnetite in the solid state or was redeposited from the iron-rich solution leached from magnetite. Martitization was carried out by the hypogene process. Under pressure and temperature conditions such as in the pyrometasomatic process, the conversion of magnetite to maghemite would be impossible: it must be carried out at lower temperatures and under lower pressure conditions, as Mason (1943) stated.

Summary

In ore from the Sasagatani mine, it is frequently observed under the microscope that the grains of pyrite are partly changed into marcasite, magnetite and hematite. The writer thinks that this change might have been carried out at the time of chalcopyrite mineralization. Also, magnetite is coated with a hematite rim along its boundary plane with chalcopyrite; magnetite was oxidized at the time of chalcopyrite mineralization. Therefore, the writer believes that chalcopyrite was deposited under oxidizing and acid conditions.

By this oxidizing agency, martite would be formed, as far as this deposit is concerned. The above martitization is different from the conversion of magnetite to maghemite. Under pressure and temperature conditions such as those in pyrometasomatic deposits, the conversion of magnetite to maghemite would be impossible.

---

* Nagata and Kobayashi (1958, 1959) studied, from the point of magnetism, the maghemite in gabbro from the Kamaishi mine, Iwate Prefecture and in iron ore from the Taishi mine, Hiroshima Prefecture, which were donated by the writer. They proposed "chemical remanent magnetism" (Nagata, T. and Kobayashi, K. (1958) Experimental studies on the generation of remanent magnetization of ferromagnetic minerals by chemical reactions: Proc. Japan Acad., **34**, 269–273; Kobayashi, K. (1959) Chemical remanent magnetization of ferromagnetic minerals and its application to rock magnetism: J. Geomagnetism and Geoelectricity, **10**, 99–117).

PART IV
STRATA-BOUND DEPOSITS

IV-1. Introduction

H. Imai

The bedded cupriferous iron sulfide (mainly pyrite) deposits simply named "Kieslager" are distributed in many places in the Sambagawa-Mikabu (metamorphic) terrain and the adjoining Chichibu-Shimanto terrains of the Outer Zone in Southwest Japan (Kanehira and Tatsumi, 1970). Most of them exist in the Sambagawa-Mikabu terrain. The Kune, Minenosawa, Iimori, Sazare, Besshi, Okuki and Makimine mines belong to this type of deposit (Fig. I-3).

In the Sangun terrain and Maizuru Folded zone of the Inner Zone in Southwest Japan, the Yanahara and Kawayama mines are examples of this type of deposit (Figs. I-1, and-3). In Northeast Japan, the Hitachi, Taro (Imai, 1963b) and Shimokawa mines occur in a similar kind of metamorphic rock (Figs. I-1 and -3).

In Taiwan, a similar metamorphic belt runs along the eastern slope of the backbone range, where bedded cupriferous pyrite deposits occur (Yen, 1959, 1974).

The ore is composed of pyrite (in rare cases pyrrhotite) and chalcopyrite. Gangue minerals are quartz, sericite, chlorite and others. The bedded or lenticular form of the ore body is concordant almost in all cases to the schistosity planes. Where the metamorphic rocks are folded, the ore bodies are also folded. The plunge of the ore body coincides with the lineation of the country rocks, folding axis or micro-folding axis (Fig. IV-1-1) (Horikoshi, 1940; Imai, 1952).

In Japan, there is another type of strata-bound deposit of copper, zinc and lead, accompained by a massive replacement or network ore deposit. It is called the Kuroko (Black Ore)-type deposit. The writer does not discuss this type in this book.

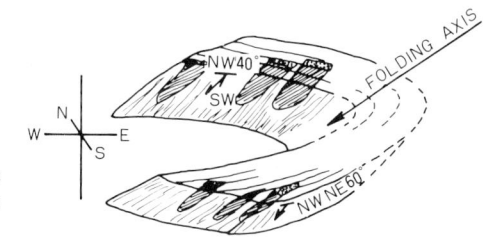

FIG. IV-1-1. Folding axis and plunge of ore bodies in bedded cupriferous pyrite deposits in the vicinity of Sata Peninsula (Fig. I-3) (after Horikoshi, 1940).

IV-2. Geology and Genesis of the Okuki Mine, Ehime Prefecture, and Other Related Cupriferous Pyrite Deposits in Southwest Japan (Imai, 1958, 1959, 1960a,b, 1963a,b)

H. Imai

The Okuki mine is situated about 45 km southwest of Matsuyama City, Shikoku Island (Fig. I-3). The ore deposit of the mine, a peculiar type of bedded cupriferous pyrite deposit, was discovered in 1935. Since then, the mine has produced about 900,000 t of ore. The tenor is 3–5% copper and 2–4 g/t gold. The mine was closed in 1971.

Geologic Setting of the Okuki Deposit

The district where the ore deposit occurs is composed of the Mikabu Metamorphic System and the Chichibu Paleozoic (Permo-Carboniferous and Triassic) System. Both systems are intruded by diabase or gabbro (Figs. IV-2-1, and -2).

The Mikabu System consists of sericite-graphite-quartz schist called black schist and epidote-actinolite-chlorite schist called greenschist*, which are intercalated with quartz schist. The massive rock composed of the same minerals as greenschist is called green rock. Towards the south of the district the Mikabu System grades into the clayslate, sandstone, schalstein, quartzite and limestone of the Chichibu System. No dislocation line is observed between Mikabu System and Chichibu System. Both systems strike E-W and generally dip

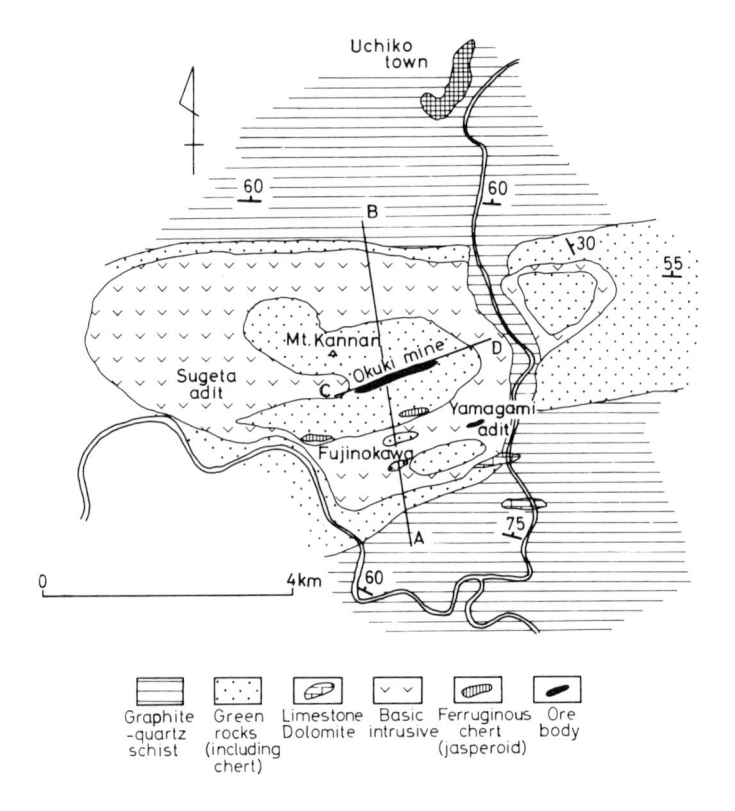

FIG. IV-2-1. Geologic map of the Okuki mining district.

---

* Watanabe *et al.* (1970) described pumpellyite, glaucophane and stilpnomelane in greenschist or basic igneous rock in this district.

N at 60°–70°, although in places they dip south. In the Mikabu System of the mining area, a large sheet of basic intrusive rock extends about 8 km E-W and 3.5 km N-S. The basic intrusive includes gabbro, diabase, metadiabase and "pyroxenite." The basic rock is gabbroic* in the core part of the mass, but toward the marginal parts gradually becomes diabase, metadiabase or "pyroxenite." The same kind of basic rock occurs in some places along nearly the same geologic zone to the east and west. The "pyroxenite" is characterized by cataclastic pyroxene, and occurs in the Mikabu System of southwest Japan. According to Suzuki (1936), the cataclastic pyroxene in this rock is a relict mineral of a basic rock. He said that it is preferable to call this rock diabase schist. In places near the basic intrusive, the black schists of the Mikabu System grade into greenschists (Fig. IV–2–1).

Above the basic intrusive in the mining district, folded greenschist, green rock and chert (manganese-bearing and ferruginous) occur as a roof-pendant (Figs. IV–2–1 and –2). Along the boundary with the basic intrusive, the green rocks of the roof-pendant become highly

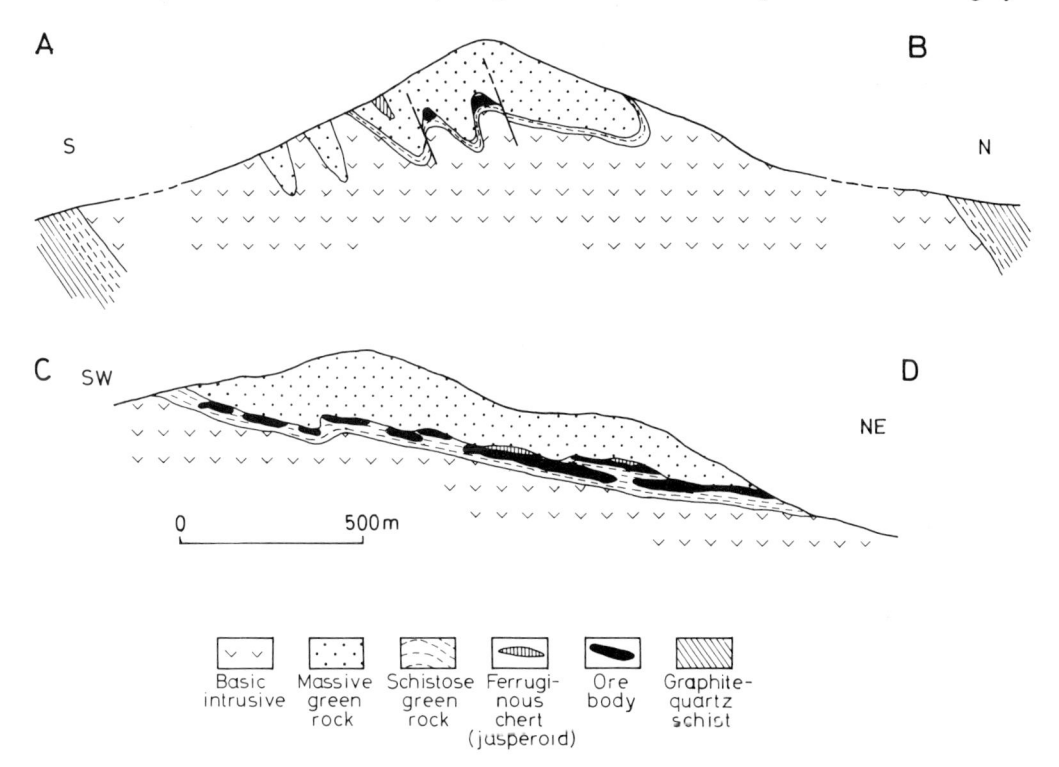

FIG. IV–2–2. Geologic sections of Fig. IV–2–1. Above, N—S section; below, E—W section.

---

* Macroscopically gray, medium-grained and equigranular. Generally massive and non-schistose, but some are schistose. Microscopically (Fig. IV–2–3), holocrystalline, medium- to coarsed-grained. Basic plagioclase transformed to sodic plagioclase by addition of soda. Optic properties of plagioclase are as follows: $\alpha(min) = 1.533 \pm 0.001$, $\gamma(max) = 1.543 \pm 0.001$ (measured by N. Isshiki), $Ab_{87}An_{13}$ (sodic oligoclase). Pyroxene crystals are 1–2 mm in size. Optic properties of pyroxene are as follows: $\alpha = 1.689 \pm 0.001$, $\beta = 1.694 \pm 0.001$, $\gamma = 1.714 \pm 0.001$, $(+) 2V = 47°$ (measured by N. Isshiki), $r > v$, $Wo_{40}En_{40}Fs_{20}$ (augite). Pyroxene is partly altered to actinolite and chlorite. The chemical analysis of the gabbro from the Kannan Adit of this mine is as follows:

SiO_2	TiO_2	Al_2O_3	Fe_2O_3	FeO	MnO	MgO	CaO	Na_2O
46.19	0.46	17.66	6.16	10.22	0.30	5.32	5.99	5.21

K_2O	$H_2O(+)$	$H_2O(-)$	P_2O_5	Total				
0.04	2.43	0.30	tr	100.28(wt. %)				

Analyst: S. Iwase (Dept. of Mining, Univ. of Tokyo).

236   2. Geology and Genesis of the Okuki Mine

FIG. IV-2-3. Photomicrograph. A thin section of gabbro. Crossed nicols.
Pl: plagioclase, Py: pyroxene.

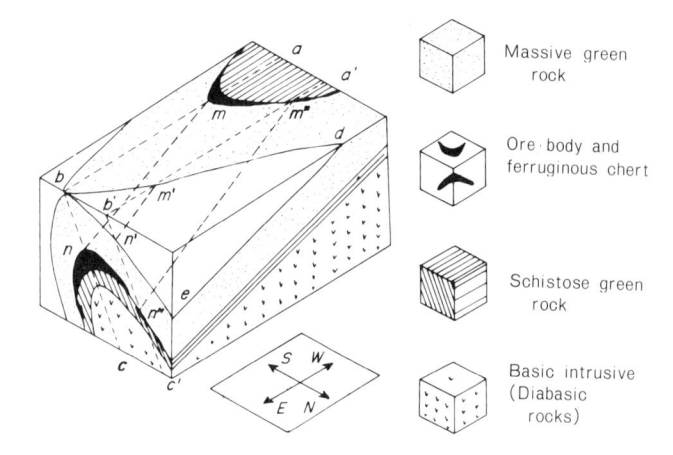

FIG. IV-2-4. Schematic block diagram of the Okuki district. The plane abc is the axial plane of the fold. The oblique hachures in the schistose green rock represent the flow cleavage which is parallel to the axial plane of the fold. The lines m'-n' and m"-n" are the intersections of the flow cleavages with bedding planes. They correspond to the lineations which are parallel to the folding axis m-n.

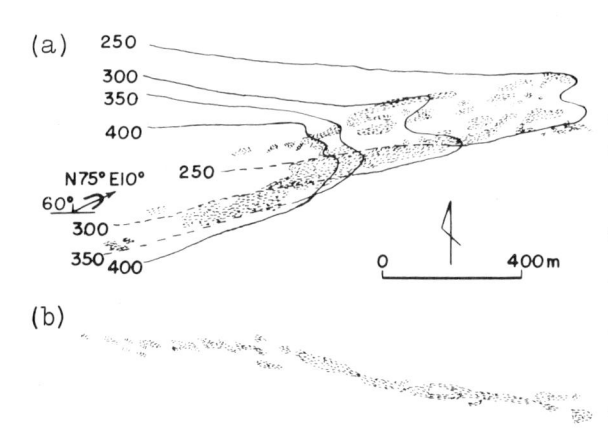

FIG. IV-2-5. Relation between the overturned and plunging fold and the ore bodies of the Okuki mine.

(a) Underground contours of the boundary surface between massive and schistose green rocks of the roof pendant. The dotted parts represent ore bodies. From the contours, the strike and dip of the boundary surface at a given place can be determined. The numeral in each contour represents the height above sea level in meters.

(b) Vertical section of ore bodies (S 75° W—N75°E section).

Part IV: Strata-Bound Deposits

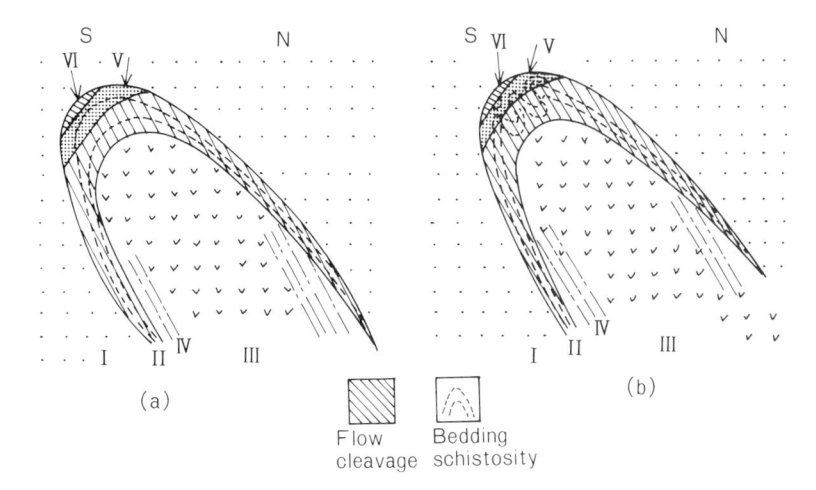

FIG. IV-2-6. Schematic geologic sections of the ore deposit of the Okuki mine. I: massive green rocks; II: schistose green rocks; III: diabasic rock; IV: schistose diabasic rocks; V: ore body; VI: ferruginous chert.

FIG. IV-2-7. Hand specimen of schistose green rock. Notice the relation between the bedding schistosity (bedding plane) (a) and the flow cleavage (b).

schistose, i.e., greenschist. Above the schistose green rocks, massive green rocks and cherts occur. Apparently, the basic intrusive does not cut the folded structure of the roof-pendant. In other words, the basic intrusive occurs as though conformable to the roof-pendant (Fig. IV-2-4), which suggests that the intrusion of the basic rock and the folding were simultaneous, accompanied by the metamorphism of the Chichibu System. It is generally believed that the majority of the rocks of the Mikabu System are metamorphosed equivalents of the Chichibu Paleozoic and Triassic System. The underground contour of the contact between the massive and schistose green rocks of the roof-pendant reveals a folded structure, as shown in Fig. IV-2-5. The axial planes of the folds of the roof-pendant strike N 80° to 90° E and

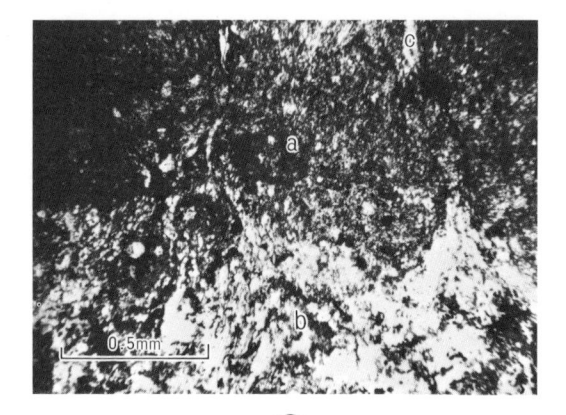

FIG. IV-2-8. Photomicrograph of the thin section of schistose green rock. One nicol. The original bedding plane is in a right-to-left direction. The upper half (a) is composed of small grains of calcite. The lower half (b) is composed of medium grains of chlorite and calcite. It is noticed that the chlorite flakes (c) are arranged in the direction of the flow cleavage (vertical direction).

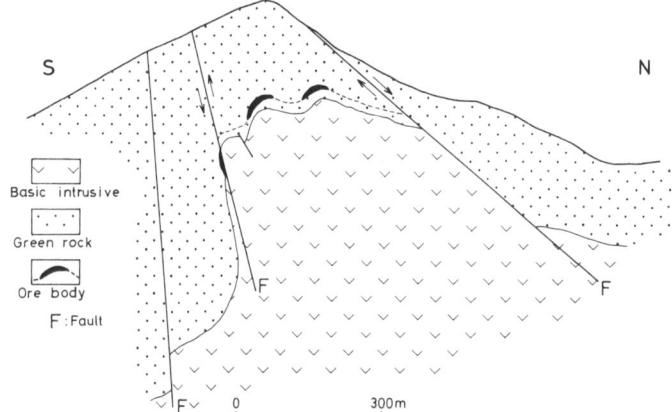

FIG. IV-2-9. Pushing up of the basic intrusive.

dip 60°–70°N, and the folding axes plunge N75°E10°; that is, the folds are overturned to the south and plunge to the east. The bedding planes dip generally north, except in the uppermost part (crest) of the south side of the folded structure where they dip south (Fig. IV-2-6).

As shown in Figs. IV-2-4 and -7, two kinds of schistosity are recognized in the schistose green rocks; one is bedding schistosity and the other is flow cleavage which resulted from compressive force exerted on the plastic matter. The boundary plane between the basic intrusive and the surrounding green rock is generally parallel to the bedding schistosity of the green rock in each point (Fig. IV-2-4), while the flow cleavage is parallel to the axial planes of the folds (Fig. IV-2-4). The intersection of the two schistosities results in a lineation parallel to the folding axes, as shown in Figs. IV-2-4 and -7. On the observation of the schistose rocks of the roof-pendant under the microscope, the thin section cut at a right angle to the lineation reveals that some flakes of chlorite are arranged parallel to the flow cleavage and that no dislocation or slip of the bedding took place along this cleavage (Fig. IV-2-8).

The above-mentioned two kinds of schistosity nearly coincide in the limbs of the folds (Figs. IV-2-4 and -6). That is, in the limbs of the folds the bedding schistosity is nearly parallel to the axial planes of the overturned folds and also to the boundary between the roof-pendant rocks and the basic intrusive. As stated before, the massive green rocks occur above the schistose green rocks, and, in most cases, the two kinds of green rocks are sharply separated from each other. That the well-developed schistosity of the roof-pendant green rocks occurred near the boundary with the intrusive might have been due to the effect of emplacement of the rigid basic igneous rock at the time of folding. Schistosity is frequently developed also in the intrusive close to the limbs of folds. This schistosity in the intrusive is also parallel to the contact plane or to the schistosity planes of the green rocks

(Fig. IV–2–6). The schistosity in the intrusive makes it difficult to distinguish the intrusive from the green rocks near the contact.

The schistose green rocks in contact with the intrusive are thick, about 20–30 m, in the anticlinal part of the folds, but are thin, 1–0.5 m, in the limbs (Fig. IV–2–6). This variable thickness may be due to flow of the incompetent schistose rocks from the limbs of the folds to the axial parts of the folds, belonging to similar fold. In the schistose green rocks adjacent to the basic intrusive, drag folds are sporadically present, especially in the axial parts (Fig. IV–2–6(b)). In the schistose green rocks just beneath the massive green rocks, cupriferous pyrite deposits are found, and the main ore bodies of this deposit occur along the axes of the overturned folds as saddle reefs (Figs. IV–2–2 and –4). The mineralized zone extends more than 1,500 m, plunging N 75°E 10°, which is the direction of the fold axes or of the lineation of the schistose green rocks (Figs. IV–2–4 and –5). In the eastern half of the main mining area, two mineralized anticlinal crests (zones) occur. Towards the west the two crests merge into each other and become one anticline (Fig. IV–2–5). In the syncline between the two anticlinal crests, trough reefs are found, although the ore bodies are smaller than the saddle reefs. Each ore body extends to N 75°E 10° for 10 m to 160 m, with a width of 5 to 10 m and a maximum thickness of 3 m.

Thinner ore bodies, which also extend in the direction of N75° E10°, occur intermittently in the limbs of the fold (Fig. IV–2–4). In most places between the ore bodies and the massive green rocks, red cherts (manganese-bearing ferruginous jasperoid) are found with widths up to 20 cm (Fig. IV–2–6). The red cherts are discussed in the next section.

The ore is mainly composed of pyrite, chalcopyrite and sphalerite, with small amounts of tetrahedrite, primary bornite, galena, cobaltite and native gold.

The massive ore bodies are surrounded by impregnated ores in many places. Around the ore bodies, chloritization, which has probably resulted from a hydrothermal alteration, is observed. Beneath the ore bodies the schistose rocks are, in some cases, impregnated with a magnetic mineral. Under the microscope, it is recognized that hematite is replaced by magnetite. The replacement would be due to reducing action, perhaps at the time of sulfide mineralization.

The two normal faults sandwiching the mineralized block extend E-W. They would be due to the pushing-up of the gabbro at the time of folding (Fig. IV–2–9), as discussed in Chapter II–1. They were partly mineralized. They would be the path of the ore-forming fluid.

As discussed in detail later, the writer interprets that deposits of this kind were formed by epigenetic replacement, closely related to regional-metamorphism and intrusion of basic rocks. But this is a minority opinion. Most of the geologists in Japan state that these deposits are of syngenetic type accompanied by basic volcanic activity at the time of Paleozoic-Triassic geosyncline. The related problems of green rocks, diabase and gabbro are discussed in the next section.

The geologic age of the metamorphism is stated in chapter I. The writer believes that the age of the main metamorphism would be late Mesozoic.

The Problems of the Basic Intrusives and the Green Rocks (Greenschists)

In the Outer Zone of Southwest Japan, a highly metamorphosed complex named the Sambagawa System borders the Median Tectonic Line (Fig. I–1). Towards the south or southeast, rocks of the Sambagawa System change gradually to rocks of lower-grade metamorphism, and are called the Mikabu System (refer to Chapter I).

In both systems the epidote-chlorite- or amphibole-bearing schists are called greenschists, and the massive rocks composed of the same minerals are called green rocks. The graphite schists are called black schists. The original rocks of the black schists are believed to have been clayslate or sandstone. The original rocks of the greenschists are discussed later.

Throughout the Sambagawa-Mikabu terrain, many deposits of cupriferous pyrite occur along the planes of schistosity, commonly as lenticular or bedded deposits. The largest deposit of this type is that of the Besshi mine.

Farther towards the south or southeast, sedimentary rocks of the Chichibu System (mostly Permo-Carboniferous) occur. It has generally been considered that the Mikabu System is in fault contact with the Chichibu System along an overthrust called the Mikabu Line, which runs nearly parallel to the Median Tectonic Line (Kobayashi, 1941). In some places this is true. To the east of the Kune, Nago and Minenosawa mines, an overthrust divides the metamorphic complex from the Paleozoic and the Mesozoic terrains (Figs. I-1 and -3). This overthrust is called the Akaishi Tectonic Line. Kimura (1954) reported a low-angle thrust between the metamorphosed and the non-metamorphosed zones in the eastern part of the Kii Peninsula.

From the writer's observations in the Okuki district, no dislocation or fault line is recognized between the Mikabu and Chichibu Systems, since the two systems gradually grade into each other. The basic rock in the Okuki district occurs as an intrusive into the folded part. The folding occurs at the transition area of Mikabu and Chichibu Systems. Horikoshi and Katano (1940) recognized that the Mikabu Metamorphic Rocks to the south of the Minenosawa mine, Shizuoka Prefecture, grade into the rocks of the Chichibu System (Figs. I-1 and -3). A similar phenomenon was observed by Tanaka (1957) near Mt. Koya, Kii Peninsula, and by Kikuchi (1950) in the Jiro mine, Shikoku (Figs. I-1 and -3).

The Chichibu System is composed of sandstone, graywacke, clayslate, chert (quartzite), limestone and schalstein; and its age ranges from Devonian to Triassic, mostly Permo-Carboniferous. In some places it is intruded by sheet-like peridotite and serpentinite. Diabasic rocks occur intercalated between the bedding planes.

Towards the south, the Chichibu System is thrust upon the Shimanto undifferentiated Mesozoic terrain along the Butsuzo Line, which is also nearly parallel to the Median Tectonic Line (Fig. I-1). From former times, the Shimanto System has been considered to be Jurassic in age, because it contains a limestone of the Jurassic Torinosu type (Kobayashi, 1941). Lately, molluscs of the Ryoseki and Monobegawa types, belonging to the Lower Cretaceous, have been found, and recently a Cretaceous ammonite has been reported from this system in Shikoku (Matsumoto et al., 1952). This system might have been deposited over a long period of geologic time. Diabasic rocks like those occurring in the Paleozoic system are also found in this system.

From the above geologic relation, it is certain that the major part of the Sambagawa-Mikabu Complex was derived from the Chichibu System.

It is necessary to discuss with what part of the Chichibu System the greenschists in the Sambagawa-Mikabu Complex should be correlated. In the Chichibu terrain, green rock or schalstein does not prevail so much as in the Sambagawa-Mikabu terrain.

Green rocks or greenschists are abundant in the terrain of the Sambagawa-Mikabu Complex, especially in the region of the Sambagawa System. In the Mikabu System, green rocks or greenschists are generally accompanied by basic intrusives which occur in sheets, as observed in the Okuki district. The highly metamorphosed parts of the basic intrusive in this district are also transformed to epidote-actinolite-chlorite schist wihch is hardly distinguishable from the surrounding greenschists. Therefore, it would be nearly impossible to discriminate the highly metamorphosed basic intrusives in the Sambagawa-Mikabu Complex from the surrounding greenschists.

Since the early days of geologic research on the Sambagawa and Mikabu Systems, it has been thought that the green rocks or greenschists were derived from basic tuffs and lava flows of submarine eruptions (Sagawa, 1910), mainly because of the intercalated occurrences and the chemical compositions of the rocks. If that is the case, they would be sediments of eugeosyncline, according to the terminology of Kay (1951), as a result of intrageosynclinal volcanism. Recently, this interpretation has been further stressed by the ophiolite problem. Most of the geologists in Japan interpret that green rocks including green or blue (glaucophane-bearing) schists and gabbroic rocks (including diabase) belong to Paleozoic to Triassic ophiolite (Uchida, 1967; Iwasaki, 1969; Watanabe *et al.*, 1970; Hashimoto *et al.*, 1970; Sugisaki *et al.*, 1970; Hide, 1972; Suzuki *et al.*, 1972; Toriumi, 1975). Suzuki *et al.* (1972) asserted that the gabbroic magma in the Okuki area intruded at the "eugeanticlinal ridge" during the sedimentation of the Chichibu Geosyncline. Recently, however, Banno *et al.* (1976) have stated that the epidote-amphibolite masses near the Besshi mine in the Sambagawa metamorphic belt were already metamorphic gabbro before the Sambagawa Metamorphism, and intruded in the solid state during the Sambagawa Metamorphism.

The differences in the interpretations of the green rocks and gabbroic rocks result in different interpretations of the genesis of the bedded cupriferous pyrite deposits.

The Chichibu System is mainly composed of sandstones, clayslates, limestones and cherts with locally distributed schalsteins. Schalsteins are generally believed to have been derived from tuffaceous rocks intercalated within pelitic sediments (Suzuki and Minato, 1952). The writer finds it difficult to determine the original rocks of the "schalsteins" in the Chichibu System from their petrographical characteristics. For example, in the so-called schalsteins, the writer has not recognized microscopically the vitroclastic texture and other decisive properties that would suggest pyroclastic rocks. There is no way to determine whether the parts now occupied by chlorite correspond to palagonites of basic tuffs or not. It is also impossible to determine from microscopical observation whether the original rocks of the greenschists in the Sambagawa-Mikabu Complex were tuffs or not. Therefore, it is only a probable interpretation to regard the original rocks as tuffs, based on the bedded occurrences and chemical compositions of the schalsteins and greenschists.

The resemblance of the chemical composition of the green rocks to the basic igneous rock is important background for the above interpretation. But, as shown in Table IV-2-1, the chemical composition of the "laminated tuff" in the Okuki mine does not necessarily correspond to that of the basic igneous rock. Also, as stated above, some parts of the greenschists of the Sambagawa-Mikabu Complex are said to have been derived from lava flows. But it is difficult to determine whether they have been derived from lava flows or from intrusive sheets.

In the Chichibu Palezoic (including Triassic) and the Shimanto undifferentiated Mesozoic terrains, there exist basic igneous rocks called diabase, in addition to the evident in-

TABLE IV-2-1. Chemical compositions of "laminated tuff" in the Okuki mine (Suzuki *et al.*, 1972). Norm was calculated by the present writer.

SiO_2	40.51		
TiO_2	1.95		Norm
Al_2O_3	17.13	Il	3.64%
Fe_2O_3	1.51	Ap	1.00
FeO	12.91	Or	8.35
MnO	0.30	Ab	13.63
MgO	13.29	An	6.12
CaO	1.38	C	10.70
Na_2O	1.62	Mt	2.08
K_2O	1.50	En-Fs	31.61
$H_2O(-)$	0.32	Ol	15.53
$H_2O(+)$	7.48		
P_2O_5	0.08	MgO: FeO (in En-Fs and Ol)	
Total	99.98	= 69:31 (in molecule).	

trusive rocks such as peridotites and serpentinites. All these igneous rocks occur in sheet-like form, intercalated between the formations. Some geologists consider the diabase to be intrusive sheets, while others think it to be lava flows. If this rock is regarded as intrusive sheets, there rises another problem to be solved, i.e., whether it was intruded at the same time as the sedimentation or after the sedimentation (probably at the same time as the orogenesis).

Mélange is not recognized in the Sambagawa-Mikabu terrain. From this geologic structure, the writer does not agree to regarding the basic rocks in the Mikabu area as entrenched oceanic crust.

The writer thinks that the Sambagawa-Mikabu and Chichibu Complexes are similar to the Franciscan Group in California. Taliaferro (1943) said, "It is of course not known if these igneous rocks (basic intrusives in the Franciscan Group) were intruded at the same time or whether they came in throughout the deposition of the Franciscan and Knoxville." According to Kay (1951), "In several instances, association of pillow lavas and radiolarian cherts are succeeded by ultrabasic intrusives. Thus the Ordovician formation has such lavas and sediments in central Newfoundland, and there are subsequent ultrabasic intrusives in the western peninsula and coast, and also near Notre Dame Bay. Ultrabasic intrusives in the Coast Range and Sierra Nevada of California adjoin the geosyncline of Late Jurassic Franciscan radiolarian cherts and pillow lavas. In each case, the evidence has been interpreted as indicating that the first downbuckling of tectogene immediately preceeded the intrusion of the ultrabasic rocks."

As stated above, most of the geologists in Japan consider that the intrusions of gabbro and diabase in the Sambagawa-Mikabu-Chichibu terrain belong to geosynclinal igneous activity during the Carboniferous-Triassic periods. The writer, however, has asserted that they were synchronously intruded with Sambagawa Metamorphism which would have occurred in the Late Mesozoic era.

From the standpoint of basic intrusion during the Chichibu geosynclinal stage, the Sambagawa Metamorphism might be considered to have begun in the Carboniferous-Triassic periods. But no data on the absolute geologic ages dated by the K-Ar method, and especially the Rb-Sr method, indicate that the metamorphism would have begun in these periods. Also, disconformity or supratenuous folding (Nevin, 1936), suggesting incipient orogenesis during the sedimentation of the geosyncline, has never been recognized in this area.

Kamiyama (1956) said that the basic rock in the the Kune mine, the country rock of the ore deposit, is intrusive. It is noticeable that this rock extends in the direction of the lineation of the surrounding black schist. This fact indicates the close relationship between the basic intrusion and the metamorphism. Imai *et al.* (1951) and Nakayama (1954) recognized the gabbroic (diabasic) intrusive in the Sambagawa-Mikabu terrain of the Minenosawa-Nago district (Fig. I–3).

In order to discuss this problem further, it is necessary to study the absolute ages of the basic rocks themselves. But they are not yet available. According to Kobayashi (1941), although some minor parts of the Sambagawa-Mikabu Complex may be Triassic, the major parts of it were originally the Paleozoic Chichibu System, and the highly pyroclastic portion belongs to the Devonian, which is said to have been a period of intense volcanism in the Chichibu Geosyncline. But, according to Kammera (1971), the main epoch of interformational volcanic activity during the Chichibu Geosyncline in this district would be Permian. In either case, the tuff and lava (schalsteins) are not so widely developed in the Chichibu terrain as are the green rocks and greenschists in the Sambagawa-Mikabu terrain.

According to Kato (1934a, b), Horikoshi (1937) and Nishio (1940), greenschists (chlorite schists) in the vicinity of the Besshi mine resulted from chloritization accompanied by the late igneous activity of the basic intrusion. Kato (1934b) and Horikoshi (1938) correlated this chloritization with propylitization in the epithermal deposits. Kato (1934a), Suzuki (1930) and Horikoshi (1937) interpreted that the albite (or glaucophane)-bearing greenschists were formed by the local introduction of sodium contained in the hydrothermal solution as related to the basic or ultrabasic intrusives. Whether or not the soda metasomatism exists in the formation of glaucophane schists has been the topic of discussion by Kojima (1953), Miyashiro and Banno (1958), Seki (1958) and Bloxam et al. (1959).

Seki et al. (1960) found jadeite in meta-gabbroic rocks of the Sambagawa Metamorphic terrain in Shizuoka Prefecture, central Japan.

Although many cases have been reported in which basaltic rocks are transformed to greenschists due to a low intensity of metamorphism and changed to amphibolites on account of a high intensity of metamorphism (Poldervaart, 1953), the writer, placing emphasis on the intimate coexistence of later basic intrusives with schalstein, green rocks and greenschists, infers that most of them may have been derived from pelitic, psammitic and calcareous sediments because of Mg-Fe metasomatism which was accompanied by the basic intrusions. As stated before, the Sambagawa-Mikabu Complex corresponds to the metamorphic facies of the Chichibu Paleozoic-Triassic terrain. If the greenschists or green rocks of the Sambagawa-Mikabu Complex were derived from tuffs or lava flows, it is difficult to explain why such formations, rich in volcanics and pyroclastics, are not more developed in the Chichibu System. Although schalsteins occur in the Chichibu System accompanied by the basic intrusives, they are not so abundant as in the Sambagawa System. If schalsteins, green rocks or greenschists represent the volcanics or pyroclastics intercalated within the pelitic sediments of the Chichibu System, the center of the volcanic activity should be found in the formation. However, no volcanic vents or necks have been reported from the Sambagawa-Mikabu or the Chichibu terrains. So the writer thinks that most of the schalsteins in the Chichibu System and some of the greenschists in the Sambagawa-Mikabu Complex were formed as the result of Mg-Fe metasomatism, which took place in association with the basic intrusion and which affected the original rocks. As stated above, Kato (1934b), Horikoshi (1938) and Nishio (1940) discussed chlorite schists adjacent to the ore bodies which were formed by the chloritization of the country rocks at the time of ore deposition. The chloritization proposed by them corresponds to part of the above-mentioned Mg-Fe metasomatism. The writer includes in the Mg-Fe metasomatism the formation of the various kinds of greenschists in the Sambagawa-Mikabu Complex. The Mg-Fe metasomatism described above is the same as the "formation of green rocks" from black schists proposed by Kojima (1948, 1951), or the metasomatic formation of green phyllite suggested by Tatsumi (1953). It is similar to the enrichment of magnesium and iron (Funahashi, 1948), Mg-Al metasomatism (Kano, 1950), Mg-Fe metasomatism (Kuroda, 1956), Mg enrichment (Iwao, 1955), Mg metasomatic phenomenon (Sundius, 1935), Mg metasomatism (Simonen, 1948), Mg-Fe metasomatism (Eskola, 1950) and Mg-Fe metasomatism (Tuominen and Mikkola, 1950; Tuominen, 1951).

The fact that the schalsteins, green rocks and greenschists extend in certain horizons may be interpreted, from the standpoint of Mg-Fe metasomatism, as the result of the replacement of certain horizons. This is the same as the sporadic occurrence of bedded cupriferous pyrite deposits in certain horizons: i.e., they are interpreted, from the replacement hypothesis, to have been deposited replacing certain horizons.

In the Okuki district an agglomeratic rock was reported. The writer thinks that it is a rock heterogeneously metasomatized by chlorite, epidote and hematite with the original texture still preserved in parts (Fig. IV–2–10), and the preserved original rock looks as if it were an agglomeratic boulder. In some basic rocks, boulders of metagabbro are found, which the writer regards as autoliths in the metadiabase (Fig. IV–2–11).

In the wall of some parts of the highest main adit of the Okuki mine, the basic rock sometimes has a rounded surface presenting a boulder-like appearance (Fig. IV–2–12). It is said to

FIG. IV-2-10. Hand specimen of the diabasic rock (polished surface). a: diabasic rock, b: part metasomatized by epidote and chlorite, c: part metasomatized by ferruginous chert (jasperoid).

FIG. IV-2-11. Hand specimen of the diabasic rock, showing gabbroic autolith (a) in diabasic rock (b).

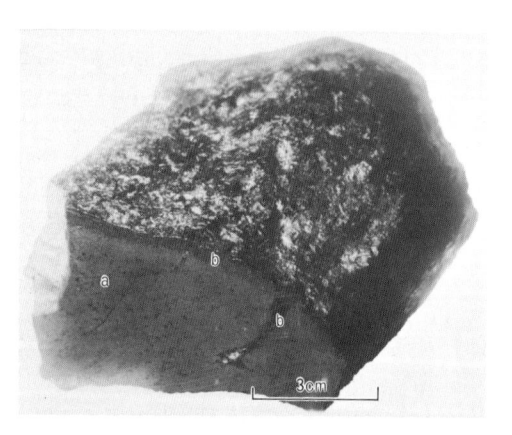

FIG. IV-2-12. Hand specimen of the diabasic rock. a: diabasic rock, b: chlorite veinlet. Notice the rounded surface controlled by the chlorite veinlet.

FIG. IV-2-13. Photomicrograph. Thin section of the schistose green rock. One nicol. Py: pyroxene.

be pillow lava, or boulders of volcanic conglomerates. But no difference in texture is recognized between the boulders and the matrix of this rock, either macroscopically or microscopically. The surface of the boulders is coated with chlorite flakes (Fig. IV-2–12). Macroscopic and microscopic observations reveal that the rounded surface is affected by chlorite veinlets.

In some schistose rocks of sedimentary origin adjacent to the basic intrusive, a small amount of porphyroblasts of pyroxene (diopside) is observed under the microscope (Fig. IV-2–13). Besides pyroxene, this rock is composed of calcite, chlorite, sericite, epidote and leucoxene. The combination of calcite, chlorite, sericite and epidote corresponds to the so-called greenschist facies or epidote-amphibolite facies. It is generally recognized as a relict mineral. The writer considers that the pyroxenes in the greenschists adjacent to the basic intrusives have been formed by pneumatolysis*, genetically related to the intrusions of the basic intrusives. For, as stated above, he interprets that the greenschists were mainly derived from sandstones, clayslates and limestones due to Mg-Fe metasomatism accompanying the basic intrusions. In other words, the pyroxenes are considered to have been deposited in the same way as the other high-temperature minerals. The tourmaline, garnet and magnetite in the Besshi mine and other mines of the same type may have been deposited by pneumatolytic action, genetically related to the intrusions of basic to ultrabasic rocks as Horikoshi (1938, 1940) stated. Taliaferro (1943) attributed the formation of glaucophane schist in the Franciscan Group to pneumatolysis. It is noticeable that pneumatolytic minerals are not found in the bedded cupriferous pyrite deposits in the Chichibu Paleozoic or Triassic Systems nor in the Shimanto undifferentiated Mesozoic System. They are found in those in the Sambagawa-Mikabu terrain, which is the core part of the regional metamorphism.

Problems of Ferruginous (Manganese-Bearing) Cherts

In the Okuki district, the ore bodies along the boundary between massive green rocks and schistose green rocks (greenschists) are in most cases accompanied by ferruginous cherts which always occur just above the ore bodies (Fig. IV-2–6). The thickness of the cherts reaches 20–30 cm. In this district, cherts of the same kind are intercalated within the greenschists. They are red or brown bedded rocks, containing iron and frequently manganese. In some places the contents of iron and manganese are high enough to have been worked before. In the southern and southwestern area of the Okuki district, iron-manganese ores of this type were worked about twenty or thirty years ago. The beds of chert in this area attain 50 m in maximum thickness. Ferruginous chert is usually reddish, but manganese-bearing ferruginous chert is brownish and is often penetrated by veinlets of rhodochrosite or rhodonite. Under a petrographic microscope, in some specimens fine-grained quartz and hematite are recognized, but in others they cannot be observed because of the submicroscopic sizes of the crystals. In the former specimens so-called radiolarian remains, which will be discussed later, are sporadically observed. In an X-ray powder diffractometer, reflections of quartz and hematite are recognized (Table IV-2–2). The content of mangenese in an ore body is variable, and the maximum content is 20% MnO. Table IV-2–3 shows the results of analyses of the iron-manganese ore from the Fujinokawa, about 3 km southwest of the Okuki mine (Fig. IV-2–1). The manganese-bearing minerals in this ore dissolve rather slowly in hydrochloric acid, which indicates that the manganese probably came from manganiferous

---

* Of course, most of the pyroxene crystals in diabase schist or metadiabase are interpreted to be relict minerals, as described earlier. The term "pneumatolysis" is used following the definition by Fenner (1933) or Graton (1940) as stated in Chapter II-17. As pyrometasomatic deposits related to diabase, the following examples have been reported: Hickok, W. O. (1933) The iron ore deposits at Cornwall, Pa.: *Econ. Geol.*, **28**, 193–255; Sobolev, V. (1935) The iron ore deposits of the Ilimpeia river, eastern Siberia: *Econ. Geol.*, **30**, 783–791.

TABLE IV-2-2. X-ray powder diffraction data for the iron-manganese ore from Fujinokawa near the Okuki mine.

Iron-manganese ore from Fujinokawa		Hematite from the Waga-Sennin mine		Hematite*		
d	I/I₀	d	I/I₀	d	I	
				4.06	0.3	
3.69	25	3.65	30	3.66	1.0	
2.70	100	2.68	100	2.69	7.0—	Coincided
2.51	75	2.50	100	2.51	4.0	with the
2.21	40	2.20	30	2.18	2.0	strongest
1.84	30	1.84	45	1.84	3.0	peak of
1.69	50	1.69	42	1.68	5.0	braunite
1.60	13	1.60	5	1.58	0.5	
1.48	13	1.48	30	1.49	1.0	
1.45	30	1.45	30	1.44	2.0	

* Harcourt G. A. (1942) Table for the identification of ore minerals by X-ray powder patterns: Am. Mineral., 27, 86.

TABLE IV-2-3. Chemical analyses of the iron-manganese ore from Fujinokawa near the Okuki mine. (Analysts: T. Nakai and H. Irokawa, National Chemical Laboratory for Industry)

Spectrochemical analysis
  Main elements presented: Fe, Mn, Si, W, Ti, Ba, Ni, (Cr)
  Elements definitely present in small amounts: Ca, Mg, Al, V, Na, Co
  Elements not detected with certainty: Zn, Bi, Sr, Sn, Pt, B, P, Be, Cu, K, rare earths

Wet analysis

SiO₂	Fe₂O₃	Al₂O₃	TiO₂	Mn₂O₃	CaO	MgO	P₂O₅	H₂O(−)	Ig.loss.	Total
					Alkaline fusion					
12.33	48.76	5.08	tr.	20.24	6.31	tr.	3.42	1.45	2.67	100.26
					Soluble part in HCl(1:1)**					
0.27	48.63			17.92						

** About 0.5 g of the sample was treated by 50 cc of HCl (1:1) for 5 hours on the water bath.

hematite. Mason (1943) recognized that MnO in hematite may occur as an isomorphous mixture attaining 15 mole %. Some of the manganese might have come from braunite, 3 MnMnO₃·MnSiO₃, because in the residue attained by heating the sample in HCl (1:1), a small amount of amorphous silica gel, which may be a decomposition product of the silicate mineral, is observed under the microscope; thus it is probable that some amount of MnO comes from the decomposition of braunite, although this mineral has not yet been recognized in X-ray charts or under the microscope. Phosphoric oxide shown in Table IV-2-3 will also be combined with calcium oxide to form tricalcium phosphate mineral or other calcium phosphate minerals, though the minerals of these compositions are not yet recognized either microscopically or in X-ray diffraction. The mode of occurrence of Al₂O₃ is unknown, but Al₂O₃ is possibly contained in hematite to some extent.

Such bedded iron-manganese ore deposits occur in the Paleozoic-Triassic (Chichibu) terrain, and also in the undifferentiated Mesozoic terrain of Japan. Typical examples of the deposits are listed below:

Name and place of the deposit	Country rock
Tokoro district, Hokkaido	Undifferentiated Mesozoic rocks (Fig. I-3)
Isakozawa, Iwate Prefecture	Paleozoic rocks (Fig. I-3)
Odake-Chibayama district, Shizuoka Prefecture	Undifferentiated Mesozoic rocks (Figs. I-1 and -3)

| Kunimiyama, Kochi Prefecture | Paleozoic rocks (Figs. I-1 and -3) |

Also, in the Chichibu terrain manganese ore composed of rhodochrosite, rhodonite, tephroite, etc., occurs in bedded form. As both types of manganese deposit occur under similar geological conditions, the origin of the ore deposits might be considered nearly the same. So the origin of these types of manganese deposits is related to the origin of the ferruginous chert in the Okuki district.

The genesis of the ferruginous chert as well as of the manganese deposit has always been a controversial subject. Three principal interpretations of the origin have been proposed:

1. Sedimentary origin, probably a submarine exhalative sedimentary origin (Takabatake, 1956; Watanabe, 1957; Watanabe et al., 1970a).

2. Replacement origin genetically related to intrusion of basic rocks (Bureau of Mines of Japan, 1932; Yoshimura, 1952, 1969; Sawamura and Yoshinaga, 1953).

3. Ferruginous cherts of sedimentary origin, enriched by diabasic intrusion. The writer calls this the enrichment origin (Kato, 1927a; Asahi, et al. 1954; Suzuki and Ohmachi, 1956).

The bases for each interpretation are as follows:

1. Sedimentary origin: Bedded or lens-like deposits of manganese intercalated in the Paleozoic or the undifferentiated Mesozoic systems occur in certain horizons, accompanied by chert or quartzite in the overlying and underlying beds.

The submarine-sedimentary origin is supported by the banded texture of ores as well as by the facts that the manganese ores and the surrounding cherts often contain radiolarian remains and the deposits are frequently in association with "schalsteins" (metatuffs).

In this interpretation the rock called diabase is regarded as a kind of lava flow or sheet of shallow intrusion which took place at nearly the same time as the deposition of the surrounding rocks. This interpretation corresponds to the "exhalative sedimentary" origin of Epprecht (1946) and Schneiderhöhn (1955), and the "volcanic-subaquatic" origin of Niggli and Niggli (1952).

2. Replacement origin: Veinlets of manganiferous hematite partly cut cherts or green rocks surrounding the ore deposits (Fig. IV-2-14). Ore bodies are apt to occur as saddle reefs in anticlines. The deposits are frequently accompanied by diabasic or gabbroic rocks which are regarded as sheet-like intrusives. These points support the replacement origin.

3. Enrichment origin: The third interpretation is an intermediate compromise between (1) and (2), since its advocators maintain that sedimentary exhalative deposits were formed by volcanic activities in geosynclines and were later enriched by diabasic intrusions.

The occurrence of radiolarian remains in manganese ores and cherts presents a perplexing problem for interpretation of the ore genesis. It may favor the sedimentary origin. In the replacement origin, the radiolarias are considered to have been silicified but have retained their original organic texture. From this standpoint, it may be conceivable that the well-retained texture of the radiolarian remains in cherts or manganese deposits are due to silicification after the sedimentation. In the Ogaki mine, Tochigi Prefecture, Yoshimura (1952) reported that radiolarian remains were replaced by rhodochrosite, which he regarded as proof of the replacement hypothesis. Watanabe (1957) asserted that siliceous remains of radiolarias would be carbonatized by diagenesis. He related the radiolarian remains composed of rhodochrosite to diagenesis.

From the standpoint of replacement origin, manganese deposits in certain horizons may be considered to have been formed along these zones as related to the intrusion of basic rocks, just as from the standpoint of the replacement hypothesis one can consider the origin of bedded cupriferous pyrite deposits which occur in certain horizons of the crystalline schists.

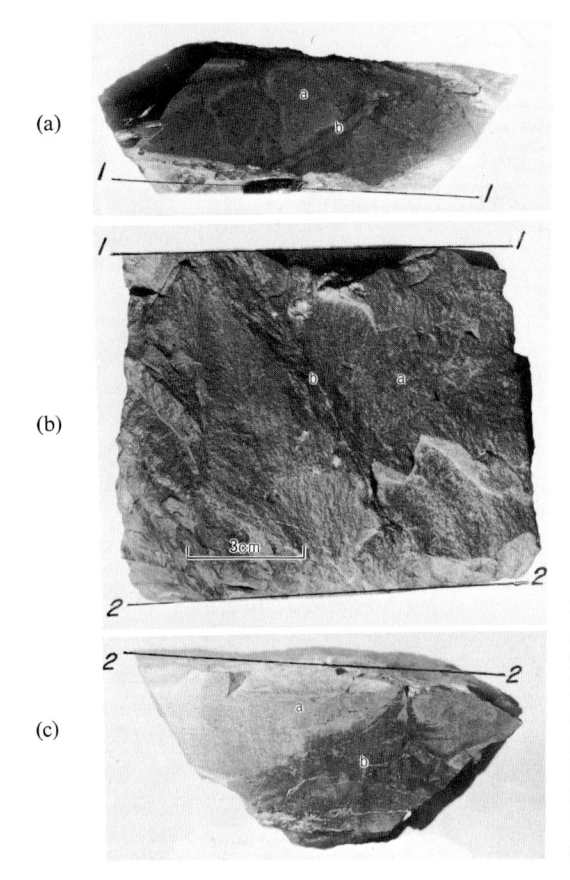

FIG. IV-2-14. Hand specimen of massive green rock. (a) Side view (polished surface), (b) front view, (c) side view (polished surface). Lines 1——1 and 2——2 in each photograph coincide with each other, respectively. The plane of photograph (b) represents the original bedding plane. a: massive green rock, b: ferruginous chert. Notice the ferruginous chert veinlet cutting the original bedding plane of the schistose green rock.

As to the frequent occurrence of manganese deposits as saddle reefs, it is necessary in the sedimentary theory to presume that the manganese or iron-manganese ores may have been more movable or plastic than the surrounding rocks and thus may have come to concentrate in the axial parts of anticlines in the course of folding. The replacement hypothesis suggests that the replacement of ore takes places more easily along the axes of the anticlines than in the limbs. Also, in regard to the existence of veinlets of manganese ore in the country rock, sedimentary origin necessitates that the ores should be movable or plastic.

At present the writer favors the replacement interpretation. He recognizes that the iron-manganese ore deposit of the Odake mine in Shizuoka Prefecture (Figs. I-1 and -3) is a replacement ore in diabase (or dolerite). In the undifferentiated Mesozoic beds in the Odake mining district, diabasic rocks accompanied by serpentinite occur in places along a zone parallel to the SW-trending bedding planes, and in these diabasic rocks masses of iron-manganese ore are found (Fig. IV-2-15). The ore bodies are massive, with dimensions of 120 m × 180 m × 25 m, and the ore bodies grade into the country rocks, distinctly indicating that the ore is of replacement origin.

Yoshimura (1952, 1969) and Togo *et al.* (1954) also concluded that the Odake deposit was formed by replacement in dolerite.

The writer thinks that the iron-manganese deposit of the Okuki district is of the same type as the Odake deposit. For this interpretation, the writer premises that the manganese or iron-manganese deposits in the Paleozoic-Triassic (including the Sambagawa-Mikabu Complex) and in the undifferentiated Mesozoic terrain of the Outer Zone in Southwest Japan are of the same origin as the Odake deposit.

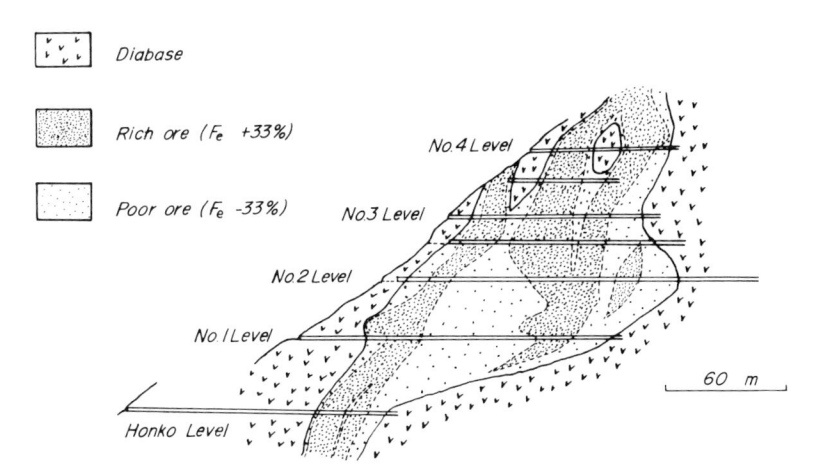

FIG. IV-2-15. Geologic section of the Odake mine (E—W section).

In the Okuki district, it is highly probable that the diabasic rocks are genetically related to the iron-manganese deposits.

As shown in Table IV-2-3, some of the minor elements detected in the iron-manganese ore, such as tungsten, nickel, chrome and cobalt, are more likely to be of replacement origin in the deeper parts of the earth rather than of exhalative sedimentary origin.

The genetical problems of bedded-type manganese deposits and (ferruginous) cherts have been discussed by many geologists throughout the world. Sampson (1923) described the ferruginous cherts of Notre Dame Bay, Newfoundland, which are found as beds of uniform thickness within a volcanic complex. He regarded the cherts as sediments of chemical precipitation. The rocks of the Notre Dame Bay region vary in age between Cambrian and Silurian, and are mostly of volcanic origin with pillow lavas predominant. Sampson considered that these widespread pillow lavas resulted from subaqueous extrusion. The bedded chert occasionally contains radiolarian fossils. Sampson thought that the silica of the cherts was mainly derived from a marine basin by magmatic emanations from submarine vents or fissures. He also described the peculiar features of the manganiferous or non-manganiferous jaspers which were formed in silica gel.

Fowler, *et al.* (1935) asserted that cherts of the Tri-State mining district were formed by silicification of limestone, and called the process "chertification." But these cherts do not contain manganese or iron.

Park (1946, 1956) reported manganese deposits from the Olympic Peninsula, Washington, which occur in volcanic rocks, red limestones and red argillite of the Early-Middle Eocene. He stated that the deposits are genetically related to submarine volcanic activity. He thinks that manganese is a common constituent of basalt, and explained the occurrence of the manganese deposits as follows. During the process of spilitization, ferro-magnesian minerals were decomposed and manganese and silica, along with other constituents, migrated to the upper part of the basalt and replaced the spilite flows and the overlying limestone or argillite.

Taliaferro (1943) thinks that at least part, if not most, of the silica in the cherts was derived from submarine springs accompanying volcanism, and that the iron and manganese commonly in association with cherts have the same origin.

Webber (1948) reported that in Costa Rica hypogene manganese minerals form deposits in diabase, chert and shale, in association with abundant jasperoidal silica. According to him, manganese may be telethermal or epithermal, situated far from its source, and it seems unlikely that the manganese mineralization is related to the diabasic intrusion.

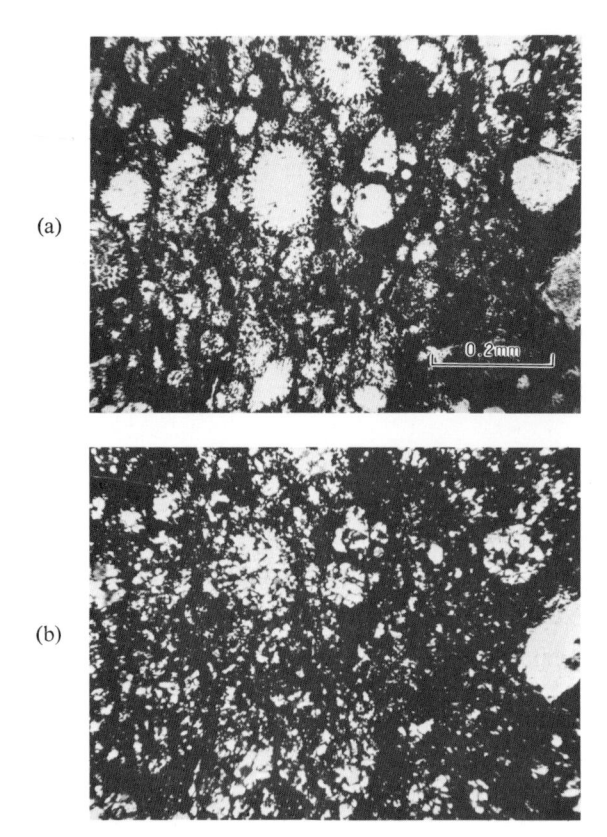

FIG. IV-2-16. Photomicrographs. Thin section of radiolarian chert. (a) One nicol, white part is quartz, black qart is hematite. (b) Same section as (a) under crossed nicols.

According to Wenk (1949), the association of radiolarian cherts with extrusive rocks of spilitic composition and with tuffs can be explained as having been affected by submarine effusions of alkali-basaltic magma rich in volatile components, by submarine exhalations (CO_2, SiO_2, etc.) and by explosions producing much tuffaceous material suspended in sea water.

In Japan, the problems of radiolarian remains have been discussed by paleontologists and geologists. Hujimoto (1938) discovered some fossils in calcareous lenses contained in the sericite schist of the Sambagawa System on the left bank of the Arakawa river, Saitama Prefecture. The fossils included thirteen species of radiolaria and one species of foraminifera. Because five of the radiolaria belong to *Cyrtoidea* and three species of *Archicapsa*, *Tricolocapsa* and *Lithocampe* have their respective allies in the Jurassic of Europe, Borneo and Japan, he concluded that the age of this fauna must be the Jurassic or at least the Mesozoic. But Suzuki (1939) considered that there is no reason to believe that the radiolarian remains in the Sambagawa schist present a marked vestige of the Mesozoic, especially not of the Jurassic type. Kobayashi and Kimura (1944) asserted that Hujimoto's claim naturally brings about a question as to whether radiolarian remains are really an indicator of geologic age, and concluded that though some minor parts of the Sambagawa-Mikabu Complex may be Triassic in age the major part was originally Permo-Carboniferous.

The so-called radiolarian remains in the ferruginous chert of the Okuki district are, when observed under the microscope by one nicol, composed of the aggregates of small-sized quartz and hematite which give a fossil-like appearance (Fig. IV–2–16(a)). Under the crossed nicols, however, the silicified aggregates do not look like fossils (Fig. IV–2–16(b)). The writer thinks that it is necessary to discuss whether the aggregates are really fossils or not. At least it is certain that they are not the remains of organisms themselves.

Either sedimentary origin or replacement origin is acceptable, both are coincident in that the manganese deposits are genetically related to the basic rocks. The frequent coexistence of (manganese-bearing) ferruginous cherts and cupriferous pyrite deposits, as observed in the Okuki mine, Asakawa mine (Kato, 1927a) and other mines, would be due to the coincidence of conditions favorable to both kinds whether in sedimentation or in replacement succession.

As stated later, the writer interprets that the cupriferous pyrite deposits were formed by replacement. The coexistence of cupriferous pyrite deposits with piedmontite or hematite-quartz schist in the Sambagawa-Mikabu Complex as observed at the Besshi or Iimori mine is also interpreted in the same way (Fig. I–1). Kamiyama (1950) reported that the plunge (or extension) of the piedmontite schist in the Iimori mine coincides with that of the ore body and the lineations of the country rocks. That is, piedmontite or hematite-quartz schist of this kind would correspond to dynamometamorphosed manganiferous or ferruginous chert, which was formed by replacement.

Genesis of the Ore Deposit

As stated in the foregoing section, the folding or dynamometamorphism in the Okuki district was accompanied by basic intrusions. The ore deposit of the bedded cupriferous pyrite was genetically related to the basic intrusion and would have been formed along folds, mainly at the anticlines and synclines of the folds, i.e., as saddle reefs and trough reefs. Folding, dynamometamorphism, intrusion of basic rock and ore deposition are a series of geologic

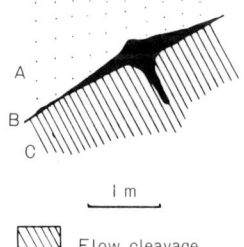

FIG. IV-2-17. Schematic geologic section of the ore body at the Okuki mine. A: massive green rock, B: ore body, C: schistose green rock. Notice the relation between the ore body and the flow cleavage.

FIG. IV-2-18. Hand specimen of copper-rich ore (polished surface). Notice the remnant structure (a) of the original bedding plane, which runs from right to left.

events. As discussed already, the writer interprets that (manganese-bearing) ferruginous chert is also genetically related to the intrusion of the basic rocks. The replacement by manganese-bearing ferruginous chert (jasperiod) preceded the deposition of the cupriferous pyrite ore, because small veinlets of the latter penetrate the former. The remnants of drag folds (bedding schistosities) are recognized in the ore bodies (Fig. IV–2–18). The forms of the ore bodies are mainly controlled by the folded structures of the green rocks and partly by flow cleavages (Fig. IV–2–17). The flow cleavage is developed in the schistose green rocks surrounding the ore bodies, but the replaced remnant of this flow cleavage has not yet been observed in the ore (Fig. IV–2–6). These facts tell that the major part of the mineralization was almost ended just before the formation of the flow cleavage. It would be difficult to produce flow cleavage in the competent ore bodies.

The genesis of the saddle or trough reef has been discussed by many geologists. The saddle reef is frequently overlain by cap rocks. The important problem of the saddle reef is whether it was deposited in an open space or deposited replacing the country rock. As for the Okuki mine, a remnant structure of drag folds is recognized in ore from the saddle reefs (Fig. IV–2–18). So the writer believes that at least some parts of the saddle reefs were formed by the replacement of the country rocks, because mineralizing solutions flowed just below the impermeable cap rock. But it is unknown whether the ores partly filled the cavities at the anticlines of the folds or not. It would be difficult for a cavity to be formed in a plastic rock such as the schistose green rock of this deposit. But it is probable that, after replacement by the sulfide ore in the schitose green rock advanced to some extent, a cavity was formed in this ore body in the course of folding, as Chace (1949) described. From the existence of the colloform structure of the ore in the saddle reef, as described later, the writer believes that an open space would occur once in the course of mineralization. It is certain that the formation of the trough reef in the Okuki mine is related to the occurrence of the thick schistose green rock in the synclinal part of a similar folding.

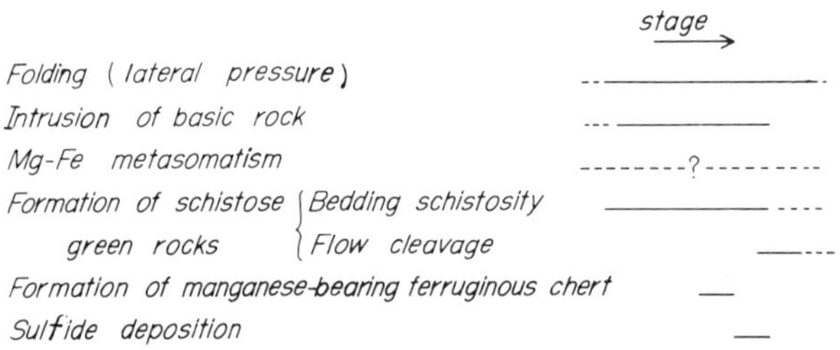

FIG. IV–2–19. Succession of events in the Okuki deposit.

From the above facts the writer has compiled the succession of events in this deposit, as shown in Fig. IV–2–19.

Many hypotheses on the genesis of the bedded cupriferous pyrite deposits in Southwest Japan, especially in the Besshi deposit, have been published. Kochibe (1892) and Nakajima (1893) accepted a sedimentary origin for the Besshi deposit in the earliest days of geologic research in Japan. Kuhara (1914) asserted that the Besshi deposit was of pyrometasomatic origin. Matsubara (1953) regarded the genesis of the Besshi deposit to be orthomagmatic. But many Japanese geologists (K. Nishio, 1910; Sagawa, 1910; Kato, 1925; Horikoshi, 1938; S. Nishio, 1940) held the opinion that deposits of this kind were formed by replacement at the time of dynamometamorphism, which was accompanied by basic igneous intrusion. Recently the theory of sedimentary origin has been revived. According to Kojima et al. (1956), the greater part of the bedded cupriferous pyrite deposits in the Sambagawa-Mikabu

Complex may have been derived from materials deposited in relation to the submarine eruption in the original geosyncline. Watanabe (1957) and Watanabe et al. (1970) stated that some of the bedded cupriferous pyrite deposits may have been formed by materials supplied from the fumaroles or hot springs accompanied by submarine volcanic activities in the original geosyncline of the Sambagawa-Mikabu Complex, and some others may have been produced by replacement at shallow depths under low-temperature conditions in the course of sedimentation. Thus, concerning the genesis of the bedded cupriferous pyrite deposits, discussions are going on in Japan similar to recent arguments on the origin of the bedded cupriferous pyrite deposit in Rammelsberg in Germany, Ergani-Maden in Turkey, Sulitelma, Outokumpu and others in Scandinavia, Rio Tinto in Spain, copper deposits in Rhodesia, Zambia and Zaire, Broken Hill in Australia, cupriferous pyrite deposits in Cyprus and massive deposits in Canada, U.S.A. and U.S.S.R. At present, the sedimentary or exhalative sedimentary hypothesis is supported by the majority.

The writer premises that all the bedded cupriferous pyrite deposits in the Sambagawa-Mikabu Complex were formed at nearly the same time by a similar mineralizing process, and the genesis is discussed as follows:

Kato (1925) reported that the Shibuki and Seki deposits in the Sambagawa terrain occur along the boundaries between greenschists and serpentinites, and the latter were replaced by the ore bodies (Fig. I–3). Kamiyama (1956) recognized that the deposit of the Kune mine in the Sambagawa terrain is found along the boundary between black schist and porphyrite (basic rock) and the porphyrite (silicified and carbonatized) was replaced by the ore bodies. He interpreted that the porphyrite belongs to intrusive rock and recognized that it extends parallel to the lineation or folding axis of the black schist and the plunge of the ore bodies.

The writer et al. (1951) believed that the ore deposit of the Nago mine, in the vicinity of the Kune mine, occurs in the sheared zone of the basic intrusive (gabbro or diabase).

As stated above, in the Okuki mine, the basic intrusive as well as the ore bodies plunge in a direction parallel to the folding axis and the lineation of the green and black schists. The ore deposit of this mine is genetically related to the basic intrusive.

In the ore from the Okuki mine, cobaltite is recognized. Yamaoka (1958, 1962) and Itoh (1976) reported that the pyrites from the bedded cupriferous pyrite deposits in Southwest Japan are relatively rich in cobalt.*

On the basis of these facts, the writer believes that bedded cupriferous pyrite deposits in the Sambagawa-Mikabu Complex belong to a replacement deposit genetically related to the basic intrusive which was synchronously formed with the dynamometamorphism.

A colloform structure is rarely recognized in ore from the Okuki mine (Fig. IV-2-20). It is certain that this structure occurs frequently in the sedimentary ore. But, more generally speaking, this structure is recognized in ores which were deposited in open cavities at lower temperatures (Edwards, 1954). As stated above, the writer thinks it probable that the cavities were formed in the ore body in the course of mineralization. The occurrence of this structure may indicate the existence of open cavities during mineralization accompanied by dynamometamorphism.

It is true that the bedded cupriferous pyrite deposits in the Sambagawa-Mikabu Complex frequently occur in certain horizons. This fact provides sound ground for the hypothesis of sedimentary origin. But, from the viewpoint of the replacement hypothesis, the situation can be interpreted as follows; igneous activities prevailed in this zone, and the bedded cupriferous pyrite deposits were formed in this zone by replacement, genetically related to these igneous activities.

---

* The content of cobalt in pyrite from the Kuroko (Black Ore) deposits is poor (Takahashi, 1963).

Fig. IV-2-20. Photomicrograph. Polished section of copper ore. One nicol. Py: pyrite, Ch: chalcopyrite. Notice the colloform texture of pyrite replaced by chalcopyrite.

In the Chichibu System to the south of the Sambagawa-Mikabu Complex, similar bedded cupriferous pyrite deposits are present, although they are much fewer in number than in the Sambagawa-Mikabu Complex. The Nanogawa and Choja deposits in Kochi Prefecture are examples (Fig. I-3). The country rocks of these deposits suffered weak metamorphism and altered to sericite phyllite and chlorite phyllite. Intrusive rocks such as diorite, diabase and serpentinite occur in the mining areas. These deposits in the Chichibu terrain may have been formed by the same series of igneous activities accompanied by metamorphism as those in the Sambagawa-Mikabu Complex.

In the Shimanto undifferentiated Mesozoic System which borders on the south of the Chichibu Paleozoic-Triassic System with a thrust fault, similar kinds of bedded cupriferous pyrite deposits to those of the Asakawa mine (Kato, 1927a) in Tokushima Prefecture or the Makimine mine (Tatsumi, 1953) in Miyazaki Prefecture occur (Fig. I-3). They may have been deposited in genetical relation to the basic intrusives. The country rocks are more or less metamorphosed. The Shimanto terrain suffered nearly the same kind of orogenic movement as that which occurred in the Chichibu terrain. It is folded, but the disturbance is not so intense as in the Chichibu terrain. According to Kobayashi (1941), a series of folds and thrusts in the Shimanto terrain probably formed the Sambagawa-Mikabu metamorphosed Complex from the Chichibu System. Kobayashi called this orogenesis the Sakawa Orogenesis. In the Shimanto System, basic intrusives occur as sheets, especially in the highly folded and metamorphosed zone where clayslates and sandstones have been changed to semischists or phyllites.

In relation to Kobayashi's conclusion, the writer concludes that the bedded cupriferous pyrite deposits in the Shimanto terrain may have also been deposited by the same series of igneous activities accompanied by dynamometamorphism as in the Sambagawa-Mikabu and Chichibu terrains.

Since the occurrence of deposits in the Sambagawa-Mikabu Complex is more frequent than in the Paleozoic-Triassic (Chichibu) and the undifferentiated Mesozoic (Shimanto) terrains, most of the deposits were probably formed at the centers (cores) of dynamometamorphism.

According to Matsumoto (1947), the pebbles of the Sambagawa-Mikabu Complex are found in the Ryoseki Formation of the Earliest Cretaceous epoch (Wealden). But some of the basic rocks in the Shimanto terrain may have been intruded after the Aptian epoch, as the Shimanto System ranges in age from the Triassic period to the Aptian epoch. From these facts, the writer agrees with the Kobayashi's concept which maintains that the geosyncline and orogenesis migrated towards the south with the lapse of geologic time. The mineralization of the bedded cupriferous pyrite deposit may also have migrated accompanied by orogenesis and metamorphism.

The writer believes that the metallogenetic epoch of the bedded cupriferous pyritic deposits in the Sambagawa-Mikabu terrain should be correlated to the time of orogenesis, i.e., metamorphism. So it would correspond to the Late Mesozoic.

Conclusions

In the Outer Zone of southwest Japan, the following types of terrain, characterized by different rock formations, are zonally arranged in a direction parallel or sub-parallel to the Median Tectonic Line. The terrains are from north to south:

(1) Sambagawa-Mikabu terrain: Zone of metamorphic rocks extending along the Median Tectonic Line which trends NE-SW or E-W. The metamorphic rocks include albite-epidote-chlorite schist, amphibole-chlorite schist, glaucophane schist, piedmontite-quartz schist, stilpnomelane-bearing greenschist, graphite-quartz schist, sericite-quartz schist, etc., accompanied by basic to ultrabasic igneous rocks. The schists containing epidote, chlorite, amphibole, etc., are called greenschists, or schistose green rocks. Massive rocks of the same mineral and chemical composition are called green rocks.

(2) Chichibu terrain: The Sambagawa-Mikabu terrain grades southward into the Chichibu terrain, which is mainly composed of Permo-Carboniferous and Triassic sedimentary rocks.

(3) Shimanto terrain: Undifferentiated Mesozoic rocks, probably Triassic to Cretaceous in age.

Throughout the Sambagawa-Mikabu terrain, many deposits of bedded cupriferous pyrite (Kieslager-type) deposits occur along the planes of schistosity, frequently as lenticular or bedded bodies. No particular type of schist constitutes the sole rock, but greenschists are the most common hosts. It is noteworthy that lenses of piedmontite-or hematite-quartz schist are frequently associated with the ore bodies.

In the Chichibu and Shimanto terrains, similar deposits occur as lenticular bodies, though they are rarer than in the Sambagawa-Mikabu terrain and most of them are accompanied by basic rocks. Frequently, ferruginous or manganiferous cherts or green rocks form the hanging-wall or the footwall.

The writer asserts that the deposits have a close relation with metamorphism.

The writer believes that the Outer Zone of Southwest Japan is the most adequate for the studies of the genesis of this kind of deposit, since it is possible to study the relation between non-metamorphosed and metamorphosed sedimentary rocks and the occurrence of the ore deposits.

The deposit of the Okuki mine in the Sambagawa-Mikabu Complex occurs in the schistose green rocks underlying massive green rocks, mainly as saddle reefs in overturned and plunging folds. The intruded basic rocks occur in the core of the folds. The writer has discussed the relations between the plunge of the deposit and the folding, and between the occurrence of flow cleavage, bedding schistosity and lineation in the schistose green rocks.

The genesis of the bedded cupriferous pyrite deposits in these terrains has been a frequent topic of discussion among Japanese geologists.

On the basis of present knowledge the writer concludes that: (a) the bedded cupriferous pyrite deposits in these terrains were formed by hydrothermal metasomatism, genetically related to intrusion of basic igneous rocks and associated folding and dynamometamorphism; (b) a large part of the greenschists in the Sambagawa-Mikabu terrain may have been derived by chloritization, epidotization and amphibolitization (Mg-Fe metasomatism) of the Paleozoic argillaceous, arenaceous or calcareous sedimentary rocks accompanied by the intrusion of the basic or ultrabasic rocks; and (c) piedmontite (hematite)-quartz schist or ferruginous (manganiferous) chert which forms either the hanging wall or footwall of the deposit is also of replacement origin, and is genetically related to the basic intrusives accompanied by dynamometamorphism.

IV–3. Besshi Mine, Ehime Prefecture (Imai, 1963b)

H. Imai

This mine is located in the rugged mountain district of the backbone range extending E-W on Shikoku Island (Fig. I-3). Opened in 1690, it was one of the largest copper deposits in Japan. It produced about 30 million tons of crude ore containing 2.45% copper. But it was closed in 1973.

The ore deposit of the Besshi mine is intercalated in the Sambagawa System which consists chiefly of graphite schist (black schist) and chlorite-amphibole schist (greenschist) with thin layers of sericite-piedmontite-quartz schist and other rocks. The rocks strike N60°W, dipping 40°–80° NE. The succession of the schists with reference to the ore deposit is shown in a profile along the main adit (Fig. IV–3–1). The occurrence of a great mass of metamorphosed basic intrusive rocks comprising amphibolite, peridotite and others on the hanging wall side is worthy of special notice.

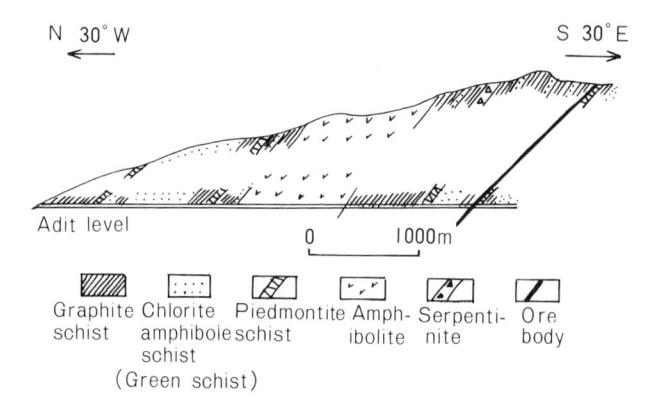

FIG. IV–3–1. Geologic profile of the Besshi mine (after Kato, 1926a).

The deposit is bedded and has a strike length of more than 1,700 m, extends over 3,500 m down the dip, and is up to 10 m thick. The ore deposit is enclosed in graphite schist and chlorite-amphibole schist, and is bordered by siliceous, light-colored schistose rocks such as piedmontite-quartz schist, sericite-quartz schist, etc.

In the upper levels of the adits the deposit plunges to N70°E 30°–50°, while below the main level of the adits the plunge changes to N35°W 60°–70° (Fig. IV–3–2). The lineation of the country rocks is parallel to the plunge of the ore body, so it changes gradually from N70°E30°–50° in the upper levels to N35°W60°–70° in the lower levels (Doi, 1961 ~ 1962)

Part IV: Strata-Bound Deposits 257

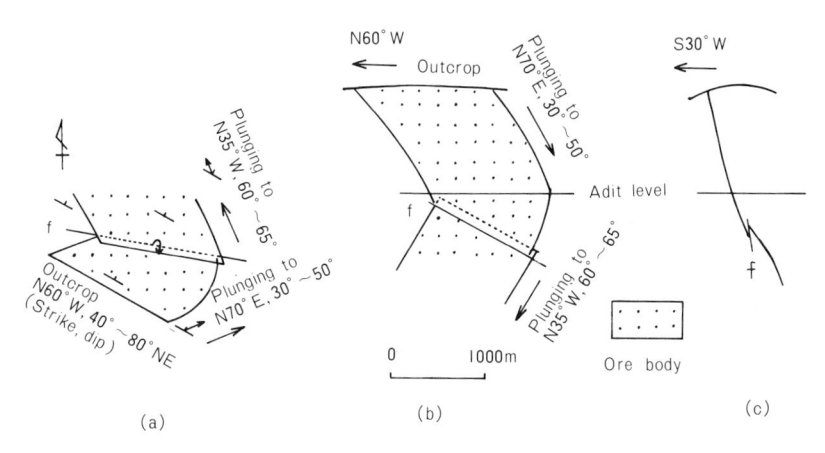

FIG. IV-3-2. The ore body of the Besshi mine. (a) horizontal projection, (b) vertical projection along the strike, (c) vertical projection along the dip. f: fault.

(Fig. IV-3-2). The reason for this change is unknown at present, but it is possible that it is due to the intrusion of the igneous mass into the deeper part of this area, as the writer discusses in Chapter IV-5.

Along the zone in which the plunge of this ore deposit changes several faults (reversed faults) exist (Fig. IV-3-2).

Two kinds of ores are recognized in the upper and middle parts of the ore deposits:

(1) Massive cupriferous pyrite ore: This is the most common ore; it consists of minute grains of pyrite and small amounts of magnetite and hematite firmly cemented by chalcopyrite. The presence of chalcopyrite as the cementing matter is well displayed on polished specimens. This ore occurs as tabular masses developed along both walls; the ores of this kind developed along the both walls are called hanging-wall sulfide ore and footwall sulfide ore, respectively.

(2) Banded ore: This ore is intercalated between the above-mentioned massive pyritic ores which are developed along the hanging- and footwalls, and ranges in thickness from 0.7 m to 6 m. It consists of magnetite (hematite)-chlorite-quartz schist thinly interbedded with layers of pyrite and chalcopyrite aggregates, and is represented by greenschist impregnated with abundant pyrite in association with chalcopyrite. It contains about 3% copper and about 30% silica, on the average.

According to Kase (1972), the mineral assemblages of the above ores vary with increasing depth from hematite-bearing pyrite-chalcopyrite ore to ilmenite-bearing pyrrhotite-chalcopyrite ore. He suggested the existence of Neogene acid igneous rock in the deeper part of this mining area. Izawa and Mukaiyama (1972) and Miyazaki *et al.* (1974) discussed the mineralogical problems of the formation of pyrrhotite due to thermal metamorphism in this deposit. Valleriite or mackinawite was described by Miyahisa (1958) and Miyazaki *et al.* (1974). Miyazaki *et al.* (1974) stated that cobaltite was produced by the transformation of pyrite into pyrrhotite.

Shimada and Tsunori (1962) recognized a small amount of stannite in the brecciated part of the above-described banded ore of the lower adit in this mine. They concluded that it was deposited by a hydrothermal solution belonging to a different igneous activity from that which was responsible for ore formation in the Besshi mine.

The writer recognized scapolite-wollastonite-diopside schist, epidote-tremolite-quartz schist, cordierite-biotite-diopside-quartz schist, biotite-garnet-quartz schist, epidote-pyroxene-calcite schist, garnet-calcite schist, etc., in the lowest level of the adits.

These facts indicate that acidic plutonic rock exists underneath this mining area. It is possible that it belongs to the Neogene igneous activity which occurred in the Outer Zone of Southwest Japan. But the writer thinks that it is also possible that the acidic igneous rock in the deeper part is affiliated with the basic and ultrabasic rocks discussed above and with the regional metamorphism in this district (Sambagawa Metamorphism) as in the case of the Hitachi mine.

As discussed in the chapter on the Okuki mine, the writer believes that the mineralizing solution in this deposit may have originated in deep-seated magma related to basic and ultrabasic rocks such as amphibolite, peridotite and others. Metamorphism would have occurred at nearly the same time as the basic igneous activity and mineralization. The absolute age of metamorphism was estimated at 102–82 m.y. by the K-Ar method (Banno and Miller, 1965) and 88–85 m.y. by the Rb-Sr method (Hayase and Ishizaka, 1967) for muscovite and biotite from the schists in this area.

IV–4. Yanahara Mine, Okayama Prefecture   (Imai, 1958, 1963b)

H. Imai

This mine is located about 35 km NE of Okayama City, west Honshu Island. Geologically, it is situated in Maizuru Folded zone Inner Zone of Southwest Japan (Figs. I–1 and –3). This district is composed of Carboniferous-Triassic sedimentary rocks and intermediate to basic intrusives, penetrated by granite, quartz porphyry and porphyrite. Unconformably overlying these rocks, Late Mesozoic rhyolitic rocks are developed (Fig. IV–4–1).

This Carboniferous-Triassic system is strongly folded with a N-S axis, plunging to the south. It is composed chiefly of black slaty rocks, sandstone, and pyroclastics (Oshima, 1964). Biotite cordierite hornfels metamorphosed from the slaty rocks are developed extensively throughout the district, passing with gradual transition into black slate. The contact metamorphism may be ascribed to the underlying granitic rock. The intermediate to basic intrusives are composed of diorite and quartz diorite in the core part and diabase in the marginal parts. In the mining area, the basic intrusive exists in the anticlinorium part of the Carboniferous-Triassic sedimentary rocks. It occurs at the core of a fold and extends in the direction of plunge of the folding axis (Figs. IV–4–1 and –2). The ore bodies exist along the boundary between sedimentary rocks and basic intrusive.

The ore bodies occurring at the depressed part of anticlinorium are the largest, i.e., the Main Yanahara ore body extends swelling and pinching to the direction of plunge of the anticlinal axis, i.e., S10°E25°. The dimensions of the proven part of the Main Yanahara ore body at the present time are as follows: extension 1,500 m, maximum width 450 m, maximum thickness 100m. (Fig IV–4–2).

On the limbs of the fold, deposits of a similar type occur, such as Shimoyanahara, Hisagi, Old Hisagi, Hidashiro and Shimodani on the west limb, and Yasumiishi and Hoden on the east limb of the fold (Fig. IV–4–1). The Shimoyanahara ore body is along a minor fold on the west limb of the major anticlinorium. The plunge of its minor folding axis coincides perfectly with that of the major fold. The plunges of the ore bodies in limbs of the major anticlinorium also extend nearly parallel to the plunge of the major folding axis, i.e., S10°E25°, as observed in the Okuki mine (refer to Chapter IV–2)

The ore bodies are surrounded by peculiar rocks called "complex zone" by the geologists of the mine (Oshima, 1964). The complex zone occurs along the boundary between the basic intrusive and the sedimentary rocks. It is composed of diabase, keratophyre, quartz keratophyre, felsite, pyroclastics, and clayslate, which are partly thermal-metamor-

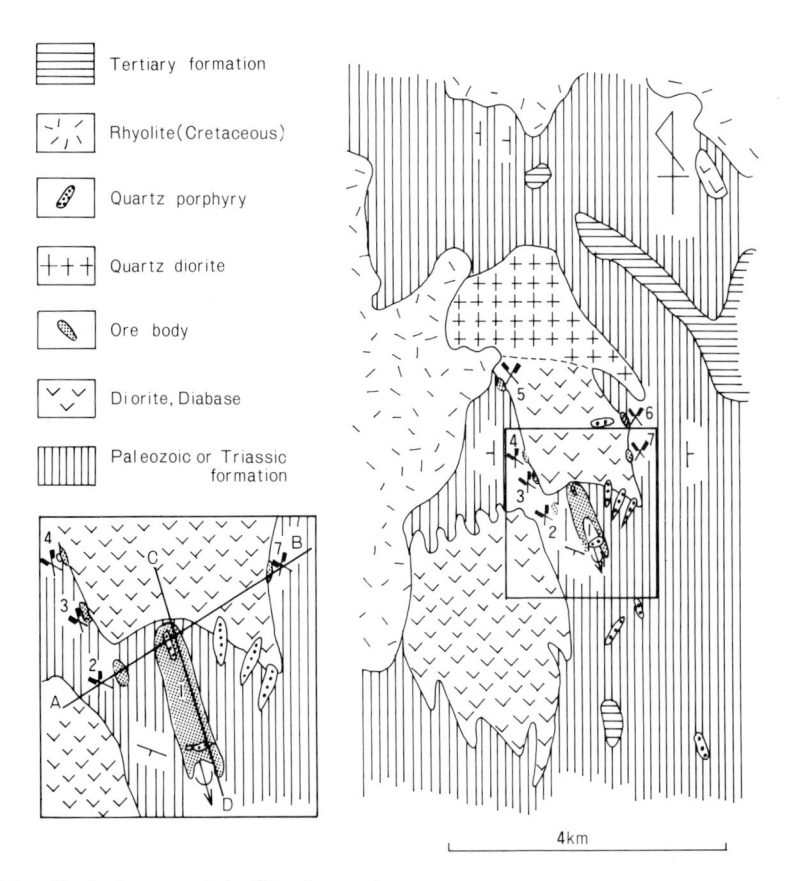

FIG. IV-4-1. Geologic map of the Yanahara mine.
1. Main Yanahara ore body, 2. Shimoyanahara ore body, 3. Old Hisagi ore body, 4. Hisagi ore body, 5. Shimodani and Hidashiro ore bodies, 6. Hoden ore body, 7. Yasumiishi ore body.

phosed and sericitized, chloritized and silicified. The genesis of the keratophyre, and felsite, is puzzling and has been a subject of controversies. The writer believes these rocks are marginal facies of the basic intrusive; that is, they have been formed by special differentiation of the basic magma and have suffered contact metamorphism and hydrothermal alterations. The hydrothermal alterations, such as sericitization, chloritization and silicification, are related to the ore deposition and are developed around the ore bodies. Thus the country rocks surrounding the ore bodies are very complicated.

The ore in the Main Yanahara ore body is composed of pyrite, with subordinate amounts of pyrrhotite and magnetite in marginal parts of the bodies. The magnetite and pyrrhotite are said to have been formed by contact metamorphism near granitic intrusive.

In the ore deposits on the west limb of the major anticlinorium, for example, the Shimoyanahara, Old Hisagi and Hisagi deposits, pyrite ore is accompanied by chalcopyrite, as indicated by 1.45% of copper content in the crude ore from the Hisagi ore deposit.

The mode of occurrence and shape of the ore deposits are similar to those of the Okuki mine, as discussed in Chapter IV-2.

The writer believes that the ore deposits were formed by the post-igneous activity of intermediate to basic rocks intruded synchronously with the folding, and recognizes that the ore bodies are controlled by the fold structure.

Dikes of granite porphyry and porphyrite penetrate the ore bodies.

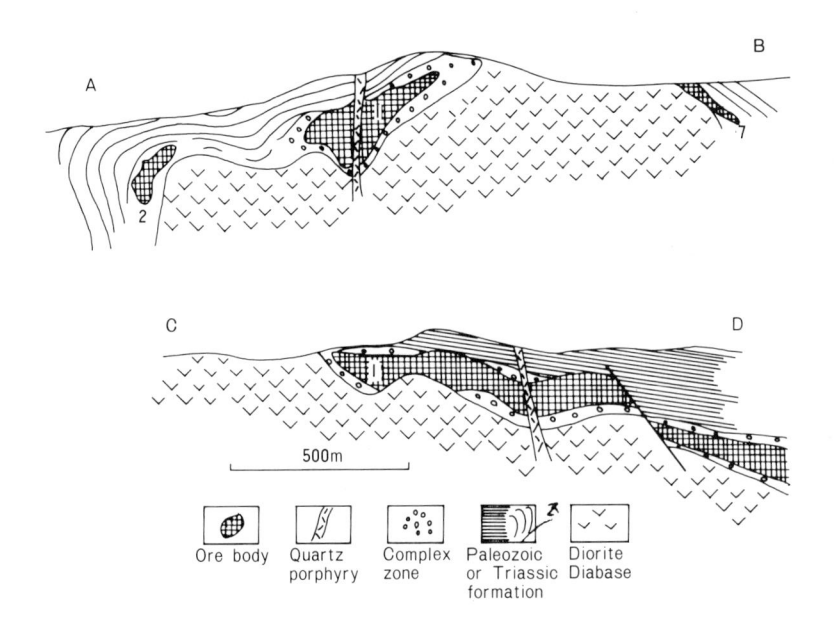

FIG. IV-4-2. Geologic sections of the Yanahara deposits. A–B and C–D sections in Fig. IV-4-1.

IV–5. Hitachi Mine, Ibaragi Prefecture (Imai, 1963b)

H. Imai

The Hitachi mine is located about 130 km NE of Tokyo. Geologically it is situated in the crystalline schists of the eastern margin in the Abukuma plateau (Figs. I–1 and –3). This mine has produced 30 million tons of ore containing 1.45% copper and 1.07% zinc.

The district is composed of metamorphosed and non-metamorphosed rocks of Paleozoic age intruded by granite, granodiorite, diorite, gabbro and diabase (epidiabase). The ore deposits occur in a metamorphosed complex called the Akazawa Formation which consists of amphibolite, amphibole schist (greenschist) chlorite schist (greenschist), sericite schist, biotite schist, etc. (Fig. IV–5–1). Limestone does not occur in this formation. Sheets of diabasic rock (epidiabase) do occur. To southeast the metamorphic complex of the Akazawa Formation gradually changes into non-metamorphosed pelitic sedimentaries called the Ayukawa Formation. The Ayukawa Formation is schistose green rock (greenschist) in the area adjoining the Akazawa Formation, but farther southeastwards it gradually merges into black sandstone and clayslate. Ottrelite schist occurs in the green rock of the Ayukawa Formation. In the pelitic sedimentary rocks of the Ayukawa Formation diabasic rocks are absent. Beds of limestone, some of which yield Visean (Lower Carboniferous) coral and other fossils, occur at many places in the Ayukawa Formation.

The general strike of the Ayukawa Formation is N30°–40°E, and the dip is about 40° to the southeast. The black clayslate and sandstone of the complex are folded with axes plunging N45°–60°E10°–20°. They belong to the recumbent fold plunging to the northeast and with an axial plane dipping to the southeast (the writer calls it NE folding) (Fig. IV–5–1). The plunge of the folding axes coincides with the direction of lineation which is formed by the elongation of prismatic minerals in the green rock and wrinkling in the pelitic rocks (the writer calls it NE lineation).

In the western area of the Ayukawa Formation the dips of the sedimentary rocks gradually become steep and then vertical, and most of the Akazawa Formation dips to the northwest by 50°–60° (Fig. IV-5-1). According to Shimada (1955), the Akazawa Formation forms a recumbent synclinal structure with an axial plane striking N70°–75°E and dipping 80°NW and with the axis plunging N80°W65°–70° (the writer calls it NW folding) (Figs. IV-5-1 and -2). In the southeast limb of the synclinal structure, the plunge of the lineation is nearly parallel to that observed in the Ayukawa Formation, i.e., N30°–50°E20°–30°. But in the northwest limb, the lineation changes to N30°–70°W40°–70°, approximately parallel to the trend of the fold axis of the above syncline (the writer calls it NW lineation). The NW lineation revealed by elongation of the amphiboles is very conspicuous. The crystalline schists having NE lineation would have suffered folding with the axis plunging N80°W65°–70° (NW folding) (Fig. IV-5-3). In other words, the folding which produced the NE axis preceded that which formed the NW axis.

Along the schistosity planes of the Akazawa Formation such as amphibolite, amphibole

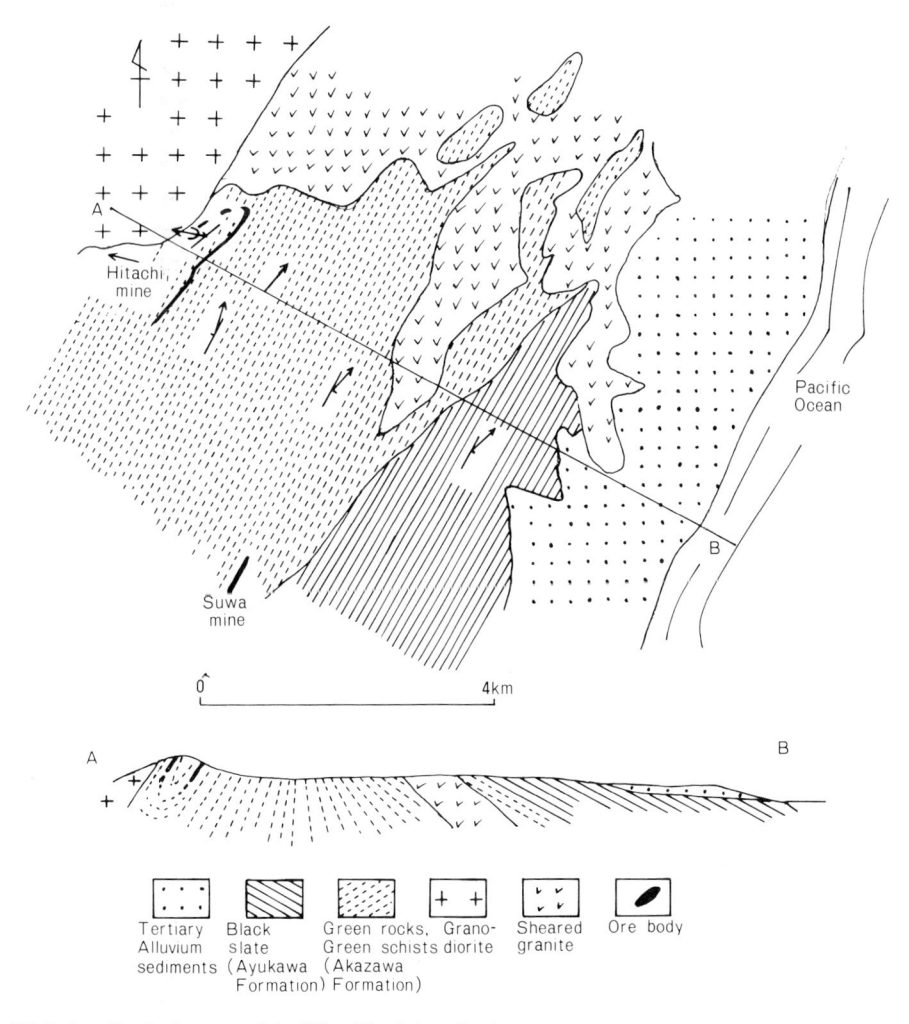

FIG. IV-5-1. Geologic map of the Hitachi mining district.

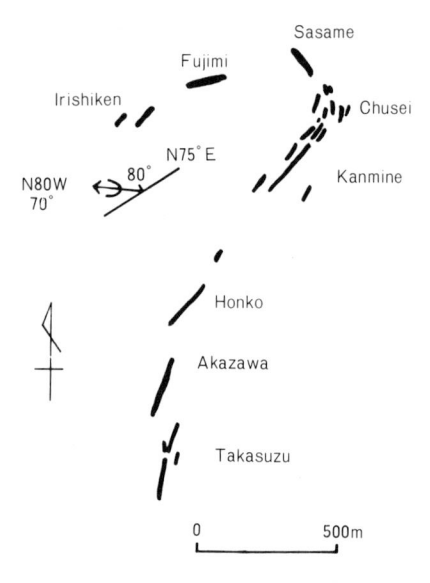

FIG. IV-5-2. The emplacement of the ore bodies in the Hitachi mine. The ore bodies are controlled by the synclinal structure.

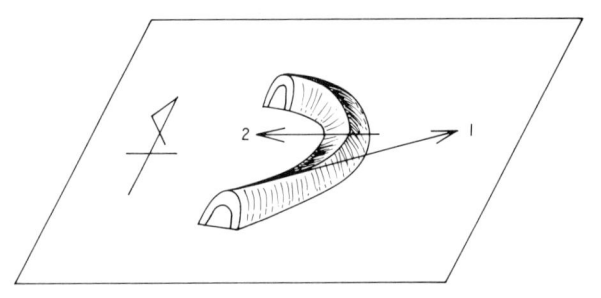

FIG. IV-5-3. The relation between NE folding and NW folding in the Hitachi mine. 1: axis of NE folding, 2: axis of NW folding.

schist (including anthopyllite-quartz schist), chlorite schist, sericite schist and biotite schist ore deposits occur. The ore deposits are controlled by the synclinal structure (Fig. IV-5-2). The Takasuzu, Akazawa, Honko, Kanmine and Chusei ore bodies are arranged along the southeast limb of the synclinal structure, and the Sasame deposit is in the apical part, whereas the Fujimi and Irishiken deposits occur along the northwest limb of the structure. The ore deposits of the southeast limb extend 2,500 m in the strike direction from the Takasuzu deposit to the Sasame deposit, and more than 600 m in the dip direction. This mineralized zone is about 90 m in width. The Irishiken ore body in the northwest limb is 200 m long; its country rocks are mainly amphibole schist having conspicuous lineation and obscure schistosity planes.

In the mineralized zone of the southeast limb the distribution of diabase (so-called epidiabase) is closely related to the ore deposits, as shown by Fig. IV-5-4. In the Chusei, Kanmine, Honko and Akazawa ore bodies, highly schistose metamorphic rocks, such as amphibole schist, biotite schist, chlorite schist and sericite schist, are developed adjacent to both sides of the sheet-like diabase. In these highly schistose rocks drag folds are commonly developed. Outward from the diabase there are many kinds of siliceous schists, such as actinolite-quartz schist, colorless amphibole*-quartz schist, chlorite-quartz schist, biotite-quartz schist, etc., which are rather massive. Ore bodies occur mainly in the schistose rocks bordering the diabases, but in some cases, they occur in sheared zones within the diabasic rocks

---

* Including anthophyllite, cummingtonite and tremolite.

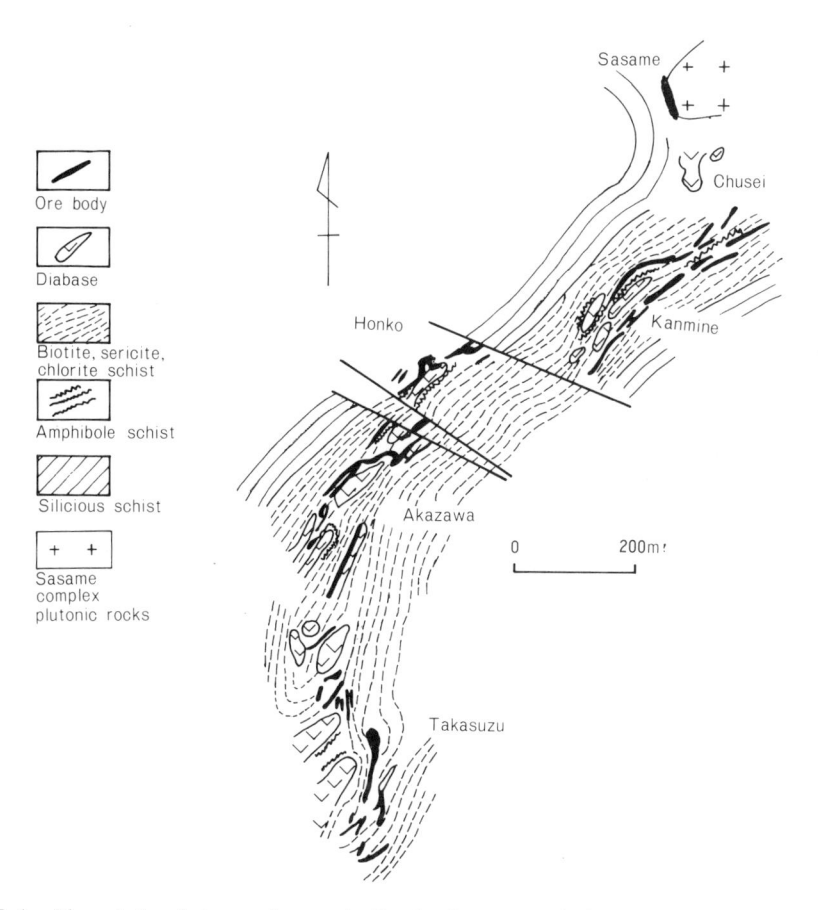

FIG. IV-5-4. The relation between the ore bodies in the eastern limb of the syncline and the country rocks of the Hitachi mine (after Shimada, 1955).

(Fig. IV-5-4). Ore bodies swell at the apex of the syncline, as exemplified by the Sasame deposit, or at curved parts of the limb like those in the northern part of the Takasuzu deposit (Fig. IV-5-4).

The ore is composed mainly of pyrite and chalcopyrite, with small amounts of pyrrhotite, magnetite, sphalerite, galena and marcasite. It is noteworthy that barite is rarely found as a gangue mineral. Rich ore bodies are always accompanied by impregnated ores. The plunge of the Irishiken ore deposit is parallel to the lineation of the surrounding country rocks, i.e., N30°-70°W, but plunges of the ore bodies of the southeast limb are complicated, probably being controlled by both NE and NW lineations. It has been observed that plunges of ore bodies are coincident with depth extensions of the diabasic rocks.

To the west of the Irishiken ore deposit, a large mass of so-called Irishiken granodiorite occurs concordantly with the Akazawa Formation (Fig. IV-5-1). Judging from the mode of occurrence, it may have been intruded at nearly the same time as the formation of the synclinal structure.

The writer thinks that the NW folding and NW lineation are closely related to the shape of the granodiorite and its mechanism of intrusion.

The highly metamorphosed rocks, such as amphibolite, anthophyllite schist or amphibole schist, which are developed near the Irishiken ore deposit, may have been formed under intense compressive force with a supply of hydrothermal solutions together with the thermal effect of granitic intrusion.

It is generally believed that the original rocks of the greenschists and green rocks of Akazawa and Ayukawa Complexes are acidic to basic volcanics and pyroclastics. The writer thinks that this is not necessarily the only appropriate interpretation. The greenschists or green rocks are composed of epidote, actinolite, chlorite, etc., and have schistose textures, but no textures characteristic of volcanic and pyroclastic rocks (such as hyalopilitic texture, vitroclastic texture, etc.). As for the origin of the diabasic rocks, there are two different opinions; one, that they are of intrusive origin; and the other, that they are of effusive origin. On the basis of studies of the Okuki and other mines in Southwest Japan, the writer thinks that the diabasic rocks in the Hitachi mine were intruded synchronously with the metamorphism, and that the greenschists and the green rocks may have formed from pelitic rocks of the Paleozoic formations by Mg-Fe metasomatism. It is probable that the diabasic rocks intruded at the same time as the folding which produced the NE axes.

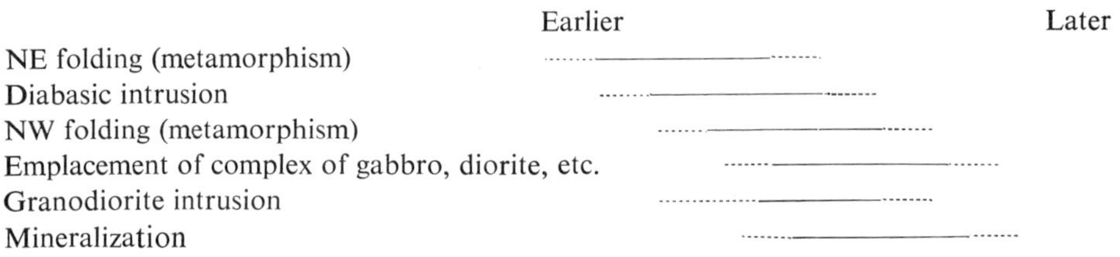

	Earlier	Later
NE folding (metamorphism)	────	
Diabasic intrusion	────	
NW folding (metamorphism)	────	
Emplacement of complex of gabbro, diorite, etc.		────
Granodiorite intrusion		────
Mineralization		────

Fig. IV-5-5. Succession of the geologic events in the Hitachi mine.

An igneous complex of gabbro, diorite, quartz diorite and fine-grained granodiorite occurs at the apex of the synclinal structure (Fig. IV-5-4). It may have intruded at nearly the same time as the NW folding (the formation of synclinal structure). Around the igneous complex of gabbro, etc., at the apex of the syncline, contaminated rocks prevail. The occurrence of sillimanite-andalusite-quartz schist in this apical part is interesting from a petrological standpoint. The existence of cordierite in the biotite-anthophyllite schist in the northwest limb of the syncline (i.e., in the vicinity of the Irishiken granodiorite) may be due to thermal metamorphism accompanying the granitic intrusion.

Summarizing the above information, the succession of events can be inferred as in Fig. IV-5-5.

The writer concludes that this deposit belongs genetically to the hydrothermal replacement type related to a series of igneous activities at the time of dynamometamorphism.

According to Ueda *et al.* (1969), the absolute age of metamorphism was determined to be 105–120 m.y. by the K-Ar method on the muscovite from muscovite-quartz schist in the Hitachi mine. The age of the Irishiken granodiorite is 90 m.y. by the K-Ar method (Kawano and Ueda, 1966a).

These values are the same as those of rocks in the Sambagawa-Mikabu and Ryoke terrains.

PART V

PORPHYRY COPPER DEPOSITS

Porphyry Copper Deposits in the Southeast Asia, with Special Reference to Fluid Inclusion Study (Nagano et al., 1977)

H. Imai, S. Takenouchi, T. Shoji and K. Nagano

Introduction

Recently, many prophyry copper deposits have been found in the southwestern Pacific region, including the Philippines and Papua-New Guinea. In the Philippines, several porphyry copper deposits have already been developed and active explorations are being carried on in many places. In Malaysia, the Mamut mine started production in May, 1975. Since the discovery of Panguna, Bougainville, in 1964, more than 25 porphyry copper mineralization have been reported from Papua-New Guinea and the Solomon Islands. In these areas, the porphyry copper mineralization took place in association with dioritic to granodioritic intrusives of Paleocene, Miocene-Pliocene or Pleistocene age. These features are very characteristic, as compared with those of the southwestern United States which are mostly of Laramide age and are associated with quartz monzonitic intrusives.

Porphyry copper deposits in the southwestern Pacific region are distributed along island arcs, and the mineralization has been discussed from the viewpoint of plate tectonics (Mitchell and Garson, 1972; Guild, 1972; Sillitoe, 1972; Titley, 1975; Pelton and Smith, 1976). It is presumed that the temporal and spatial distribution of the porphyry copper deposits are related to the activity of old and active subduction zones.

The Japanese islands are situated on island arcs similar to those of the southwestern Pacific region. Geophysical and seismological data indicate that the westward subduction of the Pacific plate is active along the Kuril, Japan and Izu-Mariana arcs. The other subduction zone is located along the Ryukyu Islands. The Chimei disseminated copper deposit in Taiwan is reported as porphyry copper. Nevertheless, no porphyry copper deposits have yet been reported in Japan. Guild (1972) stated that there is no reason to consider that porphyry copper deposits may not be present in Japan.

Fluid inclusions of porphyry copper deposits have been studied by Roedder (1971a), Nash and Theodore (1971), Moore and Nash (1974) and Hall et al. (1974). In general, three types of fluid inclusions are recognized in quartz from porphyry copper deposits: (1) polyphase fluid inclusions of high salinity, (2) gaseous inclusions having various degrees of filling and (3) liquid inclusions of high degree of filling and of intermediate salinity. The distribution of these fluid inclusions in deposits is very characteristic. Polyphase inclusions mostly occur in the cores of deposits, whereas the liquid inclusions predominate outside the ore-shell. Distribution of highly saline inclusions corresponds to the potassic alteration zone.

Studies of the stable isotopes of water in minerals and fluid inclusions revealed that the fluids in the core were rich in heavy water which is considered to have been in contact with magmas or rocks at high temperatures, while in the outside of the ore shell the fluids contain much light water of meteoric origin (Sheppard et al., 1971; Hall et al., 1974; Sheppard and Taylor, 1974).

Fluid inclusions from some porphyry copper deposits in the Philippines and East Malaysia (Sabah) have been studied to compare the results with those in the United States and with granitic rocks and porphyries in Japan, and to consider the differences in the mineralization and ore-forming fluids between prophyry copper deposits and ordinary hydrothermal ore deposits (Takenouchi, 1976b).

Mamut Mine (including Nungkok), Sabah, Malaysia

The Mamut porphyry copper deposit, which is located about 65 km to the east of Kota Kinabalu, Sabah, Malaysia, has been developed recently through the cooperation of Malaysia and Japan. Operations at the mine started in May, 1975, on the scale of 16,000 t of crude ore (0.61% of copper) per day.

The ore body is located on the southeastern flank of Mt. Kinabalu at height of about 4,300 m above sea level. The deposit was found in 1965 during geochemical prospecting undertaken by the Labuk Valley Project of the United Nations, begun in March, 1963. The ore reserve is estimated at 180 million tons of 0.476% copper.

Geology: The geology of the Kinabalu area was reported by Collenette (1958), Kirk (1967, 1968) and Jacobson (1970) (Fig. V-1). This area lies in the Northwest Borneo Geosyncline which developed in the Late Cretaceous and Early Tertiary age and is about 300 km in width and 800 km in length, trending NE-SW.

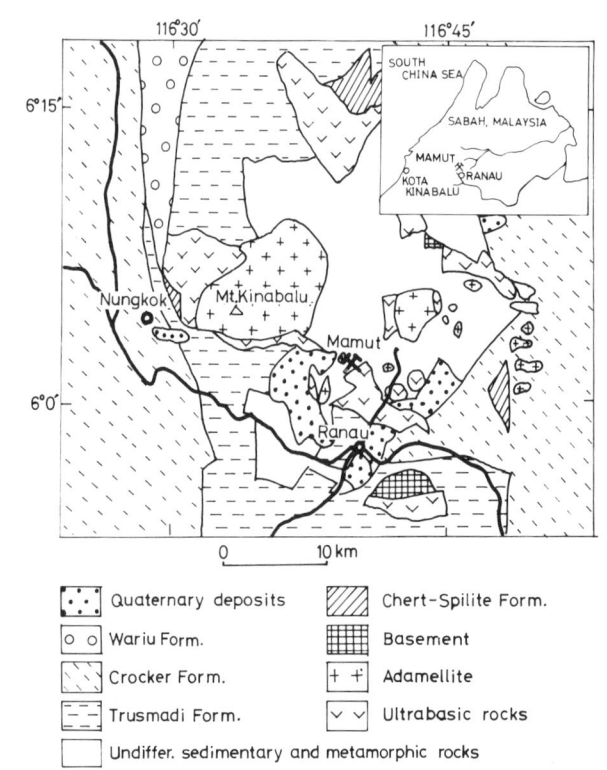

FIG. V-1. Geology of the Mt. Kinabalu area, Sabah, Malaysia (after Jacobson, 1970).

Part V: Porphyry Copper Deposits

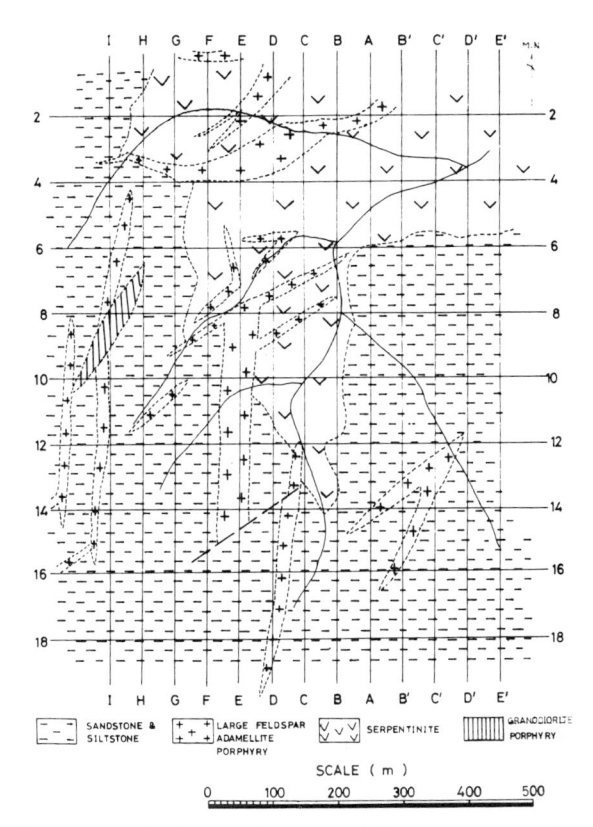

FIG. V-2. Geologic map of the Mamut porphyry copper deposit (after Kosaka and Wakita, 1975).

At the earliest stage of the geosyncline, chert, shale, sandstone and limestone were deposited associated with extrusion and intrusion of basic rocks, forming the Chert-Spilite Formation. Eugeosynclinal flysch-type sediments were deposited in the Paleocene and Early Miocene. These sedimentary rocks are divided into the Trusmadi (lower) and Crocker (upper) Formations. The former is more argillaceous than the latter.

In the Early Miocene age, folding and faulting took place, and peridotite and serpentinite emplaced along faults.

The intrusion of the Kinabalu batholith, accompanied by small satellitic stocks, took place in the Late Miocene age. The main intrusive body consists mainly of adamellite, the K-Ar age of which is reported as 9 m.y. (Jacobson, 1970). The mineralized porphyries at Mamut Valley and Mt. Nungkok belong to these satellitic stocks (Lewis, 1967). The peridotite and serpentinite were intruded by the porphyries. In serpentinite, spinel was formed by thermal metamorphism due to the intrusion of porphyry.

Major faults belong to the N-S system. The combination of the N-S faults and the NW-SE faults makes this area complicated in geologic structure. Most of these faults are pre-ore but some of them show post-ore movement.

Ore deposit: The geology of this mine has been described by Kosaka and Wakita (1975).

The country rocks of the ore deposit consist of sandstone and siltstone of the Trusmadi Formation, serpentinite and adamellite porphyry (Fig. V-2). The adamellite porphyry intrudes in sedimentary rocks and serpentinite, extending generally N-S and dipping 40° eastward. The elongation of the stock is conformable to the tectonic trend of this area. The stock branches to many dikes, forming complicated boundaries. Some of these dikes are recognized on the surface trending in the N-S and NE-SW directions. The adamellite porphyry consists of phenocrysts of K-feldspar, plagioclase and hornblende, and groundmass of fine-grained quartz and K-feldspar.

268   Mamut Mine, Sabah, Malaysia

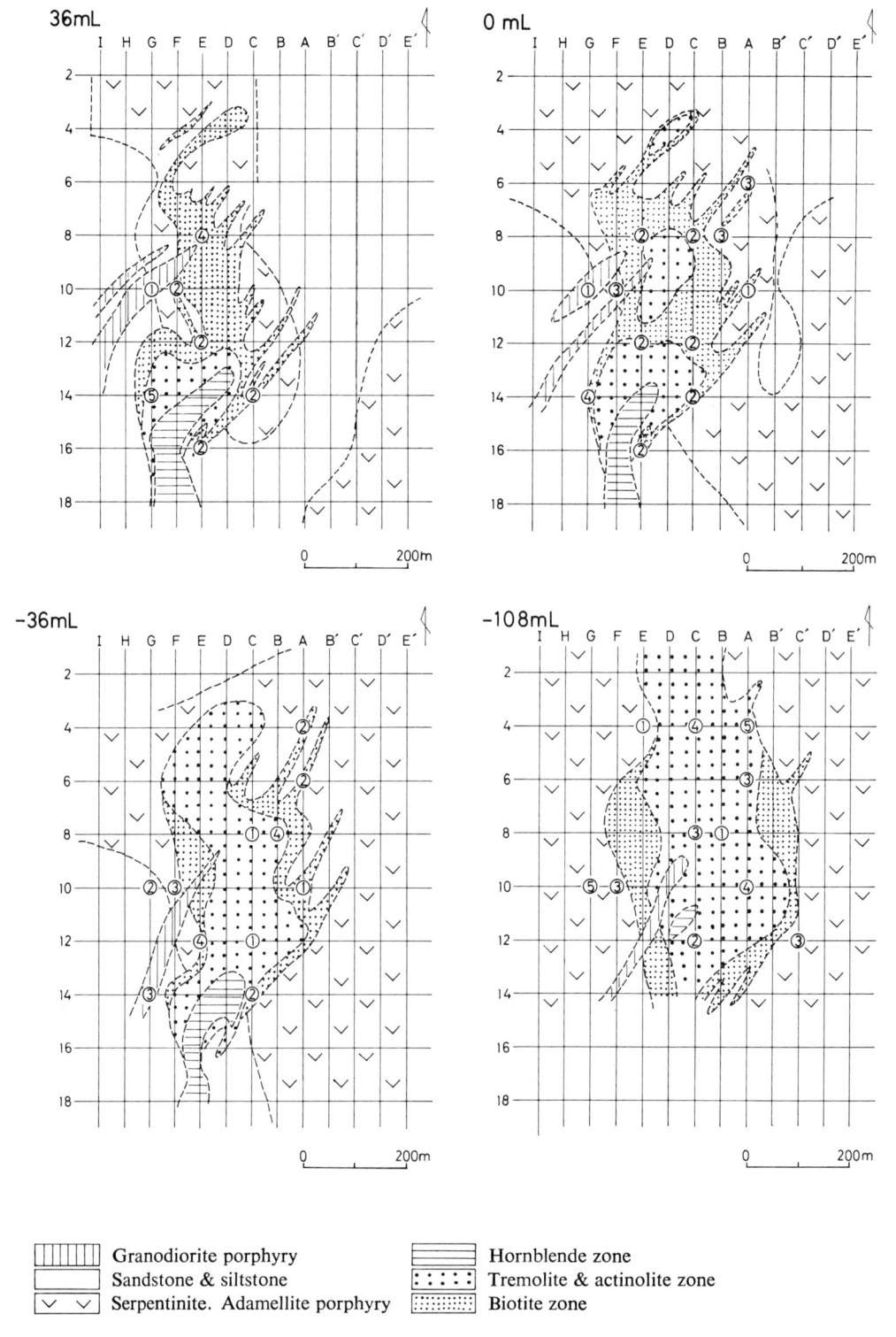

		Granodiorite porphyry		Hornblende zone
	Sandstone & siltstone		Tremolite & actinolite zone
	Serpentinite. Adamellite porphyry		Biotite zone

FIG. V-3. Horizontal planes of the Mamut deposit (after Kosaka and Wakita, 1975), and the distribution of polyphase inclusions. The number in a circle is the same as in Fig. V-4.

Part V: Porphyry Copper Deposits

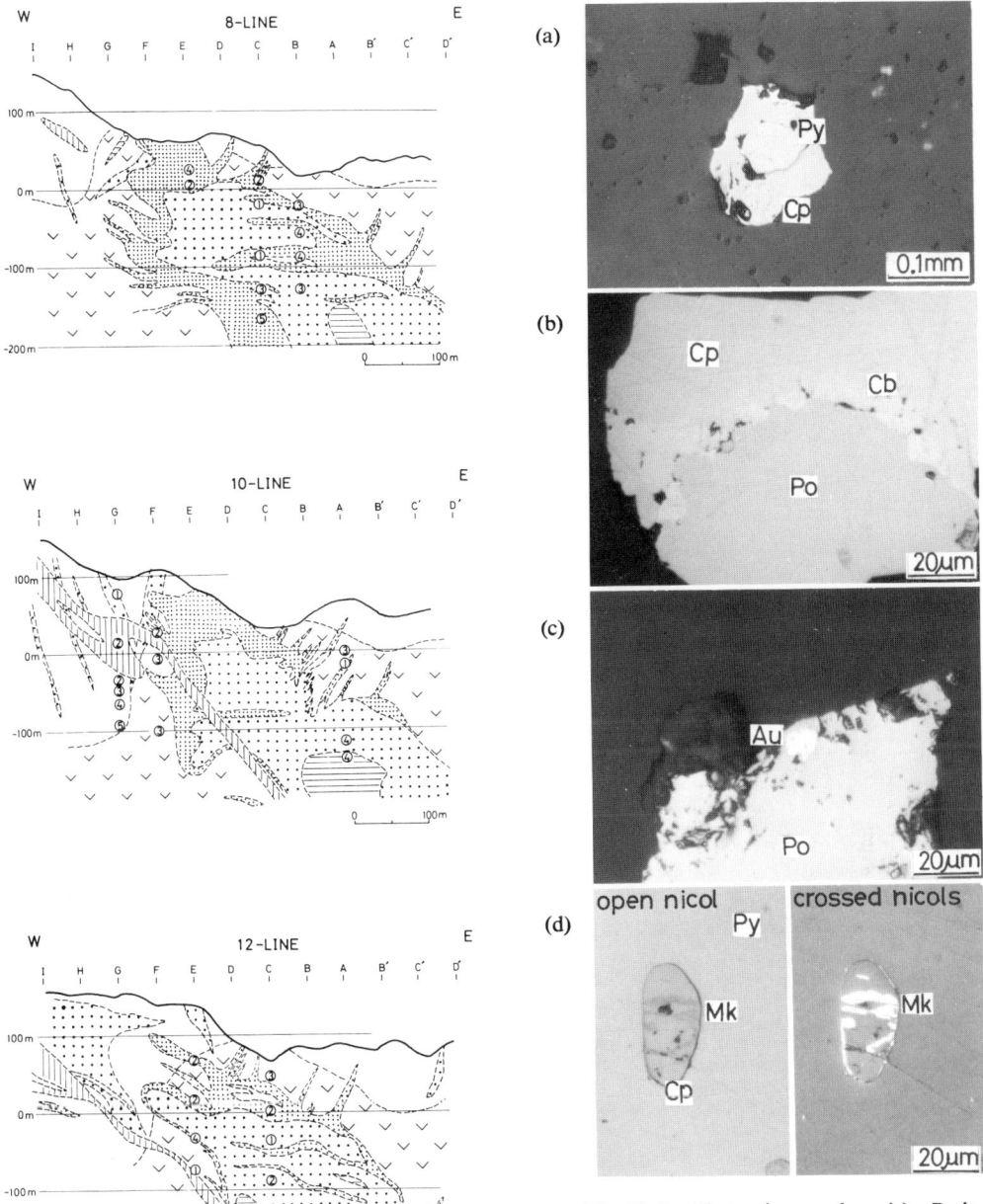

Fig. V-4. Vertical sections of the Mamut deposit (after Kosaka and Wakita, 1975), and the distribution of polyphase inclusions. The number in a circle represents the relative abundance of polyphase inclusion in parts per ten.

Fig. V-5. Photomicrographs. (a) Pyrite grain wrapped by chalcopyrite (G-10, 160 m). Cp: chalcopyrite, Py: pyrite. (b) Cubanite at the boundary between chalcopyrite and monoclinic pyrrhotite (E-8, 53 m). Cb: cubanite, Po: pyrrhotite. (c) Native gold in pyrrhotite. Au: native gold. (d) Chalcopyrite bleb in pyrite associated with exsolved mackinawite. Mk: mackinawite.

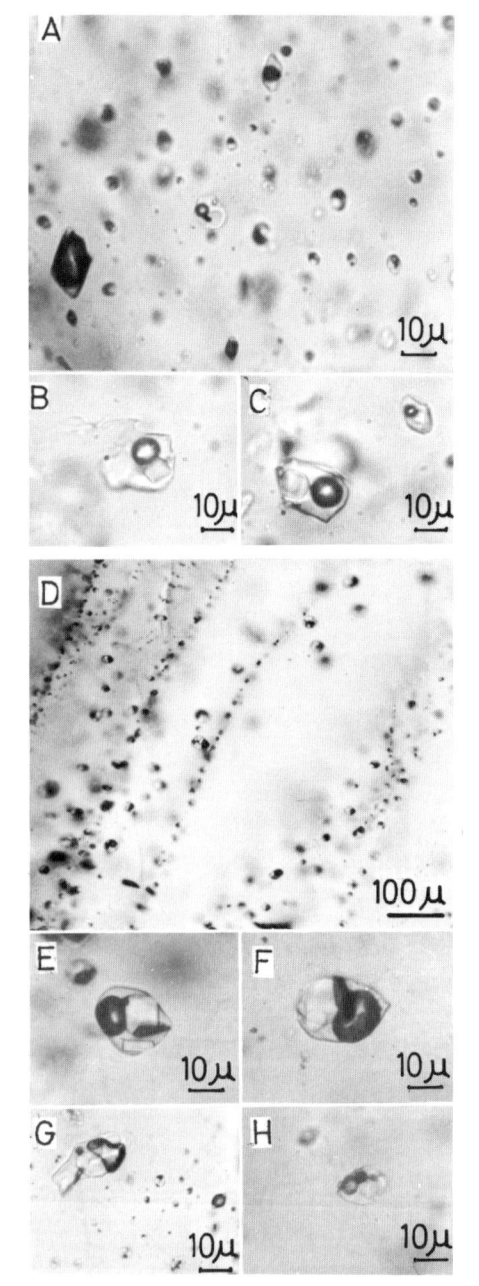

FIG. V-6. Photomicrographs of polyphase fluid inclusions in quartz of porphyry copper ores.
  (A) Distribution of polyphase fluid inclusions and gaseous inclusions (Mamut, Sabah, Malaysia).
  (B) Highly saline inclusion (Mamut, Sabah, Malaysia).
  (C) Highly saline inclusion (Nungkok, Sabah, Malaysia).
  (D) Distribution of polyphase fluid inclusions (Santo Tomas II, Luzon, Philippines).
  (E) and (F) Highly saline inclusion (Santo Tomas II, Luzon, Philippines).
  (G) Distribution of polyphase fluid inclusions (Marcopper, Marinduque, Philippines).
  (H) Highly saline inclusion (Marcopper, Marinduque, Philippines).

A fine-grained granodiorite porphyry (granophyre) dike intrudes the stock of adamellite porphyry, sedimentary rocks and serpentinite in the western part of the ore body. The dike strikes N35°E and dips 45°SE. Granodiorite porphyry (granophyre) is altered but poorly mineralized in copper and is regarded as an intrusive at a late stage of mineralization.

The copper grade of the ore is high in the marginal parts of the adamellite porphyry stock but low in the core, forming a tilted ore-shell as shown in Figs. V-3 and -4. Within the ore-shell, the copper grade is high even in siltstone and serpentinite, suggesting that the copper mineralization is independent of the kinds of wall-rocks. In the ore-shell, biotitization and silicification are predominant and the low-grade core corresponds to the tremolite-actinolite zone, where K-feldspar is also recognized in the alteration products.

The following ore minerals were identified microscopically: chalcopyrite, cubanite, pyrite, pyrrhotite, mackinawite, galena, sphalerite, molybdenite, magnetite and native gold as primary minerals; and bornite, chalcocite (digenite) and covellite as secondary minerals (Fig. V–5). Among them, chalcopyrite, pyrite and pyrrhotite are the principal primary minerals. They occur as disseminated grains rather than veinlets. Chalcopyrite and pyrite occasionally occur as separated single grains, but pyrrhotite occurs intimately associated with chalcopyrite. In some cases, a core of pyrite is coated by chalcopyrite, suggesting that the crystallization of some pyrite preceded the chalcopyrite deposition (Fig. V–5(a)).

Magnetite occurs mostly as a single grain: sporadically it is coated by sulfide minerals.

Cubanite is recognized at the boundaries between chalcopyrite and pyrrhotite (Fig. V–5(b)), and mackinawite rarely occurs with chalcopyrite dots in pyrite Fig. (V–5(d)). Existence of cubanite might be related to peridotite or serpentinite, as discussed in Chapters II–5 and III–5.

Wall-rock alteration: The alteration of adamellite porphyry is identified by the alteration products of hornblende phenocrysts. The distribution of zoning is hornblende, tremolite-actinolite, and biotite from the lower core to the outermost parts of the stock. Silicification is remarkable in the ore-shell regardless of the kinds of country rocks.

In the tremolite-actinolite zone, hornblende phenocrysts are replaced by the aggregates of tremolite, actinolite and K-feldspar, while in the biotite zone, hornblende is replaced by flakes of biotite. Epidote, carbonate and chlorite, which are the characteristic minerals of the propylitic zone, are occasionally recognized in both the above-stated alteration zones, suggesting the telescoping of alteration. Sericitization is locally recognized but not pervasive.

In serpentinite of the ore-shell, talc, phlogopite and tremolite are recognized. These minerals would have been formed by the activities of alteration and mineralization.

In sedimentary rocks of the ore-shell, aggregates of biotite flakes and, in some parts, mosaic aggregates of K-feldspar are recognized. It is likely that the biotite-K-feldspar alteration of sedimentary rocks corresponds to that of adamellite porphyry. Outside the ore-shell, sericitization is predominant.

Though the copper mineralization is poor, the alteration is intensive in fine-grained granodiorite porphyry (granophyre). Biotite, chlorite, epidote and carbonate replace hornblende phenocrysts. In some parts, pyritization is remarkable.

Fluid inclusion studies: Three types of fluid inclusions are recognized in quartz from the Mamut deposit, that is, polyphase, gaseous and liquid inclusions (Nagano *et al.*, 1977).

Polyphase fluid inclusions: Polyphase fluid inclusions generally contain a cubic crystal of halite and occasionally minute opaque minerals. Sylvite is often observed as a small rounded crystal in association with halite. Some inclusions occur in the shapes of negative crystal but others are irregular and are distributed along a supposed plane, suggesting their formation along healed fractures (Fig. V–6). In many samples, polyphase inclusions coexist intimately with gaseous inclusions. Besides the above-stated chloride crystals, a solid having a rhombohedral shape and high birefringence color, and a prismatic solid having a parallel extinction and low birefringence color are occasionally observed in inclusions. It is supposed that the former phase is probably carbonate and the latter anhydrite. Small opaque minerals are common in these polyphase inclusions, and are considered to be sulfide minerals. Several other minerals are have observed rarely in polyphase inclusions but they have not been identified owing to their small sizes.

Gaseous inclusions: Gaseous inclusions are the most common in quartz crystals. They occur with polyphase inclusions and show negative crystal or simpler shapes. In some cases, they are found along curved planes. The bubble in gaseous inclusions occupies $70 \sim 80\%$ of the volume of the inclusion at room temperature. Small opaque minerals are sometimes observed in the aqueous phase in inclusions of this type.

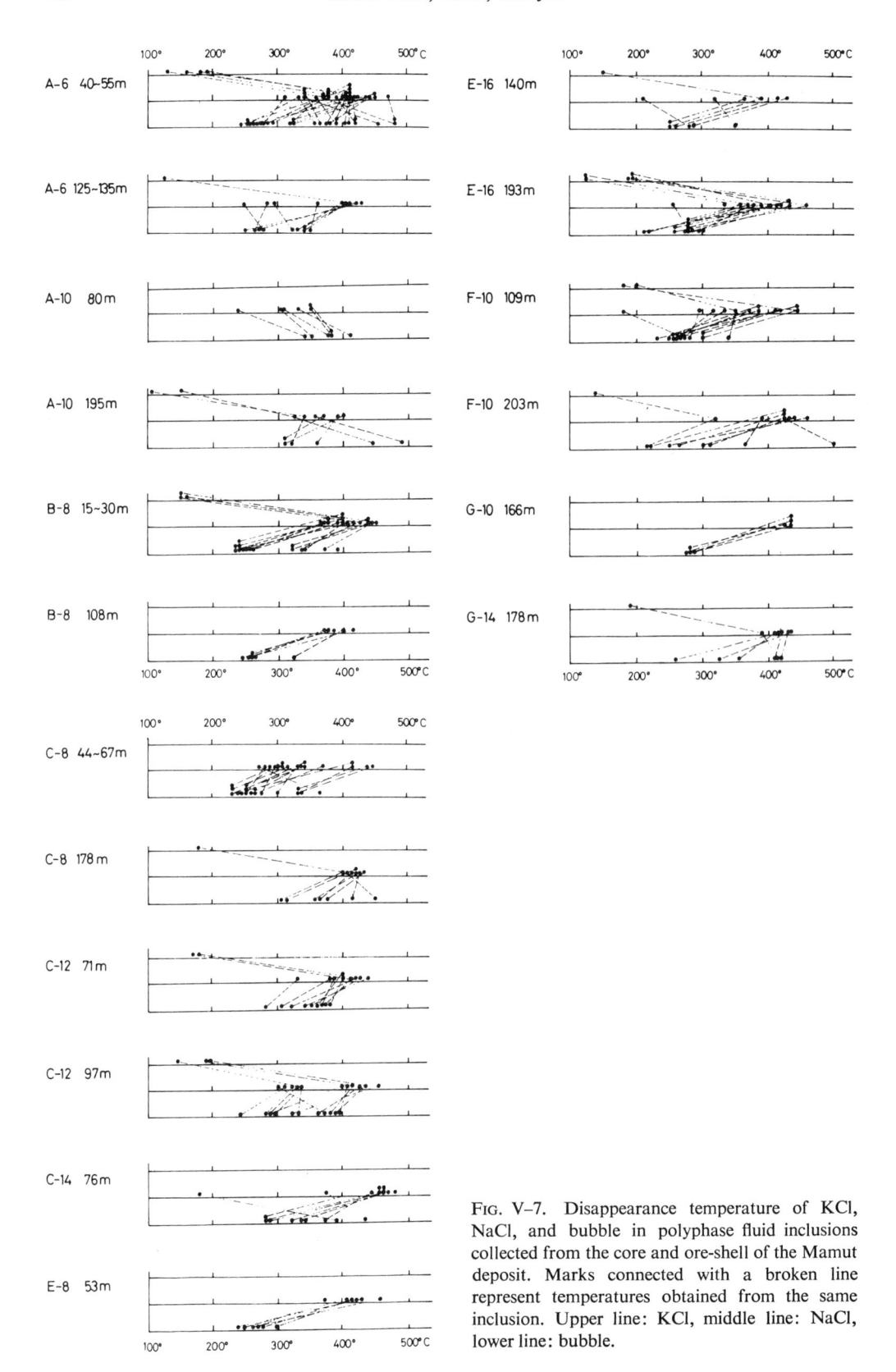

FIG. V-7. Disappearance temperature of KCl, NaCl, and bubble in polyphase fluid inclusions collected from the core and ore-shell of the Mamut deposit. Marks connected with a broken line represent temperatures obtained from the same inclusion. Upper line: KCl, middle line: NaCl, lower line: bubble.

Liquid inclusions: The degree of filling of liquid inclusions is generally from 0.75 to 0.90. The shapes of inclusions are generally irregular, and they are distributed along supposed healed fractures, suggesting secondary origin. This type is the least common of the three types of inclusions.

Homogenization temperatures and salinity: Homogenization temperatures were determined on liquid and polyphase inclusions.

Liquid inclusions generally homogenized at temperatures between 280° and 325°C. On the other hand, homogenization temperatures of polyphase inclusions ranged from 300° to 450°C, much higher than those of liquid inclusions (Fig. V–7).

The homogenization of a polyphase inclusion was caused by the disappearance of either the bubble or halite. In some cases, the disappearance of the bubble preceded that of halite, even at a careful slow heating rate. The salinity estimated from the disappearance temperature of halite indicates a range of 37–53 wt. %. An intimate coexistence of gaseous and polyphase inclusions suggests the "boiling" condition of fluids at the formation of inclusions, and it is possible to infer that the pressure at formation would have been equal to the vapor pressure of the system. In such cases, the dissolution of halite should precede the disappearance of the bubble upon heating.

Data on the samples taken from the core and ore-shell of the ore body show no remarkable difference in the disappearance temperature of halite, but the disappearance temperature of the bubble is slightly higher at the core than in the ore-shell.

Distribution of fluid inclusions: The numeral ratio of polyphase inclusions to the total number of inclusions, expressed in parts per ten, was determined under the microscope in order to discover the distribution of polyphase inclusions in the ore body. As the number of liquid inclusions is small, the ratio shows an approximate relation between polyphase and gaseous inclusions. Though the data are not sufficient and the ratios are various, the general trend of distribution is shown in Figs. V–3 and –4.

The ratio is higher in the marginal zone of adamellite porphyry (that is, in the ore-shell or biotite zone) and at the lower levels. A low value at the core is attributed to an abundance of gaseous inclusions. Data from samples which represent the outside of the ore-shell are very limited.

The volume ratio of chalcopyrite to the total volume of sulfide minerals was determined

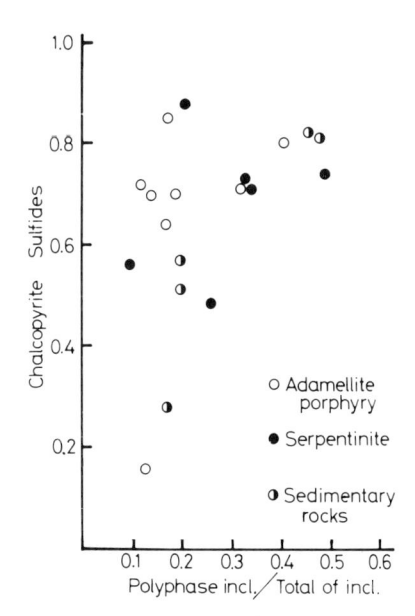

FIG. V–8. Relation between the abundance of chalcopyrite and polyphase inclusion in the same specimen.

by modal analysis of polished thin sections, on which determination of the numeral ratio of polyphase inclusion had been carried out (Fig. V–8). The figure shows that at a high-volume ratio of chalcopyrite, the numeral ratio of polyphase inclusions is distributed in a wide range from 1 to 5 in parts per ten, but at a low chalcopyrite ratio, the inclusion ratio is limited to a lower range.

Though the scarcity of data on samples from the low-copper grade zone makes the result indistinct, it is inferred from Fig. V–8 that the high saline fluids represented by polyphase inclusions might have played an important role in the copper mineralization.

Santo Tomas II Mine, Philippines

The Santo Tomas II porphyry copper deposit is located about 20km to the south of Baguio City in the Mountain Province, Luzon Island, Philippines (Fig. II–20–1). The mine was developed in July, 1958, and the present production is 21,000 t per day of 0.42% copper. The ore reserves are estimated at 120 million tons at an average grade of 0.466% copper and 0.925 g/t gold (Retardo, 1972).

Geology: The deposit is located in the southern part of the Baguio Mineral District, which comprises a geanticlinal mountain range resulting from Miocene orogenesis and accentuated by later tectonisms of the Pliocene and Pleistocene ages (Peña, 1970).

Figure V–9 shows the geology of this area. The Pugo Formation, which consists of volcanic and sedimentary rocks of the Cretaceous-Paleogene age, is the oldest formation in this area. Towards the top of the formation, sedimentary members become more dominant. The Pugo Formation is overlain by the Late Miocene Zigzag Formation, which consists of siltstone and arkosic sandstone with minor conglomerate and limestone. Kennon Limestone of the Middle Miocene age overlies the Zigzag Formation conformably.

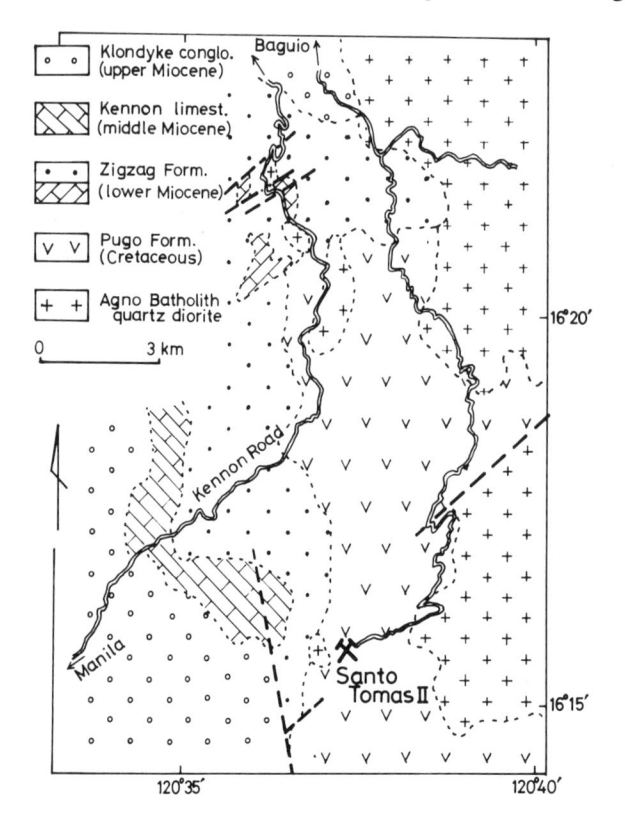

FIG. V–9. Geologic map of the southern part of the Baguio Mineral District (redrawn from Quinto, 1970).

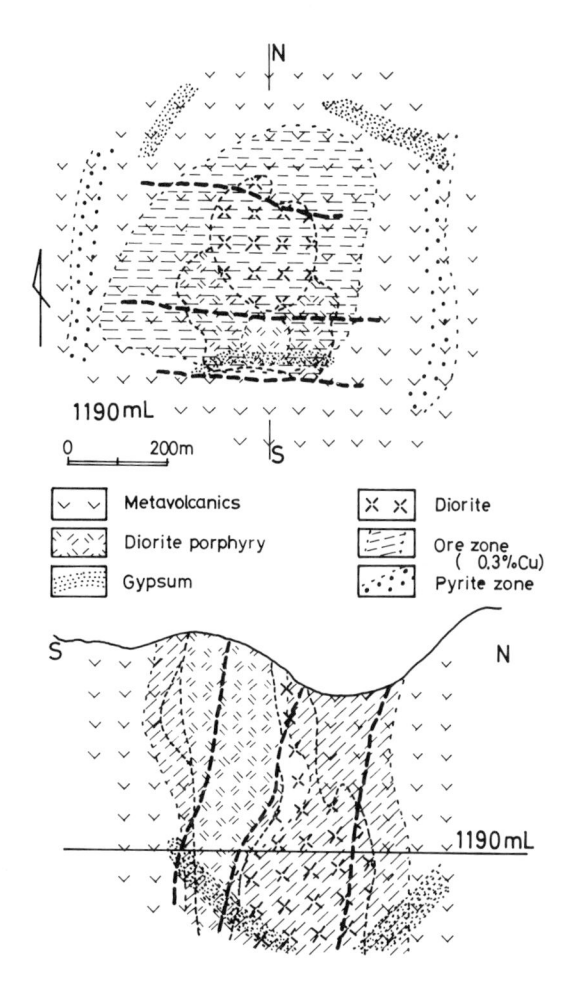

FIG. V-10. Geologic map of the Santo Tomas II porphyry copper deposit (redrawn from Retardo, 1972).

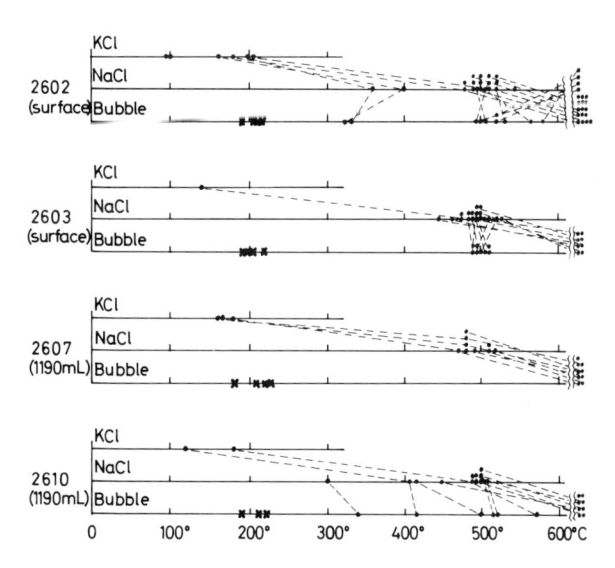

FIG. V-11. Disappearance temperature of KCl, NaCl and bubble in polyphase fluid inclusion from the Santo Tomas II deposit. Cross mark shows the filling temperature of liquid inclusion.

In the Middle-Late Miocene age, the above-stated sedimentary and volcanic rocks were intruded by a plutonic mass of quartz diorite known as the Agno Batholith. In the vicinity of the mine, the Agno Batholith is exposed in an area of 20km (N-S) by 10km (E-W).

The Klondyke Conglomerate, which covers the above-stated rocks, is a post-orogenic molasse of Late Miocene age.

Ore deposit: In the mineralized area, metavolcanics of the Pugo Formation are intruded by a stock of dioritic complex, the size of which is about 700m(E-W) by 500m(N-S). Included in the complex are diorite, diorite porphyry and porphyritic andesite. The textures of the groundmass range from aphanitic to phaneritic, and the phenocrysts comprise plagioclase, hornblende, and pyroxene. The groundmass and phenocrysts are more or less altered to biotite, sericite, epidote and chlorite.

A pipe-like ore body is formed in both the metavolcanics and dioritic complex. It extends almost vertically from the outcrops at 1,770m above sea level to the lowest explored level at 1,000m, and has a horizontal section of 370m(E-W) by 230m(N-S) (Fig. V–10) (Hidaka et al., 1965; Bryner, 1969).

At upper levels, the ore occurs mostly in the metavolcanics, but at lower levels, it occurs in the dioritic complex. Veins and networks are prominent in the metavolcanics, but networks with dissemination are remarkable in the diorite complex. Silicification is intensive in high-grade ores.

Ore minerals are chalcopyrite, bornite, pyrite, molybdenite and magnetite. Chalcopyrite, bornite and pyrite occur in veins and disseminated ores. Magnetite and molybdenite occur mainly in veins. From the paragenesis of minerals, it is inferred that magnetite would have been formed in early stages, and molybdenite in late stages. Pyrite halo is identified within 330–500m from the center of mineralization. Outside the pyrite halo, epidotization is prominent. In the lower levels, gypsum occurs in fractures at the NE and NW parts of the ore body. Quinto (1970) and Retardo (1972) reported the presence of a low-grade pipe-like core surrounded by the ore zone.

Fluid inclusion studies: Three types of fluid inclusions are found in quartz collected from the outcrops and the underground 1,190m level. Polyphase fluid inclusions of high salinity are the most abundant, succeeded by gaseous inclusions and a lesser number of liquid inclusions.

A polyphase inclusion generally contains a large cubic crystal of halite (NaCl), a small round-shaped crystal of sylvite (KCl) and occasionally a flake of hematite. In addition to these solid phases, an anisotropic prismatic crystal, anisotropic granular crystal and two

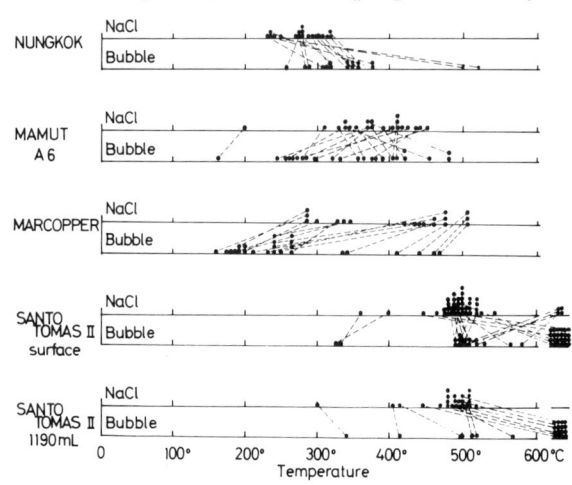

Fig. V–12. Comparison of the temperatures obtained from the Nungkok, Mamut, Marcopper and Santo Tomas II deposits.

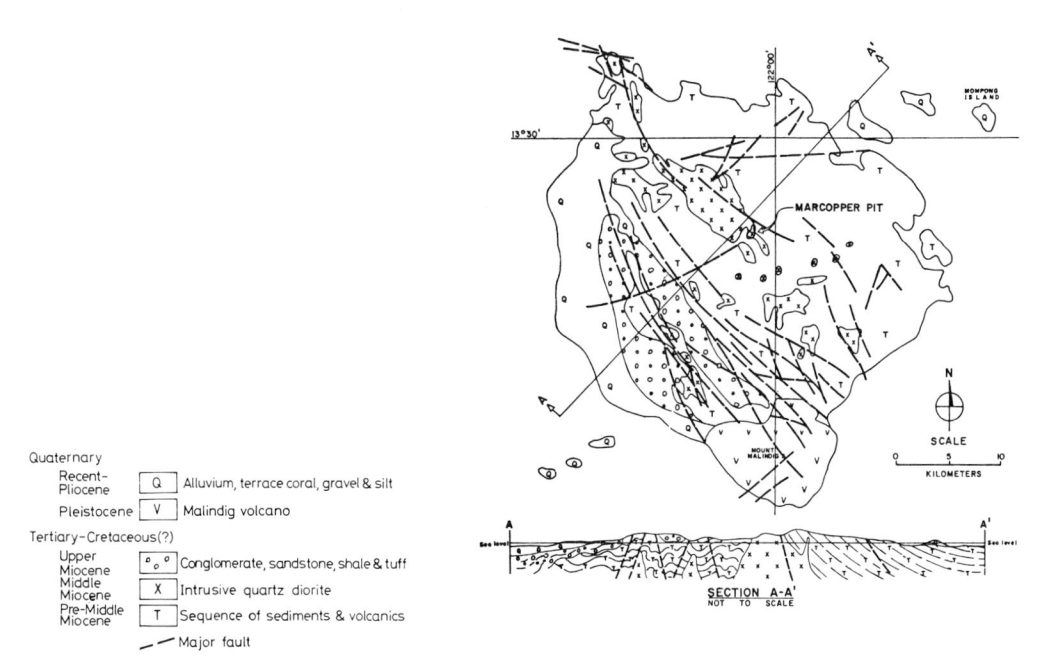

FIG. V–13. Geologic map of Marinduque Island (after Loudon, 1976).

kinds of opaque minerals are often observed. These polyphase inclusions distribute randomly or along planes in quartz, indicating their primary, secondary and pseudosecondary origin (Fig. V–6).

Upon heating, sylvite dissolved at a temperature between 100°C and 200°C, and halite at a temperature between 450°C and 550°C (Fig. V–11). It is inferred from the NaCl-KCl-H$_2$O ternary system that the approximate concentration of NaCl is around 40 to 50 wt. % and that of KCl about 20 wt. %.

Gaseous inclusions, which mostly show negative crystal shapes, occur intimately with the polyphase inclusions, suggesting the "boiling" condition of high saline fluids.

Liquid inclusions with high degrees of filling rarely occur in irregular and platy shapes. These inclusions probably represent low-temperature hydrothermal solutions of late stages, because the homogenization temperature of some of these liquid inclusions is distributed in a temperature range around 200°C.

Hydrothermal activities: From the results of a fluid inclusion study at the Santo Tomas II deposit, it was concluded that the ore-forming fluids were very high in temperature and rich in materials, and that hydrothermal activities of the late stages were deficient. As shown in Fig. V–12, the homogenization temperatures of polyphase inclusions from the Santo Tomas II are the highest, succeeded by those of Marcopper, Mamut and Nungkok. High temperature and high salinity of the inclusions suggest that the ore-forming fluids came up from origins very close to the present deposit. Unfortunately, as the number of samples was very limited, the change in temperature and salinity of fluids upwards and outwards from the deposit is unknown. But, at least, no remarkable difference is observed in the results obtained from the surface (about 1,700m above sea level) and the underground 1,190m level (Fig. V–11).

Marcopper Mine, Philippines

The Marcopper porphyry copper deposit is located near the center of Marinduque Island,

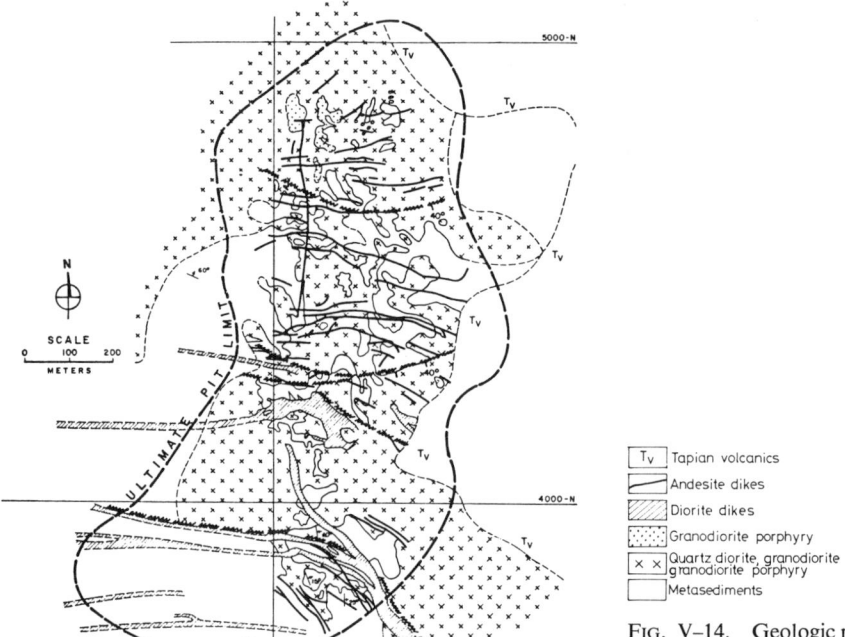

FIG. V-14. Geologic map of the Marcopper porphyry copper deposit (after Loudon, 1976).

180 km south of Manila. The Marcopper Mining Corporation began production in September, 1969, and in 1975 the milling capacity was extended to 29,000 t per day from 18,000 t. The minable ore reserves are reported as 102 million tons at an average grade of 0.58% copper.

General geology: According to Loudon (1972, 1976), the island is geologically divided into two blocks (Fig. V-13). The northeastern block consists mainly of quartz diorite and Eocene-Miocene formations, whereas the southwestern block consists mainly of Miocene, Pliocene and Pleistocene formations. The basement of Marinduque Island consists of basic lava flows intercalated with siltstone and graywacke of Cretaceous age (Loudon, 1972, 1976; Motegi, 1975).

The Eocene Tumicob Group consisting of sediments and volcanics overlies unconformably the basement rocks. The group comprises a lower formation of clastic and limy sediments and an upper dacitic and andesitic pyroclastic formation. The Tumicob Group is overlain by the Oligocene San Antonio Formation, which comprises andesitic lava flows, volcanic breccias and tuffs.

Lower Miocene limestone, clastic sediments and volcanics uncomformably overlie the San Antonio Formation.

Numerous intrusive bodies of Middle Miocene quartz diorite occur in a NW-SE direction. The geologic structure of Marinduque is characterized by the NW-SE trend shown by major faults and intrusive bodies. The largest stock is called the Mahinhin Stock, on the southeastern margin of which the Marcopper deposit is located. The stock consists of medium-grained hornblende-biotite-quartz diorite. In the vicinity of the deposit, however, the stock consists of granodiorite, diorite and their porphyritic facies and is intensively altered hydrothermally. The porphyries cut off most copper-bearing quartz veinlets at their contact, suggesting that the intrusion of porphyries occurred later than the main stages of fracturing and mineralization which occurred in quartz diorite. A swarm of post-ore andesite dikes cut the deposit in the general direction of NW-SE (Fig. V-14).

The emplacement of intrusives, mineralization and post-ore dikes suggests that the NW-SE direction of the geologic structure was dominant before and after the mineralization.

Ore deposit: The mineralization occurred in an intense stockwork of quartz veinlets and dissemination during a multiple-intrusive event (Loudon, 1976). The copper grade decreases with the decrease of fracturing away from the intrusive rocks to sedimentary wall rocks. Though the horizontal distribution pattern of the copper grade is irregular, the vertical pattern shows a mushroom shape. Chalcopyrite and pyrite are the main primary sulfide minerals. The content of magnetite varies from less than 1 % in high-grade ore to several percent in low-grade sedimentary rocks. The average MoS_2 content is about 0.005–0.01 % and Au and Ag content are 0.34 and 1.71 g/t, respectively (Loudon, 1976).

The alteration zoning changes from an inner potassic zone to an outer propylitic zone through a phyllic zone, but the low-grade potassic core proposed by Lowell and Guilbert (1970) is not clear. The horizontal zoning pattern of alteration is assymetric or irregular, while the vertical zoning pattern is a mushroom shape, as is that of the copper grade.

The copper mineralization is generally accompanied by biotite, K-feldspar, quartz, sericite and chlorite. High-grade copper ores are associated with intensive silicification and biotitization. The copper mineralization also occurs in the phyllic zone which consists of quartz, sericite and chlorite. Argillic alteration occurs in the phyllic-propylitic boundary zone. Loudon (1972) suggested a moderate depth of burial during the mineralization, inferred from the predominance of stockwork over dissemination and from the absence of the inner potassic core of low-grade copper mineralization.

Fluid inclusion studies: In quartz from the Marcopper deposit, gaseous inclusions, liquid inclusions and polyphase fluid inclusions occur, in successive order of abundance.

Liquid inclusions show irregular shapes and are distributed along planes or in clusters of

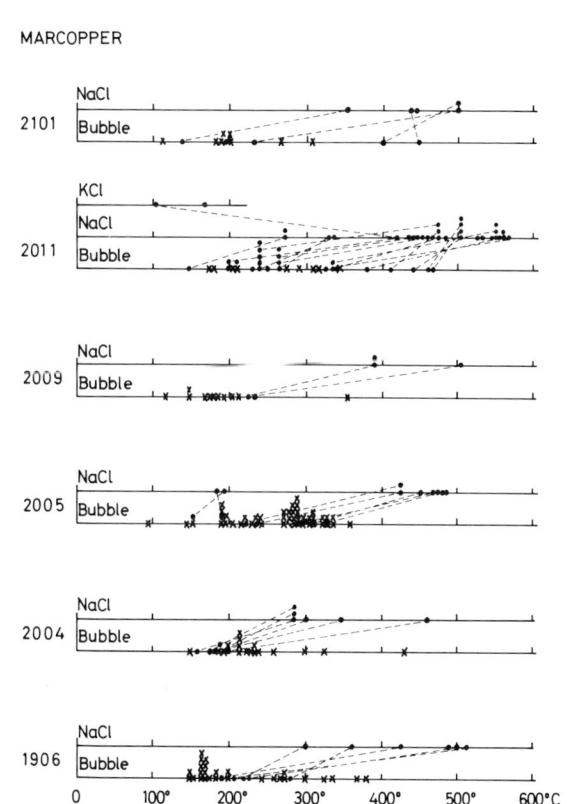

Fig. V-15. Disappearance temperature of NaCl and bubble in polyphase fluid inclusion from the Marcopper deposit. Cross mark shows the filling temperature of liquid inclusion.

dense population. The degree of filling is various from high to as low, as is that of gaseous inclusions.

Gaseous inclusions, which have various degrees of filling from 0.1 to 0.3, occur in shapes of simpler or negative crystal and, occasionally, with polyphase inclusions in an intimate relation. This type of inclusion was the most abundant in any sample.

Polyphase inclusions often carry a prismatic or round transparent mineral, hematite flake, and minute opaque mineral, besides a cubic crystal of halite. The size of the gas bubble of a polyphase inclusion is smaller than that of the other types of inclusions, suggesting high densities of fluids at the time of trapping. The number of polyphase inclusions is small, occupying less than 10% of the total number of inclusions. The size is generally smaller than 10 microns (Figs. V–6(g), and(h)).

Upon heating, the bubble in polyphase inclusions disappears first at a temperature between 160°C and 460°C, and then halite disappears at a temperature between 300°C and 550°C (Fig. V–15). From the disappearance temperature of halite, it is inferred that the salinity of fluid would have been approximately 35–50 wt. %.

The homogenization temperature of liquid inclusions is distributed in a temperature range of 150°–380°C, which is generally lower than that of polyphase inclusions.

Hydrothermal activities: From the results of fluid inclusion studies at the Marcopper deposit, it is inferred that the ore deposit was subjected to activities of various types of hydrothermal fluids, from a high-temperature and highly saline fluid in a boiling condition to a low-temperature hydrothermal solution originating from meteoric water.

The scarcity of polyphase inclusions and the abundance of liquid inclusions in Marcopper would suggest retrogressive hydrothermal activities during and after the formation of the ore deposit, which is probably indicated by the irregular or sporadic development of sericitization and argillization superimposed on the potassic alteration. This is contrasted with the results obtained for the Santo Tomas II porphyry copper deposit, where the polyphase inclusions are the most abundant and the potassic alteration (biotitization) is prominent, suggesting the scarcity of retrogressive alteration.

General Considerations

Studies of fluid inclusions in some porphyry copper deposits in southeast Asia show that the ore-forming fluids of these deposits were generally high in temperature and salinity, and similar to those of the United States (Takenouchi, 1976b).

On the other hand, the salinities of fluid inclusions in hydrothermal ore deposits are apparently lower than those of porphyry copper deposits, generally less than 10 wt. %, though polyphase inclusions of high salinities are occasionally found in porphyries or granitic rocks genetically related to the mineralization (Takenouchi and Imai, 1975).

Burnham (1967) discussed the formation of porphyry copper deposits as the result of a rapid and shallow intrusion of granitic magmas undersaturated with water. Recent experiments on the chlorine concentration of the aqueous phase equilibrated with granitic melts indicate that chlorine increases in the aqueous phase released in the late stages of crystallization of melts, when the pressure is lower than 500 bars (Kilinc and Burnham, 1972), or when the initial content of water in melts is low (Holland, 1972).

From the above-stated experimental results and the mode of occurrences of fluid inclusions in porphyry copper deposits and hydrothermal deposits, it is concluded that the porphyry copper deposits were formed from highly saline fluids released rapidly from magmas which intruded into shallow places and were deficient in water, whereas the fluids of hydrothermal deposits were intermediate in salinity and released gently from water-rich magmas, especially in the later stages of crystallization.

PART VI
SEDIMENTARY DEPOSITS

The Peculiar Phosphate and Copper Deposits in Noto Peninsula
(Imai and Yamadera, 1952)

H. Imai and H. Yamadera

Japan is not favored with sources of phosphate ore, and must import all that is consumed. However, in Noto Peninsula, central Japan, there are several localities of phosphate rocks. Although the ores are very poor in phosphate content (P_2O_5 content is 7–10% on the average), and the ore bodies are small in scale, they were prospected and worked during every period of shortage, especially during times of wars.

Several localities of phosphate deposits are found in Noto Peninsula (Figs. I–2, and –3).

Even in the most productive area, the Hiuchidani area, the total production since its discovery to the present time is said to be only about 50,000 t of phosphate ore containing 10–25% P_2O_5. In the Notojima area, total production is about 16,000 t of ore of similar content. In the Shichimi and Usetsu areas it is about 9,000 t.

Geology and Ore Deposit of the Hiuchidani and Adjacent Areas

Geological background: This area is composed of two groups of rocks. The lower formation of Lower or Middle Miocene age, is called the Noto Group, and is unconformably covered by the Nanao Group, belonging to the Middle or Upper Miocene (Fig. VI–1).

The Noto Group is composed of olivine basalt, rhyolite, andesite (two pyroxene andesite) andesitic agglomerate and andesitic tuff breccia, with intercalated thin beds of sandstone and shale. In some places lignite beds occur in this group. Gold-silver veins, manganese veins, etc., are also found in these rocks, but metalliferous veins do not exist in the Nanao Group which unconformably covers the Noto Group.

The Nanao Group is composed of fossiliferous calcareous sandstone (or limestone), arkosic sandstone and mudstone rich in diatom fragments. The group commonly contains thin intercalated beds of volcanic material which is white, soft and clayey; it is composed of fine and elongated volcanic glass shards of microscopic size. The Nanao strata are generally horizontal, or dip gently away from the areas of the Noto Group. In some places, the sequence of deposition is, from lower to upper, calcareous sandstone, arkosic sandstone and mudstone. But in other places, these rocks grade into one another laterally—that is, they are merely different facies, as already noted by Takai (1944). This relation is shown in Fig. VI–2. According to Otsuka (1935), the geologic age of the Nanao Group is correlated with the Ogashima Stage* (oil-bearing Tertiary formation), Middle or Upper Miocene epoch. He

---
* It corresponds to Onnagawa Stage; refer to Chapter I.

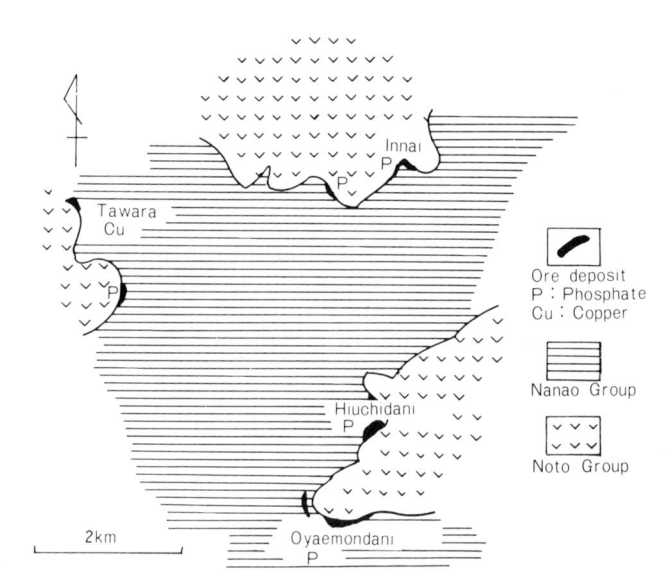

FIG. VI-1. Geologic map of the Hiuchidani area.

recognized that the rocks of the Noto Group composed islands at the time when the sediments of the Nanao Group were deposited. This inferred condition is reflected by the present topography. In this district, hills or mountainous lands 150–200m above sea level are composed of andesitic rocks of the Noto Group, while the lower areas are occupied by sedimentary rocks of the Nanao Group, which are loose and soft compared with the Noto Group. This topography is attributable to differential erosion (Figs. VI–2(a) and –3).

Phosphate Ore Deposits

The phosphate ore deposits in this area exist in the Nanao Group along the boundary with the andesitic rocks of the Noto Group. The P_2O_5 content of the ore is generally 7–15%, but in rarer cases it reaches 20% or so. The modes of occurrence of the ore bodies are as follows:

(1) Some deposits occur in lenticular form along the boundary plane between the andesitic rocks of the Noto Group and the loose sedimentary rocks of Nanao Group. The dimensions of a lenticular ore body are, at largest, 10 m in diameter and 1 m in thickness (Fig. VI–2(a)).

(2) Apparently conformably intercalated in the Nanao Group, there exist small lenses of sandy phosphate. In some cases, a lens of sandy phosphate is superimposed on impervious volcanic tuff which is intercalated in the arkosic sandstone (Fig. VI–2(b)). In such cases, it is frequently observed that cracks filled with phosphate which branch from the phosphate lens penetrate into the underlying tuff (Figs. VI–4(a),(b)). The lenses of phosphate are generally horizontal, but in some places they dip gently away from the areas of the Noto Group.

These lenses are less than 1m thick and 10–20m in diameter. Such lenses are distributed horizontally and laterally here and there. The P_2O_5 content in a lens is somewhat variable, up to 25%. Generally the P_2O_5 content is low at the margin and increases toward the central part of the lens. The ore body grades into the surrounding barren arkosic sandstone. The average content is generally 7–8% of P_2O_5. Accompanying the phosphate ore there are nodules of opaline rock in some places.

(3) At Oyaemondani (Fig. VI–1), a massive ore body 25m long, 15m wide and 5m thick occurs in the arkosic sandstone near the boundary with andesitic rocks (Fig. VI–2(a)). The P_2O_5 content of this ore body is somewhat variable as in type (2). The content is, at best,

Part VI: Sedimentary Deposits

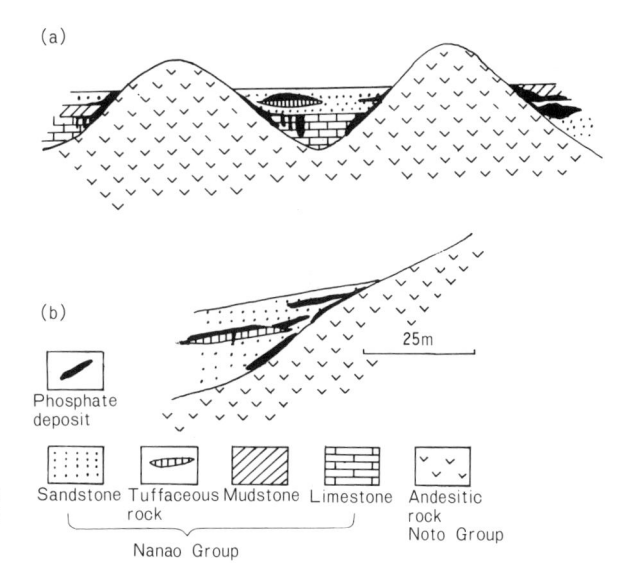

FIG. VI-2. Diagrammatic section of the phosphate ore desposits in Hiuchidani area.

FIG. VI-3. View of the working site at Oyaemondani (Hiuchidani area). The foreground is the mining site, and the hill in the background is composed of andestic rocks.

15–20% of P_2O_5 in the central part, and it becomes poor towards the margin, gradually changing into barren sandstone.

(4) Irregular ore bodies occur in the limestone. They are sharply bordered by barren limestone along very irregular boundary planes (Fig. VI-2(a)).

Phosphate deposits of all four types described above occur along or near the boundary between the sedimentary rocks of the Nanao Group and the andesitic rocks of the Noto Group; topographically they are found in the valleys which adjoin hills consisting of andesitic rocks. The valleys containing the deposits are, at present, mostly occupied by paddyfields (Fig. VI-3).

(a) Occurrence of phosphate ore at Oyaemondani (Hiuchidani area). P: phosphate ore, S: arkosic sandstone, T: tuffaceous rock.
The part enclosed in the rectangle is illustrated in Fig. VI–4(b).

(b) Occurrence of phosphate ore. This part is enclosed by the rectangle in Fig. VI–4(a). The phosphate ore penetrates the impervious tuffaceous bed.

Fig. VI–4.

Genesis of the Phosphate and Copper Ore Deposits

From the occurrences described above, the writers infer that the phosphate once rested on islands of andesitic rocks at the time of sedimentation of the Nanao Group. It was probably the excrement of seabirds. Surface water containing excrement (or its products) entered the sea. The phosphate partly settled on the bottom of the sea as lenticular beds, and partly infiltrated into the loose Tertiary sediments of the Nanao Group along the boundary with the underlying andesitic rocks of the Noto Group.

In the case of deposits of type (1) above, infiltrated water containing phosphate probably flowed along the boundary between the impervious andesitic rocks of the Noto Group and the loose sedimentary rocks of the Nanao Group, and deposited the phosphate. In type (2), some phosphatic lenses might be the phosphate deposited at the bottom of the sea, and others might be sandstone impregnated by phosphate contained in water that infiltrated below the sea bottom. In some cases, infiltrating water containing phosphate would flow in the pervious sandstone which is superposed on the impervious volcanic tuff, and would deposit the phosphate in the sandstone. As described above, the cracks filled with phosphate penetrating into the impervious bed would be due to pouring of water containing phosphate into the small cracks of the impervious bed (Fig. VI–4).

In types (3) and (4), water containing phosphate probably infiltrated into the sedimentary rocks and deposited the phosphate.

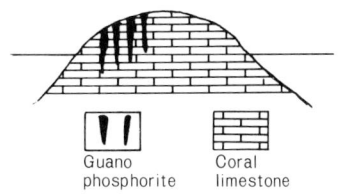

FIG. VI–5. Diagrammatic section of phosphorite in a coral reef.

The guano phosphorite in oceanic islands exists, in most cases, on reef limestone and, in very rare cases, on igneous or metamorphic rocks. It is thought that excrement on the coral reefs reacts with limestone, becoming guano phosphorite (Fig. VI–5). On the other hand, excrement on islands composed of igneous rocks or metamorphic rocks, in most cases, is probably carried into the sea.*

This would happen in the case of the Noto Peninsula deposits.

The writer found thin crustal aggregates of vivianite ($Fe_3P_2O_8 \cdot 8H_2O$), coating andesitic boulders in the phosphate deposit. They were derived from the Noto Group and rest on the unconformity surface between the two stratigraphic groups. Phosphate ion was probably combined with ferrous ion derived from the andesitic rocks.

The genesis of the opaline rock described above is not yet known.

The peculiar copper deposit at Tawara, in the western part of this area (Figs. VI–1 and 6), apparently had an origin similar to that of the phosphate deposits. It occurs in the fossiliferous mudstone of the Nanao Group at the boundary zone with the andesitic rocks of the Noto Group. It is a lenticular deposit, conformably intercalated in mudstone rich in fossils of *silicispongia*, etc. The ore body is 200m long, 15–20m wide and 1–3m thick. The ore is greenish or bluish mudstone or sandstone containing 0.3–2% copper. The copper minerals are mainly chrysocolla, with small amounts of malachite, azurite and native copper. In the chemical analysis of one sample sulfur was recognized, perhaps due to the existence of chalcocite or covelline. Opaline silica occurs as a lenticular mass beneath the ore body, as in the case of some of the phosphate ore bodies. The andesitic rocks of the Noto Group are correlated with Lower or Middle Miocene, a time of ore mineralization in Northeast Japan. As stated above, there are some gold deposits, manganese deposits, etc., in the Noto Group.

---

* It is said that the phosphorus content of sea water is a necessary condition for the breeding of plankton. The diatom earth in the Nanao Group might be related to the phosphate which poured into the sea at that time.

Andesitic rocks Mudstone Copper deposit

Opaline rock

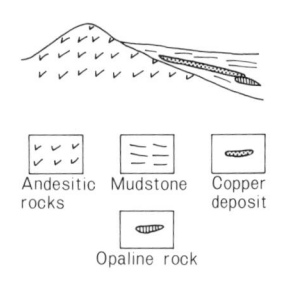

FIG. VI-6. Diagrammatic section of the copper deposit near Tawara.

The genesis of the copper deposit might be interpreted as follows. Water containing copper leached from copper deposits in the Noto Group probably poured into the sea at the time of sedimentation of the Nanao Group, and the copper was reduced and precipitated as copper minerals. The copper deposits in the Noto Group are eroded out at present. Thus the origin could be similar to that of the phosphate deposits in this region. The explanation for the opaline silica is not known, as in the case of the phosphate deposits.

Other Phosphate Deposits in the Noto Peninsula

Notoijima area: Notoijima is a small island 4km from north to south, and 8km east to west. The western part of the island is mainly composed of andesitic rocks of the Noto Group, partly fringed by calcareous sandstone, arkosic sandstone and mudstone of the Nanao Group along the beach. The phosphate ore deposit is similar to that of Hiuchidani. This deposit was worked on a comparatively large scale from 1905 to 1915.

TABLE VI-1. Chemical analyses of the phosphate ores (1).

	(a) Nanao 40 A 1	(a') Nanao 40 A 1	(b) Nanao 40 A 2	(c) Nanao 70	(d) Nanao 8**	(e) Nanao 47
H_2O (−)	3.48	3.48*	3.52	7.62	6.72	8.03
H_2O (+)	Ig. loss 7.45	Ig. loss* 7.45	Ig. loss 5.01	Ig. loss 6.75	10.04	Ig. loss 6.41
CO_2	2.14***	2.14***	2.50***	—	1.90	—
Insol. in conc HCl: SiO_2	21.39 ⎫	84.78 Insol. in acetic acid (10%)	19.30 ⎫	37.92 Insol. in HCl (1:10)	23.76 Insol. in HNO_3 (1:10)	35.44 Insol. in HCl (1:10)
$Al_2O_3(+Fe_2O_3)$	3.49 ⎬ 25.51		3.86 ⎬ 23.23			
CaO	0.61 ⎬		— ⎬			
P_2O_5	0.02 ⎭		0.07 ⎭			
Sol. in conc HCl: SiO_2	0.34	0.34 ⎫	0.44	2.22	—	1.85
Al_2O_3	5.66	⎬ 0.64 (10%)	5.73	8.48	10.15	6.49
Fe_2O_3	7.83	⎭	4.06	3.41	4.30	2.51
CaO	29.64	2.68	32.10	16.29	23.78	21.26
MgO	0.0	—	0.0	0.0	0.39	—
MnO	0.21	—	—	—	—	—
$Na_2O(+K_2O)$	3.09	—	—	—	—	—
P_2O_5	16.08	0.93	21.68	15.01	15.95	18.40
SO_4	tr.	tr.	—	—	—	—
Cl	tr.	tr.	—	—	—	—
F	—	—	—	0.09	—	—
Total	99.29	100.30	95.77	97.79	96.99	100.47
Analyst	Yamadera	Yamadera Dept. of Chemistry, University of Tokyo	Yamadera	Yamadera	Analyst in Nippon Seitetsu Co. Ltd.	Yamadera

* Cited from (a).
** This sample was analyzed in the laboratory of the Nippon Seitetsu Co. Ltd., by courtesy of Dr. T. Ikeno.
*** The analysis of CO_2 was carried out in the laboratory of the Department of Chemistry, University of Nagoya, under the guidance of Dr. S. Oana and Dr. T. Koyama.
— represents n. d. (not determined).

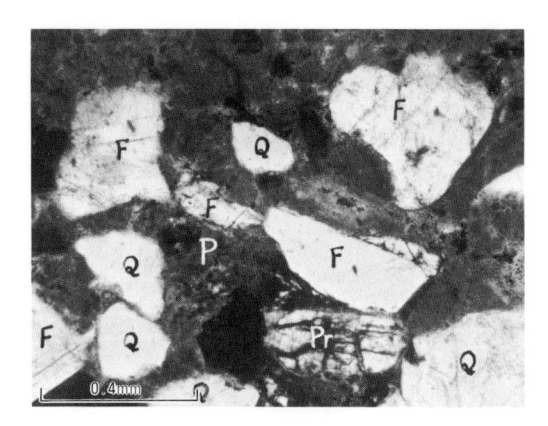

FIG. VI-7. Photomicrograph. One nicol. Thin section of phosphatized arkosic sandstone (Oyaemondani). One nicol. Q: quartz, Pr: pyroxene, F: feldspar, P: phosphatized matrix.

Shichimi and Usetsu areas: At the coastal terraces of Shichimi and Usetsu, phosphate deposits occur in massive form in the mudstone of the Nanao Group. The geological occurrences and genesis of the ore deposit are the same as in the Hiuchidani area.

Takai (1944) found a molar and incisor of *Desmostylus* in the phosphate deposits of the Notojima and Shichimi areas, respectively. The occurrence of the *Desmostylus* fossil would tell the circumstances at the time of sedimentation of the Nanao Group, i.e., at the time of phosphate deposition.

The Properties of the Phosphate Ores

The results of the chemical analyses of the phosphate ores are shown in Table VI-1.

(a) Sample No. Nanao 40 A 1. This sample was collected at Oyaemondani (in the Hiuchidani area) (Fig. VI-1). It is arkosic sandstone containing phosphorus. Under the microscope, grains of quartz and feldspar are observed surrounded by the matrix (Fig. VI-7). Sometimes crystals of diopside and hornblende are found, which would be derived from the andesitic rocks. In the matrix part, green minerals such as glauconite and unknown black minerals are recognized (refer to Chapter I). Frequently the remains of foraminifera and diatoms are found in it. The matrix was permeated by phosphate, and the remains of foraminiferas and diatoms are entirely phosphatized. The phosphate contents of samples (phosphatized sandstones) are 8–15% of P_2O_5. The variation in content is due to differences in the volume of the grains and the degree of permeation of phosphate into the matrix.

The sample was crushed and sieved by a 200-mesh screen on Tyler's scale. The undersize was mostly composed of the matrix of the ore, and grains of quartz and feldspar were concentrated in the oversize. The matrix part thus prepared was analysed. Under the microscope small quantities of quartz and feldspar crystals were recognized in it, but they were negligible in amount.

(b) Sample No. Nanao 40 A 2. This sample was collected at the same place as (a). It is a phosphatized arkosic sandstone. The permeation of phosphate into the matrix part is more marked than in (a), so the sample is much harder than (a). The method of sampling was the same as (a).

(c) Sample No. Nanao 70. This is phosphatized diatom earth, which was collected at Innai, 3km north of Hiuchidani (Fig. VI-1).

Under the microscope, fragments of diatoms, spicules, etc., are recognized (Fig. VI-8), and there exist small amounts of quartz and feldspar. Separation of those materials from the matrix is difficult, and they are quantitatively negligible, so the sample was crushed and analyzed without separating.

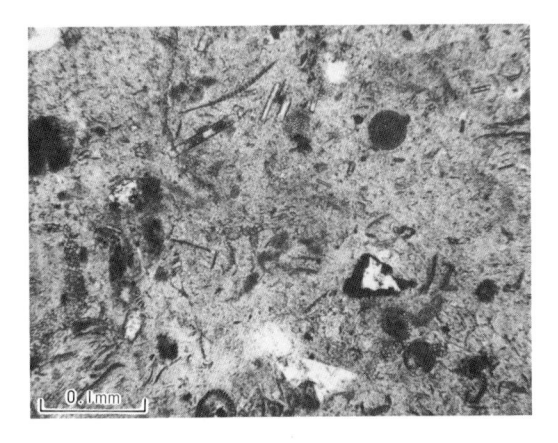

FIG. VI-8. Photomicrograph. One nicol. Thin section of phosphatized diatom earth (at Innai). Remains of diatoms and sponge spicules are observed.

(d) Sample No. Nanao 8. This sample was collected at Nanao City (Takai area). The ore is similar to (a) and (b), containing many grains of quartz and feldspar. The limestone surrounding the pockets of phosphate ore is poorer than the ore itself in contents of quartz and feldspar crystals. The abundance of the grains in the ore is probably due to the flowing of the superposed loose arkosic sand into limestone pockets at the time of solution of limestone and of deposition of phosphate. The method of sampling for analysis was the same as (a) and (b).

(e) Sample No. Nanao 47. This sample is a phosphate nodule (concretion) which was collected from mudstone in the Hamada area 9 km NE of the Hiuchidani area. Macroscopically it is a rather hard and compact rock, and microscopically it is devoid of quartz and feldspar grains and is homogeneous. This sample was crushed and analyzed.

The writer took X-ray powder photographs of samples (a) and (b) using copper or cobalt anticathode. The photographs show rings of the apatite group, although they are very weak.

TABLE VI-2. Chemical analyses of phosphate ores (2).

	(a) Nanao 40 A 1			(b) Nanao 40 A 2			(c) Nanao 70		(d) Nanao 8			(e) Nanao 47	
	Ph	Ca	Cl	Ph	Ca	Cl	Ph	Cl	Ph	Ca	Cl	Ph	Cl
CaO	19.09	2.74	8.42*	25.76	3.19	3.15	16.29		18.89	2.41	2.48	21.26	
P_2O_5	16.10			21.75			15.01		15.95			18.40	
CO_2		2.14			2.50					1.90			
SiO_2			21.73			19.74		2.22			—		1.85
Al_2O_3			}16.98			}13.65		8.48			10.15		6.49
Fe_2O_3								3.41			4.30		2.59
MgO			0.00			0.0		—			0.39		—
MnO			0.21			—		—			—		—
$Na_2O(+K_2O)$			3.09			—		—			—		—
SO_4			tr.			—		Insol. in HCl (1:10) 37.92			Insol. in HNO_3(1:10) 23.76		Insol. in HCl (1:10) 35.44
Cl			tr.			—		—			—		
$H_2O(+)$			5.31			2.51		Ig. loss 6.75			10.04		Ig. loss 6.41
$H_2O(-)$			3.48			3.52		7.62			6.72		8.03
	35.19	4.88	59.22	47.51	5.69	42.57	31.30	66.40	34.84	4.31	57.84	39.66	60.81
	99.29			95.77			F, 0.09 97.79		96.99			100.47	

* Ca^{++} which is exchangeable by K^+ corresponds to CaO 0.70%.
Ph: calcium phosphate, Ca: calcium carbonate, Cl: clay.

TABLE VI-3. X-ray powder photographs of apatite group.

(A)		(B)		(C)		(A)		(B)		(C)		(A)		(B)		(C)	
d.	I	d.	I	d.	I	d.	I	d.	I	d.	I	d.	I	d.	I	d.	I
4.08	2					2.62	5	2.61	m	2.64	m	1.89	3				
3.90	2					2.52	1					1.83	7	1.84	m	1.83	s
3.44	8	3.39	m			2.25	6	2.24	m	2.20	m	1.80	4				
3.18	1					2.14	1	2.14	w			1.77	4				
3.08	2	3.04	m			2.06	1					1.75	4				
2.82	10	2.76	s	2.74	s	2.00	1					1.64	2			1.50	m
2.71	9	2.72	s			1.92	7	1.93	m	1.91	m					1.46	m

d: measured spacing Å. I: intensity, s: strong, m: medium, w: weak.
(A) Carbonate apatite, after Frondel (1943). Cu radiation filtered through Ni foil.
(B) Phosphorite from Angaur. This photograph was taken by Dr. H. Sawada. Camera radius 28.75 mm. Co radiation, Unifiltered.
(C) Phosphate ore from Noto Peninsula (Oyaemondani). Sample No. Nanao 40 A 1. Same as (B).

The data are shown in Table VI–3, with reference to the data of Prien and Frondel (1947).

All the samples were decomposed by dilute HCl or HNO_3, leaving residues. When these residues are observed with the microscope, amorphous silica gel is recognized. In Table VI–1, it is understood that the major part of each residue is silica. But in the matrix parts of the original samples, amorphous silica is not recognized under the microscope.

Aluminum and iron exist in the matrix parts of the samples shown in Table VI–1, and a large fraction of these elements is dissolved by dilute HCl or HNO_3.

From these facts, it is interpreted that amorphous silica residues are derived from the decomposition of the hydrous silicate of the clay minerals by acids. Aluminum and iron in these minerals would be extracted into acids.

The writers attempted to arrange the components of the total analyses in Table VI–1 in order to represent the modes of minerals. The most probable combinations are shown in Table VI–2.

Calcium oxide is combined with phosphate and carbonate radicals in order to make tricalcium phosphate and calcium carbonate. As stated above, the existence of the apatite group is confirmed by X-ray powder photograph. The apatite may be a carbonate apatite or a hydrous carbonate apatite, according to Frondel (1943). The remaining components are attributable to clay minerals. The clay minerals represent the matrix part of the original rocks into which phosphate permeated and in which it was deposited.

In (c), calcium oxide is insufficient to make tricalcium phosphate. It is uncertain whether it is due to the existence of the other calcium phosphates such as monetite ($HCaPO_4$), brushite ($HCaPO_4 \cdot 2H_2O$) metabrushite ($2HCaPO_4 \cdot 3H_2O$), etc., or is due to the existence of iron or aluminum phosphates. Generally, most of the calcium oxide in the matrix of sedimentary rocks is due to calcite. But, as shown in Table VI–2, the clays of (a), (b) and (d) contain more CaO than is accounted for by the calcium phosphate and calcium carbonate. Minami (1935) published the average values of chemical analyses of clayslates in Japan and Europe, as shown in Table VI–4. The samples he analyzed are very poor in calcite. Compared with these values, the CaO content of the above samples are very high. Clayslate containing 9.91% CaO, 0.08% P_2O_5 and 0.55% CO_2 has been reported in the Carboniferous formation of Kentucky (Clarke, 1924), but a mineralogical explanation is not given.

In the present case, on the assumption that the base exchange would have been carried out at the time of permeation of the solution rich in calcium and phosphorus, the writers made a quantitative analysis of the calcium which was exchangeable with potassium ion. But it was only CaO of 0.70%.

Also, as shown in Table VI–1(a′), the CaO content which is exchangeable with hydrogen

TABLE VI-4. Chemical analyses of clayslates (after Minami, 1935).

	1	2	3
SiO_2	58.35%	61.42%	65.66%
TiO_2	0.83	0.62	0.68
Al_2O_3	18.74	17.68	14.43
Fe_2O_3	0.73	1.79	2.85
FeO	5.53	4.52	2.33
MnO	0.08	0.09	0.07
MgO	2.00	2.41	1.93
CaO	0.38	1.66	1.47
Na_2O	1.09	2.97	2.17
K_2O	3.84	2.70	2.60
$H_2O(+)$	4.25	3.05	3.15
$H_2O(-)$	1.09	0.71	1.46
P_2O_5	0.12	0.17	0.12
CO_2	0.53	n.d.	n.d.
C	1.91	n.d.	n.d.
	99.47	99.79	98.92

1. Paleozoic clayslates in Europe (average of 36 samples).
2. Paleozoic clayslates in Japan (average of 14 samples).
3. Mesozoic clayslates in Japan (average of 10 samples).

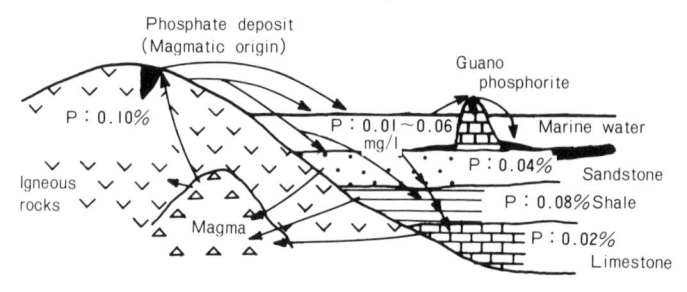

FIG. VI-9. Phosphorus cycle in the earth crust (data from Van Wazer, 1961).

ion was analyzed, using 10% acetic acid solution. However, the experiment was unsuccessful, as the amount of CaO which was removed was 2.68%, which is nearly equal to the CaO content of the $CaCO_3$ as shown in Table VI-2. Apparently the calcium carbonate and part of the calcium phosphate in the sample were dissolved. Concerning this problem, no conclusion has been reached.

In sample (c), the undissolved residue is rich in silica, which is due to the remains of siliceous spicules and diatoms.

The Phosphorus Cycle in the Earth's Crust (Fig. VI-9)

Blackwelder (1916) and Lindgren (1923) discussed this problem, and the writers will merely add some facts to their conclusions.

The average content of phosphorus in igneous rocks is $0.10 \pm 0.03\%$. The phosphate deposits related to igneous activity are orthomagmatic (including carbonatite) or pegmatitic. This kind of deposit occurs in the Kola Peninsula, USSR. Released by weathering, the phosphorus of these rocks is transported into the sea. Phosphorus in seawater amounts to about $0.01 \sim 0.06$ mg/l, and the phosphorus contents of sedimentary rocks are 0.04% in sandstone, 0.08% in shale and 0.02% in limestone (Rankama and Sahama, 1950; Goldschmidt, 1954; Van Wazer, 1961). A part of the phosphorus in seawater would be derived from the land.

Plankton concentrate the phosphorus in the seawater, fish devour the plankton, large fish eat the small, and sea-birds swallow the fish. The excrement of the seabirds accumulates on ocean islands. This is guano.* If the island is a coral reef, rainwater dissolving guano permeates into and reacts with the limestone of the coral reef, producing guano phosphorite. The phosphate ore deposits of Angaur, Ocean, Nauru, etc., belong to this type (Rogers, 1948). The writers think that the lime in guano phosphorite is attributable to two sources. One is the calcium phosphate contained in the droppings of the seabirds, which was originally derived from fish. The other is the limestone of the coral reef, which reacts with the soluble phosphates in the droppings such as ammonium phosphate, primary or secondary calcium phosphate. Phosphorite ore occurs among bizarre pinnacles of limestone (Fig. VI-5). The exceedingly irregular boundary surface between the ore body and underlying limestone is believed to have been formed by a solution of limestone. It resembles closely the weathered surface that is commonly present on limestone beneath the soil. Imai (1944) discussed these problems in a previous paper.

Excrement which rests on rocks other than limestone in most cases is carried into the sea. Such a case is fortuitously represented by the phosphate deposits in the Noto Peninsula.

Geologically, the phosphate ore in coral limestone is ultimately liable to be eroded and deposited on the sea bottom. As a result, limestone containing tricalcium phosphate would be formed.

But most of the enormous calcium phosphate deposits in the world may have been deposited from sea water on the sea bottom, though the conditions and circumstances have not been fully explained. It is reported that in some cases shells of brachiopoda rich in phosphorus content are accumulated in sedimentary beds (Correns et al., 1940). In these various ways, sedimentary rocks rich in calcium phosphate are deposited.

These sediments are folded and uplifted, and become land. Erosion of the rocks is begun again. Otherwise, the sediments may sink to a deeper part and become absorbed in magmas, in which case the phosphorus returns to the initial stage.

Summary

(1) The phosphate ore deposits in the Noto Peninsula occur in the sedimentary rocks of the Nanao Group (Upper Miocene) along or near their contacts with the andesitic rocks of the Noto Group (Lower or Middle Miocene). Topographically, they are found in valleys which are the marginal parts of hills composed of andesitic rocks. Some ore bodies are lenses intercalated in the sedimentary rocks; others are massive or pocket-like ore bodies.

(2) It is believed that the excrement of seabirds was once deposited on islands composed of andesitic rocks at the time of sedimentation of the Nanao Group and was carried into the sea. The transported phosphate was partly deposited with the sedimentary rocks of the Nanao Group, and partly infiltrated into the underlying Nanao Group beneath the sea bottom.

(3) The phosphate ores are sedimentary rocks permeated by tricalcium phosphate.

(4) Fossils of *Desmostylus* were found in the phosphate deposits in the Noto Peninsula, indicating circumstances at the time of phosphate deposition.

---

* Frondel (1943) said that the composition of guano itself is complex and not well known. The principal constituents are calcium phosphates, ammonium phosphate and ammonium acid phosphate, together with uric acid, sodium acid urate, ammonium and calcium oxalate, ammonium and alkali sulfate and sulfate-oxalates, guanine, xanthine, and other organic substances (Prien and Frondel, 1947).

PART VII

CONTRIBUTIONS

VII-1. Magma Genesis in a Dynamic Mantle

Keisuke Ito

Introduction

According to the plate tectonics hypothesis, the mantle is tectonically divided into three regions: lithosphere, asthenosphere and mesosphere. The lithosphere or plate is created from the asthenosphere at mid-ocean ridges, spreads away and subducts into the asthenosphere at trenches. The mesosphere is stable, does not move and lies below the asthenosphere.

Three types of volcanic rocks occur at different tectonic settings. The mid-ocean ridge basalt erupts at mid-ocean ridges where two plates are diverging. The oceanic island basalt occurs at oceanic islands located away from plate margins. Arc volcanic rocks characterized by calc-alkali series are restricted to the continental side of trenches where two plates are colliding.

Mid-ocean ridge basalts occur where new plates are being formed, and are not contaminated by a pre-existing plate material. Oceanic island basalts consist of tholeiites and alkali-rich basalts (alkali basalt and nephelinite). Alkali-rich basalts erupt explosively; some contain nodules of the deep mantle material and indicate that their magmas passed through the plate very rapidly. Therefore, alkali-rich basalts may be little contaminated. Calc-alkali rocks may be seriously contaminated by a thick continental crust. We therefore chose the mid-ocean ridge basalt and alkali-rich basalt as the least contaminated materials of the asthenosphere and used them as chemical keys for understanding the nature of the asthenosphere.

These two types of basalts are geochemically distinctive. The mid-ocean ridge basalt is tholeiitic, depleted in large ion lithophile elements, such as alkali, light rare earth elements, U and Th. The alkali-rich basalt is enriched in these elements. Lead and strontium isotopes in the alkali-rich basalt are more radiogenic and variable than those in the mid-ocean ridge basalt.

These geochemical features suggest that the mid-ocean ridge basalt is formed by a large degree of partial melting of a part of the asthenosphere which is depleted in the large ion lithophile elements and has low Rb/Sr, U/Pb and Th/Pb ratios. The alkali-rich basalt is formed by a small degree of partial melting of a part of the asthenosphere which is not depleted or may be enriched in the large ion lithophile elements. The mantle is required to be heterogeneous (Gast, 1968).

The scale and degree of heterogeneity of the mantle should reveal the structure and history

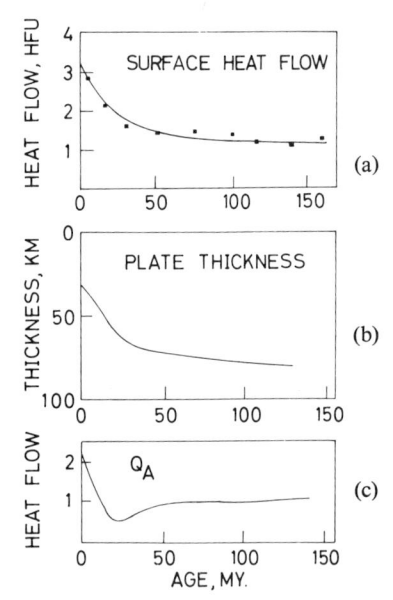

FIG. VII-1-1. Geophysical data vs. crustal age of the ocean for a thickening plate model (Ito, 1976). (a) Surface heat flow. Solid squares are averaged values for the North Pacific given by Sclater and Francheteau (1970). (b) Plate thickness. The temperature of the lithosphere-asthenosphere boundary is assumed to be 1100°C. (c) Heat flow from the asthenosphere to the lithosphere. The latent heat of freezing is not taken into account.

of the mantle convection. By studying the geochemical characters of oceanic basalts, we can estimate the structure and evolution of the mantle. I have presented a model of the lithosphere and asthenosphere and suggested that a liquid layer may exist at the lithosphere-asthenosphere boundary and that this liquid layer may be the source of the alkali-rich basalt (Ito, 1976, 1977). The present paper is a geochemical extension of my geophysical model summarized in the following section.

Model of the Lithosphere and Asthenosphere

The lithosphere is a solid slab overlying the partially molten asthenosphere. In the present model, the lithosphere is assumed to grow from the asthenosphere as the liquid in the asthenosphere freezes (Parker and Oldenburg, 1973; Yoshii, 1973). In order to obtain a thermal model of the lithosphere, I used the decrease of heat flow values as a function of the age of the ocean floor and estimated the nature of the lithosphere-asthenosphere boundary (Fig. VII-1-1). The conclusions drawn from this model relevant to the present discussion are the following:

(1) The heat flow from the asthenosphere to the lithosphere Q_A is high beneath the mid-ocean ridge, and indicates the upwelling convective flow of the asthenosphere beneath the mid-ocean ridge.

(2) The Q_A decreases rapidly with increased distance from the mid-ocean ridge, has a minimum at about 200 km away from the ridge, and suggests downwelling convective flow.

(3) The Q_A beneath older ocean basins is rather high, relatively constant at about 1 μcal cm^{-2} sec^{-1}, and requires effective heat transfer by active convection in the asthenosphere.

(4) The temperature at the lithosphere-asthenosphere boundary seems to be controlled by the amount of H_2O at about 1,100°C. The low melting temperature requires that some H_2O exists at the top of the asthenosphere.

The primary convection under the ocean involves plate motion and return flow at depth. The horizontal scale is on the order of plate dimension. Conclusion (3) requires local small-scale convection in addition to large-scale convection. According to Richter and Parsons (1975), the form of small-scale convection in the asthenosphere is spout-like beneath a slow-moving plate, but roll-like beneath a fast-moving plate. The axes of rolls are aligned in the

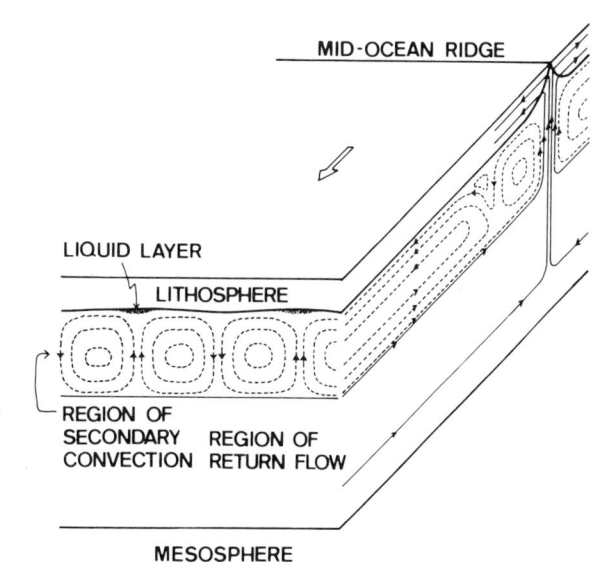

FIG. VII-1-2. Convective motion in the mantle. A white arrow indicates the plate motion. Solid lines with arrows represent flow lines of the primary convection and dashed lines are flow lines of the secondary convection.

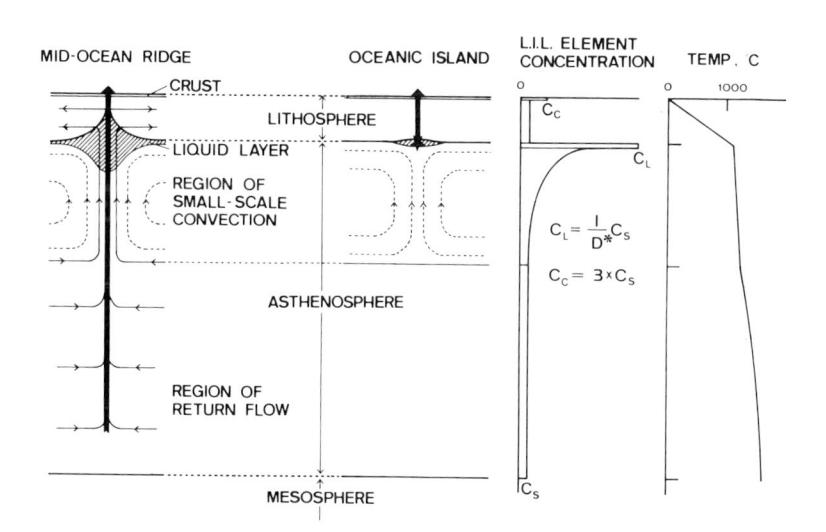

FIG. VII-1-3. Dynamic structure, concentration of large-ion lithophile (L. I. L.) elements, and temperature of the mantle. The asthenosphere is subdivided into three regions: liquid layer (shaded part); region of small-scale, secondary convection (dashed lines with arrows); and region of return flow for the primary convection (solid lines with arrows). Solid black arrows indicate the sources of mid-ocean ridge basalt magmas and oceanic island alkali-rich basalt magmas, respectively. C_c is the concentration of L. I. L. elements in the oceanic crust, C_L the concentration in the liquid layer, and C_S the concentration in the region of return flow or the depleted asthenosphere.

$$C_L = \frac{1}{D^*} C_c, \qquad C_L = 3 \times C_S$$

direction of plate movement. The horizontal scale of convective rolls is on the order of the depth of the convecting region.

The primary convection is believed to extend about 650 km because earthquakes occur down to that depth. If downwelling convective flow exists about 200 km away from the mid-ocean ridge, as conclusion (2) suggests, the vertical scale of small-scale convection may be only 200 km. The region of small-scale convection roughly coincides with the low-velocity

TABLE VII–1–1. Abundance of large ion lithophile element (ppm) in the mid-ocean ridge basalt and alkali-rich basalt.

	K	Rb	Cs	Sr	Pb	Ba	U	Th
Alkali-rich basalt[a]	12000	33	(2.0)[c]	930	(5.22)	808	(2.14)	(7.35)
Mid-ocean ridge basalt[b]	600	0.61	0.007	190	0.42	6.62	0.049	0.084
	Yb	Er	Dy	Gd	Eu	Sm	Nd	Ce	La
Alkali-rich basalt[a]	1.44	2.16	5.80	9.31	3.35	11.0	53.1	109.6	56.1
Mid-ocean ridge basalt[b]	3.02	2.93	4.54	4.07	1.17	3.09	9.33	10.3	2.99

[a]K23, olivine nephelinite from Kauai, Hawaii, analyzed by Kay and Gast (1973).
[b]C10D3, from Explorer Seamount, Juan de Fuca Ridge, analysed by Kay et al. (1970) and Church and Tatsumoto (1975).
[c]Data in parentheses are estimated from analysis data for other similar rocks.

zone or low-viscosity zone. A deeper region down to 650 km may be the region of return flow for the primary convection. The preferred model of the convecting mantle is schematically shown in Fig. VII–1–2.

Magmas of mid-ocean ridge basalts are formed by partial melting of the upwelling material suggested by conclusion (1). They are essentially anhydrous, containing 0.2% or less H_2O (Moore, 1970). Conclusion (4) requires that more H_2O, about 1%, exists beneath the old lithosphere. The H_2O content at the lithosphere-asthenosphere boundary must gradually increase from less than 0.2% at the spreading center to about 1% beneath the older ocean basin. The increase in H_2O is considered to be caused by accumulation of volatiles as well as low melting temperature components from the entire asthenosphere to the top by means of small-scale convection.

A liquid layer formed in this way is enriched in volatiles such as H_2O, CO_2 and Cl_2 and large ion lithophile elements. As a result, the residual asthenosphere is depleted in these elements. The liquid layer is an open system. Ascending liquid enters into the system. As the overlying lithosphere cools, solids crystallize and are removed from the system. Some sink into the asthenosphere and others accrete onto the base of the lithosphere to thicken the plate.

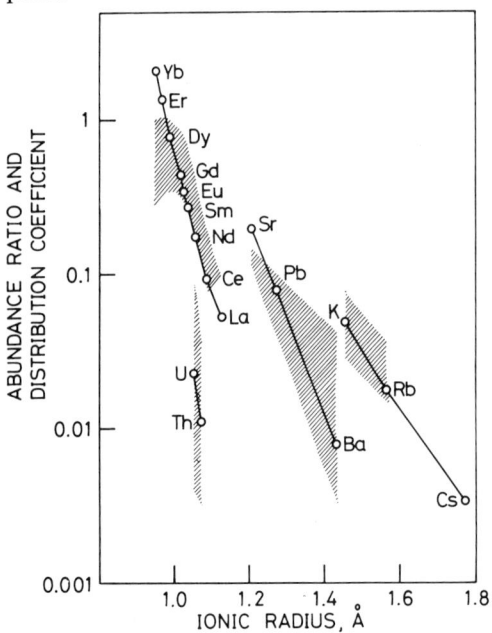

FIG. VII–1–4. Abundance ratios D* of large ion lithophile elements between the depleted asthenosphere and the enriched liquid layer. Shaded parts are ranges of the distribution coefficients between clinopyroxene and liquid (Schnetzler and Philpotts, 1970; Onuma et al., 1968).

The asthenosphere may be divided into three regions, though their boundaries are gradual: (1) the liquid layer enriched in volatiles and large ion lithophile elements, (2) the intermediate region of the secondary, small-scale convection and (3) the region of return flow for the primary convection, which is depleted in large ion lithophile elements. I propose that the liquid layer is the source of the alkali-rich basalt and the region of return flow is the source of the mid-ocean ridge basalt. It should be noted that the liquid layer cannot be thick, nor can it exist continuously over the asthenosphere, because seismic shear waves pass through the asthenosphere. A diagrammatic representation of the present model is given in Fig. VII–1–3.

Distribution of Large Ion Lithophile Elements

Trace element abundances in mid-ocean ridge basalts and alkali-rich basalts are distinctly different. Mid-ocean ridge basalts are depleted in large ion lithophile elements. Alkali-rich basalts are enriched in these elements (Kay *et al.*, 1970; Kay and Gast, 1973). Representative values of trace element abundances in these two types of basalts were tentatively selected from analysis data by Church and Tatsumoto (1975) and Kay and Gast (1973), and are given in Table VII–1–1. The selected values are not the averages but are taken from a depleted side of the whole range for the mid-ocean ridge basalt and from an enriched side for the alkali-rich basalt.

In the present model, the trace element abundances in the alkali-rich basalt represent those in the enriched liquid layer at the top of the asthenosphere. The mid-ocean ridge basalt is formed by partial melting (about 30% or more) of the depleted asthenosphere. Because large ion lithophile elements are concentrated in liquid by partial melting, abundances of these elements in the mid-ocean ridge basalt should be about three times as much as those in the depleted asthenosphere (Fig. VII–1–3). From data given in Table VII–1–1, the abundance ratios of trace elements between the depleted asthenosphere and the enriched liquid layer are estimated and plotted as a function of ionic radii in Fig. VII–1–4.

The concentration mechanism of trace elements in the liquid layer by small-scale convection is comparable to concentration of impure elements by zone refining. In zone refining at the final stage, the concentration ratio for a trace element between the concentrated liquid and the refined solid is equal to the distribution coefficient of the element between liquid and solid, regardless of the degree of partial melting at each refining process (Pfann, 1957).

Accordingly, the abundance ratios of large ion lithophile elements between the depleted asthenosphere and the liquid layer should approach the bulk distribution coefficient given by $D^N = \Sigma X_i D_i^N$, where X_i is the proportion of mineral i in the solid asthenosphere and D_i^N the distribution coefficient of element N between the mineral i and liquid (Fig. VII–1–3). In the solid asthenosphere, large ion lithophile elements are contained mostly in clinopyroxene, and the amount of clinopyroxene is estimated to be 10–30%. Therefore, the abundance ratios of these elements between the depleted asthenosphere and the liquid layer are expected to be 1/10–1/3 of the distribution coefficients of these elements between clinopyroxene and liquid.

The distribution coefficients of large ion lithophile elements between clinopyroxene and liquid have been estimated from phenocryst-groundmass data (Schnetzler and Philpotts, 1970; Onuma *et al.*, 1968). They are superimposed in Fig. VII–1–4 to be compared with the ratios between the depleted asthenosphere and the enriched liquid layer. Agreement between the two sets of data is striking. The only significant disagreement is that the distribution coefficients for rare earth elements have a maximum around dysprosium but the abundance ratios show a straight increase in heavier (smaller) rare earth elements. The difference suggests that garnet enriched in heavy rare earth elements participates in the genesis of alkali-rich basalt magmas (Kay and Gast, 1973).

1. Magma Genesis in a Dynamic Mantle

TABLE VII–1–2. Decay systems.

Parent	Daughter	Decay constant	Initial ratio
^{238}U	^{206}Pb	0.155125×10^{-9}/yr	^{206}Pb/^{204}Pb $= 9.307$
^{235}U	^{207}Pb	0.98485	^{207}Pb/^{204}Pb $= 10.294$
^{232}Th	^{208}Pb	0.049475	^{208}Pb/^{204}Pb $= 29.48$
^{87}Rb	^{86}Sr	0.0139	^{87}Sr/^{86}Sr $= 0.6990$
^{147}Sm	^{143}Nd	0.00654	^{143}Nd/^{144}Nd $= 0.50598$

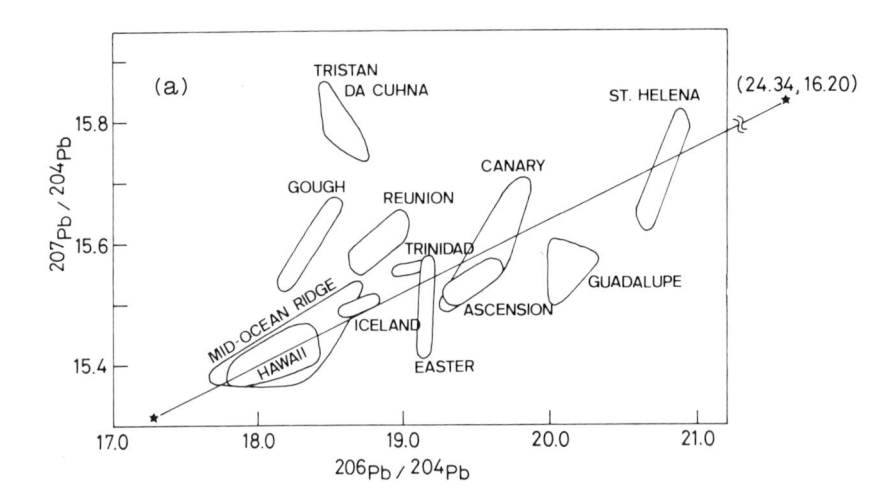

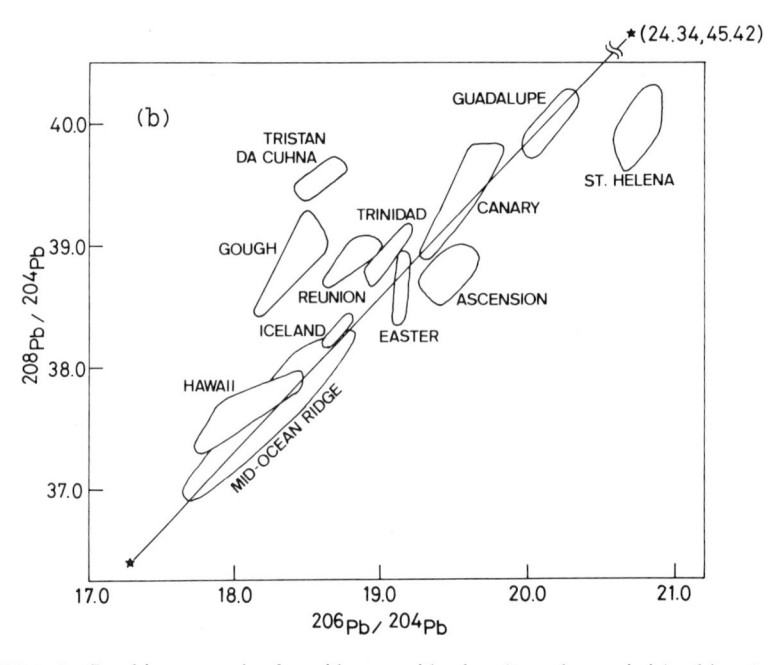

FIG. VII–1–5. Lead isotope ratios for mid-ocean ridge basalts and oceanic island basalts: (a) ^{207}Pb/^{204}Pb versus ^{206}Pb/^{204}Pb, and (b) ^{208}Pb/^{204}Pb versus ^{206}Pb/^{204}Pb. Star marks represent end-member isotope ratios generated in the two source regions (see text for explanation).

Radiogenic Isotope Evidence

Radiogenic isotopes used for studying the evolution of the mantle are listed in Table VII–1–2. Lead isotope ratios in oceanic basalts are widely variable, but they tend to lie in a straight line on both ^{207}Pb/^{204}Pb versus ^{206}Pb/^{204}Pb and ^{208}Pb/^{204}Pb versus ^{206}Pb/^{204}Pb plots (Fig. VII–1–5(a) and –5(b)) (Sun and Hanson, 1975; Church and Tatsumoto, 1975). Basalts of such islands as Gough, Tristan da Cuhna, Réunion and Azores are anomalous. They have strontium isotope ^{87}Sr/^{86}Sr ratios higher than ordinary island basalts and should represent another local heterogeneity of the mantle (Meijer, 1976). Intra-island variations for some islands are oblique to the slope and are probably caused by crustal contamination. Mid-ocean ridge basalt data have a slope which is subparallel to the slope for all the data.

The straight-line relationship of lead isotope data is interpreted in terms of either a two-stage model (Gast, 1969; Tatsumoto, 1966) or a mixing model (Sun and Hanson, 1975). In a two-stage model, the slope of the straight line gives the time when U and Pb were differentiated in the source region. The model age for the slope shown in Fig. VII–1–5(a) is 2.0 b.y.

In a mixing model, the least radiogenic material that is the source of the mid-ocean ridge basalt is mixed with more radiogenic material which is the source of the alkali-rich basalts. Some authors, accepting the mantle-plume hypothesis for the genesis of oceanic island chains (Morgan, 1972; Schilling, 1973), consider that the radiogenic lead comes from the deep mantle plume material from the mesosphere (Sun and Hanson, 1975).

In the present model of magma genesis, the source of the mid-ocean ridge basalt is the depleted asthenosphere with low U/Pb and Th/Pb ratios and therefore it has the least radiogenic lead isotopes. The source of the alkali-rich basalt is the enriched liquid layer with high U/Pb and Th/Pb ratios and therefore it has more radiogenic lead. Separation of the two source regions is maintained by small-scale convection. By means of convection accompanied by partial melting, parent and daughter elements are kept fractionated, but isotopes tend to be mixed. Mixing at present is not complete, as we see linear arrays of lead isotope variations in Fig. VII–1–5(a), (b).

Though the slope of the straight line needs not have a time significance in a mixing model, we interpret that the slope shown in Fig. VII–1–5(a) gives the time when separation of the two source regions began. By that time, 2.0 b.y. ago, the mantle is assumed to have been strongly convecting, and either the mantle was homogeneous or the isotopes were completely mixed.

Growth of ^{206}Pb/^{204}Pb in the present model is shown in Fig. VII–1–6(a). End-member growth curves at the second stage were obtained for a hypothetical case where isotopes were not mixed between the two source regions. The primary ^{238}U/^{204}Pb ratio (as the present value) was estimated to be 7.8. This was determined from a requirement of the two-stage model that the extension of primary stage growth curve and the geochron should intersect on the straight line of observed data. The ^{238}U/^{204}Pb ratios for the depleted asthenosphere and the enriched liquid layer at the second stage were obtained from U and Pb abundances in the two source regions, estimated from trace element abundance data for the mid-ocean ridge basalt and the alkali-rich basalt given in Table VII–1–1.

End-member lead isotope ratios, which might have been built up in the two source regions if the isotopes were not mixed, are shown by star marks in Fig. VII–1–5(a). It should be remembered that the end-member isotope ratios do not represent isotope ratios in the present two source regions. Lead isotope ratios in all oceanic basalts lie between the two end-member ratios and suggest that they were formed by mixing of the two end-members. The mid-ocean ridge basalt data are closer to the end-member generated in the depleted asthenosphere. The alkali-rich basalt data spread toward the other end-member generated in the

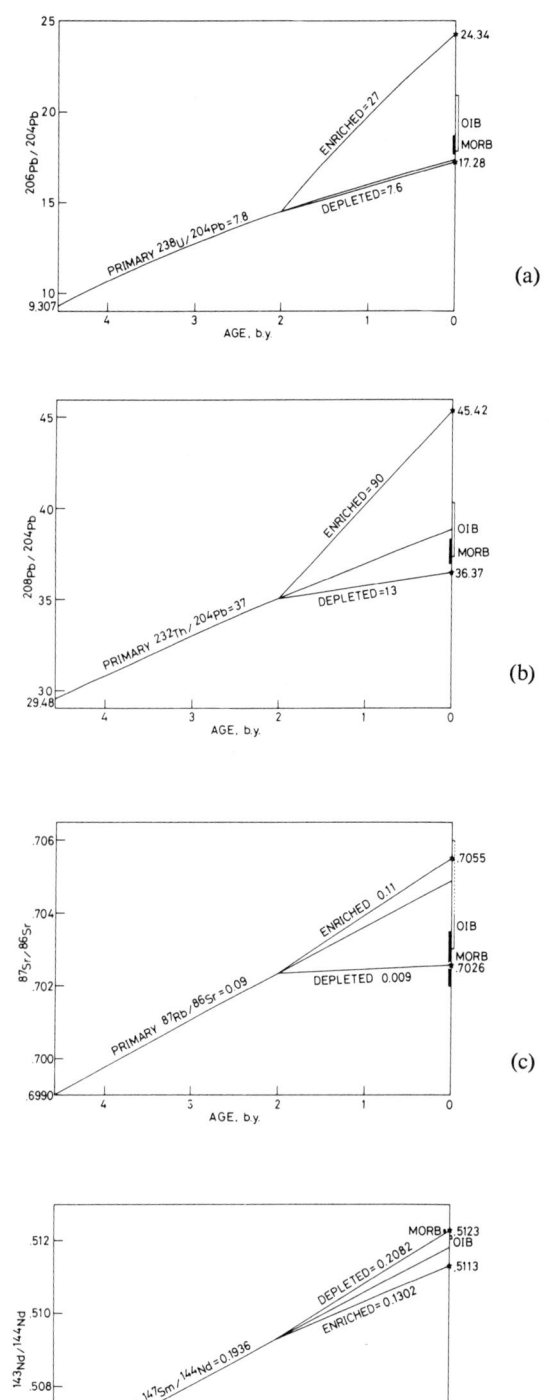

FIG. VII-1-6. Growth curves of radiogenic isotope ratios: (a) ^{206}Pb/^{204}Pb, (b) ^{208}Pb/^{204}Pb, (c) ^{87}Sr/^{86}Sr, and (d) ^{143}Nd/^{144}Nd. Solid bars at the age of zero represent the ranges of isotope ratios for mid-ocean ridge basalts (MORB), and open bars those for oceanic island basalts (OIB).

enriched liquid layer. Mixing ratios of the two end-members to make observed isotope ratios in oceanic basalts indicate that most of uranium in the asthenosphere is contained in the depleted asthenosphere, in spite of the fact that uranium is strongly enriched in the liquid layer. This means that the volume of the liquid layer is extremely small, as geophysical data require.

Growth of ^{208}Pb/^{204}Pb, ^{87}Sr/^{86}Sr and ^{143}Nd/^{144}Nd was calculated in the same way as the growth of ^{206}Pb/^{204}Pb, and is shown in Fig. VII–1–6(b), (c) and (d), respectively. The ^{232}Th/^{204}Pb, ^{87}Rb/^{86}Sr and ^{147}Sm/^{144}Nd for the depleted asthenosphere and the enriched liquid layer are obtained from Th/Pb, Rb/Sr and Sm/Nd ratios in the two source regions calculated from data given in Table VII–1–2. The primary ^{232}Th/^{204}Pb ratio was determined to satisfy the straight-line relationship of ^{208}Pb/^{204}Pb versus ^{206}Pb/^{204}Pb shown in Fig. VII–1–5(b). The primary ^{87}Rb/^{86}Sr is determined so that the primary-stage growth curve may run close to Hart and Brooks' estimate for the ^{87}Sr/^{86}Sr value in the mantle 2.6 b.y. ago (Hart and Brooks, 1970) and so that the present end-member value of the depleted asthenosphere may not be higher than the average ^{87}Sr/^{86}Sr of the mid-ocean ridge basalt. The primary ^{147}Sm/^{144}Nd is Lugmair's value used by DePaolo and Wasserburg (1976) and is in accord with the average Sm/Nd ratio of chondrite.

As ^{206}Pb/^{204}Pb ratios, all isotope ratios in oceanic basalts lie between the corresponding two end-member ratios generated in the depleted asthenosphere and the enriched liquid layer. Enrichment factors of U/Pb, Th/Pb, Rb/Sr and Sm/Nd in the liquid layer relative to the depleted asthenosphere are 3.5, 7.0, 11 and 0.6, respectively, as calculated from data given in Table VII–1–2. Variations of relative positions of observed isotope ratios and the end-member ratios seen in Fig. VII–1–6(a) ~ (d) are due to the different enrichment factors of parent and daughter elements for each decay system.

The ^{208}Pb/^{204}Pb, ^{87}Sr/^{86}Sr and ^{143}Nd/^{144}Nd ratios in oceanic basalts also show, as do ^{206}Pb/^{204}Pb data, that the source of mid-ocean ridge basalt is only weakly contaminated by the end-member radiogenic isotopes formed in the liquid layer, while the source of alkali-rich basalt is strongly contaminated by the other end-member generated in the deep asthenosphere. Estimated mixing ratios of the two end-members suggest that the mass of the liquid layer is very small relative to the entire asthenosphere.

The present model is obviously too simple. It does not take into account the evolution of continents. The formation of continents at a certain time during the primary stage of the present model should affect the parent-daughter element ratios in the asthenosphere. This effect is probably significant for the Rb/Sr ratio because Rb is extremely concentrated in the continental crust. The ^{87}Rb/^{86}Sr ratio in the asthenosphere may have gradually decreased during the primary stage. Then, the slope of the growth curve for ^{87}Sr/^{86}Sr given in Fig. VII–1–6 (c) should be steeper at the beginning and flatter near the end of the primary stage.

Not only trace element abundances but also Pb, Sr, and Nd isotope ratios in the mid-ocean ridge basalt and the alkali-rich basalt can be explained within a framework of this simple model without further complicated assumptions. The present model does not require recycling of crustal lead and strontium (Armstrong, 1968; Russel, 1972), nor does it assume the radiogenic, deep mantle plume (Sun and Hanson, 1975). It is geochemically similar to the dynamic, open-system model proposed by Church and Tatsumoto (1975). They suggested that the observed lead isotopic data arrays are the result of mixing as a direct consequence of mantle convection. The structure of the convecting mantle, however, is not clear in their model.

The present model clearly defines the location of the two source regions and the mechanism of separation. The two source regions in the present model are chemically comparable with the solid-type and liquid-type materials defined by Masuda (1966). It should be noted that the source of alkali-rich basalts exists as liquid at the top of the asthenosphere. No mineral phases for the source of large ion lithophile elements are needed to explain trace element abundances in alkali-rich basalts.

Conclusions

(1) In addition to the primary convective cells involving plate movement and return flow at depth, a secondary, small-scale convection is active in the asthenosphere.

(2) As a result of the small-scale convection, the asthenosphere is fractionated into a liquid layer enriched in volatiles and large ion lithophile elements at the top, and a residual layer depleted in these components at the base.

(3) The mid-ocean ridge basalt is formed by partial melting of the depleted layer upwelling beneath the mid-ocean ridge from the base of the asthenosphere. The source of the alkali-rich basalt is the enriched liquid layer at the top of the asthenosphere.

(4) Elements are kept fractionated between the two source regions by small-scale convection. Abundance ratios between the two source regions are determined by bulk-distribution coefficients of elements between solid and liquid.

(5) Abundances of large ion lithophile elements in the alkali-rich basalt are those in the liquid layer, and those in the mid-ocean ridge basalt are three times as great as those in the depleted layer.

(6) The enriched liquid layer has higher U/Pb, Th/Pb and Rb/Sr but lower Sm/Nd ratios than the depleted layer. More radiogenic Pb and Sr isotopes but less radiogenic Nd isotopes are generated in the liquid layer than in the depleted layer.

(7) Isotopes tend to be mixed, though elements are fractionated, by convection. Mixing is not complete and results in linear arrays of isotope ratio variations in the mid-ocean ridge basalt and alkali-rich basalt. The slope of linear array of Pb isotope ratios indicates that the separation of the two source regions occurred about 2 b.y. ago.

(8) Observed ^{206}Pb/^{204}Pb, ^{207}Pb/^{204}Pb, ^{208}Pb/^{204}Pb, ^{87}Sr/^{86}Sr and ^{143}Nd/^{144}Nd in the mid-ocean ridge basalt and alkali-rich basalt lie between the computed two end-member isotope ratios generated in the two source regions. Estimated mixing ratios of the two end-members require that the liquid layer mass be extremely small in comparison to the entire asthenosphere.

VII-2. Tectonophysics as Related to the Structural Features of Ore Deposits: The Role of the Fluid Intrusion in the Formation of Vein Fractures

Hitoshi Koide

Introduction

Ore deposits are the result of many kinds of tectonic processes, but most metallic ore deposits are formed directly or indirectly as the result of igneous activities. Imai (1966a, b) has recognized the following types of vein fractures after study of many ore deposits in Japan:
  (A) Fractures related to magmatic intrusion
    (a) Fractures due to magmatic upheaval
    (b) Fractures due to subsidence (including cauldron subsidence) related to magmatic intrusion
    (c) Joints in or around the igneous rocks due to solidification of magma
  (B) Fractures due to lateral force (horizontal compression or shear)

The early fracturing and later hydrothermal mineralization were conventionally treated as genetically unrelated events. However, recent investigations (Phillips, 1972; Raybould, 1974, 1976) suggest that hydrothermal solution under pressure is responsible for fracturing as well as for the successive process of mineralization. Mechanical study of fracturing by magmatic or hydrothermal intrusion is necessary for an understanding of the tectonic features of ore deposits.

Effect of Fluid Pressure on Fracturing of Rocks in the Crust and Mantle

A fracture system in rock provides the main passageway for extensive percolation of underground fluids (Brace, 1972). Fractures are generally classified as tension fractures or shear fractures. Tension fractures form open spaces in rock, while shear fractures in a narrow sense provide genetically no open spaces. However, pure shear fractures are rather rare. A fault zone—that is, a macroscopic shear fracture—usually contains numerous microscopic tension fractures (Koide, 1971) and shows dilatancy due to the opening of constituent microcracks (Frank, 1965). Tensile microcracks in fault zones provide the passageways for underground fluid percolation.

Usually, underground rocks are compressed in all directions due to the load of overburden rocks (Koide, 1976a). Tension cracks cannot be formed in completely homogeneous materials under total compression. However, rocks are heterogeneous and contain numerous planes of weakness and foreign inclusions. The Griffith theory is widely accepted as the model for the fracturing of rocks at about atmospheric pressure. In the Griffith theory, it is assumed that a new tension fracture is formed under locally high tensile stress which is induced around the tip of a pre-existing open crack.

Open penny-shaped cracks are closed under the following compressive stress (σ_c):

$$\sigma_c = \frac{\pi G}{2(1-\nu)s} + p \quad \quad (1)$$

where G is the rigidity of rock, ν is the Poisson's ratio, s is the aspect ratio of the penny-shaped crack (that is, the ratio of length to width) and p is the pressure of fluid within the crack. Cracks in rocks are closed under high lithostatic pressure but can remain open only if the interstitial fluid pressure exceeds or is close to the lithostatic pressure.

Local tensile stress in a macroscopically compressional stress field can be induced around boundaries of heterogeneity or planes of weakness as well as around open pre-existing cracks. The maximum tensile stress (σ_A) is induced at about the tip of various penny-shaped inclusions (Koide, 1976b, c):

$$\sigma_A = \tfrac{1}{2}(\sigma_t - \sqrt{\sigma_t^2 + b^2\tau^2}) + \sigma_L \quad \quad (2)$$

$$\sigma_t = \frac{8(1 - \nu G'/G)\{(1 - K'/K)(\sigma_L - p) - K'\Delta V'\}}{2\pi/s + 4(1-\nu)K'/G + 16(1-\nu)G'/3G} \quad \quad (3)$$

$$b = \frac{8(1 - G'/G)}{(2-\nu)\pi/s + 4(1-\nu)G'/G} \quad \quad (4)$$

where tensile stress is negative, σ_L is the lithostatic stress or tectonic stress normal to the plane of elongation of an inclusion, p is the internal pressure of the inclusion, τ is the tectonic shearing stress along the plane of elongation of the inclusion, s is the aspect ratio of the penny-shaped inclusion, ν is the Poisson's ratio of host rock, K is the bulk modulus of host rock, G is the rigidity of host rock, K' is the effective bulk modulus of the inclusion, G' is effective rigidity of the inclusion and $\Delta V'$ is the relative volume expansion of inclusion over host rock due to temperature change, phase change or dilatancy by microfracturing.

Equations (2) ~ (4) show that high local tensile stress is induced at about the tip of a penny-shaped inclusion when tectonic shearing stress, internal pressure and/or volume expansion are large enough to overcome lithostatic pressure with the help of stress concentration at the inclusion tip (Fig. VII–2–1). The coefficient of stress concentration due to shearing stress (b, Eq. (4)) is very large for an inclusion of small effective rigidity and becomes maximum in the case of a fluid inclusion ($G' = 0$). If the inclusion is a vacuum (then, $K' = G' = $

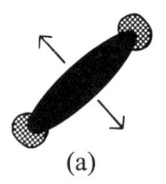

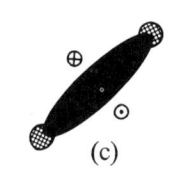

FIG. VII–2–1. Three types of stress release due to the expansion (a) and the sliding (b and c) of a Griffith crack or inclusion. The direction of displacement in (c) is perpendicular to the plane of the paper.

TABLE VII–2–1. The minimum value of volume expansion of a Griffith inclusion which can cause absolute local tensile stress in the mantle. The values of elastic moduli and lithostatic pressure in the mantle are after Clark and Ringwood (1964). Let the elastic moduli of a solid inclusion be the same as those of the mantle, for convenience.

Depth	Elastic properties of the mantle		Lithostatic pressure	Minimum volume expansion due to phase change which can cause local tensile stress(%)	
(km)	Rigidity (G,Mb)	Poisson's ratio(ν)	(kb)	From solid into solid	From solid into liquid
100	0.642	0.266	−31	4.2	1.77
200	0.674	0.285	−64	7.9	3.4
300	0.756	0.286	−98	10.8	4.6
400	0.871	0.283	−133	12.8	5.5
500	1.003	0.283	−169	14.1	6.0
600	1.182	0.282	−208	14.8	6.3
700	1.433	0.269	−248	14.9	6.3
800	1.649	0.267	−290	15.2	6.4

$p = 0$), Eqs. (2) ∼ (4) indicate the stress concentration around the conventional Griffith crack, which is unrealistic beneath the earth's surface. If fluids can easily flow out of or into a fluid inclusion through capillary tubes which are connected with an extensive fluid channel system, the pressure of internal fluid (p) is held nearly constant regardless of deformation of the individual inclusion. For such a drainable fluid inclusion, practically, $K' = G' = 0$, but p is controlled by regional fluid pressure. Underground fractures respond to quick deformation as undrainable fluid inclusions, but many fractures behave as drainable fluid inclusions under slow deformation.

The Griffith theory of fracture is based on a simple criterion: that new tension cracks begin where local tensile stress exceeds the tensile strength of the material (Griffith, 1925). Incompetent inclusions such as serpentine, mica or graphite are favorable fracture loci under compressive stress, because stress concentration due to shearing stress is high (G' is very small in Eq. (4)) but the effect of lithostatic pressure is relatively small (K' is not so small in Eq. (3)). Undrainable fluid inclusions extend most easily even under high lithostatic pressure. However, an extensive drainable fracture system cannot stay open unless interstitial fluid pressure is close to lithostatic pressure.

Eq. (2) indicates that relative expansion of a penny-shaped inclusion in comparison with the host rock induces tensile stress at the inclusion tip. The temperature change causes heterogeneous volume change due to the difference of thermal-expansion coefficients in rocks (Koide, 1976c). The quick large volume change by partial phase change could initiate tension cracks (Koide, 1974a, b). Table VII–2–1 shows the minimum volume expansion which causes absolute tensile stress at various mantle depths. As the tensile strength of rock is very low, tensile fractures can initiate in the mantle if the volume expansion slightly exceeds the value indicated in Table VII–2–1.

The calculations in Table VII–2–1 are based on Eq. (3) using the pressure and elastic moduli of the mantle estimated by Clark and Ringwood (1964). Phase transformations such as coesite-quartz ($\Delta V' = 10.3\%$), stishovite-quartz ($\Delta V' = 61.6\%$), spinel-olivine ($\Delta V'$

= 10.8%) and jadeite plus quartz-albite ($\Delta V' = 19.8\%$), can form tension fractures and trigger deep earthquakes at various mantle depths (Koide, 1974a, b).

Besides volume expansion, reduction of inclusion rigidity due to phase change also causes pressure release around a Griffith inclusion under the existence of tectonic differential stress (Eq. (4)). Partial melting or breakdown of hydrous minerals creates a thin film of fluids in the mantle. Fluid inclusions become very efficient nuclei of pressure release. Volatile materials tend to migrate into tension cracks. If temperature is sufficiently high, pressure release causes melting of the rock (Yoder, 1952; Uffen, 1959). The volatile materials concentrated in fractures also reduce the melting temperature of wall rock (Kushiro, 1976).

The criteria for formation of open cracks (Eqs. (1)~(4)) suggest that fluid phase and fracture systems coexist in the lower crust and mantle. Lithostatic pressure and fluid pressure in the fracture system increase with depth due to their own load (Fig. VII–2–2). The rate of pressure increase is proportional to their density. If the fluid density is lower than that of the host rock, the effective pressure (that is, lithostatic pressure minus interstitial fluid pressure) is lower at the top of a fracture system than at its bottom (Fig. VII–2–3). As fractures begin more easily under lower effective pressure, a fluid-filled fracture system tends to extend upwards ("buoyancy effect").

Sudden volume expansion by partial phase change forms fractures or pockets of melts in the uprising part of the mantle convection. The sinking oceanic plate takes hydrous minerals and carbonates into the mantle. The breakdown of hydrous minerals and carbonates in the sinking plate creates fractures which contain aqueous solution or CO_2 gas. The fluid-filled fracture system propagates upwards under shearing stress due to plate subduction. Volatile fluids intrude through the fracture system into the high-temperature mantle over the sinking plate. The melt-contained fracture system is formed in the low-velocity zone of the mantle. By the buoyancy effect, diapir of magma or crystal mush, which is a mixture of melt and fragments of rocks, intrudes from the mantle into the lithospheric plate.

Control of Fracture Pattern by Effective Pressure

Brace and Bombolakis (1963) have described: "Whereas in tension the critical crack, once started, will grow in the long axis until a free boundary or obstruction is reached, the growing critical crack in compression curves out of the direction of its initial long axis, approaches the direction of compression and finally stops after having travelled the distance of a few crack lengths or less." Fractures develop discontinuously under compression and form a zone of *en echélon* fractures (Koide and Hoshino, 1967; Koide, 1971). A fracture zone consists of numerous tensile and shearing microcracks (Koide and Hoshino, 1967). Tension fractures grow longer under lower effective pressure. Microscopic tension fractures are formed around the tip of a Griffith inclusion but cannot extend longer under high effective pressure (Koide, 1971; Koide *et al.*, 1974). Fractures in rocks show great variety in size. The size of an original Griffith inclusion affects the length of a fracture. However, a longer tension fracture is formed under lower effective stress from the same Griffith inclusion. As the width of a fracture zone is controlled by the length of constituent tensile microcracks, a wider fracture zone is formed under lower effective pressure (Koide, 1971). Tension fractures are predominant in a fracture zone which is formed under lower effective pressure, whereas an *en echélon* set of shear fractures is formed under high effective pressure (Fig. VII–2–4). Many vein fractures show *en echélon* arrangement of tension fractures (McKinstry, 1948; Koide, 1971). On the other hand, large tension fractures are rare in ordinary fractures which were formed under little or no influence of hydrothermal or igneous activity. An *en echélon* set of shear fractures is observed in many earthquake faults (Koide and Bhattacharji, 1977). The difference in fracture patterns between vein fractures and earthquake faults is due to the difference in effective pressure at the time when the fractures were formed. The pres-

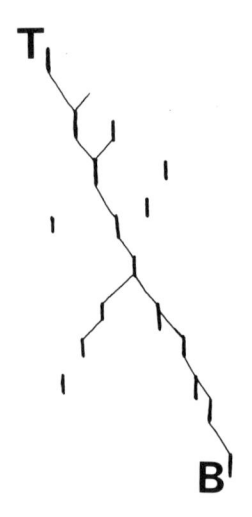

306   2. Tectonophysics as Related to the Structural Features of Ore Deposits

Fig. VII-2-2. Fluid-filled fracture system.

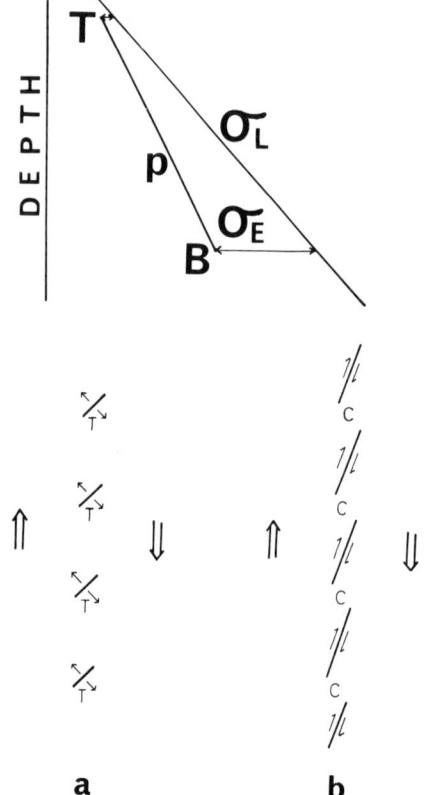

Fig. VII-2-3. Effective pressure (σ_E) at the top (T) and at the bottom (B) of a fluid-filled fracture system (cf. Fig. VII-2-2).

Fig. VII-2-4. (a) An *en echélon* set of tension fractures. Arrows show the direction of displacement. (b) An *en echélon* set of shear fractures.

ence of hydrothermal solution under high pressure reduced effective pressure and caused the formation of dilatant vein fractures.

As fracture zones are weaker than flawless surrounding rocks, further deformation and fracturing concentrate into fracture zones (Koide and Hoshino, 1967). Fractures which are

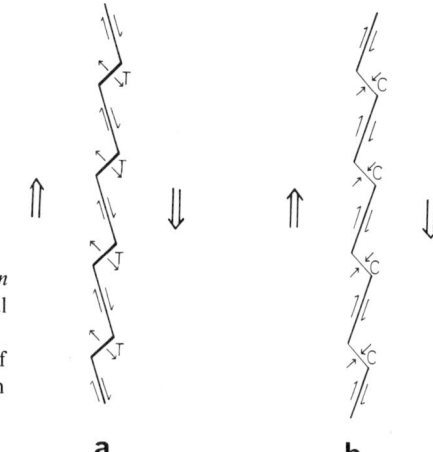

FIG. VII–2–5. (a) Zigzag macrofault resulting from an *en echélon* set of tension fractures. "T" indicates the extensional segments.
(b) Zigzag macrofault resulting from an *en echélon* set of shear fractures. "C" shows the compressional segment which is a thrust or secondary shear fracture.

arranged *en echélon* within a fracture zone are connected by further fracturing and form a larger shear fault. The large shear fault shows a zigzag or feather pattern due to the connection of *en echélon* fractures (Koide, 1968). Segments of two different directions are aligned alternatively to form a zigzag macrofault (Fig. VII–2–5). Segments at a lower angle to the general trend of the macrofault are characteristically shear fractures ("main shear segments"). Segments at a higher angle to the general trend are originally tension fractures in a macrofault which has been formed from an *en echélon* set of tension fractures (Fig. VII–2–5). On the other hand, thrusts or secondary shear faults create higher-angle segments in a macrofault system resulting from an *en echélon* set of shear fractures (Fig. VII–2–5). Gaps among *en echélon* shear fractures are highly compressed due to displacement along shear fractures (Figs. VII–2–4 and –5).

Wide veins are formed in extensional segments of one type of zigzag fault ("normal step-like pattern") which was derived from *en echélon* tension fractures (Fig. VII–2–5). However, wide veins cannot be expected in compressional segments of another type of zigzag fault ("reverse step-like pattern") which was derived from *en echélon* shear fractures (Fig. VII–2–5). As rocks are stronger under compression than under tension, great strain energy is accumulated before the fracturing of compressional gaps of *en echélon* shear fractures. The final break of a compressional gap in earthquake faults generates a large earthquake (Koide and Bhattacharji, 1977).

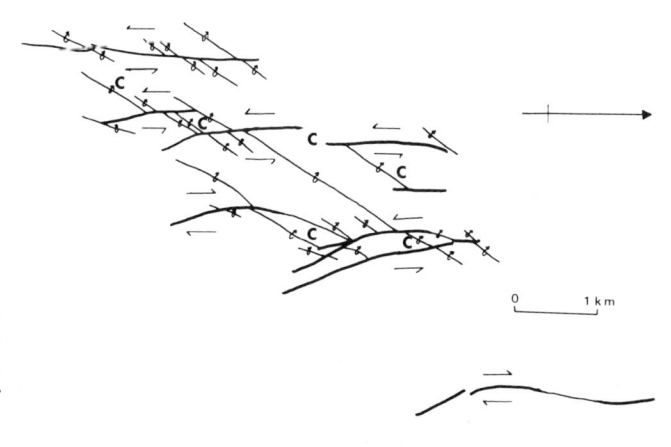

FIG. VII–2–6. Simplified mechanical relations among vein fractures at the Taishu mine. Reverse faults are formed in compressional zones (c) due to displacement of strike-slip faults (after Matsuhashi, 1968) (refer to Chapter II–6).

Vein fractures usually show *en echélon* tension or a normal step-like pattern. However, the larger regional structure of vein fractures sometimes shows reverse step-like patterns. Narrow mineralized thrust faults are observed in compressional gaps of the main vein fractures of the Taishu mine (Fig. VII–2–6).

Fractures Related to Underlying Stock-like Intrusion

Diapirs of magma or crystal mush intrude and deform the earth's crust. Some physical constraints in the crust prevent the uplift of magmatic intrusions. Shallow crustal rocks are usually porous and have low bulk density. The buoyancy effect is reversed and instead prevents upward intrusion of magma if the density of the host rock is lower than that of the magma. Magma is cooled by low-temperature crustal rocks and tends to be solidified or to increase in viscosity.

Usually, a porous shallow crustal layer contains a lot of water. A water-logged layer presents an efficient "surface barrier" against the passage of magma (Walker, 1974), because ground water cools the magma body. Ground water convection occurs around the magmatic intrusion and forms hydrothermal ore deposits (White, 1968). It is likely that many magmatic intrusions are prevented from reaching the ground surface by surface barriers. The subsurface magmatic intrusions form cryptobatholiths, crypto-stocks, concealed dikes or sills.

Koide and Bhattacharji (1975a, b) studied the mechanical effect of subsurface magmatic intrusion by the three-dimensional theory of elasticity. It is assumed that stress in rock mass is isotropic and homogeneous except for the vicinity of intrusion.

If the horizontal projection of the magma body is circular, radial and concentric fractures are formed over the intrusion. If the magma body is nearly spherical, the excess magma pressure over lithostatic pressure forms a domal uplift and radial fractures in the roof rock (Fig. VII–2–7). Wisser (1960) has given many examples of vein fractures associated with doming in the North American Cordillera region. Predominant radial fractures accompany discontinuous concentric fractures in the Sunlight mining district, Wyoming (Parsons, 1937).

Local stress and pore pressure control zonal distribution of fractures around the magma body. Pore pressure is very high near the magma reservoir wall due to volatile emanation, heated ground water and magma pressure. The extremely high pore pressure induces a zone of continuous tension fractures around a magma reservoir (Fig. VII–2–8). Abnormally high pore pressure decreases to normal pressure at a distance from the magma reservoir. A zone of "brittle faults", which is characterized by *en echélon* sets of tension fractures or a normal

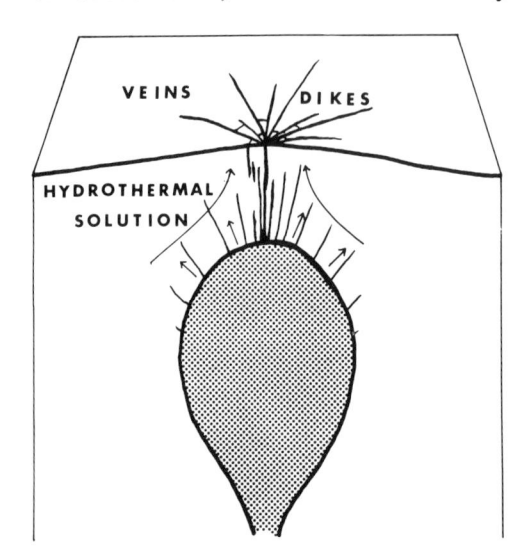

FIG. VII–2–7. Radial fractures and dome above nearly spherical magma reservoir. Short straight arrows indicate stress directions. Long curved arrows show probable flow directions of hydrothermal solutions.

FIG. VII-2-8. Zonal distribution of modes of fractures around a stock-like intrusion (Koide and Bhattacharji, 1975a). RT, CT, RS and CS denote zones of radial continuous tension fractures, concentric continuous tension fractures, radial brittle faults and concentric brittle faults, respectively. RT·CT indicates that radial tension fractures are predominant over concentric fractures; and CT·RT, vice versa. The CS·RS and RS·CS zones are favorable sites for hydrothermal veins.

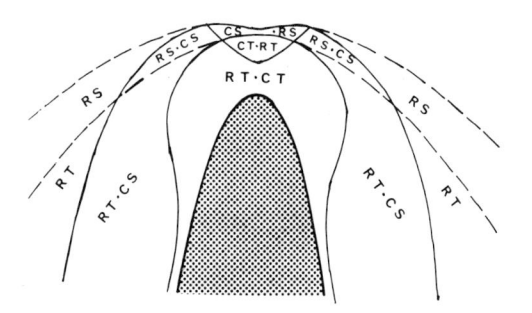

step-like pattern of faults, takes place around the zone of continuous tension fractures and on the inside of the fracture-free zone (Fig. VII-2-8). Magma can penetrate only into those continuous tension fractures with wide openings, while hydrothermal solution or gas can penetrate even into narrow capillary tubes. Dikes are usually formed in the continuous fracture zone, while hydrothermal veins are likely to be formed in the brittle fault zone (Fig. VII-2-8). Radial dikes and outer zone of hydrothermal mineralization around a top of igneous stock are observed in the Climax molybdenum deposit, Colorado, U.S.A. (Wallace et al., 1968).

If the magma reservoir is circular in horizontal sections but extends deep—that is, if it is of prolate spheroidal stock-like shape—concentric fractures are formed predominantly over radial fractures on the inside of a funnel-shaped zone above the top of the intrusion (Fig. VII-2-9). Radial fractures are predominant on the outside of the central funnel-shaped concentric fracture zone.

The crust is dilated outwards by the stock-like intrusion, because magma pressure pushes the side wall of the reservoir outwards. The outward spreading of the base causes extension in the center of the overlying rocks. Gregory (1921) explained this wedging effect by a "keystone analogy." The extension due to wedging of an underlying stock-like intrusion forms vertical ring tension fractures and cauldron subsidence along funnel-shaped ring normal faults in the center of domal uplift (Fig. VII-2-9).

Anderson (1936) analyzed stress distribution only around a circular magma reservoir in two dimensions. The stock-like magma reservoir is a more realistic model of diapiric intrusion. Billings (1943, 1945) has concluded from the study of ring dikes in New Hampshire that initial ring fractures are essentially vertical tension fractures due to the upward pressure of magma. Reynolds (1956) and Yokoyama (1974) suggested that cauldron subsidence is bounded by funnel-shaped faults. These field studies support the theory that ring fractures are formed due to wedging of an underlying stock-like intrusion of magma.

Imai (1966a, b) has found that many vein-type ore deposits in Japan are genetically related to cauldron subsidence and underlying igneous intrusion. The vein-type deposits of the Ashio mine are located within and in the vicinity of a funnel-shaped rhyolite body. Imai et al. (1975) have found an inward dipping concentric normal fault which partly surrounds the funnel-shaped rhyolite body of the Ashio mine. It is likely that the mineralized concentric normal fault in the Ashio mine is a remnant of funnel-shaped faults bounding a cauldron which has been occupied by the successive intrusion and outflow of rhyolitic magma. In the Yaso mine, steeply inward-dipping rhyolitic dikes surround a cauldron subsidence (Takenouchi, 1962a). Lead, zinc, copper, gold and silver mineralization took place in and near calderas in the San Juan volcanic field, southwestern Colorado, U.S.A. Steven et al. (1974) noticed in the San Juan volcanic field that the mineralization most commonly takes place in fractures extending outward from the caldera walls, and that a zone several kilometers wide around the periphery of a mineralized caldera generally contains the most, the largest and the highest-grade deposits.

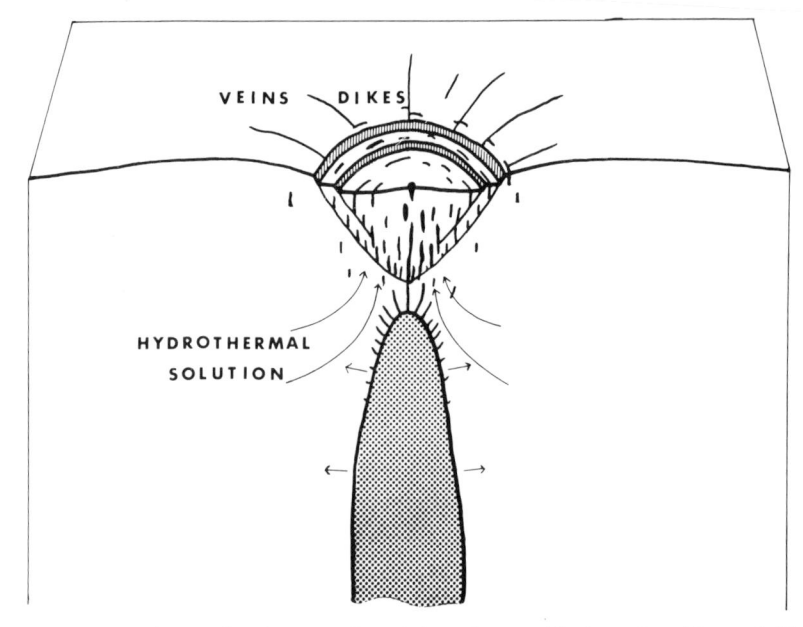

FIG. VII–2–9. Vertical ring tension fractures, funnel-shaped normal faults and cauldron subsidence over a stock-like intrusion of magma with outer radial fractures and domal uplift. Short arrows indicate stress directions. Long curved arrows show flow directions of hydrothermal solutions.

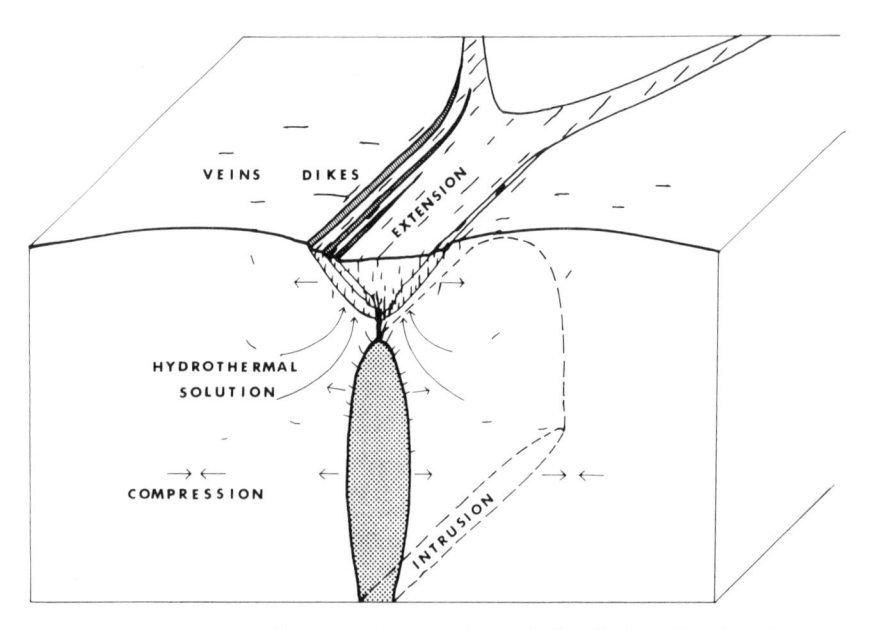

FIG. VII–2–10. Formation of a rift system due to wedging of dike-like intrusion of an igneous mass into the lithospheric plate (Koide and Bhattacharji, 1975b). Graben subsidence occurs inside the funnel-shaped normal faults. Subsurface dike-like intrusion makes areas of dike intrusion parallel and perpendicular to the main rift valley and areas of compression and extension, domal uplift and central graben subsidence. Short arrows indicate stress directions. Long curved arrows show flow directions of hydrothermal solution.

Variation of pore pressure or multiple intrusions often disturbs the ideal zonal distribution of mineralization. Tectonic differential stress, heterogeneity and anisotropy in the rocks cause deviations from the ideal fracture pattern. Many vein fractures are irregular but formed in

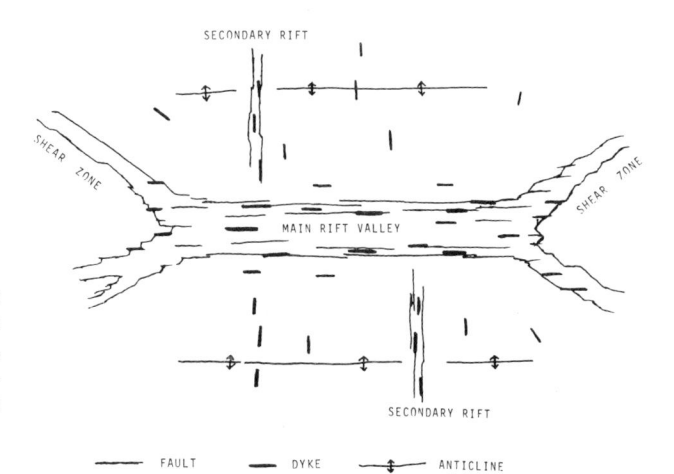

FIG. VII–2–11. Plan view of rift system. The rift system bifurcates into a set of strike-slip fault zones. The parallel anticline and perpendicular secondary rift to the main rift valley are shown.

and around domal uplift or cauldron subsidence due to an underlying stock-like intrusion of magma. The carbonatite complex often shows a typical concentric or radial fracture pattern (Takenouchi, 1973). The ideally concentric or radial fracture pattern suggests that the fluid pressure of the carbonatite intrusion was very high in comparison with lateral tectonic stress.

Graben and Rift Structure Related to Underlying Dike-like Intrusions

If the horizontal section of a stock-like intrusion is not exactly circular or if the stress is not isotropic, an elliptical basin may be formed over a diapiric intrusion of magma. The vein fractures of the Ikuno mine are formed in an oval-shaped volcanic depression (Imai et al., 1975). Thompson (1959) suggested that subsidence in the Basin and Range province, U.S.A., is bounded by a set of funnel-shaped normal faults. Koide and Bhattacharji (1975b) analyzed mechanically a graben subsidence due to an underlying dike-like intrusion of magma which is elongated vertically and horizontally (Fig. VII–2–10). The extension due to the wedging effect of an underlying dike-like intrusion forms a long trough-like subsidence bounded by a set of funnel-shaped normal faults and a surrounding gentle wide domal uplift (Fig. VII–2–10).

A graben depression bifurcates into a pair of strike-slip fault zones (Fig. VII–2–11). The vertical dike-like intrusion of magma splits and pushes aside wall rocks and causes transverse horizontal compression, which causes gentle domal uplift on the crust. The transverse compression produces sills or transverse dikes in the wall rock. A large subsurface transverse dike may form a secondary graben which is perpendicular to the trend of the main graben (Fig. VII–2–11). Such a graben-dome structure is observed on many rift systems such as Rhine graben, Red Sea rift, East African rift and Icelandic rift (Koide and Bhattacharji, 1975b; Bhattacharji and Koide, 1975).

Kutina (1972) has pointed out that hypogene mineralization tends to concentrate in belts along rift structures. Hydrothermal mineralization has been observed in the trough of the Red Sea rift (Degens and Ross, 1969). Hydrothermal deposits and mouths of the hydrothermal sources were directly observed in a transform fault zone by a recent undersea study of the Mid-Atlantic Ridge (ARCYANA, 1975). Hydrothermal circulation occurs through fractures around subsurface magmatic intrusions, and hydrothermal mineralization takes place in and around grabens or rift valleys.

Summary and Conclusion

The theory of fracture in rocks suggests the coexistence of fluids and fractures in the lower crust and mantle. The fracture pattern is controlled by the effective pressure. Continuous tension fractures are formed in the zone where fluid pressure exceeds lithostatic pressure. *En echélon* sets of tension fractures or normal step-like patterns of faults are formed under low effective pressure. Small tension fractures in the brittle fault zone provide passageways for hydrothermal solution and favorable sites for the formation of wide hydrothermal veins. *En echélon* sets of shear fractures or a reverse step-like pattern of faults are formed under high effective pressure and often trigger large earthquakes due to fracturing in compressional segments.

Phase changes such as dehydration, decarbonating and melting initiate fractures in the upper mantle. A deep fracture system which contains hydrothermal solutions or magma extends upwards due to the buoyancy effect. Magma or crystal mush intrudes into the crust through a deep fracture system.

A nearly spherical magma reservoir with excess pressure over lithostatic pressure produces radial fractures and domal uplift over the intrusion. A stock-like intrusion forms ring fractures and outer radial fractures. Cauldron subsidence takes place on the inside of funnel-shaped normal ring faults in the center of domal uplift. A subsurface vertical dike-like intrusion produces graben subsidence within a set of funnel-shaped normal faults in the center of gentle domal uplift.

The convection of hydrothermal solution is induced over a subsurface igneous intrusion. Hydrothermal mineralization takes place along fractures around a magmatic intrusion. Many hydrothermal ore deposits are found in and around a dome, caldera, graben or rift system which was formed by the intrusion of magma beneath it.

The structure of ore deposits is usually complicated due to heterogeneity of crustal rocks and later tectonic deformation. The underground fluids, hydrothermal solution and magma, play a major role in the genesis of ore deposits. More detailed investigation is necessary in order to understand the tectonic control of ore deposits and to find helpful guides to the prospecting of ore deposits.

VII-3. Fluid Inculsion Studies

Sukune Takenouchi

Introduction

The studies of fluid inclusions in minerals have developed recently into an important field in mineralogy, geology and geochemistry research. Many years have passed since H. Davy carried out the first analytical study of a large fluid inclusion in 1823 and H.C. Sorby mentioned the principle of fluid inclusion geothermometry in 1858 (Smith, 1953). Since then, many scientists and geologists have engaged in the study of inclusions in minerals, especially from the point of view of geothermometry. These reports and results are reviewed or summarized elsewhere, for instance, by Smith (1953), Yermakov (1965), Roedder (1967, 1972), and, in Japan, by Tatsumi (1951), Imai (1956a), Takenouchi (1968, 1975b, 1976a), Yajima (1969), Enjoji (1970) and Enjoji and Takenouchi (1976).

The first study of fluid inclusions in Japan was done by Kinoshita (1924) on gypsum from the Ishigamori Kuroko deposit. Later, Nishio *et al.* (1953) and Takahashi *et al.* (1955b) studied the decrepitation method, and Kashiwagi *et al.* (1955) and Takenouchi (1962b) compared the decrepitation temperature with the filling temprature measured by the heating-stage method.

Recent studies on fluid inclusions are not only measurements of homogenization temperature but are also concerned with studying the salinity of fluid inclusions by means of the freezing-stage microscope.

Furthermore, the recent development of isotopic studies of water and carbon dioxide has stimulated geochemists to turn their attention to studies of fluid inclusions, and some geochemical laboratories in Japan have started experimental work to obtain information on the origin of fluids in inclusions.

Though the volume of fluid inclusions in minerals is generally very small, inclusions offer a variety of information on the temperature, pressure, chemical composition and origin of fluids in the earth's crust. At present, however, it is very hard to obtain precise information from an individual inclusion because of the extremely small volume and the complicated occurrences of fluid inclusions. Many problems in the study of fluid inclusions are left to be solved in the future.

Homogenization Temperatures

The homogenization temperature of fluid inclusions is defined as the temperature at which the constituent phases homogenize to a single phase upon heating. In the case of liquid inclusions, the gaseous phase shrinks upon heating and the inclusion homogenizes to a high-density liquid phase, whereas in the case of gaseous inclusions, the liquid phase disappears and the inclusion homogenizes to a low-density gaseous phase.

In some cases of polyphase fluid inclusions, it is difficult to obtain a homogenized state because some solids do not dissolve at an ordinary heating rate. In the case of highly saline inclusions, there are two cases of homogenization: one is by the disappearance of the bubble and the other is by the solution of salts, generally halite. The homogenization temperature of a liquid inclusion, which is also called the filling temperature, is one of the important data obtained from fluid inclusions, along with their salinity.

Heating-stage method

At present, the heating-stage method is the most reliable technique for measuring the homogenization temperature of fluid inclusions, from which it is possible to infer the formation temperature of crystals. Since H.C. Sorby suggested the principle of fluid inclusion geothermometry, there have been many discussions and reports on the premises and reliability of the method.

Roedder and Skinner (1968) discussed the leakage of fluid inclusions and concluded that in most cases inclusions retain all their original materials except hydrogen. Though there are many premises for the estimation of the formation temperature of an inclusion, the pressure correction of the homogenization temperature is the most important and difficult problem. In special cases, such as inclusions trapped in a "boiling" condition, the measured temperature will directly indicate the formation temperature without any pressure correction. Therefore, most of the data show homogenization temperature without correction for the pressure at the formation of crystals.

Enjoji and Takenouchi (1976) summarized the data on the homogenization temperature and salinity of fluid inclusions in minerals from vein-type deposits in Japan. The number of vein deposits studied is as high as 68. A photograph and design of the heating-stage designed by Prof. H. Imai and made by Union Optical Co. are shown in Fig. VII–3–1.

Generally speaking, the homogenization temperatures of fluid inclusions from gold-silver, lead-zinc and copper veins of Tertiary age are nearly in the same range, showing that no characteristic difference in temperature exists among these types. The temperatures of tin, tungsten and molybdenum hypothermal veins of Cretaceous age are apparently higher than those of the above-mentioned three types of veins. Some examples of homogenization temperatures which were determined by the author are shown in Table VII–3–1.

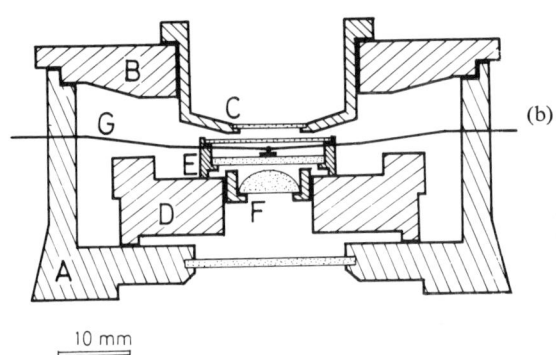

FIG. VII-3-1. Photograph (a) and section (b) of MHS-3 type heating-stage.
A: cell, B: lid, C: observation window, D: heating plate, E: inner cell, F: condenser lens, G: chromel-alumel thermocouple.

TABLE VII-3-1. Homogenization temperatures of fluid inclusions (liquid inclusions) from various ore deposits.

Deposit	Location	Miner.	Temp. range	Ore	Geol. age
Chitose	Hokkaido	Qz	190–300°C	Au–Ag	Tertiary
Sado	Niigata	Qz	245–305	Au–Ag	Tertiary
Oppu	Aomori	Qz	200–270	Pb–Zn	Tertiary
Asahi	Niigata	Sp	260–270	Pb–Zn	Tertiary
Toyoha	Hokkaido	Qz	150–260	Pb–Zn	Tertiary
Toyoha	Hokkaido	Sp	180–220	Pb–Zn	Tertiary
Taishu	Nagasaki	Qz	210–370	Pb–Zn	Tertiary
Yaso	Fukushima	Qz	180–270	Cu	Tertiary
Osarizawa	Akita	Qz	245–270	Cu	Tertiary
Suzuyama	Kagoshima	Qz	200–360	Sn	Tertiary
Hirase	Gifu	Qz	190–310	Mo	Cretaceous
Takatori	Ibaragi	Qz	225–350	W	Cretaceous
Takatori	Ibaragi	Cs	300–350	W	Cretaceous
Takatori	Ibaragi	Tp	330–350	W	Cretaceous
Takatori	Ibaragi	Fl	215–270	W	Cretaceous
Ohtani	Kyoto	Qz	225–360	W	Cretaceous
Ohtani	Kyoto	Cs	285–340	W	Cretaceous
Ohtani	Kyoto	Sh	270–330	W	Cretaceous
Kaneuchi	Kyoto	Qz	225–315	W	Cretaceous
Kaneuchi	Kyoto	Sh	276–318	W	Cretaceous
Kaneuchi	Kyoto	Wf	286–337	W	Cretaceous
Kuga	Yamaguchi	Qz	190–267	W	Cretaceous
Kiwada	Yamaguchi	Qz	200–265	W	Cretaceous
Ashio	Tochigi	Qz	200–350	Cu	Cretaceous(?)

Qz: quartz, Sp: sphalerite, Tp: topaz, Cs: cassiterite, Fl: fluorite, Sh: scheelite, Wf: wolframite.

FIG. VII-3-2. Photograph (a) and section (b) of FS type freezing-stage.
A: cell, B: inner cell, C, D: observation window with defrosting heater, E: inner glass, F: pipe for coolant, G: thermister thermometer.

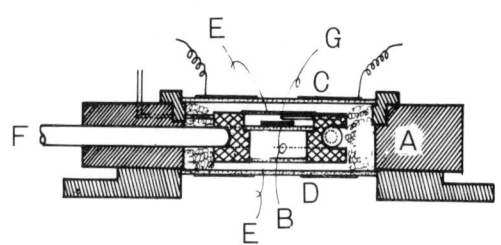

Decrepitation method

The decrepitation method was first reported by Scott (1948), and the decrepitoscope method was advocated by Deicha (1951). These methods are not as reliable as the heating-stage method for obtaining the formation temperatures of fluid inclusions. Takenouchi (1962b) discussed the relationship between the homogenization temperature and decrepitation temperature of some minerals, and recognized some relations between them, but as there are so many problems to be solved in order to use the decrepitation temperature for the determination of the formation temperature of minerals, it is difficult to correlate the decrepitation temperature with the formation temperature, or even with the homogenization temperature. Shiobara (1961) determined the decrepitation temperature of garnet, hedenbergite and sphalerite from the Kamioka mine, as described later.

Salinity

The salinity of liquid inclusions (high-density fluid inclusions) can be measured by the freezing-stage method, and that of highly saline inclusions which contain crystals of halite and sylvite can be determined by the heating-stage method.

Freezing-stage method

The freezing-stage method has been developed to measure the salinity of aqueous solution in an individual inclusion without destroying the inclusion. This can be done under the microscope by measuring the melting point of ice formed in the aqueous solution of the inclusion. Though the salt contained in the aqueous solution is mainly NaCl, other chlorides such as KCl or $CaCl_2$ are also involved. Therefore, the melting point of the ice is not that of the pure NaCl aqueous solution, and the salinity of fluid inclusions measured by the freezing-stage method should be expressed as NaCl equivalent concentration.

In the United States, the freezing-stage method was developed by Cameron *et al.* (1952), Weis (1953) and Roedder (1962). Roedder (1963) reported the salinity of fluid inclusions in sphalerite from the Ani mine, Akita Prefecture, as 12% NaCl equivalent. In Japan, the first type of freezing-stage was made by the Union Optical Co. in cooperation with Professor H. Imai of the University of Tokyo. In this stage, the Peltier effect was utilized for the cooling mechanism, but the stage was not suitable for the study of fluid inclusions. Later, the company made a second type (FS type), in which alcohol cooled by dry ice was used as the coolant. The Nippon Optical Co. has also provided a freezing-stage (NE type) which uses liquid nitrogen as the coolant.

The FS type freezing-stage consists of a freezing cell, storage tank and circulation pump for the coolant, as shown in Fig. VII–3–2. The freezing cell consists of an outer cell made of stainless steel and an inner cell of brass. A loop of copper pipe for the circulation of coolant is set between the outer and inner cells. The observation glass of the cell has a silver-plated line heater to prevent frost on the glass surface. The volume of the inner cell is kept small to keep the temperature gradient caused by the convection as small as possible. A thermister thermometer having a temperature range between $-55°$ and $+55°C$ with a scale interval of $0.5°C$ is used for temperature measurement. The temperature control of the cell is carried out manually by adjusting the flow rate of the coolant.

The salinity of aqueous solutions can be read as NaCl equivalent concentrations from the H_2O-NaCl phase diagram at low temperatures (Roedder, 1962). From the phase diagram, it is known that the lowest temperature of the melting point of ice is $-21.1°C$, and that the maximum salinity of aqueous solution which can be measured by the freezing method is 23.3 wt. % of NaCl.

Salinity of polyphase inclusions

High saline inclusions carry a cubic crystal of halite at room temperatures. This means that the salinity of the inclusion is much higher than 26 wt. %, the saturated solubility of NaCl in aqueous solution at room temperature. In such cases, the salinity can be measured by the heating-stage instead of the freezing-stage, or by calculation from the volume of halite and saturated aqueous solution.

If the temperature at which the halite crystal dissolves to aqueous solution is known, it is possible to determine the salinity from the temperature using the data of Keevil (1942). However, Keevil's data are only applicable to the disappearance temperature of halite which coexists with the saturated aqueous solution and vapor phase. Therefore, when the halite crystal disappears at a higher temperature than the bubble does, the obtained salinity from Keevil's data will be higher than the true salinity; but no data on the pressure effect on the solubility of NaCl in this condition are available.

Besides a crystal of halite, a small crystal of sylvite (KCl) is occasionally found in highly saline inclusions. In this case, it is possible to measure the concentration of KCl as well as NaCl by the heating-stage method. Using the ternary phase diagram of the H_2O-NaCl-KCl system (Roedder, 1971a), the approximate concentrations of NaCl and KCl in the aqueous solution of the inclusion can be obtained from their disappearance temperature, but, as in the former case, the assemblage in the inclusion should include the vapor phase—that is, the vapor bubble in the inclusion should exist at the disappearance temperatures of these two salt crystals. These data are especially useful for inferring the ratio of K and Na in ore-forming fluids of post-magmatic ore deposits.

Temperature and Salinity Measurements at the Takatori Tungsten-Quartz Veins

The geology of the Takatori mine is described in Chapter II–3.

Fluid inclusions are found in quartz, fluorite, cassiterite, topaz and rhodochrosite. Most inclusions are liquid inclusions, but some of them are CO_2-rich fluid inclusions or polyphase fluid inclusions containing a minute transparent mineral.

Quartz crystals especially contain a large number of liquid inclusions which are suitable for the heating-stage and freezing-stage methods. The results of temperature and salinity measurements are shown in Figs. VII–3–3 and –4, respectively. The temperature-salinity relation which is determined on the same fluid inclusion, is shown in Fig. VII–3–5. From these three figures, it is inferred that the temperature and salinity of fluid inclusions in quartz vary in a wide range but are generally lower than those of cassiterite. Figure VII–3–5 also shows

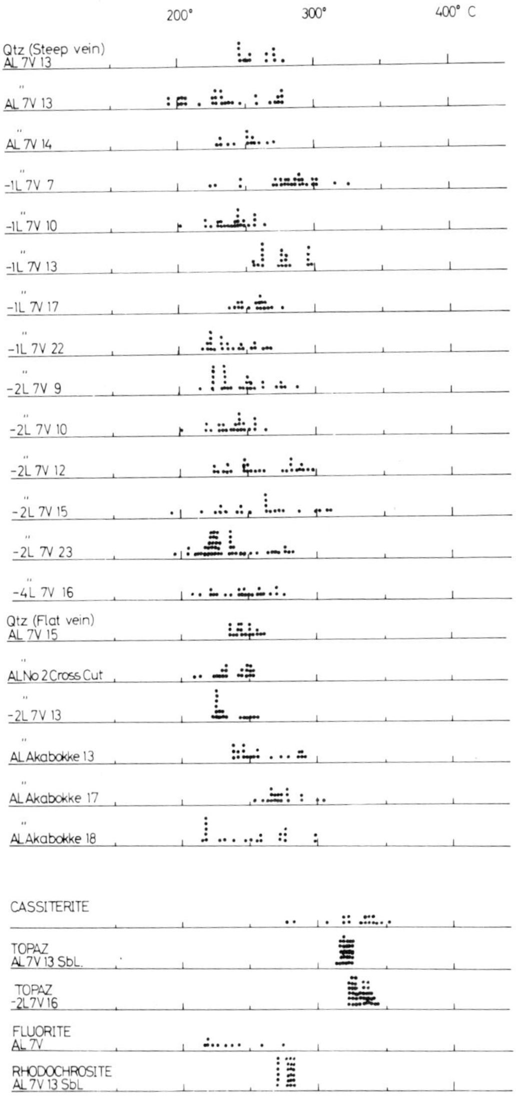

FIG. VII–3–3. Filling temperatures of fluid inclusions in quartz and other minerals from the Takatori wolframite-quartz veins.

that the general trend of distribution is inclined to both the temperature and salinity axis, showing that the higher the temperature, the higher the salinity.

Topaz occurs in flat veins with quartz forming comb-structures. The temperature of topaz is generally higher than that of quartz, but the salinity distributes in a lower range than quartz.

CO_2-rich Fluid Inclusions

The presence of CO_2 in fluid inclusions has been known since the time of the early fluid inclusion studies. According to the review of Smith (1953), in 1823 Brewster recognized the second liquid phase, which had a larger coefficient of expansion than that of water, besides the aqueous phase in fluid inclusions in topaz. Later, in 1869 Sorby identified this second liquid as liquid CO_2 from the physicochemical data on CO_2 at temperatures near the critical point. In 1876, Zirkel recognized liquid CO_2 in granites and gneisses from the western United States, and in 1869, Vogelsang also recognized it in granites and gneisses from the Alps and Ireland.

318   3. Fluid Inclusion Studies

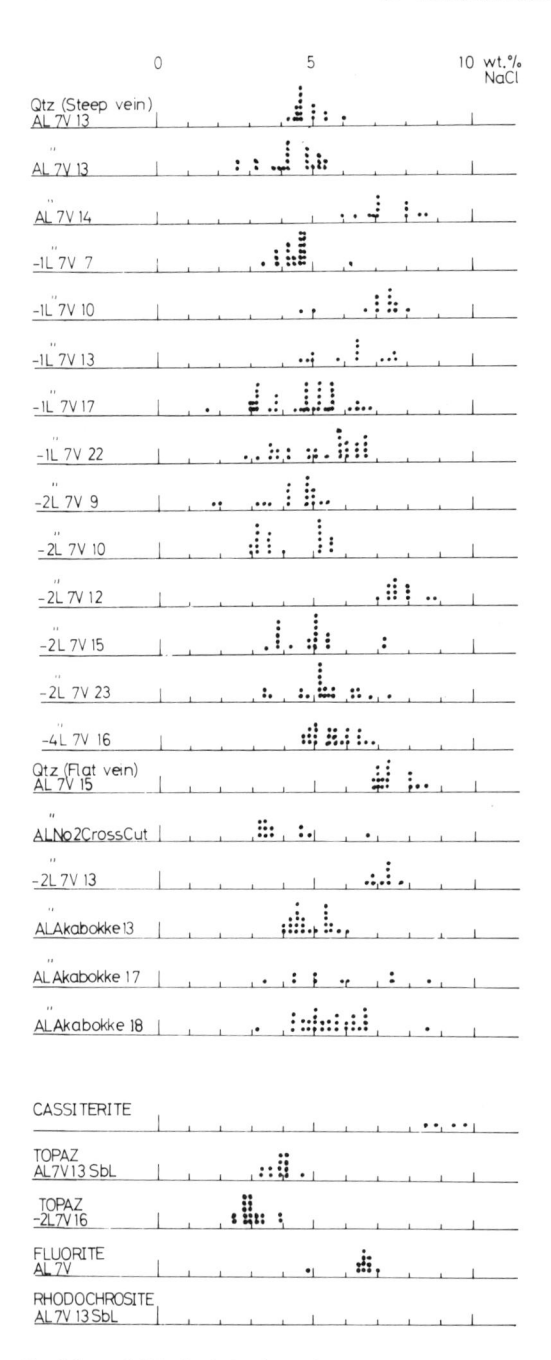

FIG. VII-3-4. Salinity of fluid inclusions in quartz and other minerals from the Takatori wolframite-quartz veins.

Smith and Little (1959) and Little (1960) studied fluid inclusions of the H_2O-CO_2 system and concluded that this type of inclusion is commonly found in minerals of the late stages of pegmatite formation or high-temperature hydrothermal veins. They discussed the physico-chemical conditions of the ore-forming fluids of hydrothermal tin deposits.

Roedder (1963) reported the presence of CO_2 in fluid inclusions in quartz from mercury deposits, pegmatites and Alpine quartz veins, and in beryl. Furthermore, Roedder (1965) investigated olivine nodules in basic and ultrabasic rocks and found that most of the samples contained CO_2. Roedder and Coombs (1967) also reported the presence of CO_2 in granitic rocks from Ascension Island in the Atlantic Ocean.

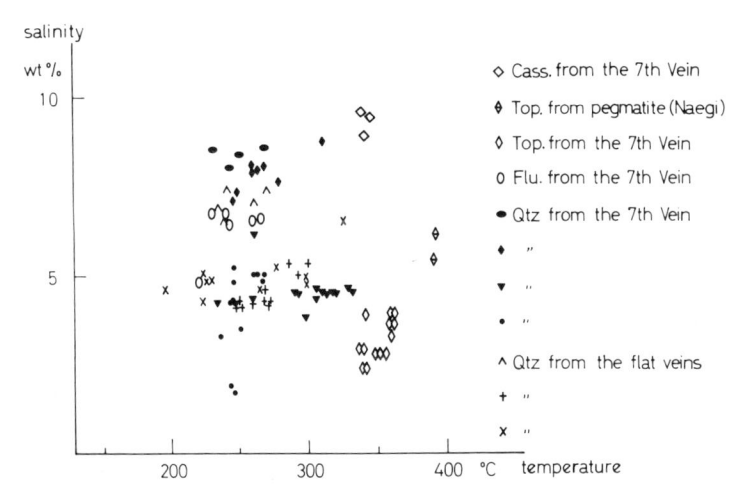

Fig. VII-3-5. Filling temperature-salinity relation of fluid inclusions in quartz and other minerals from the Takatori wolframite-quartz veins.
Cass: cassiterite, Top: topaz, Flu: fluorite, Qtz: quartz.

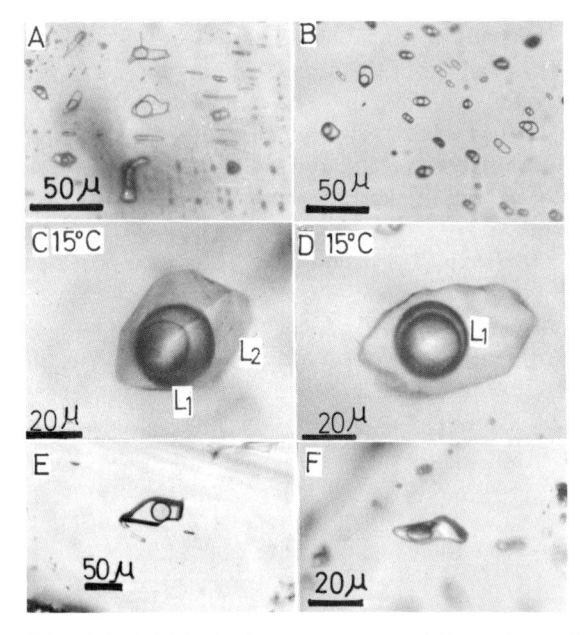

Fig. VII-3-6. Microphotographs of fluid inclusions in various minerals.
(A) Primary liquid inclusion in topaz (Takatori).
(B) Primary liquid inclusion in rhodochrosite (Takatori).
(C) CO_2-rich fluid inclusion (type C) in quartz (Ohtani).
(D) CO_2-rich fluid inclusion (type B) in quartz (Ohtani).
(E) Liquid inclusion in cassiterite (Takatori).
(F) Liquid inclusion in scheelite (Kaneuchi).

CO_2-rich fluid inclusions are reported from tin and tungsten deposits at various places of the world. Sushchevskaya and Ivanova (1967) and Naumov and Malinin (1968) have studied tungsten quartz veins in Transbaikal; Elinson et al. (1969) studied tungsten quartz veins in East Tian-Shan, and Pomirleanu (1968) studied scheelite from Tincova, Rumania. Groves and Solomon (1969) studied tin deposits in Tasmania, and Kelly and Turneaure (1970) studied tin deposits in Bolivia.

From these results, it can be said that CO_2-rich fluid inclusions are frequently observed in cassiterite or scheelite, suggesting an intimate relation between the H_2O-CO_2 ore-forming fluids and tin and tungsten mineralization.

H_2O-CO_2 system at low temperatures

The appearance of liquid CO_2 or CO_2 hydrate at low temperatures will indicate the presence of some amounts of CO_2 in a fluid inclusion, and it is possible to discuss semi-

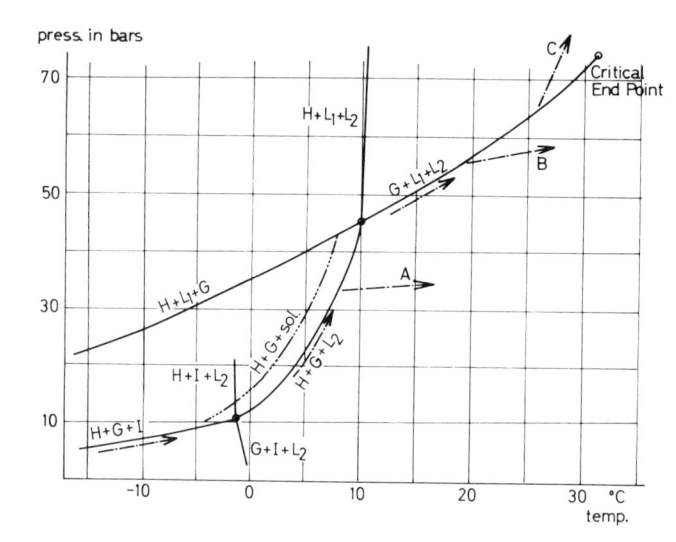

FIG. VII-3-7. Phase diagram of the system H_2O-CO_2 at low temperatures.
L_1: liquid CO_2, L_2: aqueous solution, G: gaseous phase, H: CO_2 gas hydrate, I: ice. Arrows shown by A, B, and C are referred to in the text.

quantitatively the content of CO_2 in an inclusion from the data on the disappearance temperature of these CO_2 phases and the phase diagram of the H_2O-CO_2 system at low temperatures.

The phase diagram of the H_2O-CO_2 system at low temperatures which was studied by Larson (1956) is shown in Fig. VII-3-7. As is known from the phase diagram, the CO_2 hydrate is stable at temperatures lower than approximately 10°C in the range of pressure in fluid inclusions—that is, at pressures lower than several tens of bars. As the solubility of CO_2 in the aqueous phase is very low at this range of temperature, the univariant curve of the gas-liquid CO_2-aqueous solution assemblage is located in almost the same position as the vapor pressure curve of CO_2.

The phase changes which occur in a CO_2-rich fluid inclusion with the increase of temperature are classified as one of three types.

The first type, represented by arrow A in Fig. VII-3-7, is less abundant in CO_2 than the other types. When a CO_2-rich fluid inclusion is frozen, the hydrate, ice and gaseous phase are formed. With the increase of temperature, the inner pressure of the inclusion—that is, the equilibrium pressure of the hydrate, ice and gas—changes along the univariant curve HGI up to the invariant point at −1.5°C and 10.3 bars. Here, ice melts to form the aqueous solution (L_2), and then the inner pressure of the inclusion changes along the other univariant curve, shown by HGL_2. Finally, at a point on the curve, the hydrate disappears completely, and the aqueous solution and gas bubble form the constituent phases of the inclusion. The inner pressure of the inclusion then changes with temperature along arrow A.

In the case of the second type, which contains more CO_2 than type A does, the hydrate exists stably up to the second invariant point at 10°C and 44.6 bars, where the hydrate melts to form liquid CO_2. In this case, as the volume of liquid CO_2 (L_1) formed is small, it forms a thin layer around the gas bubble of the inclusion. At a point on the univariant curve GL_1L_2, the thin layer of liquid CO_2 disappears, and the aqueous solution and gas bubble consitute the inclusion. At temperatures higher than this point, the inner pressure of the inclusion will change along arrow B.

The third type contains much CO_2, and when the hydrate melts at the second invariant point, the formed volume of liquid CO_2 is great. In this case, the gas bubble is smaller and seems to be floating in a globule of liquid CO_2. At a point on the univariant curve GL_1L_2, the gas bubble homogenizes to the liquid CO_2 phase. Above this temperature, the increase of the inner pressure with temperature is very rapid compared with the former two cases.

In the case of types B and C, the temperature at which the liquid CO_2 phase disappears

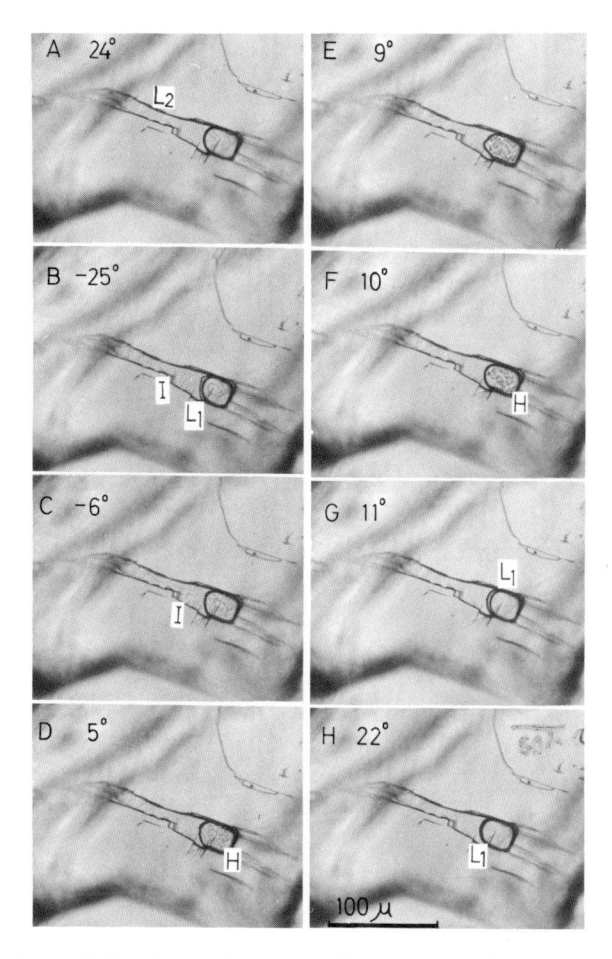

FIG. VII-3-8. Phase change of a CO_2-rich inclusion at low temperatures (rhodochrosite from the Takatori mine). The abbreviations are the same as those in Fig. VII-3-7.

indicates semi-quantitatively the content of CO_2 in the inclusion. In the case of type C, a lower temperature shows a higher content of CO_2, but in the case of type B, a lower temperature shows a lower content of CO_2.

Formation of CO_2 hydrate in inclusions

A fluid inclusion which contains several wt. % of CO_2 is generally two-phase (liquid and gas), and the liquid CO_2 phase is not observed at room temperature. When it is cooled, however, a small amount of liquid CO_2 or CO_2 hydrate will be formed around the gas bubble. The phase change of a CO_2-rich fluid inclusion in rhodochrosite from the Takatori mine is shown in Fig. VII-3-8. When the fluid inclusion is frozen at $-25° \sim -30°C$, ice and liquid CO_2 are observed, besides the deformed gas bubble. After a while, liquid CO_2 disappears and CO_2 hydrate is formed around the bubble at $-15° \sim -6°C$. The hydrate exists stably even above $0°C$, but at $10°C$ it disappears rapidly to form liquid CO_2, as is known from the phase diagram of the H_2O-CO_2 system. This case belongs to type B. In general, the formation of CO_2 hydrate is stagnant, and it is most frequently observed at temperatures between $-6°$ and $-8°C$ during the cooling of an inclusion.

Estimation of CO_2 content in inclusions

It is possible to estimate the approximate content of CO_2 in a fluid inclusion under the microscope, if the volume of each phase in the inclusion is known. The content of CO_2 in each phase can be estimated from the phase diagram if the temperature is fixed. However, as the estimation of volume under the microscope is very error-prone, the data obtained by this method give us only approximate values.

322   3. Fluid Inclusion Studies

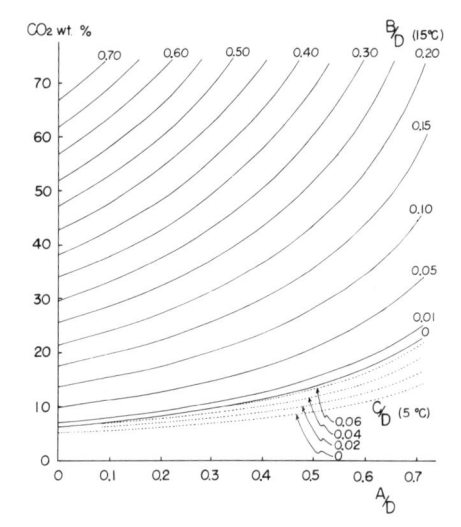

FIG. VII-3-9. Diagram showing the relation between the CO_2 concentration and the volume ratios of the constituent phases in a CO_2-rich fluid inclusion.
 A: volume of bubble; B: volume of liquid CO_2 at 15°C; C: volume of CO_2 hydrate at 5°C; D: volume of inclusion.

An approximation of the CO_2 content in a fluid inclusion can be obtained graphically using the diagram shown in Fig. VII-3-9. To use this diagram, it is necessary to know the volume of inclusion and gas bubble, and the volume of liquid CO_2 at 15°C or that of CO_2 hydrate at 5°C.

Occurrences of CO_2-rich fluid inclusions

Occurrences of CO_2-rich fluid inclusions have been reported from some ore deposits in Japan. Most of them are tungsten-quartz veins (Takenouchi and Imai, 1971).

Takatori mine: CO_2-rich fluid inclusions are found in quartz, topaz and rhodochrosite.

In quartz, inclusions of this type are both primary and secondary in origin. Some of them are liquid inclusions, but the others are gaseous inclusions. The CO_2 concentration varies, but, in general, inclusions of high CO_2 concentration are found in quartz of late stages, especially in rock crystals from druses (Fig. VII-3-6).

Topaz also contains CO_2-rich primary inclusions, but CO_2 hydrate is the only phase which indicates the presence of CO_2 when it is cooled. The homogenization temperatures of the inclusions in topaz are in the range between 320° and 350°C, and the salinities range between 2.5 and 5.0 wt. % (Fig. VII-3-4).

In rhodochrosite, two kinds of fluid inclusions are recognized. The first group includes the primary inclusions, in which liquid CO_2 is not recognized at low temperatures. They show negative crystal shapes and are distributed parallel to growth zones of the crystal (Fig. VII-3-6). The homogenization temperatures range from 260° to 300°C. Fluid inclusions of the second group are secondary in origin. They show irregular flat shapes, and, when cooled, a small amount of liquid CO_2 appears around the bubble. The phase change of this type of inclusion is shown in Fig. VII-3-8. From the volume of liquid CO_2, it is inferred that the CO_2 concentration is 7 ∼ 9 wt. %.

Ohtani and Kaneuchi mines: The geology of the area and its fluid inclusion problems are described in Chapter II-4.

CO_2-rich inclusions are found in quartz of late stages. It is supposed, from their shape and occurrence, that the inclusions are mostly primary in origin, but the CO_2 concentration varies greatly (Fig. VII-3-6). Some inclusions consist of the aqueous solution and liquid CO_2 without having the bubble at room temperature. From the volume of the CO_2 phase, it is estimated that the CO_2 concentration of the inclusion is approximately 11 ∼ 20 wt. %. The inclusions which accompany liquid CO_2 are often found in quartz from the drusy parts of veins but not in massive vein quartz. The homogenization temperatures of CO_2-rich inclusions are within 220° ∼ 280°C and the salinity is within 0.9 ∼ 3.5 wt. %. The above-stated homogenization temperatures are lower than those of vein quartz, and the salinities are within the lowest range of values of the deposit. Compared with the ore deposit of the Takatori mine, inclusions in quartz from druses contain more CO_2 at the Ohtani mine.

In the Kaneuchi mine, the occurrence of CO_2-rich inclusions is similar to that of the Ohtani mine, suggesting that the ore-forming fluids were rich in CO_2 in the late stages of mineralization.

Taishu mine: The geology and fluid inclusion problems of this mine are described in Chapter II–6.

CO_2-rich fluid inclusions are found in veins which are located far from the granitic rock genetically related to the mineralization. Some inclusions in places furthest from the granitic rock consist of the aqueous solution and liquid CO_2 without the bubble, suggesting a remarkably high CO_2 concentration. It is concluded from the study of CO_2-rich inclusions that the CO_2 concentration of inclusions increases while the salinity decreases with distance from the granitic rock.

Akagane mine: The geology and fluid inclusion study of the Akagane mine is described in a later section; the occurrence of CO_2-rich inclusions in granitic rocks and porphyries and in a scheelite-bearing quartz breccia pipe is described here (Takenouchi, 1975a).

CO_2-rich inclusions occur in both the Sanjinsha and the Ohatano granodioritic stocks with polyphase fluid inclusions and liquid inclusions. They are generally so small that an oil immersion objective of 100 magnification is necessary for observation. These inclusions, belonging to type B or C of CO_2-rich fluid inclusions, show simple shapes and are distributed randomly in quartz grains (Fig. VII–3–10).

The numeral ratio of CO_2-rich inclusions to the total number of inclusions found in the same crystal is estimated at about 30 ~ 40% in the Sanjinsha stock and 10 ~ 20% in the Ohatano stock.

In porphyries near the granodioritic stocks, CO_2-rich fluid inclusions are commonly found, though the numeral ratio of CO_2-rich inclusions is generally 10 ~ 20%. Inclusions of type B and C are observed along presumed fractured planes in quartz phenocrysts.

The Koganetsubo gold-scheelite deposit is a breccia pipe formed in the gabbroic complex. It is located about 500m to the south of the Sanjinsha granodioritic stock. The breccia pipe has an areal size of about 50 m × 30m and plunges steeply to the southeast as deep as about 160m from the surface (Fig. VII–3–11).

The breccia pipe consists of a great number of breccias of gabbroic rocks and matrices of

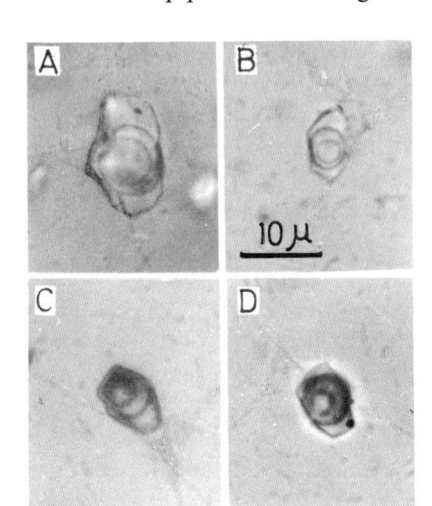

FIG. VII–3–10. CO_2-rich fluid inclusions in quartz and scheelite from the Akagane mine.
   (A) CO_2-rich fluid inclusion (type B) in quartz from Ohatano granodiorite.
   (B) CO_2-rich fluid inclusion (type C) in quartz from Ohatano granodiorite.
   (C) CO_2-rich fluid inclusion (type C) in scheelite from the Koganetsubo breccia pipe.
   (D) CO_2-rich fluid inclusion (type C) in scheelite, bearing an opaque mineral (Koganetsubo breccia pipe).

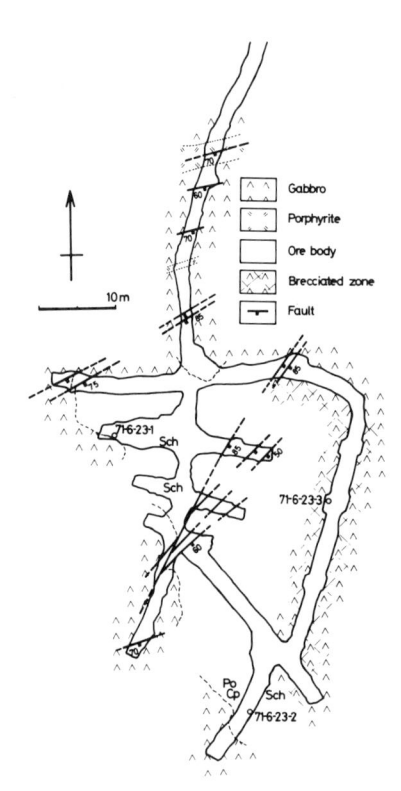

FIG. VII-3-11. Underground geologic map of the Koganetsubo breccia pipe, Akagane mine.

quartz. In the marginal zone of the pipe, the structure of brecciation gradually changes to a network structure of quartz veins in the gabbroic rock. In the marginal zone, it is clear that the displacement of breccias is very small because the walls of quartz veins are nearly concordant.

Several faults trending NE-SW are recognized especially in the northwestern and central parts of the pipe. Brecciation is especially remarkable between these faults and their southeastern side. From field observations, it is inferred that the Koganetsubo breccia pipe was formed by an intensive replacement of gabbro caused by the hydrothermal solutions which passed through fissures and joints developed along the faults. The brecciation might have been stimulated by a slight subsidence caused by the solution of gabbro.

Scheelite is found in quartz along the walls of gabbroic breccias with some pyrrhotite, chalcopyrite, bismuthinite, native bismuth and a small amount of native gold.

CO_2-rich fluid inclusions are observed in both quartz and scheelite. In general, the size of inclusions is smaller than 10 micrometers, and the shapes are simple or negative crystal. CO_2-rich inclusions belonging to type C are distributed randomly in the crystal of quartz and scheelite, suggesting that they are primary in origin, but those of type B are distributed in quartz along planes, indicating that they are secondary in origin. The estimated concentration of CO_2 in inclusions of type C is approximately 35 ~ 50 wt. %, and that of type B is less than 30 wt. %.

From fluid inclusion studies of the Koganetsubo breccia pipe of the Akagane mine, it is inferred that the ore-forming fluids varied in fluid density and that the CO_2 concentration was high in the early stages of mineralization. It is also deduced that the CO_2-rich ore-forming fluids of the Koganetsubo deposit would have been supplied from some deeper places in the granodioritic complex, because CO_2-rich fluid inclusions are found in the two granodioritic stocks, especially in the Sanjinsha stock, suggesting an intimate genetic relation between the breccia pipe and stock.

Crushing-stage method

In the case of fluid inclusions which do not show any CO_2 phases, it is difficult to check the presence of CO_2 by means of the freezing-stage method. In general, fluid inclusions contain some gases, and, in some cases, the pressure reaches as high as a few tens of bars. CO_2 occupies a large percentage of the total amount of gases.

For the detection of gas pressure in fluid inclusions, the crushing-stage method is a very sensitive technique. Various types of crushing-stages were reported by Deicha (1950), Shugurova (1968) and Roedder (1970). A slightly modified design after Shugurova's is shown in Fig. VII-3-12. A small chip of sample containing a fluid inclusion is inserted in the glycerin-filled space between the upper and lower observation glasses and crushed gently by pressing the upper glass. As the surface of the upper observation glass is convex downward and that of the lower is concave upward, the sample plate is easily crushed by the slight advance of the upper glass. When a fracture reaches the inclusion, the gases are released into the glycerin and form a bubble in it (Fig. VII-3-13). Measuring the diameters of the bubble before and after crushing under the microscope, the volume of released gases and the gas pressure in the inclusion can be calculated. Some data on fluid inclusions from the Takatori and Ohtani mines are shown in Table VII-3-2.

The apparatus for this technique is simple. Measurement of the gas pressure of an individual inclusion is possible and sensitivity is very high. However, the accuracy of this method is fairly low because relative error in the calculation of gas volume is as high as three times that in the measurement of the diameter of the bubble. Therefore, data obtained by means of the crushing-stage method give us merely the order of gas volume or inner pressure of the inclusion.

When a bubble released in glycerin is transmitted to a KOH aqueous solution, it is possible to calculate the volume of CO_2 in the bubble from the decrease in the bubble's diameter owing to the absorption of CO_2 into the solution.

Fluid Inclusions in Acidic Igneous Rocks

Acidic volcanic, hypabyssal and plutonic rocks often occur in proximity to hydrothermal ore deposits. Some of these rocks are considered to be genetically related to the mineralization from their geological settings.

Many fluid inclusions are found in quartz phenocrysts of rhyolite and porphyry, and in quartz grains of granite from mining areas (Takenouchi and Imai, 1975). Some of them are

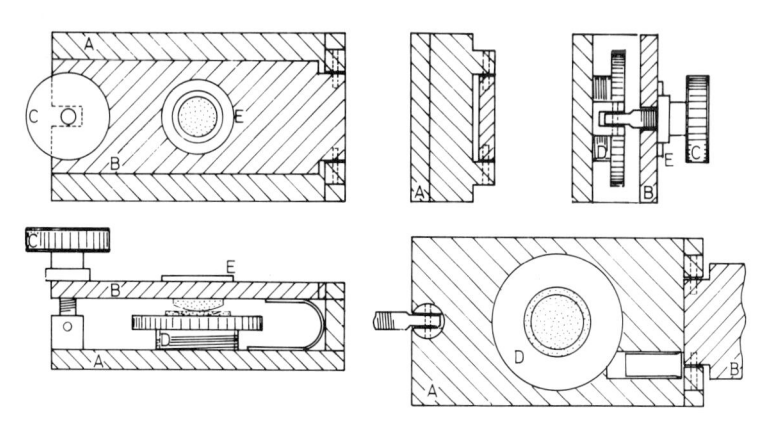

FIG. VII-3-12. Sketch of crushing stage.
A: base, B: pressure plate, C: screw, D: lower concave observation glass, E: upper convex observation glass.

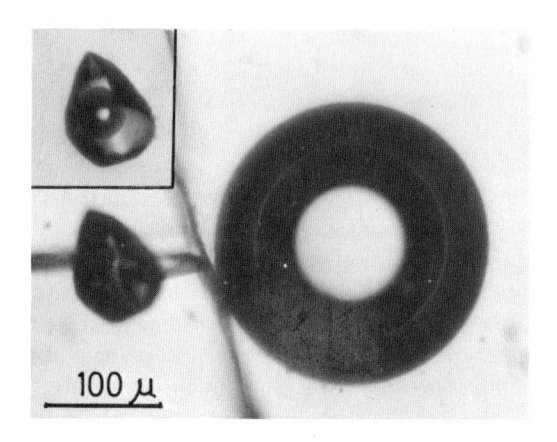

FIG. VII-3-13. Comparison of bubble before and after crushing of inclusion (liquid inclusion in quartz from the Takatori mine).

TABLE VII-3-2. Results of the crushing method.

Sample		Diameter I mm	Diameter II mm	Volume μl	Vol. ratio
Takatori No. 7 Vein					
Quartz	1–A	2.2×10^{-2}	4.7×10^{-2}	5.6×10^{-5}	10
	–B	1.6	3.6	2.4	12
	–C	4.0	7.2	19	6
	–D	2.0	4.2	3.7	9
	–E	1.4	2.3	0.6	4
	2–A	1.2	5.1	6.8	69
	–B	1.2	4.4	4.3	52
	–C	1.0	3.6	2.5	52
	–D	1.5	4.9	6.3	36
	–E	1.4	4.7	5.5	44
Rhodochrosite					
(primary)		1.4	5.1	6.8	51
		4.7	15.1	170	32
Ohtani					
Quartz	–A	1.6	4.7	5.3	24
	–B	1.5	4.4	4.5	26
	–C	2.1	5.9	11	21
	–D	1.7	3.1	1.5	58
	–E	1.8	3.9	3.1	11
Taishu, Shintomi					
Quartz	–A	1.3	2.6	0.9	8
	–B	2.0	5.0	6.6	17
	–C	1.8	4.3	4.1	14
	–D	1.8	4.6	5.2	16
	–E	2.6	7.8	25	27
Osarizawa, Tanosawa					
Quartz	–A	0.7	2.1	0.5	27
	–B	1.8	2.9	1.2	43

Diameter I and II: diameter of bubble in fluid inclusion and that released in oil, respectively.
Volume: volume at 1 atm. and room temperature.
Vol. ratio: volume ratio of bubbles before and after crushing.

highly saline inclusions. These inclusions are generally small in size, ranging from 10 to a few microns in diameter. They are distributed randomly or along planes in quartz crystals, and their shapes vary from irregular forms to negative crystals. The modes of occurrence of these inclusions suggest their secondary origin. Most highly saline and gaseous inclusions, and liquid inclusions which homogenize at relatively high temperatures, have shapes of negative crystal or partly faceted forms.

RHYOLITE, CHITOSE

RHYOLITE, ASHIO

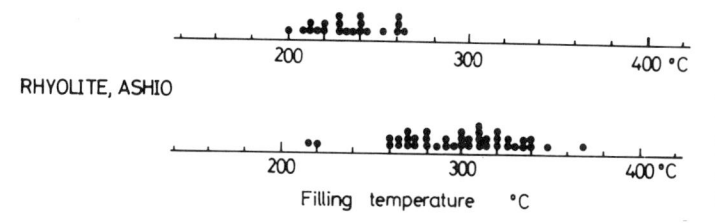

Fig. VII–3–14. Filling temperatures of liquid inclusions in phenocrysts of rhyolite from the Chitose, Hokkaido, and Ashio mines, Tochigi Prefecture.

If these fluid inclusions represent trapped ore-forming fluids which passed through the rocks, the study of them, correlated with studies of fluid inclusions from the associated ore deposits, would give us significant information concerning the nature of ore-forming fluids. Fluid inclusions in these rocks are so small that ordinary microscopic methods of fluid inclusion study are generally inapplicable. However, microscopic observation of the fluid density, concentration of solid phases and CO_2 content in the inclusions could give us some information on the physicochemical conditions of the fluids.

Fluid inclusions in rhyolite

Rhyolites from the Chitose and Ashio mines are characterized by the presence of liquid and/or gaseous inclusions and the absence of highly saline inclusions. Liquid inclusions in the Chitose rhyolite, having a degree of filling of about 0.8, homogenize at 200° ～ 260°C. No gaseous inclusions were found with liquid inclusions.

On the other hand, liquid inclusions in the Ashio rhyolite often occur with gaseous inclusions. In some parts, the degree of filling of fluid inclusions varies from that of liquid to gaseous inclusions, suggesting the "boiling" condition of fluids. Liquid inclusions having the highest degree of filling in the same group homogenize at 215° ～ 375°C (Fig. VII–3–14). In general, the fluid inclusions showing negative crystal shapes homogenize at higher temperatures than the inclusions of irregular shapes.

The difference in filling temperatures of these fluid inclusions would represent the difference in the conditions of hydrothermal fluids between these two ore deposits. The ore-forming fluids in gold-silver-quartz veins of the Chitose mine were of high density and intermediate temperature. On the other hand, the fluids of the Ashio xenothermal copper veins were at high temperatures and intermittently under a "boiling" condition.

Fluid inclusions in granitic rocks

The occurrence of fluid inclusions in granitic rocks genetically related to the mineralization are classified into two types. The first type is characterized by the presence of liquid inclusions. The second type is characterized by having highly saline inclusions as well as liquid and gaseous inclusions.

The granitic rocks of the Fujigatani and Kiwada area, Yamaguchi Prefecture, are an example of the first type. The geology and the results of fluid inclusion studies are described in Chapter II–4. Fluid inclusion studies show that the thermal condition of this area was kept nearly constant during the hydrothermal activity, though the salinity of fluids varied. Recent drilling exploration at the Fujigatani mine revealed that the ore deposit is located above the shoulder of a flat-topped batholith. The homogeneous thermal condition of the area probably depends on this geologic setting. It is presumed that the ore-forming fluids were fed from deeper places.

The granitic rock of the Taishu mine, Nagasaki Prefecture, is a good example of the second type. The results of a fluid inclusion study are described in Chapter II–6. Here, the granitic rock contains a great number of polyphase fluid inclusions, and their temperature and salinity are in the highest ranges in this area. In the ore deposit, the temperature and salinity decrease in accordance with distance from the granitic stock, suggesting that the ore-forming fluids were diluted by fluids of low salinities during their migration.

Fluid inclusions in porphyries

A large number of polyphase, gaseous and liquid inclusions is found in quartz phenocrysts of porphyries collected from mineralized areas. In general, they are so small that ordinary heating and freezing methods are not adequate. Microscopic observation, however, reveals some information on the variations of ore-forming fluids ascending through the rocks.

Suzuyama mine: The Suzuyama tin deposit is located to the southwest of Kagoshima City in Kyushu. Cassiterite-bearing quartz veins occur in the Cretaceous formation, consisting of sandstone and shale. It generally strikes NE-SW and dips 50°N. At several places, the formation is intruded by stocks of Miocene granodiorite porphyry (12 ~ 15 m.y., Nozawa, 1968b), in a NW-SE direction. In the center of the Suzuyama mineralized area, a small stock of the porphyry is situated in an area of 200 m × 300 m. The Cretaceous formation is thermally metamorphosed to hornfels for a horizontal distance of several hundred meters from the stock (Fig. VII–3–15).

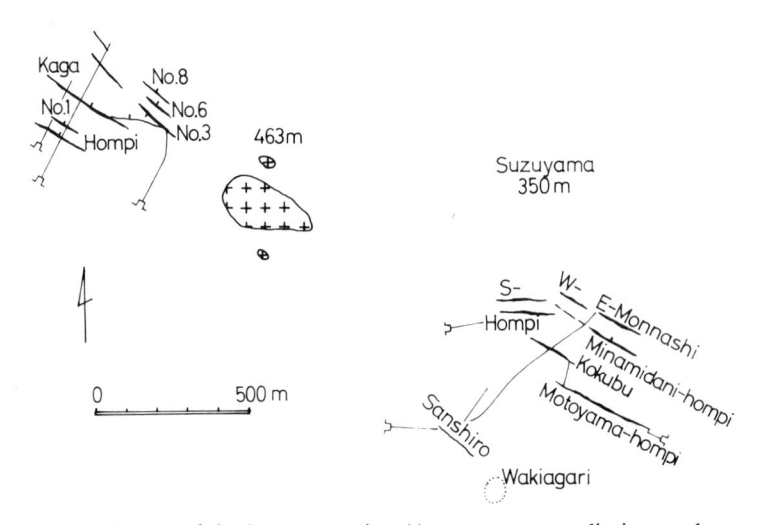

FIG. VII–3–15. Geologic map of the Suzuyama mine. ‡‡ represents granodiorite porphyry.

More than ten veins running parallel to the direction of the outcrops of porphyry occur in the Cretaceous formation, in both the eastern and western sides of the stock. The eastern group of veins consists of several quartz veins with cassiterite, while the western group consists of several sulfide-quartz veins carrying cassiterite. Cassiterite occurs as aggregates and streaks in quartz veins as well as in networks and disseminations in ores and silicified or chloritized country rocks.

In the porphyry, sulfide minerals such as pyrrhotite, pyrite and chalcopyrite are found replacing mafic minerals. With these minerals, cassiterite, ilmenite, sphene, apatite and fluorite are also found. Thin quartz veinlets carrying arsenopyrite occur in the stock, though no productive vein is found.

In quartz phenocrysts of the porphyry, many small fluid inclusions, representing liquid, gaseous and highly saline types, are recognized. Liquid inclusions of irregular shapes occur along curved planes or as clusters. The majority of gaseous and highly saline inclusions fill negative crystals or cavities with partly faceted shapes, and they coexist along planes as well as in clusters. The degree of filling of highly saline inclusions is generally constant, but that of gaseous inclusions is variable. In the highly saline inclusions, four or five kinds of solid phases are recognized. Identification of these solid phases is generally difficult because of their small sizes, but the transparent and optically isotropic cubic solid occupying the largest volume among the solid phases is probably halite (Fig. VII–3–16). As the volume

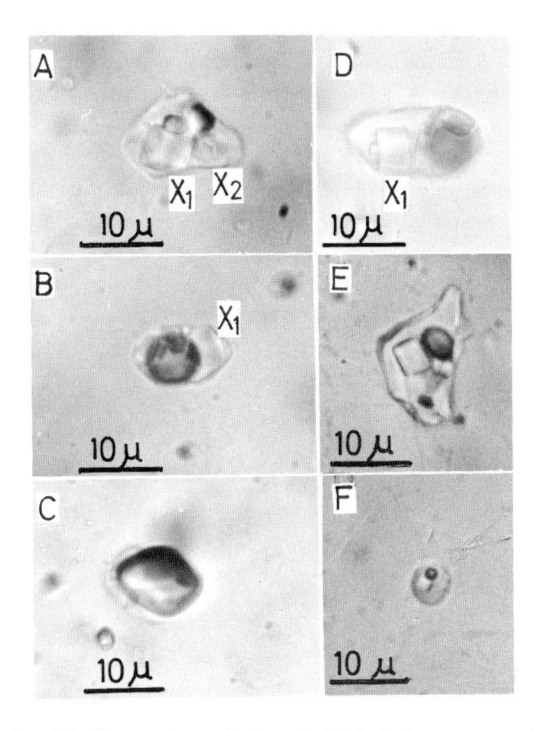

FIG. VII-3-16. Polyphase fluid inclusions in quartz of granitic rocks, porphyries, and veins.
- X_1: NaCl crystal, X_2: KCl crystal, X_3: high birefringent mineral, X_4: unknown transparent mineral, X_5: opaque mineral.
- (A), (B) and (C) High-salinity fluid inclusions in granite porphyry from the Kamioka mine.
- (D) High-salinity fluid inclusion in granite porphyry from the Suzuyama mine.
- (E) High-salinity fluid inclusion in quartz from the Ohatano granodiorite of the Akagane mine.
- (F) Minute polyphase fluid inclusion in quartz porphyry of the Jishakuyama ore body, the Akagane mine.

ratio of halite to the solution is high, it is presumed that salinity of the inclusions was higher than 50 wt. % at the time of trapping.

Besides halite, another transparent solid with cubic form, considerably smaller than halite, is often observed. This phase is presumed to be sylvite, although there is no positive way to prove it. The third phase is an optically anisotropic, minute, irregularly shaped mineral. The fourth is a transparent, prismatic, low birefringent crystal, and the fifth is a very minute opaque solid. The solid phases are not necessarily found in every polyphase inclusion.

In some gaseous inclusions a small transparent cubic phase (probably halite) often occurs in the liquid phase. Since the ratio of solid to liquid is similar to that in the highly saline inclusions, it is probable that the variation in degree of filling of gaseous inclusions is caused by trapping of the gas and liquid phases under boiling conditions. Quartz veinlets cutting the porphyry are especially rich in these gaseous and highly saline inclusions.

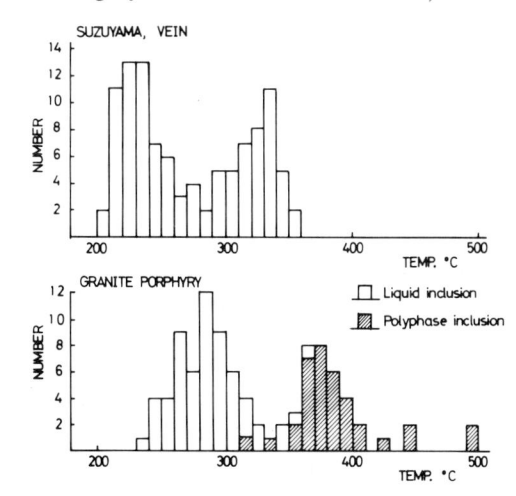

FIG. VII-3-17. Homogenization temperature of fluid inclusions in quartz from veins and granite poprhyry from the Suzuyama mine.

3. Fluid Inclusion Studies

Homogenization temperatures of some polyphase and liquid inclusions in the porphyry were measured. The temperatures of polyphase inclusions cover a range between 310°C and 500°C, with the maximum frequency at about 370°C. Filling temperatures of liquid inclusions are in the range of 230° ~ 350°C with a peak at 280°C. (Fig. VII–3–17).

Fluid inclusions in quartz collected from the veins of the eastern and western groups were also investigated. Most vein quartz is milky white. Some inclusions suitable for measurement, however, were found in quartz from comb-structured or vuggy parts of veins. The size of the inclusion was so small that it permits only the determination of filling temperature. The filling temperatures of liquid inclusions in veins range from 200°C to 350°C with two frequency peaks at 230°C and 350°C.

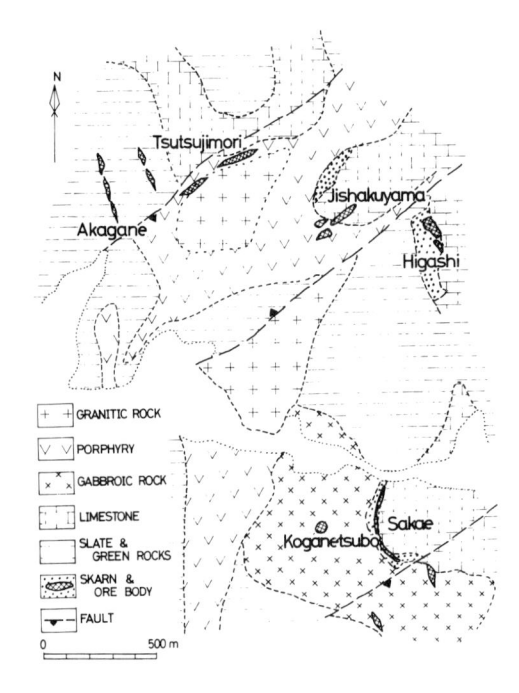

Fig. VII–3–18. Geologic map of the Akagane mine.

The difference of temperature distribution between porphyry and veins is several tens of degrees. Salinity of fluids should be lower in veins than in porphyry, as no halite crystal is recognized in fluid inclusions from vein quartz. It is considered that the highly saline inclusions in porphyry represent a part of the original ore-forming fluids which ascended through fissures in porphyry, and that the fluid inclusions in veins represent the solution trapped during or after the deposition of ore.

Akagane mine: The Akagane mine, Iwate Prefecture, is located about 110 km to the north-northeast of Sendai City in northeastern Japan.

The oldest rocks of this area are slate, limestone and green rocks of Late Paleozoic age. The Paleozoic rocks generally strike NNW-SSE, but in the mineralized area the structure is complicated with folding and faulting. The Paleozoic rocks are intruded by gabbroic rocks in the southern part of the area. These rocks were later intruded by quartz porphyry and granodioritic rock of Late Cretaceous age (Fig. VII–3–18).

Quartz porphyry forms a funnel-shaped body elongated in a NNE-SSW direction and branches to many small dikes showing intricate boundaries. Two small stocks of granodiorite are intruded into porphyry. The texture of the granodiorite is either fine-grained or porphyritic, and the boundary between the granodiorite and porphyry is gradational or sharp.

The ore bodies are mostly pyrometasomatic deposits formed in Paleozoic limestones and gabbroic rocks. They are the Akagane, Tsutsujimori, Jishakuyama, Higashi and Sakae bodies. The principal ore mineral is chalcopyrite, but the ore contains some scheelite, bismuthinite and native bismuth. Some parts of the Jishakuyama ore body are disseminated vein types formed along fissures and joints in silicified quartz porphyry.

The Koganetsubo breccia pipe carries scheelite and native gold.

Fluid inclusions in quartz from the Jishakuyama ore body and those in quartz porphyry and granodiorite were studied to correlate mineralization with ore-forming fluids. All investigated fluid inclusions were so small that neither the homogenization temperature nor the salinity could be determined. Recently, however, Muramatsu and Nambu (1976) have measured the homogenization temperature and salinity of fluid inclusions in the Hozumi and Sakae skarn ore bodies.

Fluid inclusions in vein quartz from the Jishakuyama ore body are mostly liquid inclusions. The degree of filling of these inclusions ranges from approximately 0.7 to 0.8. Polyphase inclusions are rarely found. On the other hand, quartz phenocrysts in the silicified porphyry of the country rocks contain a larger number of polyphase inclusions than veins. The solid phase in polyphase inclusions has a cubic form (Fig. VII–3–16). The numeral ratio of polyphase inclusions to the total number of inclusions is estimated to be about 30 %. No gaseous inclusions are recognized in either the veins or the porphyry.

In quartz from the Koganetsubo breccia pipe, CO_2-rich fluid inclusions are commonly found in addition to liquid inclusions, as described earlier.

Three types of fluid inclusions are recognized in quartz from the granodiorite. The liquid inclusions are by far the most abundant and show different shapes and modes of distribution. Polyphase and CO_2-rich inclusions show rather simpler shapes. The mode of occurrence of CO_2-rich inclusions in granodiorite is similar to that of the Koganetsubo ore body. However, in the northern stock of granodiorite, polyphase inclusions are common and CO_2-rich inclusions are less abundant. On the other hand, in the southern stock, polyphase inclusions are fewer than CO_2-rich inclusions. The CO_2 concentration of these inclusions is generally high, as indicated by the homogenization of the gas phase to liquid CO_2. CO_2-rich inclusions are much less common in the porphyry.

The volume of halite in polyphase inclusions from the Akagane area is generally smaller than those reported from other localities where polyphase inclusions occur with gaseous inclusions. It is assumed from the volume ratio of halite to inclusions that the salinity would be about 30 to 35 wt. %. As stated already, quartz phenocrysts from the silicified porphyry which forms the country rock of the Jishakuyama ore body are rich in polyphase inclusions. Some samples of quartz porphyry collected from the adit between the Jishakuyama ore body and the northern granodiorite stock contain a few tiny polyphase inclusions besides many liquid inclusions. On the other hand, samples collected from the northern and southern parts of the quartz porphyry dikes contain scarcely any polyphase inclusions.

The abundance of polyphase inclusions in and near the granodioritic stocks suggests that the copper mineralization of the Jishakuyama ore body, and probably of the other ore bodies, was caused by ore-forming fluids released from the granodioritic stocks. The predominance of liquid inclusions in vein quartz from the Jishakuyama ore body and the presence of polyphase inclusions in the wall rocks indicate that at early stages the fluids would have had high salinities, but at the main stages of mineralization and the sites of deposition the salinities would have been lowered to intermediate concentrations either by the mixing of meteoric waters of low salinities or by a decrease in salinity of the fluids released from magmas.

The genetic relation between the Koganetsubo breccia pipe and the granodiorite stocks is manifested in a study of CO_2-rich fluid inclusions. The breccia pipe would have been formed by hydrothermal fluids produced in depth in the granodiorite body. Burnham (1967) suggested that the enrichment of CO_2 in hydrothermal fluids separates at early stages from a felsic magma at pressures below 2,000 bars. Sushchevskaya and Ivanova (1967) reported a close relation between CO_2-rich hydrothermal fluids and tungsten mineralization. These facts are consistent with the present observation of the Koganetsubo ore body and granodiorite stocks.

Kamioka mine: The geology of this mine is described in Chapter III-2.

The zonal arrangement of ore minerals, which is generally molybdenite, chalcopyrite, sphalerite and galena outward from the center, is reported in the Tochibora-Maruyama area (Sakai, 1963) and in the Mozumi area (Nitta and Fukabori, 1969; Nitta *et al.*, 1971). One of the centers of mineral zoning is located in the area between Tochibora and Maruyama, where dikes of granite porphyry are intruded. In the Mozumi area, a horizontal and vertical zoning which is centered in the granite porphyry stock intruded into the lower part of the area is reported.

Shiobara (1961) reported on the decrepitation temperatures measured on more than 1,000 samples of sphalerite, hedenbergite and garnet from ore bodies of the Tochibora and Maruyama groups. According to his results, the decrepitation temperature of hedenbergite is distributed in the range between 320° and 455°C with a maximum frequency at 375°C. Sphalerite distributes in the range of 200° to 390°C with two peaks at 250° and 320°C, while garnet is in the range of 410° to 475°C with a peak at 420°C. High decrepitation temperatures are found in the area between Tochibora and Maruyama, where several dikes of granite porphyry are intruded. The decrepitation temperature is reported to decrease toward the SE and NW.

In the area between Tochibora and Maruyama, one quartz porphyry dike and seven granite porphyry dikes are intruded. The granite porphyry dikes are numbered from the south to the north. No. 2 is the largest. Molybdenite is recognized in quartz veinlets along joints in this dike.

Fluid inclusions in porphyry are classified into three types: polyphase, gaseous and liquid inclusions. These inclusions are found in quartz phenocrysts and quartz veinlets in porphyry. The size of the inclusions, however, is too small to permit the use of heating-stage and freezing-stage methods for most of them.

Polyphase inclusions contain various combinations of the following solid phases: transparent cubic solid (halite), smaller transparent cubic solid (sylvite), optically isotropic rounded solid, long prismatic anisotropic solid, high birefringent solid (calcite) and opaque solid (Fig. VII-3-16). All of these solid phases are not necessarily included in every poly-

TABLE VII-3-3. Relative abundance of polyphase fluid inclusions in porphyries from the Kamioka mine.

		Q.P	G.P.I	G.P.II	G.P.III	G.P.IV	G.P.V	G.P.VI	G.P.VII
Surface	Veinlet					+++	++	++	
	Phenocryst								++
0mL	Veinlet			+++		+++			
	Phenocryst			++		+	+		
−200mL	Veinlet			+++	+	+++			
	Phenocryst	++	+	+++		+	+		
−370mL	Veinlet			+++					
	Phenocryst			+++					

+++: abundant, more than 60%.
++: 60 ~ 30%.
+: scarce, less than 30%.
Q.P.: quartz porphyry.
G.P.I.: No. 1 granite porphyry dike.

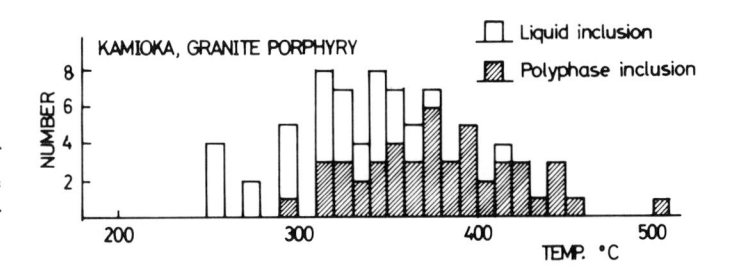

FIG. VII-3-19. Homogenization temperature of fluid inclusions in granite porphyry from the Kamioka mine.

phase inclusion. The opaque mineral is principally found in polyphase inclusions from the No. 2 granite porphyry dike, which carries molybdenite veinlets, especially in the lower parts. Most of the polyphase inclusions, however, contain halite crystal, the volume of which indicates a salinity of 35–40 wt. %.

The relative abundance of polyphase inclusions varies in different porphyry dikes and is roughly classified into three groups (Table VII-3-3). In general, polyphase inclusions are more abundant in quartz veinlets than in quartz phenocrysts. No. 2 granite porphyry dike is rich in varieties of solids and has the greatest abundance of polyphase inclusions. Both the abundance and variation of solids in polyphase inclusions decrease upward in the dike and outward from it.

Homogenization temperatures of polyphase inclusions from No. 2 granite porphyry dike are shown in Fig. VII-3-19, along with those of liquid inclusions. Homogenization temperatures reported here represent the temperatures of disappearance of either the halite crystal or the gas bubble.

Gaseous inclusions with a negative crystal form or partially faceted shape are abundant in the porphyry. These gaseous inclusions have different degrees of filling and coexist with polyphase inclusions. Small halite crystals are occasionally found in the liquid phase of gaseous inclusions. This suggests that the inclusion has trapped the highly saline phase as well as the gaseous phase.

From these data, it is presumed that the ore-forming solutions with high salinity have been released from deeper places in this area and have ascended through fissures in porphyries. The boiling of fluids is strongly manifested in inclusions collected from the investigated levels.

A preliminary observation was carried out on some samples of granite porphyry and adjacent highly silicified Mesozoic formation in the Mozumi area. Polyphase, gaseous and liquid inclusions similar to those in the Tochibora-Maruyama area were observed.

Many tiny fluid inclusions are recognized in quartz phenocrysts of the porphyry. In general, gaseous inclusions predominate over liquid inclusions. Polyphase inclusions are also present, but fewer. The number of polyphase inclusions amounts to about 10% of the total. From the volume of cubic halite crystals, it is inferred that the salinity would be around 30 to 40 wt. %. The second solid is a tiny anisotropic phase, and only a few polyphase inclusions contain this phase.

A remarkable feature of inclusions in the quartz porphyry is the predominance of gaseous inclusions and the variation in degree of filling of fluid inclusions. It is considered that the porphyry principally trapped a heterogeneous mixture of gas and liquid phases, and that the temperature and pressure of hydrothermal fluids were near the immiscibility—that is, boiling—condition.

In some specimens the porphyry is cut by pyrite-bearing quartz veinlets, which are in turn cut by molybdenite-bearing quartz veinlets. In quartz of the former veinlets and in quartz phenocrysts of the porphyry adjacent to the veinlets, polyphase inclusions are common.

This type of inclusion, however, is scarce in quartz of the molybdenite-bearing veinlets. It appears that in the early stages of mineralization the ore-forming fluids ascending through thin fissures in the prophyry were highly saline, and that in the later stages, represented by the molybdenite mineralization, the salinity of the fluids decreased to a value less than about 26 wt. %.

Geneses of Ore-forming Fluids from Felsic Magmas

Porphyries genetically related to the mineralization are characterized by the presence of highly saline inclusions. On the other hand, no highly saline inclusions are found in rhyolites. Granitic rocks are generally devoid of highly saline inclusions, but small granitic bodies occasionally contain a large number of polyphase inclusions.

The studies of fluid inclusions in these igneous rocks give us significant information on the evolution of ore-forming fluids during their migration from the sources. In some ore deposits, such as the Taishu and Suzuyama, the changes in temperature and salinity of fluids from the centers of mineralization to the peripheries are particularly shown by the study of fluid inclusions. Microscopic examinations of fluid inclusions also reveal compositional changes of inclusions outward from the related igneous rocks.

Recently, highly saline inclusions have been reported from some postmagmatic ore deposits (Kelly and Turneaure, 1970; Roedder, 1971a; Nash and Theodore, 1971; Sillitoe and Sawkins, 1971; Nash, 1973). These reports indicate that highly saline inclusions are rather common in porphyries genetically related to the mineralization. In porphyry copper deposits, the coexistence of polyphase inclusions and gaseous inclusions is characteristic of the core, while the marginal ore zones are characterized by the presence of liquid inclusions (intermediate salinity and high-density fluids). Nash (1973) suggested that the deposition of base metal sulfides is associated with relatively dilute fluids, although brines are important as carriers of metals, and that marked changes in temperature and salinity of fluids from early to late stages is attributable to the mixing of fluids having different temperatures and salinities. In ordinary hydrothermal deposits, it is shown that highly saline inclusions are rare in ore zones, but they are common in porphyries and granitic rocks genetically related to the mineralization.

The generation of hydrothermal fluids from acidic magmas was discussed by Burnham (1967), Kilinc and Burnham (1972), Holland (1972) and Whitney (1975). It was experimentally shown that chlorides concentrate more in the aqueous phases than in coexisting silicate melts, and that partition ratios of metals between the aqueous phase and the coexisting silicate melt phase increase as the chloride concentration of the aqueous phase increases. Holland (1972) concluded from these results that highly saline hydrothermal solutions are excellent scavengers for metals from magmatic melts.

These facts are strongly supported by the study of polyphase inclusions, which often carry small opaque minerals. The concentration of these opaque minerals estimated from their volume reaches as high as 1 gram per liter of fluid (Takenouchi, 1962c).

It is highly probable that highly saline inclusions found in igneous rocks genetically related to the mineralization represent a part of the original ore-forming fluids, and the dilution of these fluids would be one of the important factors controlling ore deposition. Since the presence of polyphase inclusions suggests the center of mineralization, the fluid inclusion study of genetically related igneous rocks as well as of ore deposits would give us useful indications for the exploration of ore deposits.

VII-4. A Thermodynamic Study of Manganese Minerals and its Applications to the Formation of Some Manganese Mineral Deposits in Japan

Taro Takahashi and Edward Schreiber

Introduction

Manganese occurs as a major constituent in a variety of mineral species which include oxides, hydroxides, silicates, carbonates and sulfides. Such occurrences of manganese in varying mineral species can be attributed partially to its three valence states, i.e., 2^+, 3^+ and 4^+, which are stable in various natural conditions existing in the crust of the earth. Because of its variable valence states and its varying chemical affinities with a number of geologically important anion species, manganese minerals are useful indicators of the physicochemical conditions which prevailed during the processes of the formation of ore deposits, metasomatism and metamorphism. Thus, a study of the thermodynamic stability of various ore minerals of manganese is pertinent to an improvement of our understanding of the ore-forming processes. In this chapter the results of the experimental determination of the univariant boundary for the reaction: 2 rhodochrosite ($MnCO_3$) + quartz (SiO_2) = tephroite (Mn_2SiO_4) + $2CO_2$ (gas) are presented first (Schreiber, 1963). Since rhodochrosite is one of the most common primary ore minerals and is often associated with quartz, this boundary gives an upper limit for the stability of rhodochrosite. In addition, the thermodynamic stability fields for various manganese mineral species, including manganosite (MnO), pyrochroite (Mn(OH)$_2$), hausmannite (Mn_3O_4), bixbyite (Mn_2O_3) and rhodonite ($MnSiO_3$), have been computed using the available thermochemical data as a function of temperature and pressure of H_2O, CO_2 and/or O_2. The results of these experimental and computational studies have been used to estimate the possible range of the temperatures and the pressures of H_2O, CO_2 and O_2 at which some of the major Japanese manganese ore deposits were formed.

An Experimental Study of Tephroite and Rhodochrosite

Since tephroite and rhodochrosite are common manganese minerals in ore deposits, a knowledge of their stability relations would increase our understanding of the ore-forming environment. Accordingly, the univariant equilibrium boundary for the reaction: Mn_2SiO_4 + $2CO_2$ = $2MnCO_3$ + SiO_2, has been determined over a CO_2 pressure and temperature range of 638 to 2,065 bars and 470° to 530°C, respectively, using a quench method followed by an X-ray diffraction examination of the specimen.

Apparatus: A conventional cold seal rod reactor (Hastelloy-C, 12″ long, 1-3/4″ o.d., 5/16″ i.d.) was employed as a container for specimens under high CO_2 pressure. Each reactor was heated externally in an electric resistance furnace, and the temperature was measured with a Pt − (Pt + 10% Rh) thermocouple with an accuracy of ± 0.5°C. The thermocouples were calibrated with an accuracy of ± 0.1°C against the melting points of high-purity tin (231.85°C), lead (327.4°C), zinc (494.45°C) and antimony (630.5°C). Temperature gradient observed within the sample chamber in a reactor was 3°C at 500°C, and that between the inside and outside walls of the same section of the reactor was found to be no more than 1°C. Therefore, the temperature of a specimen in the first 3/4 inch of the reactor can be estimated within ± 2°C. The high-pressure CO_2 was generated by pumping liquefied CO_2 with a Sprague pump which is capable of producing 30,000 psi. High purity CO_2 (99.99%) was dried with magnesium perchlorate and then filtered by a micron filter before use. The pressure was measured by a Heise precision gauge with an accuracy of 0.2%.

Starting Materials: The rhodochrosite specimen used for the experiments was crystallized from reagent grade $MnCO_3$ (the Fisher Scientific Co.; Mn, 46.5%; insoluble in HCl, 0.004%; Cl, 0.01%; SO_4, 0.001%; heavy metals as Pb, 0.003%; Fe, 0.001%, Ni, 0.01%) at 450°C and 200 bars CO_2 pressure. The specimen thus formed is white with a faint pinkish cast, and has lattice parameters of $a_0 = 4.77$ Å and $c_0 = 15.67$ Å, which is in excellent agreement with the values of Swanson and Tatge (1957) and Graf (1961).

Fine-grained tephroite crystals were synthesized from a stoichiometric mixture of the reagent grade MnO and (purified natural quartz) SiO_2 (impurity less than 0.03%) at 1,150°C. The mixture was placed in a platinum-lined graphite container to prevent oxidation. After the first 12 hours, the sample was taken out of the furnace, ground thoroughly, and put back into the furnace. This process was repeated until no changes in line intensities of the X-ray diffraction peaks were observed. The lattice parameters of this synthetic tephroite are $a_0 = 5.00$ Å, b_0, $c_0 = 6.15$ Å at room temperature, and are in agreement within 1% with the values of O'Daniel and Tscheischwili (1944) and Liedau et al. (1958).

Experimental methods: Two platinum envelopes (3/32″ dia., 3/4″ long) containing samples were placed in a high-pressure reactor. One was filled with approximately 200 mg. of the synthetic tephroite (Mn_2SiO_4) and the other was filled with a mixture of rhodochrosite ($MnCO_3$) and quartz (SiO_2). These capsules were lightly clamped at their ends so that the gases in the envelopes could equilibrate with the gases in the reactor. Since the reaction was extremely sluggish at temperatures below 500°C, water was added to the specimen as a reaction catalyzer. For most of the "wet" runs, 0.02 *ml* of distilled water was introduced into each capsule and soaked into the powdered sample. Two capsules thus prepared were then set side by side in a thin-walled graphite cylinder, one end of which was loosely packed with a small quantity of steel wool. The graphite and steel wool both served as an oxygen buffer for maintaining the oxidation state of manganese to divalent. This assembly was, in turn, placed on the bottom of a high-pressure reactor. Following this, a 1/4″ diameter stainless steel rod approximately 10″ long was inserted into the reactor to fill the dead space and minimize convection. The reactor was sealed and pumped with CO_2 until it was filled with liquid CO_2. Most of the runs were made by filling a reactor with CO_2 at room temperature to an estimated pressure that would yield the desired pressure at the temperature of the run. This method was useful for minimizing the removal of water from the neighborhood of the sample prior to establishing the run temperature.

At the end of the run, the reactor was quenched in a water bath. The temperature of the reactor dropped to 40° ~ 60°C within 1 minute, and to room temperature within 3 minutes. After the quenching, the sample capsules were removed from the reactor and were carefully examined for signs of oxidation of divalent manganese and capsule distortion that would indicate unsatisfactory communication of the gases in and out of the capsule. None of the runs produced such conditions.

Results: The extent of reaction of the samples was examined by an X-ray diffractometer. The amount of each mineral species was estimated semi-quantitatively by using the intensities of the following diffraction lines: d = 3.34 Å ($I/I_0 = 100$) for quartz, d = 2.84 Å ($I/I_0 = 100$) and 1.76 Å ($I/I_0 = 33$) for rhodochrosite, and d = 2.86 Å ($I/I_0 = 90$), 2.66 Å ($I/I_0 = 30$), 2.60 Å ($I/I_0 = 80$), 2.56 Å ($I/I_0 = 100$) and 1.81 Å ($I/I_0 = 80$) for tephroite. The reactions in lower-temperature runs did not proceed to completion. However, a pair of capsules in a reactor, each containing reactants and products respectively at the beginning of runs, always produced consistent results, as indicated by growth or decline of the X-ray diffraction peak intensities for the three respective crystalline phases. The results of the experiments are shown in Fig. VII–4–1, and those of the detailed examination of each specimen are listed in Schreiber (1963).

The univariant boundary shown in Fig. VII–4–1 can be expressed by:

$$\log P_{CO_2} \text{ (bars)} = 9.107 - 4724/T(°K)$$

or
$$\ln f_{CO_2} \text{ (bars)} = 27.315 - 15.317 \times 10^3/T(°K)$$
for a temperature range of 740° to 810°K or 465° to 535°C. For the present reaction, the Gibbs free energy for reaction is given by:
$$\Delta F°_T = -2RT \ln f_{CO_2}$$
along the univariant equilibrium curve. Thus the present data yield the free energy value of -15.1 Kcal at 700°K and -26.0 Kcal at 800°K, respectively. Using those values, the Gibbs free energy of formation for tephroite (from the elements) has been computed to be -356.1 Kcal/mole at 700°K and -350.9 Kcal/mole at 800°K. Robie and Waldbaum (1968) computed the Gibbs free energy of formation for tephroite using the thermodynamic data of Jeffes et al. (1954), Mah (1960) and Kelley and King (1961), and obtained -358.56 (± 0.82) Kcal/mole at 700°K and -350.84 Kcal/mole at 800°K. The results of the present study are in excellent agreement with the basic thermodynamic data.

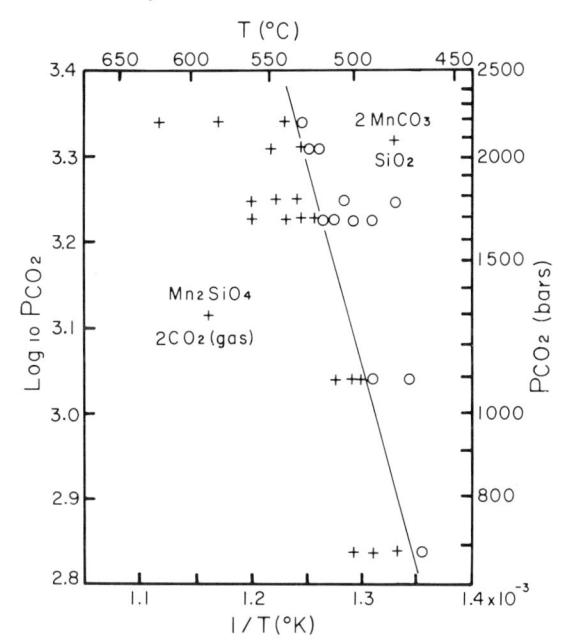

FIG. VII-4-1. The results of quench experiments for the determination of the univariant equilibrium curve for the reaction: $2MnCO_3 + SiO_2 = Mn_2SiO_4 + 2CO_2$. The boundary line, along which ΔF for the reaction is zero, is expressed by: $\log P_{CO_2} = 9.107 - 4724/T(°K)$. The circles indicate the $MnCO_3$ and SiO_2, and the crosses indicate the Mn_2SiO_4 stable conditions.

Stability Calculations for Manganese Minerals

Although some of the manganese minerals, such as hausmannite and rhodochrosite, usually contain substantial amounts of other ionic species such as Fe^{2+}, Fe^{3+} and Mg^{2+} in solid solution, those mineral species have been assumed to be pure end members as represented by the simple chemical formulae shown below. This assumption is necessitated by the lack of thermochemical data for non-ideal mixing in solid solution. The following chemical reactions are considered for the present study:

(A) Reactions involving H_2O:
$Mn(OH)_2 = MnO + H_2O(gas)$ ....(1)
pyrochroite manganosite

(B) Reactions involving CO_2:
$MnCO_3 = MnO + CO_2(gas)$ ....(2)
rhodochrosite manganosite
$2MnCO_3 + SiO_2 = Mn_2SiO_4 + 2CO_2(gas)$ ....(3)
rhodochrosite quartz tephroite

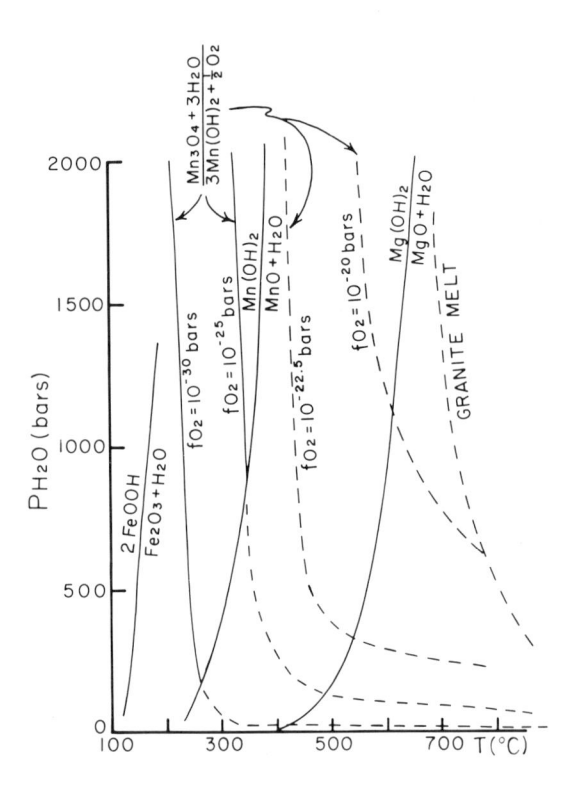

Fig. VII-4-2. The univariant equilibrium curves for the reactions: $2FeOOH = Fe_2O_3 + H_2O$, $Mn(OH)_2 = MnO + H_2O$, and $Mg(OH)_2 = MgO + H_2O$, as a function of water pressure and temperature. The equilibrium boundary for the reaction: $Mn_3O_4 + 3H_2O = 3Mn(OH)_2 + 1/2 O_2$, is shown for the three values of oxygen fugacity. The dashed line portion of the curves indicates that, since one of the solid phases is no longer stable, it has no physical meaning. Pyrochroite stability field is the triangular area bordered by the $Mn(OH)_2 - MnO$ boundary and the solid line portion of the $Mn_3O_4 - Mn(OH)_2$ boundary curves. Note that the stability field for pyrochroite vanishes at an oxygen fugacity greater than 10^{-23} bars.

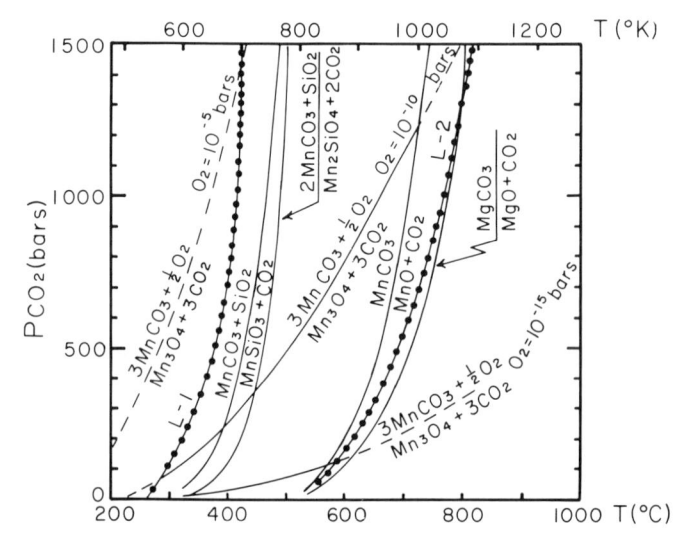

Fig. VII-4-3. The univariant equilibrium curves for the reactions between $MnCO_3$, SiO_2, $MnSiO_3$, Mn_2SiO_4, MgO and $MgCO_3$ as a function of CO_2 pressure and temperature. The equilibrium curves for the reaction: $3MnCO_3 + 1/2 O_2 = Mn_3O_4 + 3CO_2$ are also shown for the three values of oxygen fugacity. The dashed line portion of those curves indicates that one of the solid phases is no longer stable and thus the curve has no physical significance. Curve L-1 is a locus of the points at which the $MnCO_3$–Mn_3O_4 curves intersect with the Mn_2O_3–Mn_3O_4 curve at various oxygen fugacity values. Curve L-2 is also a locus of the points at which the $MnCO_3$–Mn_3O_4 curves intersect with the Mn_3O_4–MnO curve at various oxygen fugacity values. Hence those two curves bound an area in which Mn_3O_4 is stable at various oxygen fugacities and the $MnCO_3$–Mn_3O_4 curve is meaningful.

$$MnCO_3 + SiO_2 = MnSiO_3 + CO_2(gas) \quad \dots\dots\dots\dots\dots\dots(4)$$
rhodochrosite quartz rhodonite

(C) Reactions involving O_2:
$$3Mn_2O_3 = 2Mn_3O_4 + {}^1/_2O_2(gas) \quad \dots\dots\dots\dots\dots\dots\dots\dots\dots(5)$$
bixbyite hausmannite
$$Mn_3O_4 + 3MnSiO_3 = 3Mn_2SiO_4 + {}^1/_2O_2(gas) \quad \dots\dots\dots\dots\dots\dots(6)$$
hausmannite rhodonite tephroite
$$Mn_3O_4 = 3MnO + {}^1/_2O_2(gas) \quad \dots\dots\dots\dots\dots\dots\dots\dots\dots\dots(7)$$
hausmannite manganosite
$$3Fe_2O_3 = 2Fe_3O_4 + {}^1/_2O_2(gas) \quad \dots\dots\dots\dots\dots\dots\dots\dots\dots\dots(8)$$
hematite magnetite

(D) Reactions involving CO_2 and O_2:
$$Mn_3O_4 + 3CO_2(gas) = 3MnCO_3 + {}^1/_2O_2(gas) \quad \dots\dots\dots\dots\dots\dots(9)$$
hausmannite rhodochrosite
$$Fe_3O_4 + 3CO_2(gas) = 3FeCO_2 + {}^1/_2O_2(gas) \quad \dots\dots\dots\dots\dots\dots(10)$$
magnetite siderite
$$CO_2(gas) = C + O_2(gas) \quad \dots\dots\dots\dots\dots\dots(11)$$
graphite

(E) Reactions involving H_2O and O_2:
$$Mn_3O_4 + 3H_2O(gas) = 3Mn(OH)_2 + O_2(gas) \quad \dots\dots\dots\dots\dots\dots(12)$$
hausmannite pyrochroite

(F) Reactions involving H_2O and CO_2:
$$MnCO_3 + H_2O(gas) = Mn(OH)_2 + CO_2(gas) \quad \dots\dots\dots\dots\dots\dots(13)$$
rhodochrosite pyrochroite

The primary sources for the thermochemical data used for the computations are Latimer (1952), Robie (1966), Robie and Waldbaum (1968), Burnham et al. (1969) and JANAF Thermochemical Tables (Stull and Prophet, 1971). The P–V–T data of Kennedy (1954) for CO_2 gas have been used to compute the fugacity of CO_2 gas at high temperatures and pressures. The univariant boundaries for simple Reactions (1) through (8), in which only one gaseous species is present, are plotted in Figs. VII–4–2 through VII–4–5 as a function of temperature and the fugacity or pressure of gaseous species. For Reactions (9) through (12), in which two gaseous species are involved, the univariant boundaries are shown in temperature vs. oxygen fugacity plots for each of the three values of water fugacity (or pressure) or CO_2 fugacity.

Stability Relations for the Oxide and Hydroxide Manganese Minerals: The univariant boundary, along which the Gibbs free energy for Reaction (1) is zero, is shown in Fig. VII–4–2 as a function of temperature and water pressure. The water pressure has been computed using the free energy and fugacity values for water tabulated by Burnham et al. (1969). The univariant stability boundaries for the reactions: 2 goethite (FeO·OH) = hematite (Fe_2O_3) + H_2O by Schmalz (1959), and brucite ($Mg(OH)_2$) = periclase (MgO) + H_2O by Barnes and Ernst (1963), as well as the melting temperature of water-saturated granite by Tuttle and Bowen (1958) are also shown. It is seen that dehydration of brucite takes place at the temperatures closest to the minimum melting temperature for granite, whereas the dehydration of pyrochroite to manganosite takes place between 300° and 400°C in a water pressure range of 500 to 2,000 bars. The dehydration of goethite to hematite occurs at temperatures below 200°C.

The univariant stability boundaries for manganosite-hausmannite (Reaction 7) and hausmannite-bixbyite (Reaction 5) are shown in Fig. VII–4–4 as a function of oxygen fugacity and temperature. The magnetite-hematite boundary (Reaction 8) is located within the stability field of hausmannite. Hence, hausmannite is a thermodynamically stable phase in

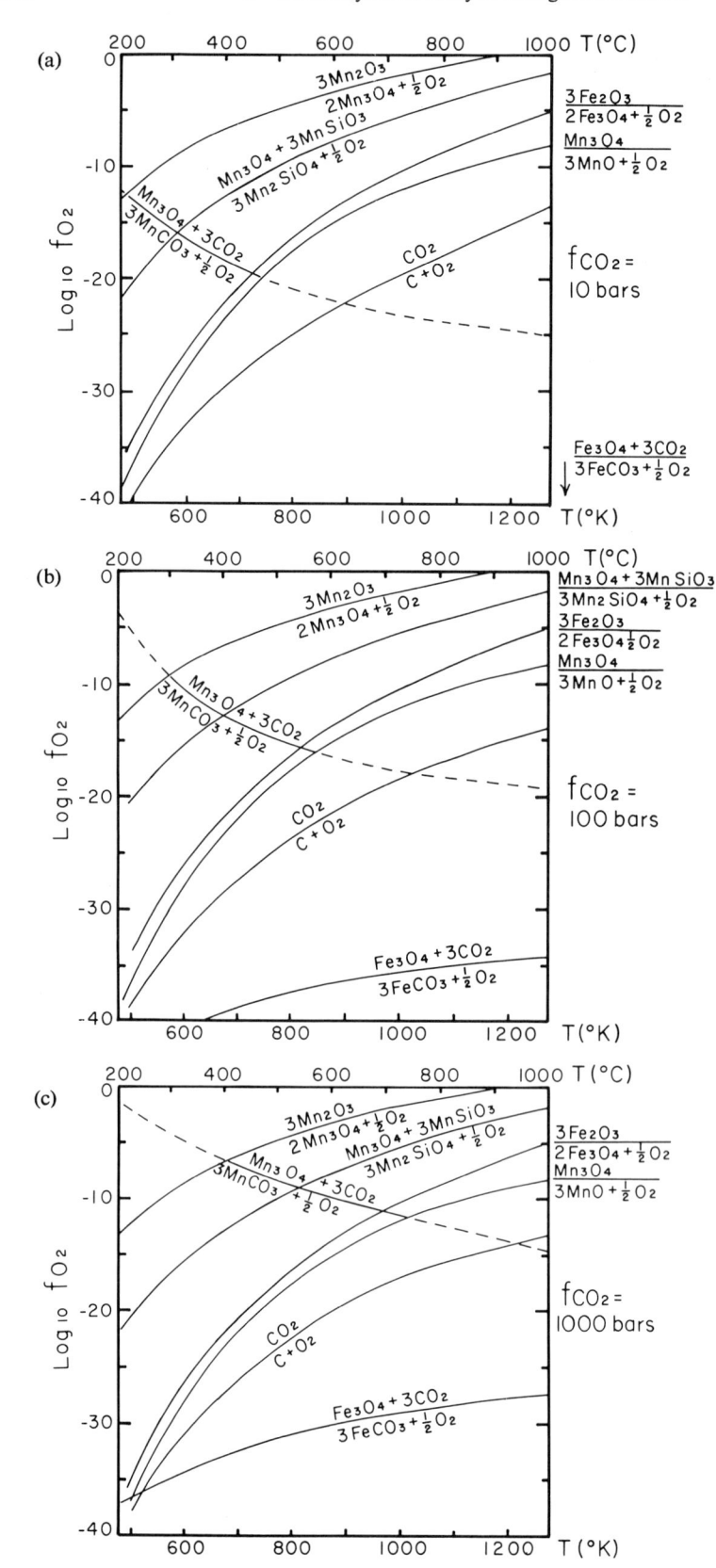

340   4. Thermodynamic Study of Manganese Minerals

Fig. VII-4-4. The equilibrium curve for reactions involving the oxidation and reduction of manganese and iron ions. The equilibrium curves for the reactions: $Mn_3O_4 + 3CO_2 = 3MnCO_3 + 1/2 O_2$, $Fe_3O_4 + 3CO_2 = 3FeCO_3 + 1/2 O_2$, and $C + O_2 = CO_2$, are also shown for the three values of CO_2 fugacity, i.e., 10 bars in (a), 100 bars in (b) and 1,000 bars in (c). The dashed line portion of those curves indicates the conditions in which Mn_3O_4 is no longer stable. Note that the stability field for $MnCO_3$ increases with increasing CO_2 fugacity. The stability field for siderite appears in the lower right corner in Fig. VII-4-5(b) as the fugacity is increased to 1,000 bars. An occurrence of siderite ($FeCO_3$) indicates a lower oxygen fugacity and higher temperature conditions compared to an environment where rhodochrosite ($MnCO_3$) is stable.

the presence of either hematite or magnetite. The hausmannite-bixbyite boundary lies in the highest oxygen fugacity range among this group of minerals. Rare occurrences of bixbyite suggest that oxygen fugacities in the stability range of bixbyite were seldom reached during the formation of manganese mineral deposits.

Reaction (12) describes the hydration of hausmannite to pyrochroite. This reaction is a function of H_2O as well as O_2 since Mn^{3+} ions in hausmannite are reduced to Mn^{2+}. The stability boundary for this reaction is plotted in Fig. VII–4–2 as a function of H_2O pressure and temperature for constant oxygen fugacity values ranging between 10^{-20} and 10^{-30} bars. The boundaries are nearly parallel to the P_{H_2O} axis at lower temperatures, due partially to the dependence of the boundary on $(P_{H_2O})^3$ and to the effect of fugacity-pressure con-

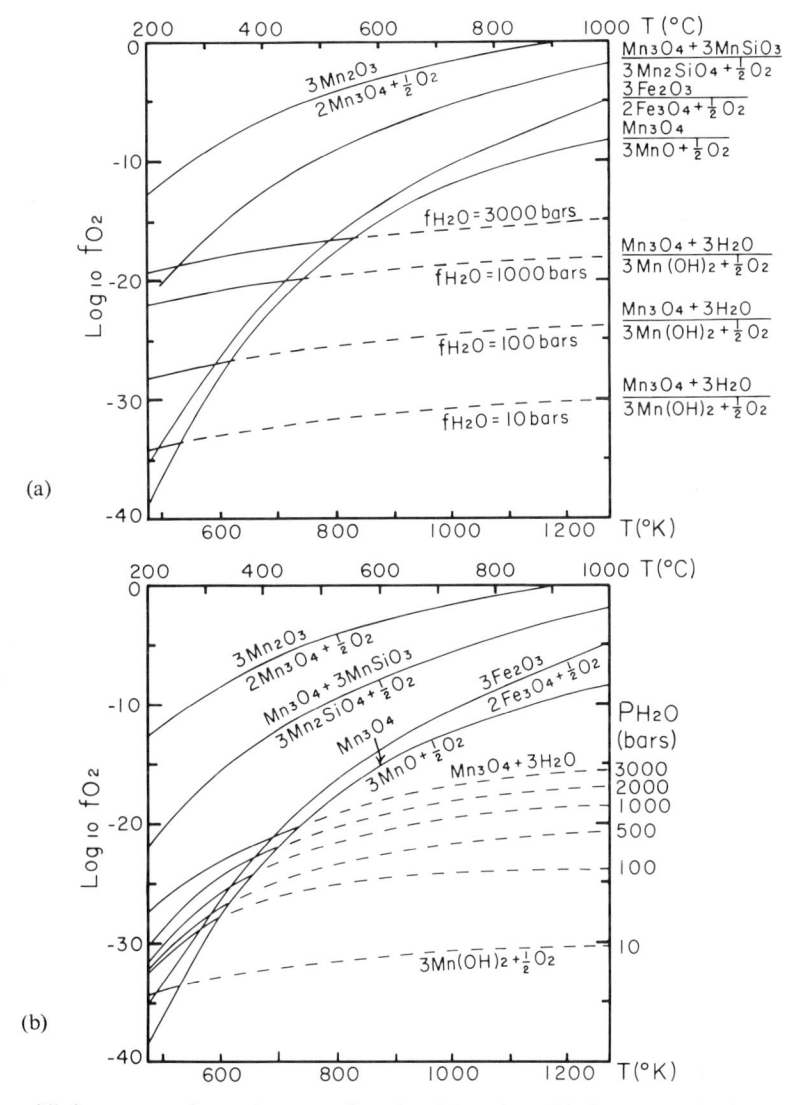

FIG. VII–4–5. The equilibrium curves for various reactions involving the oxidation and reduction of manganese and iron ions. The equilibrium curves for the reaction: $Mn_3O_4 + 3H_2O = 3Mn(OH)_2 + 1/2 O_2$ are also shown for the four values of H_2O fugacities in (a), and for the six values of H_2O pressures in (b). The dashed line portion of those families of curves indicates the conditions in which Mn_3O_4 is no longer stable. Note that the stability field of pyrochroite ($Mn(OH)_2$) is defined by the Mn_3O_4-MnO and Mn_3O_4-Mn(OH)$_2$ curves, and increases with increasing water pressure. However, its stability field is limited to temperatures below 450°C and to oxygen fugacities below 10^{-20} bars, even at a water pressure as high as 3,000 bars.

version. Because of this steepness of the boundary curves in a range of water pressures greater than 500 bars, the hausmannite-pyrochroite boundary curves no longer intersect with the pyrochroite-manganosite boundary curve when oxygen fugacities exceed 10^{-23} bars. This means that in normal hydrothermal conditions, in which the water pressure does not exceed 2,000 bars, manganosite becomes unstable with respect to hausmannite under oxygen fugacities greater than 10^{-23} bars. Under oxygen fugacities smaller than 10^{-23} bars, manganosite is stable on the high-temperature (or right-hand) side of the pyrochroite-manganosite boundary curve in Fig. IIV–4–2, and pyrochroite is stable in a triangular area bordered by the pyrochroite-manganosite and hausmannite-pyrochroite boundary curves. Hausmannite becomes stable on the low-temperature (or left-hand) side of the hausmannite-pyrochroite boundary curve. This stability relationship is also depicted in Fig. VII–4–5(a) and (b), where the hausmannite-pyrochroite boundaries are plotted as a function of temperature and oxygen fugacity for constant water fugacity or pressure values. Since those boundary curves are valid only in the stability field of hausmannite, broken lines are used to indicate the extension of the boundary curves.

On the basis of the stability boundary curves presented above, the following constraints for physicochemical environments which existed during the formation of the manganese minerals can be estimated. It is assumed that those minerals were formed under thermodynamic equilibrium conditions and at a total pressure of less than 2,000 bars. A total pressure of 2,000 bars is equivalent to lithostatic pressure at approximately 8 km deep in the crust of the earth. On the basis of stratigraphic and mineralogical reasons, this is considered to be a reasonable depth for the formation of mineral deposits associated with granitic intrusive rocks.

1) An occurrence of pyrochroite indicates that the temperature was no higher than 450°C. At a temperature higher than this, manganosite should be the stable mineral.

2) An occurrence of pyrochroite also indicates that the fugacity of oxygen was smaller than 10^{-23} bars. At oxygen fugacities greater than this, hausmannite should be the stable mineral.

3) Hausmannite is stable over a wide range of oxygen fugacities, and can, thus, stably coexist with either hematite or magnetite.

Stability Relations for the Silicate and Carbonate Manganese Minerals: The univariant stability boundary curve for rhodochrosite-manganosite (Reaction (2)) experimentally determined by Goldsmith and Graf (1957) and that for (rhodochrosite + quartz) − tephroite (Reaction (3)) reported earlier in this paper are shown in Fig. VII–4–3 as a function of temperature and CO_2 pressure. In addition, the stability boundary for magnesite-periclase determined by Harker and Tuttle (1955) and that for (rhodochrosite + quartz) − rhodonite (Reaction (4)) are shown. According to those stability curves, an observed tephroite-rhodonite assemblage without free quartz suggests that Reaction (3) had progressed to the right until all the available quartz was reacted, and that the conditions for the formation of the tephroite-rhodonite assemblage can be represented by the area on the right-hand side of the univariant curve in Fig. VII–4–3. The upper stability limit for rhodochrosite is defined by the rhodochrosite-manganosite univariant curve. A similar conclusion can be drawn for the rhodonite-rhodochrosite assemblage. Thus, the rhodochrosite-quartz assemblage can be stable to 400°C if the CO_2 pressure in the environment is 100 bars, and to 500°C if the CO_2 pressure is 1,000 bars. The rhodochrosite-tephroite (and/or rhodonite) assemblage (without quartz) should be stable up to 600°C at 200 bars and 710°C at 1,000 bars P_{CO_2}.

A complete phase diagram for the MnO-SiO_2 system has been determined by Glasser (1958) under a dry atmosphere of 1 bar total pressure. Tephroite and rhodonite exhibit simple eutectic melting at 1,251°C, below which these minerals coexist as thermodynamically stable phases. In a SiO_2-rich composition, rhodonite becomes stable in association with silica, whereas in a MnO-rich composition, tephroite becomes stable in association with manganosite. However, when hausmannite is associated with those silicate minerals, the stability of rhodonite and tephroite becomes a function of temperature and oxygen fugacity as expressed by Reaction (6). The univariant boundary for this reaction, which is shown in Fig. VII–4–4, indicates that the hausmannite-rhodonite assemblage is stable at conditions above this curve but below the univariant curve for the bixbyite-hausmannite reaction. Below this curve, the tephroite-hausmannite assemblage becomes stable if an excess of hausmannite is present, or the tephroite-rhodonite assemblage (free of hausmannite) becomes stable when not enough hausmannite is present to react with rhodonite.

Stability Reactions for the Oxide and Carbonate Minerals of Manganese and Iron: Reactions (9) and (10) describe the stability relationships for hausmannite-rhodochrosite and for magnetite-siderite. The univariant curves for those reactions are plotted in Fig. VII–4–4(a), (b) and (c) as a function of temperature and oxygen fugacity for three CO_2 fugacity values, i.e., 10, 100 and 1,000 bars. Hausmannite is stable in the temperature-oxygen fugacity conditions above the curve, and rhodochrosite is stable below the curve. The rhodochrosite stability field increases with increasing CO_2 fugacity. Since this univariant curve is only valid for the conditions in which hausmannite is stable, it is terminated at the intersection with the bixbyite-hausmannite and hausmannite-manganosite univariant curves. The stability field for siderite is found in the lower right corner of Fig. VII–4–4(b) and (c), and increases rapidly with increasing CO_2 fugacity. Thus, occurrences of siderite indicate an exceedingly low oxygen fugacity environment, in which graphite (Reaction (11)) is more stable than CO_2.

The hausmannite-rhodochrosite stability boundary is also shown in Fig. VII-4-3 as a function of P_{CO_2} and temperature for three oxygen fugacity values, i.e., 10^{-5}, 10^{-10} and 10^{-15} bars. Those stability curves are only valid in the stability field of rhodochrosite and hausmannite, and hence they are terminated at the rhodochrosite-manganosite boundary curve for high temperatures and at the hausmannite-bixbyite stability boundary curve for low temperatures. A locus of the termination conditions for the rhodochrosite-hausmannite reaction at varying oxygen fugacities is also shown in Fig. VII–4–3 and is marked L–1 for the bixbyite-hausmannite termination and L–2 for the rhodochrosite-manganosite termination condition.

Stability Relations for Rhodochrosite and Pyrochroite: Since pyrochroite occurs abundantly as a primary mineral in the Noda-Tamagawa deposits while it rarely occurs in the stability field for pyrochroite, it will be investigated in more detail here. Reaction (13) describes the reaction of pyrochroite with CO_2 gas, and its univariant boundary curve is shown in Fig. VII–4–6 as a function of H_2O fugacity and temperature for three values of CO_2 fugacity, i.e., 1, 10 and 100 bars. The univariant stability boundary curve for pyrochroite-manganosite is also shown. The dash-dot-dash curve indicates a locus of CO_2 fugacity and temperature points, where rhodochrosite decomposes to manganosite and CO_2. At a CO_2 fugacity of 1 bar, rhodochrosite is stable at temperatures below 330°C, and pyrochroite is stable at temperatures below 355°C as seen in Fig. VII–4–6. Hence, the boundary curve for the rhodochrosite-pyrochroite is valid at temperatures below 330°C, as indicated by a solid curve. Similarly, the boundary curve of the range in which the rhodochrosite-pyrochroite curve is valid at a CO_2 fugacity of 10 bars is indicated by a solid curve. Pyrochroite is stable in wedge-shaped areas located on the left-hand side of the pyrochroite-manganosite boundary curve and on the right-hand side of those for each of the given CO_2 fugacities. Thus, it is seen that the stability field for pyrochroite decreases rapidly with increasing CO_2 fugacity from 1 to 10 bars, and it disappears entirely at a CO_2 fugacity of approximately 20 bars when the H_2O fugacity is less than 2,000 bars. This means that rhodochrosite is a far more stable mineral than pyrochroite, and the latter can become more stable only when the H_2O fugacity exceeds the CO_2 fugacity by 500 to 1 at 327°C and 50 to 1 at 527°C. Therefore, pyrochroite can form only when the environment is nearly or totally depleted of CO_2 gas.

Stability Relations for the Manganese Oxide, Hydroxide and Carbonate Minerals: The stability fields for manganosite, hausmannite, bixbyite, pyrochroite and rhodochrosite can be defined as a function of temperature and the fugacities (or pressures) of oxygen, CO_2 and H_2O by assembling the information presented in Figs. VII–4–2 through VII–4–6. Two cases, in which the H_2O pressure is 2,000 bars and the CO_2 pressure is 1 and 10 bars, respectively, are presented in Fig. VII–4–7. For the conversion of fugacity to pressure of water, the tabulation of Burnham et al. (1969) was used. It was also assumed that the fugacity of CO_2 is approximately the same as the pressure and that those two gaseous species are mixed ideally. Since the CO_2 pressure is 10 bars or less and the gaseous solution is predominantly H_2O, those assumptions are considered to be reasonable.

In Fig. VII–4–7, it is seen that the stability field for pyrochroite does not exist when the CO_2 pressure is 10 bars, and it appears as an irregular area in the middle of the diagram when the CO_2 pressure is reduced to 1 bar. The pyrochroite field disappears when the H_2O pressure is lowered to 500 bars at 1 bar CO_2 pressure, or when the CO_2 pressure exceeds about 7 bars at 2,000 bar H_2O pressure. If CO_2 is absent from the system, pyrochroite should be stable in the conditions represented by an irregularly shaped area defined by the Mn_3O_4-$Mn(OH)_2$, Mn_3O_4-MnO and $Mn(OH)_2$-MnO boundaries in the lower left-hand corner of the diagram. If a small amount of silica is present in association with manganese minerals, it will occur as quartz with bixbyite, and as rhodonite or tephroite in association with hausmannite and rhodochrosite in their respective stability fields, indicated by a chained curve. Manganosite will be associated with tephroite in such a silica-deficient environment. It must be noted here that due to the lack of thermochemical data for other manganese minerals such as braunite ($3Mn_2O_3 \cdot MnSiO_3$), bementite ($Mn_8Si_7O_{27}H_{10}$), jacobsite ($MnFe_2O_4$) and galaxite ($MnAl_2O_4$), the stability fields for those minerals could not be computed. Hence, the stability diagrams presented in Fig. VII–4–7 apply only to those minerals considered in the computation, and does not exclude the stability fields of others whose thermodynamic data are not available.

Effect of Mixing of H_2O and CO_2 on the Stability Field: At high temperatures and pressures, two major gaseous species considered here, H_2O and CO_2, do not mix ideally, due mainly to extra interactions between these two molecular species. The P-V-T relation-

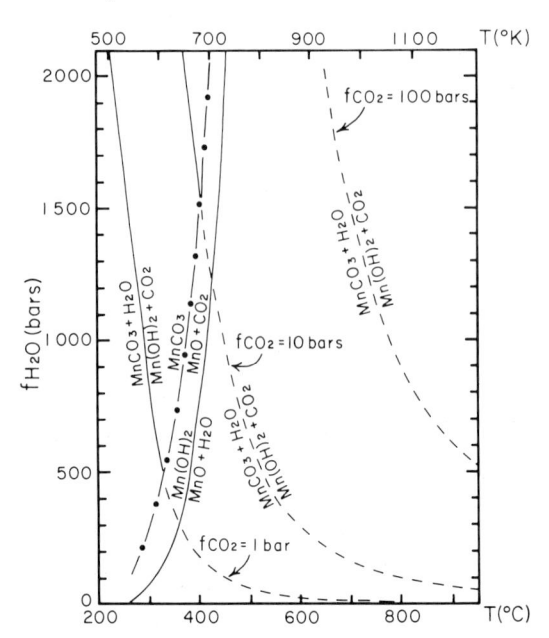

FIG. VII–4–6. The equilibrium curves for the reaction: $Mn(OH)_2 = MnO + H_2O$, and $MnCO_3 + H_2O = Mn(OH)_2 + CO_2$, at three values of CO_2 fugacity. The stability field for pyrochroite is the triangular area bordered by $Mn(OH)_2$-MnO curve and the $MnCO_3$-$Mn(OH)_2$ curves. It is seen that the pyrochroite field decreases with increasing CO_2 fugacity from 1 bar to 10 bars. This indicates that, with the presence of rhodochrosite ($MnCO_3$), pyrochroite can become stable only when the CO_2 fugacity is less than 20 bars at a water fugacity as high as 2,000 bars.

Part VII: Contributions

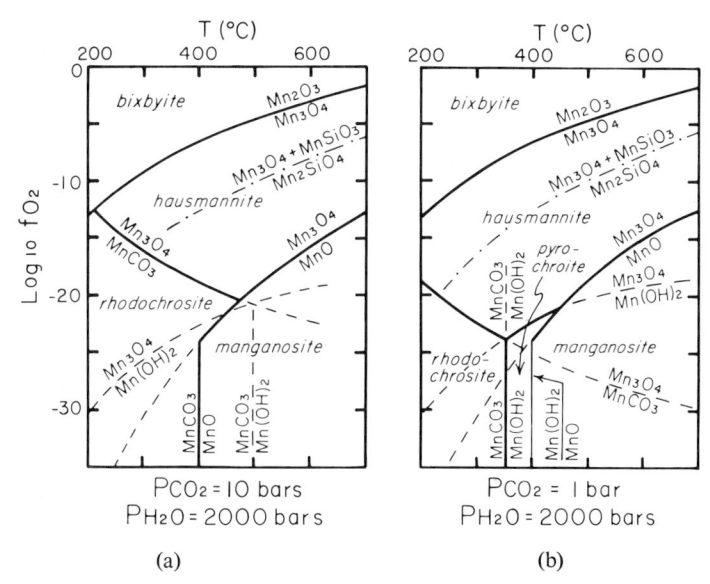

Fig. VII-4-7. Stability fields of various manganese minerals as a function of the oxygen fugacity and temperature at a CO_2 pressure of 10 bars and a H_2O pressure of 2,000 bars in (a), and at a CO_2 pressure of 1 bar and a H_2O pressure of 2,000 bars in (b). Note that the stability field for pyrochroite appears between the rhodochroiste and manganosite fields as the CO_2 pressure is reduced to 1 bar. The stability field for rhodochrosite increases as the CO_2 pressure is increased to 10 bars.

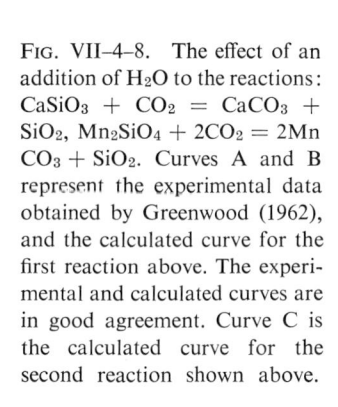

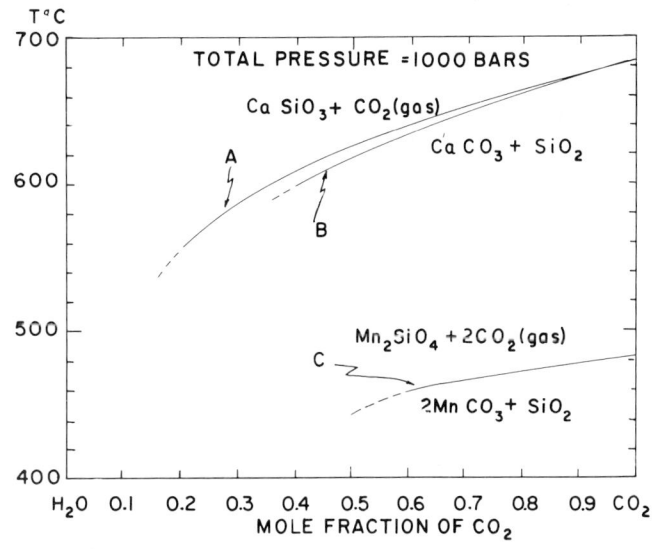

Fig. VII-4-8. The effect of an addition of H_2O to the reactions: $CaSiO_3 + CO_2 = CaCO_3 + SiO_2$, $Mn_2SiO_4 + 2CO_2 = 2MnCO_3 + SiO_2$. Curves A and B represent the experimental data obtained by Greenwood (1962), and the calculated curve for the first reaction above. The experimental and calculated curves are in good agreement. Curve C is the calculated curve for the second reaction shown above.

ships for the H_2O-CO_2 mixtures have been extensively investigated by Franck and Todheide (1959), Takenouchi and Kennedy (1964) and Greenwood (1969). Takahashi and Schreiber (1965) computed the partial molar fugacity of CO_2 in the H_2O-CO_2 gas mixtures using the data of Franck and Todheide (1959). The effect of H_2O on the reaction: $CaSiO_3 + CO_2(gas) = CaCO_3 + SiO_2$ was computed using the partial molar fugacity of CO_2 of Takahashi and Schreiber (1965). In Fig. VII-4-8, the results (Curve B) thus computed are shown to be in good agreement with the experimental results (Curve A) of Greenwood (1962). The effect of H_2O on the reaction: $Mn_2SiO_4 + 2CO_2$ (gas) $= 2MnCO_3 + SiO_2$

has been similarly calculated, and is shown in Fig. VII–4–8 (Curve C). It is seen that the temperature of stability boundary curves may be lowered as much as 60°C by replacing 50% of CO_2 with H_2O while the total pressure is kept constant. It is thus important to know probable H_2O–CO_2 ratios in ore-forming or metamorphic pore fluids. Such information may be obtained by either microgasometric analysis or microscopic observation of fluid inclusions in minerals. When such information becomes available, more refined thermodynamic computations which take the non-ideal gas mixing and the effect of dissolved electrolytes into consideration should be made in order to further delineate the condition of mineral formation.

The Noda-Tamagawa Manganese Deposits, Iwate Prefecture (Fig. I–3)

General Geology and Mineralogy: The Noda-Tamagawa mine is one of the major manganese producers in Japan. In this mine, pyrochroite, rhodochrosite, hausmannite, braunite and others occur in a thermally metamorphosed section of the Akkagawa Formation (pre-Cretaceous sediments) which includes slate, graywacke, thin bedded quartzite, small lenses of limestone and basic tuff. This section of the Akkagawa Formation is steeply inclined (60°–85°W) and is underlain by the Tanohata granite mass (Cretaceous). The geology and mineralogy of this mining area has been extensively studied by Watanabe and his associates (e.g., Watanabe, 1959; Watanabe *et al.*, 1970a, b). They observed sillimanite, andalusite and cordierite in some hornfels adjacent to the manganese ore bodies, and interpreted that the area was subjected to a high-grade thermal metamorphism caused by the Cretaceous granite intrusion.

The genesis of the bedded rhodochrosite deposits in Japan has always been a controversial subject, whether syngenetic sedimentary origin or epigenetic replacement origin (Yoshimura, 1967; Watanabe *et al.* 1970a), as partly discussed in Chapter IV–2. The deposits at the Noda-Tamagawa were subsequently thermally metamorphosed at the time of the Cretaceous granite intrusion.

The manganese ore bodies in the Noda-Tamagawa mine are in general stratified in the following sequence from the center of the ore body to the surrounding country rocks, according to Watanabe *et al.* (1970b):

1) Pyrochroite ore (center section)
    Principal minerals = pyrochroite and manganosite
    Minor minerals   = tephroite, galaxite, rhodochrosite, alabandite
2) Rhodochrosite ore (center section)
    Principal minerals = rhodochrosite
    Minor minerals   = tephroite, galaxite, manganosite
3) Hausmannite ore (lenses in the rhodochrosite ore)
    Principal minerals = hausmannite and rhodochrosite
    Minor minerals   = tephroite, manganosite, galaxite, Mn-micas
4) Tephroite ore (along the pyrochroite or rhodochrosite ores)
    Principal minerals = tephroite
    Minor minerals   = rhodochrosite, spessartite, rhodonite, bustamite
5) Rhodonite ore (most common ore in the mine)
    Principal minerals = rhodonite
6) Braunite ore (rare)
    Principal minerals = braunite
    Minor minerals   = hausmannite, tephroite, rhodonite

Thermodynamic Interpretation: The physicochemical conditions for the formation of the observed mineralogical assemblages in the Noda-Tamagawa mine can be estimated using the thermodynamic stability information presented above. For the following discussion, all the minerals are assumed to be equilibrium products. It must also be noted that the hausmannite found in the mine is not always the pure manganese end-member and instead contains other molecules such as galaxite, jacobsite and magnetite as solid solutions (Watanabe and Kato, 1966). Thus, the calculated stability field for hausmannite would be altered according to the extent of solid solutions.

A. General P-T Conditions: Considering the possible thickness of sedimentary rocks and the minimum melting temperature for water-saturated granite (see Fig. VII–4–2), the P-T condition which existed during the thermal metamorphism in the vicinity of the granite intrusion in the Noda-Tamagawa area may have been temperatures below 700°C and pressures no greater than 2,000 bars. The occurrences of sillimanite and andalusite (but not kyanite) in the hornfelsic rocks in the area appear to be consistent with the upper bounds for temperature and pressure, although the P-T stability range for those aluminosilicates are not well understood (e.g. Zen, 1969).

B. The Pyrochroite Ore: Figure VII–4–7(b) shows that the stability field for pyrochroite is limited to a narrow range of CO_2 pressure and temperature. For a condition of H_2O pressure less than 2,000 bars, pyrochroite cannot be stable at a CO_2 pressure greater than 7 bars. It is also seen in Fig. VII–4–7(b) that the upper temperature limit for pyrochroite is 450°C if the oxygen fugacity is high enough for the formation of hausmannite, and it is 400°C if manganosite is the stable oxide phase. Since pyrochroite in the Noda-Tamagawa mine is associated mainly with manganosite (but not with hausmannite) and with a minor quantity of tephroite, rhodochrosite and alabandite, it appears that the oxygen fugacity was smaller than 10^{-21} bars and the temperature was no higher than 400°C when it was formed.

On the basis of textural evidence, Watanabe et al. (1970b) interpreted that pyrochroite was probably formed by hydration of manganosite, which was, in turn, a thermal decomposition product of rhodochrosite. Their explanation may be stated in terms of the thermodynamic variables as follows:

1) As shown in Figs. VII–4–2 and –7, manganosite is always stable at higher temperatures than pyrochroite for a given water pressure. Hence, manganosite will be less stable than pyrochroite as the temperature of the ore-forming environment falls below 400°C.

2) Manganosite is always more stable than rhodochrosite at higher temperatures and/or lower CO_2 pressures, as shown in Figs. VII–4–3 and –7. If manganosite was indeed formed by decomposition of rhodochrosite as speculated by Watanabe et al. (1970b), the environment must have been warming and/or losing CO_2 pressure. Since the stability field for rhodochrosite is located in a lower temperature range than that for pyrochroite (Fig. VII–4–7), the suggested sequence of mineral alterations from rhodochrosite to manganosite and then to pyrochroite means an initial temperature increase from the rhodochrosite field to the manganosite field followed by a temperature decrease to the stability field of pyrochroite, while the pressures of H_2O and CO_2 remained more or less constant. Alternatively, it may be interpreted that the temperature and water pressure in the ore-forming environment stayed in a range of 350° to 400°C and at 2,000 bars, respectively, while the CO_2 pressure was decreased from above 10 bars to below 1 bar. Referring to Fig. VII–4–7, an area between 350° and 400°C and oxygen fugacity less than 10^{-25} bars is within the stability field of rhodochrosite when the CO_2 pressure is greater than 10 bars; then it becomes a part of the manganosite field as the CO_2 pressure is reduced below 10 bars. When the CO_2 pressure falls to 1 bar and lower, it becomes a part of the stability field for pyrochroite. The occurrence of tephroite and rhodochrosite as minor accessory minerals is consistent with either interpretation discussed above.

C. The Rhodochrosite and Hausmannite Ores: Those two ore types will be discussed here together since hausmannite ore occurs as lenticular bodies in close association with rhodochrosite ore. As shown in Figs. VII–4–4,–5 and –7, the occurrence of hausmannite indicates a greater oxygen fugacity condition than those for rhodochrosite, pyrochroite and manganosite. A close association of the hausmannite and rhodochrosite ores suggests that the condition was in the vicinity of the Mn_3O_4-$MnCO_3$ stability boundary for Reaction (9). The occurrence of manganosite as an accessory mineral suggests that the condition was probably not far from the Mn_3O_4-MnO stability boundary. The absence of bixbyite from the mineral assemblage indicates that the oxygen fugacity did not exceed the Mn_2O_3-Mn_3O_4 boundary. The occurrence of tephroite and the absence of rhodonite as an accessory mineral in the hausmannite ore suggests that the oxygen fugacity of the environment was below the (hausmannite + rhodonite) − tephroite boundary shown in Figs. VII–4–4,–5 and –7. Furthermore, the absence of hematite and the presence of magnetite molecules in solid solutions with hausmannite (Watanabe and Kato, 1966) appear to indicate that the oxygen fugacity of the environment did not exceed the Fe_2O_3-Fe_3O_4 boundary (Fig. VII–4–4). However, this boundary cnnnot be taken with confidence as the upper limit of oxygen fugacity until thermochemical data for jacobsite ($MnFe_2O_4$) become available. On the other hand, the absence of siderite (or $FeCO_3$ molecules in solid solutions with other carbonates) in this deposit suggests that the oxygen fugacity was greater than the Fe_3O_4-$FeCO_3$ (Reaction (10)) boundary. The lack of a conspicuous occurrence of graphite in hornfels may also indicate that the oxygen fugacity was greater than the CO_2-graphite (Reaction (11)) boundary (Figs. VII–4–4(a), (b) and (c)).

D. Tephroite Ore: According to Watanabe *et al.* (1970b), the tephroite ore is associated neither with hausmannite nor with manganosite. Hence, it appears that this ore is not saturated with manganese oxides. Since the rhodonite-tephroite boundary shown in Figs. VII–4–4,–5 and –7 is only applicable to a system supersaturated with hausmannite, this boundary cannot be applied to the interpretation of the environment for the formation of tephroite ore. The observed association of tephroite with a minor quantity of rhodochrosite and the absence of quartz indicate that Reaction (3) had progressed all the way to the tephroite side, and an excess of rhodochrosite had been left unreacted after quartz was consumed by the reaction. Thus, the univariant boundary for the ($2MnCO_3 + SiO_2$) − Mn_2SiO_4 reaction (Figs. VII–4–1 and –3) defines the minimum temperature for the formation of the tephroite ore, which is about 246°, 310°, 392° and 500°C at the CO_2 pressures of 1, 10, 100 and 1,000 bars, respectively. If the maximum CO_2 pressure is taken to be 7 bars as estimated from the occurrence of pyrochroite, the minimum temperature of formation for the tephroite ore is 300°C.

Summary

If various ore types occurring in the Noda-Tamagawa ore bodies were formed under smoothly varying physicochemical parameters in space and time, the observed mineral assemblages can be explained by the following conditions:

a. At a temperature of 350°–450°C and an oxygen fugacity smaller than 10^{-21} bars, the central pyrochroite zone was formed. The pore fluids should be depleted of CO_2 with respect to H_2O (i.e., $P_{H_2O}/P_{CO_2} > 1,000$).

b. As temperatures decrease outward to below 350°–400°C, rhodochrosite becomes stable even in such a CO_2-depleted environment.

c. An increase in oxygen fugacity or, alternatively, a further decrease in temperature accompanied by a decrease in water and CO_2 pressures, would cause hausmannite to become stable. An increase in oxygen fugacity may be attained by decomposition of water to oxygen, hydrogen, and subsequent loss of hydrogen from the immediate ore-forming areas. As shown in Figs. VII–4–4 and –5, a decrease in temperature and in the pressures of water and CO_2 would also cause the stability fields of pyrochroite and rhodochrosite to be replaced by the stability field of hausmannite. Considering the absence of hematite from the entire ore body, the latter explanation appears to be more plausible.

Funds for experimental study were given by N.Y. State College of Ceramics and U.S. National Science Foundation. We gratefully acknowledge their financial support.

Appendix: Thermal metamorphism of the manganese ore deposits in the Ashio mountainland, Gunma Prefecture (Refer to Chapter II–15)

The geology and mineralogy of the stratified manganese deposits occurring in the Ashio Mountainland (Fig I-3) have been summarized by Watanabe et al.* (1967) and Bunno** (1973). According to Bunno (1973), the mineral assemblages in the primary manganese ores in this area vary systematically as the metamorphic grade of the host rocks is increased from shale to biotite hornfels and to cordierite hornfels. The ore minerals occurring in shale are mainly jacobsite ($MnFe_2O_4$) and rhodochrosite with a minor amount of rhodonite (including pyroxmangite), tephroite and manganosite. No hausmannite has been reported in this class of deposits. The manganese ore deposits situated in biotite hornfels are mainly composed of jacobsite ($MnFe_2O_4$), hausmannite ($MnMn_2O_4$), galaxite ($MnAl_2O_4$), rhodochrosite, rhodonite, tephroite and spessartite. The ratio of rhodochrosite/jacobsite appears to be reduced compared to the previous class of deposits. A small quantity of alabandite, manganosite and other sulfide minerals have been also found. In the ore deposits found in cordierite hornfels, rhodonite, spessartite, and alabandite are common and those manganese minerals are associated with a fair quantity of sulfide minerals including pyrite, pyrrhotite, sphalerite and galena. A lesser quantity of rhodochrosite, tephroite and hydrous manganese silicate minerals have been found. Those observations suggest that the environment, in which the observed mineral assemblages were formed, became increasingly more reducing with increasing degree of the metamorphism. As shown in Figs. VII–4–5 and–7, the mineral assemblages become dominated by Mn^{+2} and Fe^{+2} at higher temperatures even if the oxygen fugacity remained constant. This is because all the stability boundary curves slope upward to higher temperatures. At high temperatures, rhodochrosite becomes less stable to manganosite, which may, in turn, react with sulfur (if present) to form alabandite. Hence, the observed trend in the manganese mineral assemblage in the Ashio area can be explained in terms of increasing temperature alone without altering the oxygen fugacity. The absence of pyrochroite in the ores of Ashio suggests that the Ashio environment was probably more enriched in CO_2 and had a smaller oxygen fugacity compared to that in the Noda-Tamagawa deposits. A study for the CO_2/H_2O ratios in fluid inclusions in the minerals occurring in those respective areas might provide critical information to test this postulate.

---

* Watanabe, T., Mukaiyama, H., Kanehira, K. and Hamada, T. (1957) Geological Map and its Explanatory Text of the Ashio Mountainland. Office of Tochigi Prefecture, 1–40 (in Japanese).
** Bunno, M. (1973) Thermal metamorphism of the bedded manganese deposits in the Ashio area, in Imai, H., Kawai, K. and Miyazawa, T. eds., Ore Deposits in Kanto District, Asakura, Tokyo, 206–209 (in Japanese).

REFERENCES

Abe, M. (1963) Zonal distribution of the ore deposits at the Akenobe mine: Mining Geol., **13**, 101–114. (in Japanese with English abstract)
Akizuki, M. (1969) Fibrous sphalerite from Hosokura mine: Sci. Rept. Tohoku Univ. Ser. 3, No. *10*, 359–369.
Akome, K. and Haraguchi, M. (1967) The characteristics of fracture and mineralization at the Toyoha mine: Mining Geol., **17**, 93–100. (in Japanese with English abstract)
Alexander, J. B. and Flinter, B. H. (1965) A note on varlamoffite and associated minerals from the Batang Pedang district, Perak, Malaya, Malaysia: Mineral. Mag., **35**, 622–627.
Allen, E. T., Crenshaw, J. L., Johnston, J. and Larsen, E. S. (1912) Mineral sulphides of iron: Am. J. Sci., **33**, 169–236.
———, ———, and Merwin, H. E. (1914) Effect of temperature and acidity in the formation of marcasite and wurtzite, a contribution to the genesis of unstable forms: Am. J. Sci., **38**, 393–431.
Allen, V. T. and Fahey, J. J. (1957) Some pyroxenes associated with pyrometasomatic zinc deposits in Mexico and New Mexico: Geol. Soc. Am. Bull., **68**, 881–896.
Aminoff, G. (1923) Untersuchungen über die Kristallstrukturen von Wurtzit und Rotnickelkies: Z. Kristallogr., **58**, 203–219.
Anderson, E. M. (1936) The dynamics of formation of cone sheets, ring-dykes and cauldron subsidences: Proc. Royal Soc. Edinburgh, **56**, 128–157.
——— (1937) Cone-sheets and ring-dikes. The dynamical explanation: Bull. Volcano., **1**, 35–40.
——— (1951) The Dynamics of Faulting and Dyke Formation with Applications to Britain, Oliver and Boyd, Edinburgh, 206 p.
ARCYANA (1975) Transform fault and rift valley from bathyscaph and diving saucer: Sci., **190**, 108–116.
Armstrong, R. L. (1968) A model for the evolution of strontium and lead isotopes in a dynamic earth: Rev. Geophys. Space Phys., **6**, 175–199.
Arnold, R. G. (1966) Mixtures of hexagonal and monoclinic pyrrhotite and the measurement of the metal content of pyrrhotite by X-ray diffraction: Am. Mineral., **51**, 1221–1227.
——— (1967) Range in composition and structure of 82 natural terrestrial pyrrhotite: Can. Mineral., **9**, 31–50.
Asahi, N., Saito, M., Kozaki, K., Togo, F., Matsumura, A., Igarashi, T. and Watanabe, Y. (1954) Manganiferous hematite deposits in the Tokoro district, Hokkaido: Bull. Geol. Surv. Japan, **5**, 219–234. (in Japanese with English abstract)
Asano, G. (1950) On the geologic structure and its relation to the mineralization in Arikoshi syncline, Ashio mine: Bull. Res. Inst. Mineral. Dressing, Metallur. Tohoku Univ., **6**, 83–92. (in Japanese with English abstract)
Balk, P. (1936) Structure elements of domes: Am. Assoc. Petroleum Geol. Bull., **20**, 51–67.
Banno, S. and Miller, J. A. (1965) Additional data on the age of metamorphism of the Ryoke and Sambagawa metamorphic belts: Japan. J. Geol. Geogr., **36**, 17–22.
———, Yokoyama, K., Iwata, O. and Terashima, S. (1976) Genesis of epidote amphibolite masses in the Sambagawa metamorphic belt of central Shikoku: J. Geol. Soc. Japan, **82**, 199–210. (in Japanese with English abstract)
Barnes, H. L., and Ernst, W. G. (1963) Ideality and ionization in hydrothermal fluids: The system MgO-H$_2$O-NaOH: Am. J. Sci., **261**, 129–150.
Bartholomé, P. (1958) On the paragenesis of copper ore: Studia Univ. Lavan, Faculté des Sciences (Léopoldville), **4**, 1–31.

Barton, P. B., Jr. and Skinner, B. J. (1967) Sulfide mineral stabilities: *in* Barnes, H. L. ed., Geochemistry of Hydrothermal Ore Deposits, Holt, Rinehart and Winston Inc., New York, 236–333.

────── and Toulmin, P., III (1964) The electrum-tarnish method for the determination of the fugacity of sulfur in laboratory sulfide systems: Geochim. Cosmochim. Acta, **28**, 619–640.

Bastin, E. S., Graton, L. C., Lindgren, W., Newhouse, W. H., Schwartz, G. M. and Short, M. N. (1931) Criteria of age relation of minerals with special reference to polished section of ore: Econ. Geol., **26**, 561–610.

Berg, G. (1928) Über den Begriff der Rejuvenation und seine Bedeutung für die Beurteilung von Mineralparagenesen: Z. prakt. Geol., **46**, 17–19.

Berman, J. and Campbell, W. J. (1957) Relationship of composition to the thermal stability in the huebnerite-ferberite series of tungstates: U. S. Bur. Mines, RI 5300, 14p.

Bernhard, H. (1972) Untersuchungen im pseudobinären System Stannit-Kupferkies: N. Jb. Mineral., Mh. **1972**, 553–556.

Bhattacharji, S. and Koide, H. (1975) Mechanistic model for triple junction fracture geometry: Nature, **225** (*5503*), 21–24.

Billings, M. P. (1943) Ring-dikes and their origin: N. Y. Acad. Sci. Trans., Ser. 2, **5**, 131–144.

────── (1945) Mechanics of igneous intrusion in New Hampshire: Am. J. Sci., **243**(A), 40–68.

────── (1972) Structural Geology, Prentice-Hall, Inc. N. J., 606p.

Blackwelder, E. (1916) The geologic role of phosphorus: Am. J. Sci., **42**, 285–298.

Bloxam, T. W. (1959) Glaucophane schist and associated rocks near Valley Ford, Calif.: Am. J. Sci., **257**, 95–112.

Borchert, H. (1934) Über Entmischungen in System Cu-Fe-S und ihre Bedeutung als geologische Thermometer: Chemie der Erde, **9**, 145–172.

Boyle, R. W. (1954) A decrepitation study of quartz from the Campbell and Negus-Rycon shear zone system, Yellowknife, Northwest Territories: Geol. Surv. Canada, Bull. **30**, 20p.

Brace, W. F. (1972) Pore Pressure in Geophysics: *in* Heard, H. C., Borg, I. Y., Carter, N. L. and Raleigh, C. B., eds., Flow and Fracture of Rocks: Geophys. Monogr., *16*, Am. Geophys. Union, 265–273.

────── and Bomborakis, E. G. (1963) A note on brittle crack growth in compression: J. Geophys. Res., **68**, 3709–3713.

Bryner, L. (1961) Breccia and pebble columns associated with epigenetic ore deposits: Econ. Geol., **56**, 488–508.

────── (1969) Ore deposits of the Philippines, an introduction to their geology: Econ. Geol., **64**, 644–666.

Buckner, D. A., Roy, D. M. and Roy, R. (1960) Studies in the system $CaO-Al_2O_3-SiO_2-H_2O$, II: the system $CaSiO_3-H_2O$: Am. J. Sci., **258**, 132–147.

Buddington, A. F. (1935) High-temperature mineral associations at shallow to moderate depths: Econ. Geol., **30**, 205–222.

────── (1959) Granite emplacement with special reference to North America: Geol. Soc. Am., Bull., **70**, 671–747.

Buerger, M. J. (1928) The plastic deformation of ore minerals. 2: Am. Mineral., **13**, 35–51.

────── (1934) The pyrite-marcasite relation: Am. Mineral., **19**, 37–61.

Bunno, M. (1971) Hessite from the Sado Mine, Niigata Prefecture: Mining Geol., **21**, 301–302. (in Japanese with English abstract)

Bureau of Mines, Ministry of Commerce and Industry of Japan (1932): The reconnaissance of the iron deposits in Japan, 170p. (in Japanese)

Burnham, C. W. (1967) Hydrothermal fluids at the magmatic stage: *in* Barnes, H. L., ed., Geochemistry of Hydrothermal Ore Deposits, Holt, Rinehart and Winston, Inc., New York, 34–76.

──────, Holloway, J. R. and Davis, N. F. (1969) Thermodynamic properties of water to 1000°C and 10,000 bars: Geol. Soc. Am. Spec. Paper, *132*, 96p.

Burt, D. M. (1971) Some phase equilibria in the system Ca-Fe-Si-C-O: Carnegie Inst. Wash., Yearbook **70**, Geophys. Lab., 178–185.

Burt, D. M. (1972) Silicate-sulfide equilibria in Ca-Fe-Si skarn deposits: Carnegie Inst. Wash., Yearbook **71**, Geophys. Lab., 450–457.

Cabri, L. J. (1967) A new copper-iron sulfide: Econ. Geol., **62**, 910–925.

—— (1973) New data on phase relations in the Cu-Fe-S system: Econ. Geol., **68**, 443–454.

—— and Hall, S. R. (1972) Mooihoekite and haycockite, two new copper-iron sulfides, and their relationship to chalcopyrite and talnakhite: Am. Mineral., **57**, 689–708.

——, ——, Szymanski, J. T. and Stewart, J. M. (1973) On the transformation of cubanite: Can. Mineral., **12**, 33–38.

—— and Harris, D. C. (1971) New compositional data for talnakhite, Cu_{18} (Fe, $Ni)_{16}S_{32}$: Econ. Geol., **66**, 673–675.

Cameron, E. N., Rowe, R. B. and Weis, P. L. (1952) Fluid inclusions in beryl and quartz from pegmatites of the Middletown district, Connecticut: Am. Mineral., **38**, 218–262.

Chase, F. M. (1949) Origin of the Bendigo saddle reefs with comments on the formation of ribbon quartz: Econ. Geol., **44**, 561–597.

Christophe-Michel-Lévy, M. (1956) Reproduction artificielle des grenats calciques: grossulaire et andradite: Bull. Soc. franç. Minéral. Cristallogr., **79**, 124–218.

Chung, C. K. (1964) Geology and origin of zoning in the main vein of the Sangdong ore deposits: J. Korean Inst. Mining, **1**, 3–8. (in Korean with English abstract)

—— (1966) Geological structure and ore deposits in Sangdong Area: J. Korean Inst. Mining, **1**, 117–121. (in Korean with English abstract)

Church, S. E. and Tatsumoto, M. (1975) Lead isotope ralations in oceanic ridge basalts from the Juan de Fuca – Gorda Ridge area, N. E. Pacific Ocean: Contr. Mineral. Petrol., **53**, 253–279.

Clark, S. P. and Ringwood, A. E. (1964) Density distribution and composition of the mantle: Rev. Geophys., **2**, 35–88.

Clarke, F. W. (1924) Data of Geochemistry: U. S. Geol. Surv., Bull., *770*, 552p.

Cloos, H. (1928) Über antithetische Bewegungen: Geol. Runds., **19**, 241–251.

Coats, R. (1940) Propylitization and related types of alteration on the Comstock Lode: Econ. Geol., **35**, 1–6.

Collenette, P. (1958) The geology and mineral resources of the Jesselton-Kinabalu area, north Borneo: Geol. Surv. Dept., British Territories in Borneo, Mem., *6*, 194p.

Correns, F. W., Barth, T. F. W. and Eskola, P. (1940) Die Entstehung der Gesteine, Springer Verlag, Berlin, 422p.

Damon, P. E. (1968) Application of the potassium-argon method to the dating of igneous and metamorphic rocks within the Basin and Range ranges of southwest: *in* Titley, S. R., ed., South Arizona Guidebook III, Arizona Geol. Soc., Tucson, 7–20.

Degens, E. T. and Ross, D. A., eds. (1969) Hot Brines and Recent Heavy Metal Deposits in the Red Sea, Springer Verlag, New York, 600p.

Deicha, G. (1950) Essais par ecrasement de fragments mineraux pour la mise en evidence d'inclusions de gaz sous pression: Bull. Soc. franç. Mineral. Cristallogr., **73**, 439–445.

—— (1951) Neue Methoden zur Erforschung der hydrothermalen und pneumatolytischen Einschlüsse in Mineralien und Gestein: N. Jb. Mineral., Mh., **9**, 193–203.

DePaolo, D. J. and Wasserburg, G. J. (1976) Nd isotope variations and petrogenetic models: Geophys. Res. Letters, **3**, 249–252.

De Sitter, L. U. (1956) Structural Geology, McGraw-Hill, New York, 552p.

De Waard, D. (1949) Diapiric structures: Kon. Ned. Akad. Wetensch. Proc., **52**, 1–14.

Doi, M. (1961–1962) Geology and cupriferous pyrite deposits (Besshi-type) in the Sambagawa metamorphic zone including Besshi and Sazare mines in central Shikoku: Mining Geol., **11**, 610–626, **12**, 1–15, 63–83. (in Japanese with English abstract)

Doelter, C. and Leitmeier, H. (1926) Handbuch der Mineralchemie, Bd. 4, Erste Hf. Theodor Steinkopf, Dresden, 1003p.

Edwards, A. B. (1954) Textures of the Ore Minerals, Revised Ed., Aust. Inst. Mining Metallur., Melbourne, 242p.

Eggler, D. H. (1973) Role of CO_2 in melting processes in the mantle. Carnegie Inst. Wash., Yearbook **72**, Geophys Lab. 457–467.

Ehrenberg, H. (1931) Der Aufbau der Schalenblenden der Aachener Blei-Zink Erzlagerstätten und der Einfluss ihres Eisengehaltes auf die Mineralbildung: N. Jb., Beil. Bd., **64**, A. 397–422.

Elinson, M. M., Polykovsky, V. S. and Shuvalov, V. B. (1969) On the gaseous composition of solutions which precipitated in the formation of greisens and quartz-wolframite veins of Maidantal: Geokhimya, **1969** 571–581. (in Russian with English abstract)

El Sharkawi, M. A. and Dearman, W. R. (1966) Tin-bearing skarns from the northwest border of the Dartmoor granite, Devonshire, England: Econ. Geol., **61**, 362–369.

El Shatoury, H. M., Takenouchi, S. and Imai, H. (1974) Fluid inclusion studies in the Toyoha mine, Hokkaido, Japan: IAGOD in Varna, Bulgaria, 1974, Abstracts of Papers, 275–277.

——, —— and —— (1975) Nature and temperature of ore-forming fluids at Toyoha mine in the light of fluid inclusions in quartz porphyry: Mining Geol., **25**, 11–25.

Emmons, W. H. (1924) Primary Downward changes in ore deposits: AIME, Trans., **70**, 964–992.

—— (1933) On the mechanism of the deposition of certain metalliferous lode systems associated with granitic batholiths: *in* Ore Deposits of the Western States (Lindgren Volume), AIME, 327–349.

—— (1935) On the origin of certain systems of ore-bearing fractures: AIME, Trans., **115**, 9–35.

—— (1938) Diatremes and certain ore-bearing pipes: AIME, Tech. Pub., No. *891*, 1–15.

—— (1940) The Principles of Economic Geology, McGraw-Hill, New York, 529p.

Enjoji, M. (1970) Analytical methods of fluid inclusions in minerals: Notes on Ore Deposits, **9**, 4–12. (in Japanese)

—— (1972) Studies on fluid inclusions as the media of the ore formation. Sci. Rept. Tokyo Kyoiku Daigaku, Sec. C, **11**, 79–126.

—— (1978) Fluid inclusions in minerals in some contact metasomatic ore deposits: *in* Prof. T. Miyazawa Memorial Volume on the Occasion of his Retirement, B49–B64. (in Japanese)

—— and Takenouchi, S. (1976) Present and future researches of fluid inclusions from vein-type deposits, *in* Nakamura, T., ed., Genesis of Vein-type Deposits in Japan, Mining Geol., Spec. Issue **7**, 85–100. (in Japanese with English abstract)

Epprecht, W. (1946) Die Manganmineralien vom Gonzen und ihre Paragenese: Schweiz. Mineral. Petrogr. Mit., **26**, 19–27.

Eskola, P. (1950) Orijarvi re-interpreted: Commission Geol. Fin. Bull., **150**, 93–111.

Feiss, P. G. (1974) Reconnaissance of the tetrahedrite-tennantite/enargite-famatinite phase relations as a possible geothermometer: Econ. Geol., **69**, 383–396.

Fenner, C. N. (1933) Pneumatolytic process in the formation of minerals and ores: *in* Ore Deposit of the Western States (Lindgren Volume), AIME, 58–106.

Fernandez, J. C. and Pulanco, D. H. (1967) Reconnaissance geology of northwestern Luzon, Philippines: First Symposium on the Geology of the Mineral Resources of the Philippines and Neighboring Countries, Geol. Surv. of Philippines, **1**, 35–44.

Fleet, M. E. (1970) Refinement of the crystal structure of cubanite and polymorphism of $CuFe_2S_3$: Z. Kristallogr., **132**, 276–287.

Fleischer, M. (1955) Minor elements in some sulfide minerals: Econ. Geol., 50th Anniv. Vol., 970–1024.

Fletcher, C. J. N. (1977) The geology, mineralization, and alteration of Ilkwang mine, Republic of Korea. A Cu-W bearing tourmaline breccia pipe: Econ. Geol., **72**, 753–768.

Folinsbee, R. E., Kirkland, K., Nekolaichuk, A. and Smejkal, V. (1972) Chinkuashih, a gold-pyrite-enargite-barite hydrothermal deposit in Taiwan: *in* Doe, B. R. and Smith, D. D., eds., Studies in Mineralogy and Precambrian Geology, Geol. Soc. Am. Mem., *135*, 323–335.

Fowler, G. M., Lyden, J. P., Gregory, F. E. and Agar, W. M. (1935) Chertification in the Tri-State (Oklahoma-Kansas-Missouri) mining district: AIME, Trans., **115**, 106–163.

Franck, E. U., and Todheide, K. (1959) Thermische Eigenshaften über kritischer Mischungen von Kohlendioxyd und Wasser bis zu 750°C und 2000 atm: Z. Phys. Chemie, N. F., **22**, 232–245.

Frank, F. C. (1965) On dilatancy in relation to seismic sources: Rev. Geophys., **3**, 485–503.
Franz, E. D. (1971) Kubischer Zinnkies und tetragonaler Zinnkies mit Kupferkiesstruktur: N. Jb. Mineral., Mh., **1971**, 218–223.
Frebold, G. (1927) Über einge Mineralien der Enargitgruppe und ihre paragenetischen Verhältnisse in der Kupferlagerstätte von Mankayan auf Luzon (Philippinen): N. Jb. Mineral. Abh., **56A**, 316–333.
Frondel, C. (1943) Mineralogy of the calcium phosphates in insular phosphate rock: Am. Mineral., **28**, 215–232.
Fujii, T. (1970) Unmixing in the system, sphalerite and chalcopyrite: *in* Tatsumi, T., ed., Volcanism and Ore Genesis, Univ. Tokyo Press, Tokyo, 357–366.
Fujiki, Y. (1963) Paragenetic relations in the ore minerals from the Komori mine. On the exsolution textures of Cu-Fe-S system minerals: Mining Geol., **13**, 339–350. (in Japanese with English abstract)
────── (1964a) Geology and ore deposit of the Komori mining district, Japan: Mining Geol., **14**, 36-47. (in Japanese with English abstract)
────── (1964b) Geochemical study on minor elements in sulphide minerals and igneous rocks from the Komori mine: Mining Geol., **14**, 48–57. (in Japanese with English abstract)
Funahashi, M. (1948) Contact metasomatism associated with the pyroxene peridotite of the Horoman region in the Hidaka metamorphic zone: J. Fac. Sci. Hokkaido Univ., Ser. IV, **8**, 1–33. (in Japanese with English abstract)
Gains, R. V. (1957) Luzonite, famatinite and some related minerals: Am. Mineral., **42**, 766–779.
Garrels, R. M. and Christ, C. L. (1965) Solutions, Minerals and Equilibria, Harper & Row, New York, 450p.
Gast, P. W. (1968) Trace element fractionation and the origin of tholeiitic and alkaline magma types: Geochim. Cosmochim. Acta, **32**, 1057–1086.
────── (1969) The isotopic composition of lead from St. Helena and Ascension Islands: Earth Planet. Sci. Letters, **5**, 353–359.
Gavelin, S. (1936) Auftreten und Paragenese der Antimon Minerale in Zwei Sulfid-Vorkommen in Skelleftefelde, Nordschweden: Sveriges Geol. Untersök., Ser. C, No. *404*, 20p.
Geijer, P. (1924) On cubanite and chalcopyrrhotite from Kaveltorp. Geol. Foren. Förh., Stockh. **46**, 354–355.
Genkin, A. D., Filimonova, A. A., Shadlun, T. N., Sovoleva, S. V. and Toroneva, N. V. (1966) On cubic cubanite and cubic chalcopyrite: Geol. Rudnykh Mestorozhdenii, **1966**(*1*), 41–54. (in Russian with English abstract)
Geological Society of China (1963) Change of chemical composition of uniform ore-bearing solution: *in* Symposium "Problems of Postmagmatic Ore Deposition", Appendix **v. I**, Geol. Surv. Czechoslovakia, Praha, 109–111.
Gilmour, P. C. (1972) The relationship between porphyry copper deposits and post-orogenic volcanism. AIME, Ann. Meeting, Preprints, 1–15.
Glasser, F. P. (1958) The system MnO-SiO_2: Am. J. Sci., **256**, 398–412.
Goldschmidt, V. M. (1954) Geochemistry, Clarendon Press, Oxford, 730p.
Goldsmith, J. and Graf, D. L. (1957) The system CaO-MnO-CO_2: Solid solution and decomposition relations: Geochim. Cosmochim. Acta, **11**, 310–334.
Gonzalez, A. (1956) Geology of the Lepanto copper mine, Mankayan, Mountain Province: *in* Kinkel, A., Santos-Yñigo, L. M., Samaniego, S. and Crispin, O., eds., Copper Deposits of the Philippines, Pt. 1., Philippine Bureau of Mines, 17–50.
Gordon, T. M. and Greenwood, H. J. (1971) The stability of grossularite in H_2O-CO_2 mixtures. Am. Mineral., **56**, 1674–1688.
Graf, D. L. (1961) Crystallographic tables for the rhombohedral carbonates: Am. Mineral., **46**, 1283–1316.
Graton, L. C. (1933) The depth zones in ore deposition: Econ. Geol., **28**, 513–555.
────── (1940) Nature of the ore-forming fluid: Econ. Geol., **35**, 197–358.
────── and Bowditch, S. L. (1936) Alkaline and acid solution in hypogene zoning at Cerro de Pasco: Econ. Geol., **31**, 651–698.

Greenwood, H. J. (1962) Metamorphic reactions involving two volatile components: Carnegie Inst. Wash. Yearbook, 61, Geophys. Lab., 82–85.

———— (1967) Wollastonite: Stability in H_2O-CO_2 mixtures and occurrence in a contact-metamorphic aureole near Salmo, British Columbia, Canada: Am. Mineral., 52, 1669–1680.

———— (1969) The compressibility of gaseous mixtures of carbon dioxide and water between 0 and 500 bars pressure and 450° and 800° centigrade: Am. J. Sci., 267-A, 191–208.

Gregory, J. W. (1921) The Rift Valleys and Geology of East Africa, Seeley Service, London, 479p.

Griffiith, A. A. (1925) The theory of rupture: Intern. Cong. Applied Mechanics, 1st, Delft, Proc., 53–63.

Groves, D. I., and Solomon, M. (1969) Fluid inclusion studies at Mt. Bishoff, Tasmania: IMM, Trans. B, 78, Bl-Bll.

Guild, P. W. (1972) Massive sulfides vs. porphyry deposits in their global tectonic settings: Joint Meeting MMIJ-AIME 1972, Tokyo, No. GI3, 12p.

Gustafson, L. B. (1963) Phase equilibria in the system Cu-Fe-As-S: Econ. Geol., 58, 667–701.

Gustafson, W. I. (1974) The stability of andradite, hedenbergite, and related minerals in the system Ca-Fe-Si-O-H: J. Petrol., 15, 455–496.

Hall, S. R. and Gabe, E. J. (1972) The crystal structure of talnakhite. Am. Mineral. 57, 368–380.

Hall, W. E., Friedman, I. and Nash, J. T. (1974) Fluid inclusion and light stable isotope study of the Climax molybdenum deposits, Colorado: Econ. Geol., 69, 884–901.

Harada, K. (1962) Skarn zone of the Doshinkubo ore deposits of the Chichibu mine, Saitama Prefecture: J. Japan. Assoc. Mineral. Petrol. Econ. Geol., 48, 60–66. (in Japanese with English abstract)

Haranczyk, C. and Jarosz, J. (1966) Secondary minerals in sedimentary copper deposits: Rudy Metale Niezelazne, 11, 290–296 [CA 66–5455].

Harcourt, G. A. (1937) The distinction between enargite and famatinite (luzonite): Am. Mineral., 22, 517–525.

Harker, R. I. and Tuttle, O. F. (1955), Studies in the system CaO-MgO-CO_2. Part 1: Am. J. Sci., 253, 209–224.

Hart, S. R. and Brooks, C. (1970) Rb-Sr mantle evolution models: Carnegie Inst. Wash., Yearbook 68, Dept. of Terrestrial Magnetism, 426–429.

Hashimoto, M., Kashima, N. and Saito, Y. (1970) Chemical composition of Paleozoic greenstones from two areas of southwest Japan: Geol. Soc. Japan, 76, 463–476.

Hayakawa, N., Nambu, M. and Aoshima, T. (1972) Studies on fluid inclusions as geothermometer (1), Decrepitation analysis: J. Mining Metallur. Inst. Japan, 88, 185–190. (in Japanese with English abstract)

————, Suzuki, M., Suzuki, S., Nambu, M. and Oka, H. (1975) Filling temperatures of the fluid inclusions in quartz from the Kaneuchi mine: Ann. Meeting Mining Metallur. Inst. Japan, Preprint, 162–163. (in Japanese)

Hayase, I. and Ishizaka, K. (1967) Rb-Sr dating on the rocks in Japan (I), Southwest Japan: J. Japan. Assoc. Mineral. Petrol. Econ. Geol., 58, 201–212. (in Japanese with English abstract)

Henderson, J. F. and Brown, I. C. (1949) Yellowknife district of Mackenzie, Northwest Territories: Geol. Surv. Canada, Preliminary map 48–17A.

Henmi, K. (1941) Wurtzite from the Okkoppe mine, Aomori Pref.: J. Geol. Soc. Japan, 48, 527–528. (in Japanese)

Hesemann, J. and Pilger, A. (1951) Der Blei-Zink Erzgang der Zeche Victoria in Mari-Hüls (Westfalen): Monographien Deutschen Blei-Zink Erzlagerstätten, 7–184.

Hidaka, S., Agui, H. and Abe, Y., (1965) Geology and ore deposit of the Santo Tomas mine, Philippines: Mining Geol., 15, 83–91. (in Japanese with English abstract)

Hide, K. (1972) Significance of finding of two recumbent folds in the Sambagawa metamorphic belt of the Nagahama-Ozu district, west Shikoku: Mem. Fac. General. Educ., Hiroshima Univ., 5, 35–51. (in Japanese with English abstract)

Higgins, J. B. and Ribbe, P.H. (1977) The structure of malayaite, $CaSnOSiO_4$, a tin analog of titanite: Am Mineral., 62, 801–806.

Hiller, J. E. (1940) Versuch einer Klassifikation der Sulfid nach strukturellen Gesichtpunkten: Z. Kristallogr., **102**, 353–376.

────── and Probsthain, K. (1956) Thermische und röntgenographische Untersuchungen am Kupferkies: Z. Kristallogr., **108**, 108–129.

Hirata, H. (1931) Experimental studies on form and growth of cracks in glass plate: Sci. Papers Inst. Phys. Chem. Res., **16**, 172–195.

Hirokawa, O., Kambe, N. and Toga, F. (1954) Explanatory Text of the Geological Map of Japan, 1/50,000, Ohyaichiba: Geol. Surv. Japan. (in Japanese with English abstract)

Holland, H. D. (1972) Granites, solutions, and base metal deposits: Econ. Geol., **67**, 281–301.

Hong, C. K. and John, Y. W. (1965) Mineralization of Sangdong, Opyung and Kumjung areas I: Sci. Eng. Rept., College Eng., Seoul Nat. Univ., **1**, 285–294. (in Korean with English abstract)

────── and ────── (1966) Mineralization of Sangdong, Opyung and Kumjung areas II: Formation temperature of quartz veins in Sangdong ore deposits by heating stage method: Sci. Eng. Rept., College Eng., Seoul Nat. Univ, **2**, 101–111. (in Korean with English abstract)

Horikoshi, Y. (1937) Geology and petrology of the vicinity of Besshi, Ehime Pref.: J. Geol. Soc. Japan, **44**, 121–140. (in Japanese)

────── (1938) Geology and ore deposit of the Kune mine, Shizuoka Pref.: J. Geol. Soc. Japan, **45**, 857–872. (in Japanese)

────── (1940) Morphological studies of the Besshi type ore deposits: Japan. Soc. Promotion of Sci. 2nd Sub-Comm. Rept., No. *1*, 23p. (in Japanese)

────── and Katano, T. (1940) Geology and ore deposit of the Minenosawa mine, Shizuoka Pref.: J. Geol. Soc. Japan, **47**, 91–102. (in Japanese)

Hosking, K. F. G. (1964) Permo-Carboniferous and later primary mineralization of Cornwall and Southwest Devon: *in* Hosking, K. F. G. and Shrimpton, G. J., eds., Present Views of Some Aspects of the Geology of Cornwall and Devon, Royal Geol. Soc. Cornwall, Truro, 201–245.

Huang, C. K. (1955) Gold-copper deposits of the Chinkuashih mine, with special reference to the mineralogy: Acta Geol. Taiwanica, No. *7* (Dec.), 1–20.

────── (1963) Factors controlling the gold-copper deposits of the Chinkuashih mine, Taiwan: Acta Geol. Taiwanica, No. *10* (Dec.), 1–9.

────── (1964) Mineralogy of the Tsaoshan deposit of the Chinkuashih mine, Taiwan: Proc. Geol. Soc. China, No. *7*, (Apr.), 31–39.

────── (1965) Further notes on the mineralogy of the Chinkuashih gold-copper deposits, Taiwan: Acta Geol. Taiwanica, No. *11* (Dec.), 31–42.

────── (1972) Crystal forms of native gold and luzonite from the Chinkuashih mine, Taiwan: Acta Geol. Taiwanica, No. *15* (Apr.), 101–104.

──────, (1973) Minor gangue minerals of the Chinkuashih gold-copper deposits, Taiwan: Acta Geol. Taiwanica, No. *16* (May), 31–38.

────── (1974) Antimony contents of enargite and luzonite-famatinite from the Chinkuashih mine, Taiwan: Acta Geol. Taiwanica, No. *17* (May), 1–5.

Huckenholz, H. G. and Yoder, H. S., Jr. (1971) Andradite stability relations in the $CaSiO_3$-Fe_2O_3 join up to 30 kb: N. Jb. Mineral. Abh., **114**, 246–280.

Hujimoto, H. (1938) Radiolarian remains discovered in a crystalline schist of the Sambagawa System.: Proc. Imp. Acad. Tokyo, **14**, 252–254.

Huebner, J. S. (1971) Buffering techniques for hydrostatic systems at elevated pressure: *in* Ulmer, G. C. ed., Research Technique for High Pressure and High Temperature, Springer-Verlag, New York, 123–177.

Hulin, C. D. (1929) Structural control of ore deposition: Econ. Geol., **24**, 15–49.

────── (1948) Factors in the localization of mineralized districts: AIME, Trans., **178**, 36–57.

Iddings, J. P. (1898) Bysmalith: J. Geol., **6**, 704–710.

Ikebe, N., Fujita, K., Nakagawa, K., Kishida, K., Sakaguchi, S. and Tsukawaki, Y. (1961) Geological Map of the Hyogo Prefecture and the Explanatory Text: Hyogo Prefecture Office. (in Japanese)

Imai, H. (1941) On "wurtzite" from the Hosokura mine, Miyagi Prefecture: J. Geol. Soc. Japan, **48**, 505–515. (in Japanese)

────── (1942) Geology and ore deposit of the Nikko mine, with special reference to the genesis of gudmundite: J. Geol. Soc. Japan, **49**, 267–278. (in Japanese)

────── (1943) The copper ores from the Mankayan mine, Luzon, Philippine Islands, with special reference to enargite and luzonite problelms: J. Geol. Soc. Japan, **50**, 253–261. (in Japanese)

────── (1944) The form of the residual deposits in limestone or dolomitic limestone, with special reference to the Nakagawa-Syogun manganese deposit in Korea: J. Japan. Assoc. Mineral. Petrol. Econ. Geol., **32**, 1–10. (in Japanese)

────── (1947) On wurtzite from the Hosokura mine, Miyagi Pref., Japan: Japan. J. Geol. Geogr., **20**, 97–110.

────── (1949a) Geology and ore deposits of the Nikko mine, with special reference to the genesis of gudmundite: Japan. J. Geol. Geogr., **21**, 39–52.

────── (1949b) The copper ore from the Mankayan mine, Luzon, Philippine Islands, with special reference to the problem of luzonite, enargite and famatinite: Japan. J. Geol. Geogr., **21**, 57–69.

────── (1951) Transition of pyrite into marcasite, magnetite and hematite by chalcopyrite mineralization at the Sasagatani mine, western Japan: J. Geol. Soc. Japan, **57**, 211–216.

────── (1952) Universal compass and the plunge of the bedded cupriferous pyritic ore deposit in Japan: Am. Mineral., **37**, 861–864.

────── (1955) On the fissure system of the Hosokura and Ohdomori mines, northern Japan: J. Mining Metallur. Inst. Japan, **71**, 1–5 and 49–51. (in Japanese with English abstract)

────── (1956a) Liquid inclusions geothermometer, *in* Watanabe, T., ed., Progress in Economic Geology: Fuzambo, Tokyo, 288–293. (in Japanese)

────── (1956b) Vein system of the Hosokura and Ohdomori mines, north-eastern Japan: Japan. J. Geol. Geogr., **27**, 21–36.

────── (1958) Saddle reef type of the cupriferous pyrite deposits in Japan: J. Mining Metallur. Inst. Japan, **74**, 70–74. (in Japanese with English abstract)

────── (1959) Some problems associated with the genesis of the bedded curpiferous pyrite deposits and manganiferous hematitte deposits in the Outer Zone of Southwest Japan: Mining Geol. **9**, 1–18. (in Japanese with English abstract)

────── (1960a) Geology of the Okuki mine and other related cupriferous pyrite deposits in southwestern Japan: N. Jb. Mineral. Abh., **94** (Festband Ramdohr), 352–389.

────── (1960b) Basic rocks in the Okuki mine: Mining Geol., **10**, 31–32. (in Japanese with English abstract)

────── (1961) Formation of the gold-quartz veins at the Ohya mine, Miyagi Pref.: Mining Geol., **11**, 66–69. (in Japanese with English abstract)

────── (1963a) Vein-type deposits around the intrusive igneous masses. Veins filling the synthetic normal faults produced by magmatic upheaval: Mining Geol., **13**, 253–260. (in Japanese with English abstract)

────── (1963b) Pre-Tertiary igneous activity, metamorphism and metallogenesis: *in* Takai, F., Matsumoto, T. and Toriyama, R., eds., Geology of Japan, Univ. of Tokyo Press, Tokyo, 197–222.

────── (1964) Formations of the vein fissures and their mineralizations: J. Fac. Eng. Univ. of Tokyo, Ser. A, No. 2, 44–45. (in Japanese with English abstract)

────── (1966a) Occurrence of ore deposits as related to structural geology: J. Geol. Soc. Japan, **72**, 131–141. (in Japanese with English abstract)

────── (1966b) Formation of fissures and their mineralization in the vein-type deposits of Japan: J. Fac. Eng., Univ. Tokyo, Ser. B, **28**, No. 3, 255–302.

────── (1970) Geology and mineral deposits of the Akenobe mine: IMA-IAGOD Meetings '70, Guidebook 8, 1–23.

────── (1973) Geologic structure and mineralization at the Taishu mine, Nagasaki Pref., Japan: J. Mining Metallur. Inst. Japan, **89**, 509–514. (in Japanese with English abstract)

────── and Fujiki, Y. (1963) Study on the nickel- and cobalt-bearing sulfide minerals from the

Komori mine by means of electron microprobe: Mining Geol., **13**, 333–338. (in Japanese with English abstract)

——, —— and Tsukagoshi, S. (1967) Late Mesozoic or early Tertiary metallogenetic province of the western Kinki district: Mining Geol., **17**, 50. (in Japanese)

—— and Hayashi, S. (1959) Characteristic geologic structure observed in several tintungsten-bearing veins, with special reference to the vein system of the Takatori mine: J. Mining Metallur. Inst. Japan, **75**, 145–150. (in Japanese with English abstract)

—— and Ito, K. (1959) Geologic structure and tungsten-copper mineralization of the Kuga mine, Yamaguchi Pref.: Mining Geol., **9**, 95–100. (in Japanese with English abstract)

——, Kim, M. S. and Fujiki, Y. (1972) Geologic structure and mineralization of the hypothermal or pegmatitic tungsten vein-type deposits at the Ohtani and Kaneuchi mines, Kyoto Pref., Japan: Mining Geol., **22**, 371–381. (in Japanese with English abstract)

——, Kimura, K., Shimomura, Y., Nishio, S., Fuchida, T. and Fujiwara, S. (1951) All-around prospectings for ore deposits, survey at the Kune mine, 1: J. Mining Metallur. Inst. Japan, **67**, 297–302. (in Japanese with English abstract)

—— and Lee, M. S. (1972) Silver minerals from the xenothermal deposits in the Western Kinki Metallogenetic Province: J. Fac. Eng., Univ. Tokyo, Ser. A, Ann. Rept. **10**, 54–55. (in Japanese with English abstract)

——, ——, Iida, K., Fujiki, Y. and Takenouchi, S. (1975) Geologic structure and mineralization of the xenothermal vein-type deposits in Japan: Econ. Geol., **70**, 647–676.

——, Saito, N., Hayashi, S., Sato, K. and Kawachi, Y. (1960) The absolute age of granitic rocks in the Miyako-Taro district, Iwate Pref., Japan: J. Geol. Soc. Japan, **66**, 405–409. (in Japanese with English abstract)

——, Takenouchi, S., and Kihara, T. (1971) Fluid inclusion study at the Taishu mine, Japan, as related to geologic structure: Proc. IMA-IAGOD Meeting '70, IAGOD Volume, (Mining Geol., Spec. Issue No. *3*), 321–326.

—— and Yamadera, H. (1952), The peculiar phosphate deposits in the Noto peninsula, Japan: J. Geol. Soc. Japan, **58**, 79–92.

Imai, N., Mariko, T. and Nakamura, T. (1967) On the stanniferous iron ores from the Sampo mine, Okayama Prefecture, Japan: J. Mining Metallur. Inst. Japan, **83**, 731–738. (in Japanese with English abstract)

Ingham, F. T. and Bradfood, E. T. (1960) The geology and mineral resources of the Kinta Valley, Perak: Federation of Malaya, Geol. Surv. District Mem. *9*, 347p.

Ishihara, S. (1971) Modal and chemical composition of the granitic rocks related to the major molybdenum and tungsten deposits in the Inner Zone of southwest Japan: J. Geol. Soc. Japan, **77**, 441–452.

—— and Shibata, K. (1972) Re-examination of the metallogenic epoch of the Ikuno-Akenobe Province in Japan: Mining Geol., **22**, 67–73.

Ishizaka, K. (1971) A Rb-Sr isotopic study of the Ibaragi granitic complex, Osaka, Japan: J. Geol. Soc. Japan, **77**, 731–740.

—— and Yamaguchi, M. (1969) U-Th-Pb ages of sphene and zircon from the Hida metamorphic terrain, Japan: Earth Planet. Sci. Letters, **6**, 179–185.

Isomi, H. and Kawada, K. (1968) Correlation of the basement rocks on both sides of the Fossa Magna: Preprints on a Symposium "Fossa Magna", 75th Autumn Meeting, Geol. Soc. Japan, 4–12. (in Japanese)

Ito, J. and Arem, J. E. (1970) Idoclase: Synthesis, phase relation and crystal chemistry: Am. Mineral., **55**, 880–912.

Ito, K. (1962) Zoned skarn of the Fujigatani mine, Yamaguchi Pref.: Japan. J. Geol. Geogr., **33**, 169–190.

—— (1976) Heat flow and thickness of the oceanic lithosphere: Earth Planet. Sci. Letters, **30**, 65–70.

—— (1977) Physical and chemical natures of the lithosphere and asthenosphere, especially at their boundary: *in* Manghnani, M. H. and Akimoto, S., eds. High-Pressure Research, Applications to Geophysics, Academic Press, N. Y., 129–150.

Itoh, S. (1976) Geochemical study of bedded cupriferous pyrite deposits in Japan: Bull. Geol. Surv. Japan, **27**, 245–377.

Iwafune, T. (1952) Kamioka mine: *in* Ore Deposits and Geologic Structures (1), Mining Metallur. Inst. Japan, Tech. Paper, 4–12. (in Japanese)

Iwao, S. (1955) Mg-enrichment around some ore deposits in Japan, particularly with reference to hydrothermal gypsum and silica deposits: J. Geol. Soc. Japan, **61**, 543–555. (in Japanese with English abstract)

Iwasaki, M. (1969) The basic metamorphic rocks at the boundary between the Sambagawa metamorphic belt and the Chichibu unmetamorphosed Paleozoic sediments: Mem. Geol. Soc. Japan, No. *4*, 41–50. (in Japanese with English abstract)

Iwasaki, T., Watanabe, T. and Ando, R. (1940) X-ray study of some Japanese minerals. 1. So-called "wurtzite": J. Chem. Soc. Japan, **61**, 719–725. (in Japanese)

Izawa, E. and Mukaiyama, H. (1972) Thermally metamorphosed sulfide mineral deposits in Japan: 24th Intern. Geol. Congr., Montreal, 1972, Sec. 4 (Mineral Deposits), 455–462.

Jacobsen, W. (1950) Die Erzgange des nordwestlichen Oberharzes: Geol. Jb., **65**, 707–768.

Jacobson, G. (1970) Gunong Kinabalu area, Sabah, Malaysia: Geol. Surv. Malaysia, Rept. *8*, 111p.

Jambor, J. L. and Ross, M. (1966) Mckinstryite, a new copper-silver sulfide: Econ. Geol., **61**, 1383–1389.

Jarosz, J. (1966) A mineral of the stromeyerite group in copper-bearing sandstone: Rudy Metale Niezelazne, **11**, 464–465 [CA 66–5455].

Jeffes, J. H. E., Richardson, F. D. and Pearson, J. (1954) The heats of formation of manganese orthosilicate and manganous sulfide: Trans. Faraday Soc., **50**, 364–370.

John, Y. W. (1963) Geology and origin of Sangdong tungsten mine, Republic of Korea: Econ. Geol., **58**, 1285–1300.

——— (1967) On the malayaite occurring in Pinyok mine, Thailand: J. Japan. Assoc. Mineral. Petrol. Econ. Geol., **58**, 116–120. (in Japanese with English abstract)

——— (1968) Studies on the genesis of some pyrometasomatic deposits in Korea and Japan: Dr. thesis, Univ. of Tokyo, 150p.

Jong, W. F. (1928) Die Enargitgruppe. Struktur des Sulvanit Cu_3VS_4: Z. Kristallogr., **68**, 522–530.

Kajiwara, Y. (1971) Sulfur isotope study of the Kuroko ores of the Shakanai No. 1 deposit, Akita Pref., Japan: Geochem. J., **4**, 157–181.

——— and Krouse, H. R. (1971) Sulfur isotope partitioning in metallic sulfide systems: Can. J. Earth Sci., **8**, 1397–1408.

Kamiyama, T. (1950) On the lineation in the country rocks of the bedded cupriferous pyrite deposits: J. Geol. Soc. Japan, **56**, 249. (in Japanese)

——— (1956) On the Kune mine: *in* Watanabe, T., ed., Progress in Economic Geology, Fuzanbo, Tokyo, 426–427. (in Japanese)

Kammera, K. (1971) Paleozoic and early Mesozoic geosyncline volcanicity in Japan: Mem. Geol. Soc. Japan, No. *6*, 97–110. (in Japanese with English abstract)

Kanada, M. (1967) A history of prospecting in the Chichibu mine: J. Mining Metallur. Inst. Japan, **83**, 155–158. (in Japanese)

——— (1968) Economic geology of the Chichibu deposits, Saitama Prefecture: Ph. D. Thesis, Kyoto Univ., 105p. (in Japanese)

———, Watanabe, T. *et al.* (1961) On the geology and prospecting of the Akaiwa and Doshinkubo deposits, Chichibu mine: Mining Geol. **11**, 481–490. (in Japanese with English abstract)

Kaneda, H., Takenouchi, S., Shoji, T. and Imai, H. (1975) On cubanite at high temperatures: Mining Geol., **25**, 49. (in Japanese)

——— Shoji, T. and Takenouchi, S. (1978) Heating experiments of cubanite: Mining Geol., **28**, 71–82. (in Japanese with English abstract)

Kanehira, K. and Tatsumi, T. (1970) Bedded cupriferous iron sulfide deposits in Japan, a review: *in* Tatsumi, T., ed., Volcanism and Ore Genesis, Univ. of Tokyo Press, Tokyo, 51–76.

Kang, J. M., Chon, H. T. and John, Y. W. (1976) Mineralization and origin of breccia pipe of Ilkwang tungsten-copper mine: J. Korean Inst. Mineral Mining Eng., **13**, 218–228. (in Korean with English abstract)

Kano, H. (1948) A treatise on metasomatism in rock metamorphism: Chikyu Kagaku [Geol. Sci.], No. *2*, 41–50. (in Japanese with English abstract)

―――― (1965) A new interpretation on the genesis of the "breccia skarn" from the Omine mine, Iwate Prefecture: J. Japan. Assoc. Mineral. Petrol. Econ. Geol., **54**, 104–108. (in Japanese)

Kase, K. (1972) Metamorphism and mineral assemblages of ores from cupriferous iron sulfide deposit of the Besshi mine, central Shikoku, Japan: J. Fac. Sci. Univ. Tokyo, Sec. 2, **18**, No. *2*, 301–323.

Kashiwagi, H. (1953) Geology and scheelite deposit in the environs of Kuwanemura, Kuga-gun, Yamaguchi Pref., Japan: Geol. Rept. Hiroshima Univ. No. *3*, 15–23. (in Japanese)

Kashiwagi, T., Nishio, S. and Imai, H. (1955) On the formation temperatures of minerals by thermal microscope (heating-stage microscope) and decrepitation method: Mining Geol., **5**, 155–161. (in Japanese with English abstract)

Kato, A. (1959) Ikunolite, a new bismuth mineral from the Ikuno mine, Japan: Mineral. J., **2**, 397–407.

―――― (1965) A new mineral, sakuraiite: Chigaku Kenkyu [Earth Science Studies], Sakurai Vol., 1–5. (in Japanese)

―――― (1969) Stannoidite $Cu_5 (Fe, Zn)_2 SnS_8$, a new stannite-like mineral from the Konjo mine, Okayama Prefecture, Japan: Nat. Sci. Mus. [Tokyo] Bull., **12**, 165–172.

―――― and Fujiki, Y. (1969) The occurrence of stannoidite from the xenothermal ore deposits of the Akenobe, Ikuno and Tada mines, Hyogo Prefecture, and the Fukoku mine, Kyoto Prefecture, Japan: Mineral. J., **5**, 417–443.

―――― and Shinohara, K. (1968) The occurrence of roquesite from the Akenobe mine, Hyogo Prefecture: Mineral. J., **5**, 276–284.

Kato, T. (1912) The tourmaline copper veins in the Yokuoji mine, Nagato Prov., Japan: J. Geol. Soc. Japan, **19**, 69–88.

―――― (1920) A contribution to the knowledge of the cassiterite veins of pneumato-hydatogenetic or hydrothermal origin. A study of the copper-tin veins of the Akenobe district in the Province of Tajima, Japan: Imp. Univ. Tokyo Coll. Sci. J., **43**, Art. 5, 60p.

―――― (1925) The cupriferous pyrite ore deposits of the Shibuki and Seki mines in the Province of Bungo, Japan: J. Fac. Sci. Imp. Univ. Tokyo, **1**, Pt. 2, 65–76.

―――― (1926a) Besshi copper mine: Pan-Pacific. Sci. Congr., Excursion Guidebook, E-2, 16–27.

―――― (1926b) Replacement copper ore deposits in quartzite, Japan: Econ. Geol., **21**, 394–396.

―――― (1927a) On the peculiar ore deposit of the Asakawa mine: Japan. J. Geol. Geogr., **4**, 73–84.

―――― (1927b) The Ikuno-Akenobe metallogenetic province: Japan. J. Geol. Geogr., **5**, 121–126.

―――― (1928) Some characteristic features of the ore deposits of Japan, related to the Late Tertiary volcanic activity: Japan. J. Geol. Geogr., **6**, 41–43.

―――― (1931) Mineralization sequence in the formation of the gold-silver veins of the Toi mine, Izu Province: Japan. J. Geol. Geogr., **9**, 71–86.

―――― (1934a) On the Sambagawa, Mikabu and Oboke systems: Geography, **2**, 797–798. (in Japanese)

―――― (1934b) On the chlorite schist: Geography, **2**, 1107–1108. (in Japanese)

―――― (1937) Geology of Ore Deposits: Fuzanbo, Tokyo, 757p. (in Japanese)

―――― and Oyama, I. (1923) The tourmaline copper veins of the Kanan mine, South Keishodo, Korea: Japan. J. Geol. Geogr., **2**, 11–17.

Kawada, K. (1966) A remarkable pyroclastic flow deposit of Cretaceous age of northeast Japan: Chikyu Kagaku [Earth Science], No. *84*, 6–13. (in Japanese)

Kawai, N. and Hirooka, K. (1966) Results of age determinations on some Cenozoic volcanic rocks in southwest Japan: Symposium "Eruption ages of Japanese acidic rocks based mainly upon results of age determinations," 73th Annual Meeting, Geol. Soc. Japan, Preprints 5. (in Japanese)

Kawano, Y. and Ueda, Y. (1964) K-A dating on the igneous rocks in Japan(1): Sci. Rept. Tohoku Univ. Ser. III, **9**, 99–122.

────── and ────── (1965) K-A dating on the igneous rocks in Japan (II). Granitic rocks in Kitakami massif: Sci. Rept. Tohoku Univ. Ser. III, **9**, 199–215.

────── and ────── (1966a) K-A dating on the igneous rocks in Japan (III). Granitic rocks of the Abukuma massif: Sci. Rept. Tohoku Univ. Ser. III, **9**, 513–523.

────── and ────── (1966b) K-A dating on the igneous rocks in Japan (IV) Granitic rocks in Northeastern Japan: Sci. Rept., Tohoku Univ., Ser. III, **9**, 525–539.

────── and ────── (1967a) K-A dating on the igneous rocks in Japan (V). Granitic rocks in Southwestern Japan: Sci. Rept., Tohoku Univ., Ser. III, **10**, 55–63.

────── and ────── (1967b) K-A dating on the igneous rocks in Japan (VI). Granitic rocks, summary: Sci. Rept., Tohoku Univ., Ser. III, **10**, 65–76.

Kay, M. (1951), North American Geosyncline: Geol. Soc. Amer. Mem. *48*, 143p.

Kay, R. W. and Gast, P. W. (1973) The rare earth content and origin of alkali-rich basalts: J. Geol., **81**, 653–682.

──────, Hubbard, N. J. and Gast, P. W. (1970) Chemical characteristics and origin of oceanic ridge volcanic rocks: J. Geophys. Res., **75**, 1585–1613.

Kayaba, K. and Ishii, K. (1953) A statistical study of branches in the Ogoya mine, Ishikawa Pref.: Mining Geol., **3**, 110–113. (in Japanese with English abstract)

Keevil, N. B. (1942) Vapor pressures of aqueous solutions at high temperatures: Am. Chem. Soc. J., **64**, 841–850.

Kelley, K. K. and King, E. G. (1961) Contributions to the data on theoretical metallurgy XIV. Entropies of the elements and inorganic compounds: U. S. Bur. Mines, Bull. *592*, 149p.

Kelly, W. C. and Turneaure, F. S. (1970) Mineralogy, paragenesis, and geothermometry of the tin and tungsten deposits of the Eastern Andes, Bolivia: Econ. Geol., **65**, 609–680.

Kennedy, G. C. (1954) Pressure-volume-temperature relations in CO_2 at elevated temperatures and pressures: Am. J. Sci., **252**, 225–241.

Kigai, I. N. (1963) Lifudzin deposit as an example of the combination of mono- and poly-ascendent zoning: *in* Symposium, "Problems of Postmagmatic Ore Deposition" Appendix to v. I, Geol. Surv. Czechoslovakia, Praha, 64–66.

Kikuchi, T. (1950) Geology and ore deposit of the Jiro mine, Tokushima Pref.: Bull. Geol. Surv. Japan, **1**, 243–246. (in Japanese with English abstract)

Kilinc, I. A. and Burnham, C. W. (1972) Partitioning of chloride between a silicate melt and coexisting aqueous phase from 2 to 8 kilobars: Econ. Geol., **67**, 231–235.

Kim, M. Y., Fujiki, Y., Takenouchi, S. and Imai, H. (1972) Studies on the fluid inclusions in the minerals from the Ohtani and Kaneuchi mines: Mining Geol., **22**, 449–455. (in Japanese with English abstract)

Kimura, T. (1954) The discovery of a low angle thrust along the Mikabu Line in the eastern Kii peninsula, western Japan: J. Earth Sci. Univ. Nagoya, **2**, 173–190.

Kinoshita, K. (1924) Gypsum crystals from Ishigamori: Japan J. Geol., Geogr., **3**, 113–118.

────── (1953) Germanium ore from the Obira mine, Ohita Pref.: J. Mining Inst. Kyushu, **21**, 385–393. (in Japanese)

Kirk, H. J. C. (1967) Porphyry copper deposit in northern Sabah, Malaysia: IMM, Trans. B, **76**, B212–B213.

────── (1968) The igneous rocks of Sarawak and Sabah: Geol. Surv. Borneo Region, Malaysia, Bull. *5*, 210p.

Kitamura, K. (1974) A consideration on deposition of PbS in skarn: 2nd Symp. "Studies on the formation temperatures of ore deposits", Tokyo, Preprint 10–11. (in Japanese)

────── (1975) Al-Fe partitioning between garnet and epidote from contact metasomatic copper deposits of the Chichibu mine, Japan: Econ. Geol., **70**, 725–738.

Kiyosu, Y. and Nakai, N. (1977) Estimation of formation temperature for the Taishu, Toyoha and Kamioka Pb-Zn deposits on the basis of sulfur isotope temperature scale: Japan. Assoc. Mineral. Petrol. Econ. Geol., **72**, 103–108. (in Japanese with English abstract)

Klepper, M. R. (1947) The Sangdong tungsten deposits, southern Korea: Econ. Geol., **42**, 465–477.

Klockmann, F. (1891) Mineralogische Mitteilungen aus Sammlungen der Bergakademie zu Clausthal: Z. Kristallogr., **19**, 265–275.

Kobayashi, T. (1941) The Sakawa orogenic cycle and its bearing on the origin of the Japanese islands: J. Fac. Sci. Imp. Univ. Tokyo, Sec. 2, **5**, Pt. 7, 219–578.

―――― (1957) The late Mesozoic Sakawa orogeny: Proc. 9th Pacific Sci. Congr., **12** (Geol. and Geogr.), 121–127.

―――― and Kimura, T. (1944) A study on the radiolarian rocks: J. Fac. Sci. Imp. Univ. Tokyo, Sec. 2, **7**, Pt. 2, 75–178.

Kochibe, T. (1892) The pyrite deposit of the Besshi mine: J. Geol. Soc. Tokyo **4**, 541–543. (in Japanese)

Koide, H. (1968) A tectonophysical study of the initiation and development of fracture in rocks: Ph. D. Thesis, Univ. of Tokyo, 183p.

―――― (1971) Fractures aligned en echelon and fracture patterns: Proc. IMA-IAGOD Meeting '70, IAGOD Vol. (Mining Geol., Spec. Issue *3*), 107–114.

―――― (1974a) Brittle fracturing by a phase change, a mechanism of deep earthquakes: EOS (Am. Geophys. Union Trans.), **55**, 430.

―――― (1974b) Fractures and their relation to the generation of magma in the depth, *in* Fracturing and Igneous Activity: Chidanken-Sempo [Assoc. Geol. Collaboration Japan, Monogr.], *18*, 87–90. (in Japanese with English abstract)

―――― (1976a) In situ measurement of stress in the crust, Chishitsu-Nyusu [Geology Monthly], No. *257*, 14–22. (in Japanese)

―――― (1976b) Griffith inclusion model of fault gouge zone and mechanical effect of creep or microcrack concentration before earthquakes: EOS (Am. Geophys. Union Trans.), **57**, 287.

―――― (1976c) Fracturing of heterogeneous materials by thermally induced residual stress with special reference to rocks: Proc. 2nd Intern. Conf. Mechan. Behav. Materials, 1345–1348.

―――― and Bhattacharji, S. (1975a) Formation of fractures around magmatic intrusions and their role in ore localization: Econ. Geol., **70**, 781–799.

―――― and ―――― (1975b) Mechanistic interpretation of rift valley formation: Sci. **189**, 791–793.

―――― and ―――― (1977) Geometric patterns of active strike-slip faults and their significance as indicators for areas of energy release, *in* Saxena, S. and Bhattacharji, S. eds., Energetics in Geological Process, Springer Verlag, New York, 44–66.

―――― and Hoshino, K. (1967) Development of microfractures in experimentally deformed rocks: Jour. Seism. Soc. Japan, 2nd ser. **20**, 85–97. (in Japanese with English abstract)

――――, ――――, and Inami, K. (1974) Fault models based on an extended Griffith's theory: Bull. Geol. Surv. Japan, **25**, 89–103. (in Japanese with English abstract)

Kojima, G. (1948) A consideration on the metasomatic formation of green rocks from black schists and the geological age of the green rocks of the Sambagawa system: J. Geol. Soc. Japan, **54**, 109–110. (in Japanese)

―――― (1951) Über das "Feld der Metamorphose" der kristallinen Schiefer, besonders in bezug auf Bildung des kristallinen Schiefergebietes in Zentral-Shikoku: J. Sci. Hiroshima Univ., Ser. C, **1**(*1*), 1–18.

―――― (1953) Contribution to the knowledge of mutual relations between three metamorphic zones of Chugoku and Shikoku, southwestern Japan, with special reference to the metamorphic and structural features of each metamorphic zone: J. Sci. Hiroshima Univ., Ser. C, **1**(*3*), 17–46.

――――, Hide, K. and Yoshino, G. (1956) The stratigraphical positions of the bedded cupriferous pyritic deposits in the Sambagawa crystalline schist zone in Shikoku: J. Geol. Soc. Japan, **62**, 30–45. (in Japanese with English abstract)

Kojima, Y. and Asada, I. (1973) The Akenobe ore deposits—their geologic structure and fracture pattern: Mining Geol., **23**, 137–151. (in Japanese with English abstract)

Konda, T. and Taguchi, K. (1976) Volcanostratigraphy of the Neogene Tertiary of Tohoku area, Japan: Intern. Congr. Pacific Neogene Stratigr., Preprints, 126–129.

Kosaka, H. and Wakita, K. (1975) Geology and mineralization of the Mamut mine, Sabah, Malaysia: Mining Geol., **25**, 303–320. (in Japanese with English abstract)

Kozu, S., Kawano, Y. and Yagi, K. (1941) Studies on the garnet from the Wada pass (2): J. Japan. Assoc. Mineral. Petrol. Econ. Geol., **25**, 1–12. (in Japanese)

―――― and Takane, K. (1938) Formation of tennantite by heating of enargite: J. Japan. Assoc. Mineral. Petrol. Econ, Geol., **19**, 253–266. (in Japanese)

Kuhara, M. (1914) On the origin of the ore deposits of the Iya and Besshi mines: J. Geol. Soc. Tokyo, **21**, 185–199, 246–253, 325–344. (in Japanese)

Kuno, H., Baadsgaard, H., Goldich, S. and Shiobara, K. (1960) Potassium-Argon dating of the Hida Metamorphic Complex, Japan: Japan. J. Geol. Geogr., **31**, 273–278.

Kuroda, Y. (1956) On the Mg-Fe metasomatism in the Hitachi district, southern Abukuma plateau, northern Japan: Sci. Rept. Tokyo Kyoiku Daigaku, Ser. C, **44**, 57–80.

Kurshakova, L. D. (1970) Redox condition of andraditization of hedenbergite: Geochem. Intern., **7**, 395–399.

―――― (1971) Stability field of hedenbergite of the log P_{O_2}-T diagram: Geochem. Intern., **8**, 340–349.

Kusanagi, T. (1955) On the geological structure and its relation to the mineralization in the rhyolitic complex of Ashio mine (1): Mining Geol., **5**, 77–88. (in Japanese with English abstract)

―――― (1957) On the relation of the geological structure to the ore deposition at the Ashio mine. J. Japan. Assoc. Mineral. Petrol. Econ. Geol., **41**, 263–312. (in Japanese with English abstract)

―――― (1963) The mineral zoning at the Ashio mine: Mining Geol., **13**, 35–40. (in Japanese with English abstract)

Kushiro, I. (1976) Carbon dioxide in the earth's interior: Kagaku, **46**, 683–688. (in Japanese)

Kutina, J. (1957) A contribution to the classification of zoning in ore veins: Univ. Carolina Geol., **3**, 197–225.

―――― (1965) The concept of monoascendent and polyascendent zoning, in Symposium "Problems of Postmagmatic Ore Deposition", Geol. Surv. Czechoslovakia, Praha, v. II, 47–55.

―――― (1972) Regularities in the distribution of hypogene mineralization along rift structures: 24th Intern. Geol. Congr., Montreal, 1972, Sec. 4 (Mineral Deposits), 65–73.

Larson, S. D. (1956) Phase studies of the two component carbon dioxide-water system involving the carbon dioxide hydrate: Doctral Dissert. Publ. No. 15235, Univ. Microfilms, Ann Arbor, Michigan, 85p.

Lasky, S. G. (1935) Distribution of silver in base metal ores: AIME, Trans. **115**, 69–80.

Latimer, W. M. (1952) Oxidation Potentials, 2nd ed., Prentice-Hall, N. Y., 392 p.

Lawrence, L. J. and Plimer, I. R. (1969) A cubanite ore from the North Mine Broken Hill, N. S. W.: Proc. Aust. Inst. Mining Metallur., No. *231*, 27–32.

―――― and Golding, H. G. (1969) A cubanite-rich sulphide ore associated with ultramafic rock near Tumut, N. S. W.: Proc. Aust. Inst. Mining. Metallur., No. *231*, 33–39.

Lee, M. S., Takenouchi, S. and Imai, H. (1974) Occurrence and paragenesis of the Cu-Fe-Sn-S minerals, with reference to stannite, stannoidite and mawsonite: J. Mineral. Soc. Japan, **11**, Spec. Issue *2*, 155–164. (in Japanese)

――――, ―――― and ―――― (1975) Syntheses of stannoidite and mawsonite and their genesis in ore deposits: Econ. Geol., **70**, 834–843.

Lee, J. Y. (1972) Experimental investigation on stannite-sphalerite solid solution series: N. Jb. Mineral. Mh., **1972**, No. *12*, 556–559.

Lesnyak, V. F. (1957) Mineralothermometric research at Tyrny-Auz skarn-ore complex, north Caucasus: in Yermakov, N. P., ed., Research on the Nature of Mineral-forming Solutions, with Special Reference to Data from Fluid Inclusions, Intern. Ser. Monogr. Earth Sci., No. *22*, Pergamon Press, New York, 458–489.

Lévy, C. (1967) Contribution à la minéralogie des sulfures de cuivre du type Cu_3XS_4: Bur. Rech. Géol. Minières, Mém. *54*, 110–124.

Lewis, D. E. (1967) Geology of the Nungkok copper prospect Kinabalu area, Sabah, Malaysia: Proc. 2nd Geol. Convention and 1st Symp. on the Geology of the Mineral Resources of the Philippines and Neighboring Coutries, 2, Geol. Surv. Philippines, 288–295.

Liedau, P., Sprung, M. and Thilo, E. (1958) The system $MnSiO_3$-$CaMn(SiO_3)_2$: Z. Anorg. Allg. Chem., *297*, 213–225.

Lindrgen, W. (1923) Concentration and circulation of the elements from the standpoint of economic geology: Econ. Geol., **18**, 431–432.

——— (1927) Paragenesis of minerals in Butte veins: Econ. Geol., **22**, 304–307.

——— (1933) Mineral Deposits, McGraw-Hill Book Co., New York, 930p.

Liou, J. G. (1974) Stability relations of andradite-quartz in the system Ca-Fe-Si-O-H: Am. Mineral., **59**, 1016–1025.

Little, W. M. (1960) Inclusions in cassiterite and associated minerals: Econ. Geol., **55**, 485–509.

Loudon, A. G. (1972) Marcopper disseminated copper deposit, Philippines: Joint Meeting MMIJ-AIME 1972, Tokyo, TI'd3, 12p.

——— (1976) Marcopper porphyry copper deposit, Philippines: Econ. Geol., **71**, 721–732.

Loughlin, G. F. and Koschmann, A. H. (1942) Geology and ore deposits of Magdalena mining district, New Mexico: U.S. Geol. Surv., Prof. Paper *200*, 168p.

Lowell, J.D., and Guilbert, J. M. (1970) Lateral and vertical alteration-mineralization zoning in porphyry ore deposits: Econ. Geol., **65**, 373–408.

MacLean, W. H., Cabri, L. J. and Gill, J. E. (1972) Exsolution products in heated chalcopyrite. Can. J. Earth Sci., **9**, 1305–1317.

Mah, A. D. (1960) Thermodynamic properties of manganese and its compounds: U. S. Bur. Mines, RI 5600, 34p.

Mariko, T., Imai, N. and Shiga, Y. (1973) A new occurrence of a argentiferous pentlandite from the Kamaishi mine, Iwate Prefecture, Japan: Mining Geol., **23**, 355–358. (in Japanese with English abstract).

———, ———, ——— and Ichige, Y. (1974) Occurrence and paragenesis of the nickel- and cobalt-bearing minerals from the Nippo and Shinyama ore deposits of the Kamaishi mine, Iwate Prefecture, Japan: Mining Geol., **24**, 335–354. (in Japanese with English abstract)

Markham, N. L., and Lawrence, L. J. (1965) Mawsonite, a new copper-iron-tin sulfide from Mt. Lyell, Tasmania and Tingha, New South Wales: Am. Mineral., **50**, 900–908.

Marumo, F. and Nowacki, W. (1967) A refinement of the crystal structure of luzonite: Z. Kristtallogr., *124*, 1–8.

Maruyama, S. (1957) The relation between ore veins and igneous intrusive at the Ikuno mine: Mining Geol., **7**, 281–284. (in Japanese with English abstract)

——— (1959) Zonal distribution of ore deposit and prospecting at the Ikuno mine: J. Mining Metallur. Inst. Japan., **75**, 673–677. (in Japanese)

Maske, S. and Skinner, B. J. (1971) Studies of the sulfosalts of copper 1. Phases and phase relations in the system Cu-As-S: Econ. Geol., **66**, 901–918.

Mason, B. (1943) Mineralogical aspects of the system FeO-Fe_2O_3-MnO-Mn_2O_3: Geol. För. Förhand., **65**, 97–180.

Masuda, A. (1966) Lanthanides in basalts of Japan with three distinct types: Geochem. J., **1**, 11–26.

Matsubara, A. (1953) Differentiation of ore magma in Tsuchikura and Besshi mines: J. Geol. Soc. Japan, **59**, 79–87.

Matsueda, H. (1973a) Iron-wollastonite from the Sampo mine showing properties distinct from those of wollastonite: Mineral. J., **7**, 180–201.

——— (1973b) On the mode of occurrence and mineral paragenesis of iron-wollastonite skarn in the Sampo mine, Okayama Prefecture: Sci. Rept. Dept. Geol. Kyushu Univ., **11**, 265–273. (in Japanese with English abstract)

Matsueda, H. (1974) Immiscibility gap in the system CaSiO$_3$-CaFeSi$_2$O$_6$ at low temperatures: Mineral. J., **7**, 327–343.

Matsuhashi, S. (1968) Analysis of structural control and results of prospecting for bedding plane veins in the Taishu mine: Mining Geol., **18**, 161–172. (in Japanese with English abstract)

Matsukuma, T. and Horikoshi, E. (1970) Kuroko deposits in Japan, a review: *in* Tatsumi, T., ed., Volcanism and Ore Genesis, Univ. of Tokyo Press, Tokyo, 153–179.

Matsumoto, T. (1947) The geologic research of the Aritagawa valley, Wakayama Pref.: A contribution to the tectonic history of the Outer Zone of southwest Japan: Sci. Rept. Fac. Sci. Kyushu Univ., Geol., **2**, 1–12. (in Japanese)

——, Kimura, T. and Katto, J. (1952) Discovery of Cretaceous ammonites from the undivided Mesozoic complex of Shikoku, Japan: Mem. Fac. Sci. Kyushu Univ., Ser. D, **3**, 179–183.

Matsuoka, M. (1976) Iron-wollastonite from the Tsumo mine, Shimane Prefecture: Joint Meeting Soc. Mining Geol. Japan., Mineral. Soc. Japan, and Japan. Assoc. Mineral. Petrol. Econ. Geol., Preprints, 45.

Maucher, A. (1940) Über die Kieslagerstätte der Grube Bayerland bei Waldsassen in der Oberpfalz: Z. angew. Mineral., **2**, 219–275.

McKinstry, H. E. (1941) Structural control of ore deposition in fissure veins: AIME, Tech. Pub. No. *1267*, 1–23.

—— (1948) Mining Geology, Prentice-Hall, New York, 680p.

—— (1953) Shears of the second order: Am. J. Sci., **251**, 401–414.

—— (1957) Phase assemblages in sulfide ore deposits: N. Y. Acad. Sci. Trans., Ser. 2, **20**, 15–26.

—— (1959) Mineral assemblages in sulfide ores. The system Cu-Fe-S-O: Econ. Geol., **54**, 975–1001.

—— (1963) Mineral assemblages in sulfide ores. The system Cu-Fe-As-S: Econ. Geol., **58**, 483–506.

—— and Kennedy, G. C. (1957) Some suggestions concerning the sequence of certain ore minerals: Econ. Geol., **52**, 379–390.

Meijer, A. (1976) Pb and Sr isotopic data bearing on the origin of volcanic rocks from the Mariana island-arc system: Geol. Soc. Am. Bull., **87**, 1358–1369.

Merwin, H. E. and Lombard, R. H. (1937) The system Cu-Fe-S: Econ. Geol., **32**, 203–284.

Miller, J. A., Shibata, K. and Kawachi, Y. (1962) Potassium-Argon ages of granitic rocks from the Outer Zone of Kyushu, Japan: Bull. Geol. Surv. Japan, **13**, 712–714.

——, Shido, F., Banno, S. and Uyeda, S. (1961) New data on the age of orogeny and metamorphism in Japan: Japan. J. Geol. Geogr., **32**, 145–151.

Minami, E. (1935) Selen-Gehalte von europäischen und japanischen Tonschiefern. Nachr. Gesellschaft Wiss. Göttingen, Math-Phy. Kl. 4. Neue Folge., **1**, 143–145.

Minato, H., Takano, Y. and Muraoka, H. (1954) Antimony-rich luzonite from the Teine mine, Hokkaido, Japan. Sci. Paper, College General Educ. Univ. Tokyo, **4**, 155–162.

Minato, M., Gorai, M. and Funahashi, M. (Eds.) (1965) The Geologic Development of the Japanese Islands; Tsukijishokan, Tokyo, 442p.

Ministry of International Trade and Industy of Japan (1973) Aerial geologic survey of the Bantan district: 1–30. (in Japanese)

Mitchell, A. H. G. and Garson, M. S. (1972) Relationship of porphyry copper and Circum-Pacific tin deposits to paleo-Benioff zones: IMM, Trans., sec. B, **81**, B10–B25.

Mitchell, R. S. and Corey, A. S. (1954) The coalescence of hexagonal and cubic polymorphs in tetrahedral structures as illustrated by some wurtzite-sphalerite groups: Am. Mineral., **39**, 773–782.

Miyahisa, M. (1953–1954) Sulfide ore deposits of Dohi vein, Obira mine, Kyushu: J. Mining Inst. Kyushu, **21**, 93–101, **22**, 149–161, 145–155. (in Japanese with English abstracts)

—— (1958) Valleriite-bearing cupriferous iron sulfide ores from the Besshi and Kōra mines: Mining Geol., **8**, 300–303. (in Japanese with English abstract)

—— (1961) Geological studies on the ore deposits of Obira-type in Kyushu. (1) General geology and metallogenetic provinces: Japan. J. Geol. Geogr., **32**, 39–54.

Miyahisa, M. (1974) Antimony-bearing luzonite from Nansatsu gold-silver Metallogenetic province: Mining Geol., **24**, 69. (in Japanese)

——, Hashimoto, I. and Matsumoto, Y. (1971) Geological age and paleogeographical behaviors of the Mitate Conglomerate in Kyushu: Professor H. Matsushita Memorial Volume, Kyushu Univ., 103–112. (in Japanese with English abstract)

——, Ishibashi, K. and Adachi, T. (1975) Mineral paragenesis and chemical composition of malayaite from Toroku mine, Miyazaki Prefecture, Japan: J. Japan. Assoc. Mineral. Petrol. Econ. Geol., **70**, 25–29. (in Japanese with English abstract)

Miyashiro, A. (1961) Evolution of metamorphic belts: J. Petrol., **2**, 277–311.

—— and Banno, S. (1958) Nature of glaucophanitic metamorphism: Am. J. Sci., **256**, 97–110.

Miyazaki, K., Mukaiyama, H. and Izawa, E. (1974) Thermal metamorphism of the bedded cupriferous iron sulfide deposit at the Besshi mine, Ehime Prefecture, Japan: Mining Geol., **24**, 1–11. (in Japanese with English abstract)

Miyazawa, T. (1959a) What is indicated by fluid inclusions?: Mining Geol., **9**, 29–30. (in Japanese)

—— (1959b) Prospecting of contact-metasomatic deposits: J. Mining Metallur. Inst. Japan, **75**, 591–594. (in Japanese with English abstract)

—— and Enjoji, M. (1972) Studies of fluid inclusions and origin of hydrothermal fluids: J. Japan. Geothermal Energy Assoc., **34**, 60–75. (in Japanese with English abstract)

——, Tokunaga, M., Okamura, S. and Enjoji, M. (1971) Formation temperatures of veins in Japan: Proc. IMA-IAGOD Meeting '70, IAGOD Vol. (Mining Geol., Spec. Issue *3*), 340–344.

Mizota, T. (1971) Phase transition of cubanite: Preprint of 1st Scientific Research Symposium (Kyoto) on "Syntheses of Sulfide Minerals and Conditions of Formation in their Natural Occurrences", 44–46. (in Japanese)

Moh, G. H. (1960) Experimentelle Untersuchungen an Zinnkiesen und analogen Germaniumverbindungen: N. Jb. Mineral., Abh., **94**, 1125–1146.

—— (1969) The tin-sulfur system and related minerals: N. Jb. Mineral., Abh., **111**, 227–263.

—— und Ottemann, J. (1962) Neue Untersuchungen an Zinnkiesverwandten: N. Jb. Mineral., Abh., **99**, 1–28.

Moody, J. P. (1956) Wrench tectonics: Geol. Soc. Am. Bull., **67**, 1207–1246.

Moore, J. G. (1970) Water content of basalt erupted on the ocean floor: Contr. Mineral. Petrol., **28**, 272–279.

Moore, W. J. and Nash, J. I. (1974) Alteration and fluid inclusion studies of the porphyry copper ore body at Bingham, Utah: Econ. Geol., **69**, 631–645.

Morgan, B. A. (1975) Mineralogy and origin of skarns in the Mount Morrison Pendant, Sierra Nevada, Calif.: Am. J. Sci., **275**, 119–142.

Morgan, W. J. (1972) Deep mantle convection plumes and plate motions: Am. Assoc. Petroleum Geol. Bull., **56**, 203–213.

Moses, A. T. (1905) The crystallization of luzonite, and other crystallographic studies: Am. J. Sci., **20**, 277–284.

Motegi, M. (1975) Mineralization of the Philippines—A geohistorical review, *in* Kobayashi, T. and Toriyama, R., eds., Geology and Palaeontology of Southeast Asia: Univ. of Tokyo Press, Tokyo, **15**, 393–417.

Mukaiyama, H. and Izawa, E. (1970) Phase relations in the Cu-Fe-S system, the copper-deficient part: *in* Tatsumi, T., ed, Volcanism and Ore Genesis, Univ. of Tokyo Press, Tokyo, 339–355.

—— and Miyazaki, K. (1973) Sphalerite and filling temperatures of fluid inclusions from the Mozumi deposit of the Kamioka mine: Joint Meetings Soc. Mining Geol. Japan, Mineral. Soc. Japan, Japan. Assoc. Mineral. Petrol. Econ. Geol., Preprints C3.

Müller, H. (1952) Die eindimensionale Umwandlung Zinkblend-Wutrzit und die dabei auftretende Anomalien: N. Jb. Mineral. Abh., **84**, 43–76.

Muramatsu, Y. and Nambu, M. (1976) Studies on the mineralization of the pyrometasomatic copper and iron ore deposits of the Akagane mine, Iwate Prefecture, Japan (I). Filling temperature and salinity of fluid inclusions in quartz and calcite of the Hozumi and Sakae ore deposits: J. Japan. Assoc. Mineral. Petrol. Econ. Geol., **71**, 264–272. (in Japanese with English abstract)

Muta, K. (1958) Minor elements in galena and sphalerite from Kyushu: J. Mining Inst. Kyushu, **26**, 247–264. (in Japanese with English abstract)

Nabetani, S., Suzuki, M., Nozawa, T. and Tainosho, Y. (1972) Gravimetric investigation of the Ibaragi granitic complex: J. Geod. Soc. Japan, **18**, 78–88 (in Japanese)

Nadai, A. L. (1963) Theory of Flow and Fracture of Solids, v. **2**, McGraw-Hill Book Co., New York, 705 p.

Nagano, K., Takenouchi, S., Imai, H. and Shoji, T. (1977) Fluid inclusion study at the Mamut porphyry copper deposit, Sabah, Malaysia: Mining Geol., **27**, 201–212.

Nagasawa, K. (1961) Mineralization at the Mikawa mine, northern Japan: J. Earth Sci. Nagoya Univ., **9**, 129–172.

Nakajima, K. (1893) Sulfide ore deposits in the Outer Zone of the south part of Japan: Bul. Imper. Geol. Surv. Japan, **1**, 126p. (in Japanese)

Nakamura, T. (1951) High temperature mineral association in a certain quartz vein at the Ashio mine: J. Fac. Sci., Univ. Tokyo, Sec. 2, **8**, Pt. 2, 89–98.

——— (1954) Tin mineralization at the Ashio copper mine, Japan: J. Inst. Polytech. Osaka City Univ., Ser. G., **2**, 35–52.

——— (1961) Mineralization and wall rock alteration at the Ashio copper mine, Japan: J. Inst. Polytech. Osaka City Univ., Ser. G., **5**, 53–127.

——— (1970) Mineral zoning and characteristic minerals in the polymetallic veins of the Ashio copper mine: in Tatsumi, T., ed., Volcanism and Ore Genesis, Univ. of Tokyo Press, Tokyo, 231–246

——— and Aikawa, N. (1973) Stannoidite-canfieldite association as related to vein mineralization at the Ashio copper mine: J. Geosci. Osaka City Univ., **16**, Pt. 1, 1–10.

Nakamura, Y. (1974) The system Fe_2SiO_4-$KAlSi_2O_6$-SiO_2 at 15 Kbar: Carnegie Inst. Wash., Yearbook **73**, Geophys. Lab., 352–354.

Nakayama, I. (1954) The lineation of the geologic structure in the Sambagawa metamorphic zone of the Tenryu river basin: Mem. Coll. Sci. Univ. Kyoto, Ser. B, **21**, 273–286.

Narita, E. (1961) Wall rock alteration in the Kanoko vein system, Hosokura mine, Miyagi Pref., northeast Japan: J. Fac. Sci. Hokkaido Univ. Ser. 4, **11**(*1*), 59–75.

Nash, J. T. (1973) Geochemical studies in the Park City district: 1. Ore fluids in the Mayflower mine: Econ. Geol., **68**, 34–51.

——— and Theodore, T. G. (1971) Ore fluids in the porphyry copper deposit at Copper Canyon, Nevada: Econ. Geol., **66**, 385–399.

Natarajan, R. and Garrels, R. M. (1965) Partial pressure diagram: in Garrels, R. M. and Christ, C. L., eds., Solution, Minerals and Equilibria, Harper & Row, New York, 166–169.

Naumov, V. B. and Malinin, S. D. (1968) New method of determination of pressure according to the gas-liquid inclusions: Geokhimiya, **1968**, 432–441. (in Russian with English abstract)

Nevin, C. M. (1936) Principles of Structural Geology, John Wiley and Sons, London, 348p.

Newhouse, W. H. (1925) Paragenesis of marcasite: Econ. Geol., **20**, 54–66.

——— (1940) Openings due to movement along a curved or irregular fault plane: Econ. Geol., **35**, 445–464.

——— and Glass, J. P. (1936) Some physical properties of certain iron oxides: Econ. Geol., **31**, 699–711.

Niggli, P. and Niggli, E. (1952) Gesteine und Minerallagerstätten, v. **2**, Verlag Birkhäuser, Basel, 557p.

Nishimura, S. and Ishida, S. (1972) Fission-track ages of tuffs of the Neogene Tertiary in Oga Peninsula, Akita Prefecture, Japan: J. Japan. Assoc. Mineral. Petrol. Econ. Geol., **67**, 166–168. (in Japanese)

Nishio, K. (1910) On the geologic occurrences of the bedded cupriferous pyrite deposits in Japan: J. Mining Inst. Japan, **26**, 211–227. (in Japanese)

Nishio, S. (1940) Study of the cupriferous pyrite ores from the Besshi mine: J. Fac. Eng. Tokyo Imper. Univ., **23**, (*1*), 1–87.

———, Imai, H., Inuzuka, S. and Okada, Y. (1953) Temperatures of mineral formation in some deposits in Japan, as measured by the decrepitation method: Mining Geol., **3**, 21–29. (in Japanese with English abstract)

Nitta, T. and Fukabori, Y. (1969) On the successful exploration at the lower part of the Mozumi mining area, Kamioka mine: Mining Geol., **19**, 147–159. (in Japanese with English abstract)

———, ——— and Mishima, H. (1971) On the successful exploration at the lower part of the Mozumi mining area, Kamioka mine: Mining Geol., **21**, 84–96. (in Japanese with English abstract)

Nozawa, T. (1968a) Isotopic ages of Hida metamorphic belt, summary: J. Geol. Soc. Japan, **74**, 447–450. (in Japanese with English abstract)

——— (1968b) Radiometric ages of granitic rocks in Outer Zone of Southwest Japan and its extension; 1968 summary and north-shift hypothesis of igneous activity: J. Geol. Soc. Japan, **74**, 485–489. (in Japanese with English abstract)

O'Daniel, H. and Tscheischwili, L. (1944) Structure investigation of tephroite (Mn_2SiO_4), glaucochroite ((Mn, Ca) SiO_4)), and willemite (Zn_2SiO_4): Z. Krtisallogr., **105**, 273–278.

Oen, S. (1970) Paragenetic relations of some Cu-Fe-Sn sulfides in the Mangualde pegmatite, North Portugal: Mineral. Deposita, **5**, 59–84.

Ogawa, K. (1975) Malayaite from the Sampo mine, Okayama Prefecture, Japan: Mining Geol., **25**, 417–426.

Ogawa, W. (1935) Sphalerite from the Ashio mine, *in* Ito, T., ed., Beitrage zur Mineralogie von Japan, Neue Folge 1, Dept of Mineralogy, Univ. Tokyo. 24–26, (in Japanese)

Onuma, N., Higuchi, H., Wakita, H. and Nagasawa, H. (1968) Trace element partition between two pyroxenes and the host lava: Earth Planet. Sci. Letters, **5**, 47–51.

Oshima, T. (1964) Geology and ore deposits of the Yanahara mine, western Japan, with reference to structural control of ore deposition: Japan. J. Geol. Geogr., **35**, 81–100.

Otsuka, Y. (1935) The Oti graben in the southern Noto peninsula, Japan, pt. 2: Bull. Earthq. Inst. Tokyo Imper. Univ., **13**, 806–845.

Ozima, M., Ueno, N., Shimizu, N. and Kuno, H. (1967) Rb-Sr and K-A isotopic investigations of the Shidara granodiorites and associated Ryoke metamorphic belt, central Japan: Japan. J. Geol. Geogr., **38**, 159–162.

Park, C. F. (1946) The spilite and manganese problem of the Olympic Peninsula: Am. J. Sci., **244**, 305–323.

——— (1956) On the origin of manganese: 20th Intern. Geol. Congr. Manganeso Symp. t. 1, Mexico City, 75–98.

——— (1963) Zoning in ore deposits, the pulsation theory and the role of structure in zoning: *in* Symposium "Problems of Postmagmatic Ore Deposition" v. **I.**, Geol. Surv. Czechoslovakia, Praha, 47–51.

——— and MacDiarmid, R. A. (1964) Ore Deposits, Freeman and Co., San Francisco, 522p.

Park, H. I. and Miyazawa, T. (1971a) Occurrence of skarn and fluid inclusions in the Chichibu mine: Mining Geol., **21**, 43. (in Japanese)

——— and ——— (1971b) On the pyrrhotite from the Chichibu mine, Saitama Prefecture, Japan: Mining Geol., **21**, 259–273. (in Japanese with English abstract)

Parker, R. L. and Oldenburg, D. W. (1973) Thermal model of ocean ridges: Nature Phys. Sci., **242**, 137–139.

Paroni, N. (1961a) Die Nordanatolische Horizontal-Verschiebung: Geol. Runds., **51**, 122–139.

——— (1961b) Faltung durch Horizontalverschiebung: Ecologae Geologicae Helvetiae, **54**, 515–534.

Parsons, W. H. (1937) The ore deposits of the Sunlight region, Park Co., Wyoming: Econ. Geol., **32**, 832–854.

Pauling, L. and Weinbaum, S. (1934) The structure of enargite Cu_3AsS_4: Z, Kristallogr., **88**, 48–53.

Peach, P. A. (1951) Geothermometry of some pegmatite minerals of Hybla, Ontario: J. Geol., **59**, 32–38.

Peacock, M. A. and Yatsevitch, G. M. (1936) Cubanite from Sudbury, Ontario: Am. Mineral., **21**, 55–62.

Pelton, W. H. and Smith, P. K. (1976) Mapping porphyry copper deposits in the Philippines with IP: Geophys., **41**, 106–122.

Peña, R. (1970) Brief geology of a portion of the Baguio mineral district: J. Geol. Soc. Philippines, **24**, 42–43.

Perry, V. D. (1961) The significance of mineralized breccia pipes: Mining Eng., **13**, 367–376.

Petersen, U. (1965) Major ore deposits of central Peru: Econ. Geol., **60**, 407–476.

Petruk, W. (1973) Tin sulphides from the deposit of Brunswick Tin Mines Limited: Can. Mineral., **12**, 46–54.

Pfann, W. G. (1957) Technique of zone melting and crystal growing: Solid State Phys., **4**, 424–521.

Phillips, W. J. (1972) Hydraulic fracturing and mineralization: J. Geol. Soc. Lond., **128**, 337–359.

Poldervaart, A. (1953) Metasomatism of basaltic rocks, a review: Geol. Soc. Am., Bull., **64**, 259–274.

Pomirleanu, V. (1968) Die Bedeutung der Flüssigkeitseinschlusse in Scheelitkristallen fur die geologische Thermometrie: Chemie der Erde, **2**, 178–186.

Prien, E. L. and Frondel, C. (1947) Studies in urolithiasis 1. The composition of urinary calculi: J. Urology, **57**, 940–991.

Quinto, P. T., Jr. (1970) Geology, mineralization and exploration program in Philex mines, Tuba, Benguet: J. Geol. Soc. Philippines, **24**, 44–46.

Ramdohr, P. (1935) Ein Zinnvorkommen im Marmor der Stiepelmanngrube bei Arandis, Sudwestafrika: N. Jb. Mineral. Beil., **70A**, 1–45.

―――― (1938a) Uber Schapbachit, Matildit, und den Silber- und Wismutgehalt mancher Bleiglanze: Abh. Preuss. Akad. Wiss. Phy-math. Kl, VII, 21p.

―――― (1938b) Antimonreiche Paragenesen von Jakobsbacken bei Sulitelma: Norsk Geol. Tskr., **18**, 275–289.

―――― (1950) Die Lagerstätte von Broken Hill in New South Wales, im Lichte der neuen geologischen Erkenntnisse und erzmikroskopischen Untersuchung: Heidel. Beitr. Mineral. Petrogr., **2**, 291–333.

―――― (1960) Die Erzmineralien und ihre Verwachsungen, 3. Aufl., Akad.-Verlag, Berlin, 1089p.

―――― (1963) The opaque minerals in stony meteorites: J. Geol. Res., **68**, 2011–2036.

―――― (1969) The Ore Minerals and their Intergrowths, Pergamon Press, Oxford, 1174p.

―――― and Strunz, H. (1967) Lehrbuch der Mineralogie, Enke, Stuttgart, 820p.

Rankama, K. and Sahama, Th. G. (1950) Geochemistry, Univ. of Chicago Press, Chicago, 912p.

Raybould, J. G. (1974) Ore textures, paragenesis and zoning in the lead-zinc veins of mid-Wales: IMM, Trans. B, **83**, B112–B119.

―――― (1976) The influence of pre-existing planes of weakness in rocks on the localization of vein-type ore deposits: Econ. Geol., **71**, 636–641.

Retardo, N. (1972) Block caving Philex Mining Corporation's Sto. Tomas II ore body in northern Luzon, Philippine islands: Joint Meeting MMIJ-AIME 1972, Tokyo, TIId4, 16p.

Reynolds, D. L. (1956) Calderas and ring-complexes: Verhand. Konink. Nederlandsch Geol. Mijnb. Gen., Geol., Ser. **16**, 355–379.

Richter, F. M. and Parsons, B. (1975) On the interaction of two scales of convection in the mantle: J. Geophys. Res., **80**, 2529–2541.

Riedel, W. (1929) Das Ausquellen geologischer Schmelzmassen als plastischer Formänderungsvorgang: N. Jb. Geol. Paläont. Beil. 62B. 151–170.

Riley, J. F. (1974) The tetrahedrite-freibergite series, with reference to the Mount Isa Pb-Zn-Ag ore body: Mineral. Deposita, **9**, 117–124.

Riley, L. B. (1936) Ore-body zoning: Econ. Geol., **31**, 170–184.

Roberts, W. M. B. (1965) The synthesis of copper-iron sulfides at low temperature and implications of their crystallization behavior: 8th Commonw. Mining and Metall. Congr. Australia and New Zealand, **6**, Proceed.-general, 1260–1274.

Robie, R. A. (1966) Thermodynamic properties of minerals, *in* Clark, S. P., ed., Handbook of Physical Constants, Geol. Soc. Am. Mem. *97*, 442–458.

────── and Waldbaum, D. R. (1968) Thermodynamic properties of minerals and related substances at 198.15°K (25.0°C) and one atmosphere (1.013 bars) pressure and at high temperatures: U. S. Geol. Surv., Bull. *1259*, 256p.

Robinson, B. W. and Morton, R. D. (1971) Mckinstryite from the Echo Bay mine, N. W. T., Canada: Econ. Geol., **66**, 342–347.

Roedder, E. (1962) Studies of fluid inclusions I, low temperature application of a dual-purpose freezing and heating stage: Econ. Geol., **57**, 1045–1061.

────── (1963) Studies of fluid inclusions II, freezing data and their interpretation: Econ. Geol., **58**, 167–211.

────── (1965) Liquid CO_2 inclusions in olivine-bearing nodules and phenocrysts from basalts: Am. Mineral., **50**, 1746–1782.

────── (1967) Fluid inclusions as samples of ore fluids: *in* Barnes, H. L., ed., Geochemistry of Hydrothermal Ore Deposits, Holt, Rinehart and Winston, Inc., New York, 515–574.

────── (1970) Application of an improved crushing microscope stage to studies of the gases in fluid inclusions: Schweiz. Mineral. Petrogr. Mit., **50**, 41–58.

────── (1971a) Fluid inclusion studies on the porphyry type ore deposits at Bingham, Utah, Butte, Montana, and Climax, Colorado: Econ. Geol., **66**, 98–120.

────── (1971b) Metastability in fluid inclusions: Proc. IMA-IAGOD Meeting '70, IAGOD Vol. (Mining Geol., Spec. Issue *3*), 327–334.

────── (1971c) Metastable superheated ice in liquid water inclusions under high negative pressure: Sci., **155**, 1413–1417.

────── (1971d) Petrology of silicate melt inclusions, Apollo 11 and Apollo 12 and terrestrial equivalents: Proc. Second Lunar Sci. Conf., **1**, 507–508.

────── (1972) Composition of fluid inclusions: *in* Fleischer, M., ed., Data of Geochemistry, 6th ed.: U. S. Geol. Surv. Prof. Paper, 440-JJ, 164p.

────── and Coombs, D. S. (1967) Immiscibility in granitic melts, indicated by fluid inclusions in ejected granitic blocks from Ascention Island: J. Petrol., **8**, 417–451.

────── and Skinner, B. J. (1968) Experimental evidence that fluid inclusions do not leak: Econ. Geol., **63**, 715–730.

────── and Weiblen, P. W. (1970) Lunar petrology of silicate melt inclusions, Apollo 11 rocks: Proc. Apollo 11 Lunar Sci. Conf., **1**, 801–837.

Rogers, J. (1948) Phosphate deposits of the former Japanese islands in the Pacific. A reconnaissance report: Econ. Geol., **43**, 400–407.

Russel, R. D. (1972) Evolutionary model for lead isotopes in conformable ores and in ocean volcanics: Rev. Geophys. Space Phys., **10**, 529–549.

Rutstein, M. S. (1971) Reexamination of the wollastonite-hedenbergite ($CaSiO_3$-$CaFeSi_2O_6$) equilibria: Am. Mineral., **56**, 2040–2052.

Sagawa, E. (1910) Resume of a report on the geology of the cupriferous pyrite deposits in the crystalline schists of the northern part of Iyo in the Island of Shikoku: Bull. Imper. Geol. Surv. Japan, **22**, No. *1*, 184p. (in Japanese)

Saigusa, M. (1958) Geology and mineralization of the Akenobe mine, Hyogo Prefecture: Mining Geol. **8**, 218–238. (in Japanese with English abstract)

Saito, N., Matsuda, A. and Nagasawa, H. (1961) Age determination of rocks by means of K-Ar method: J. Geol. Soc. Japan, **67**, 425–426. (in Japanese)

Sakai, H. (1968) Isotopic properties of sulfur compounds in hydrothermal process: Geochem. J., **2**, 29–50.

Sakai, S. (1963) Study on zonal distribution of ore deposits in Kamioka mine: Mining Geol., **13**, 115–120. (in Japanese with English abstract)

Sakai, Y. and Ohba, M. (1970) Geology and Ore Deposits of the Sado Mine: Mining Geol., **20**, 149–165. (in Japanese with English abstract)

Sales, R. H. (1948) Wall rock alteration at Butte, Montana: AIME, Trans., **178**, 9–35.

Sales, R. H. and Meyer, C. (1949) Results from preliminary studies of vein formation at Butte, Montana: Econ. Geol., **44**, 465–484.

Sampson, E. (1923) The ferruginous chert formation of Notre Dame Bay, Newfoundland: J. Geol., **31**, 571–598.

——— (1941) Note on the occurrence of gudmundite: Econ. Geol., **36**, 175–184.

Sasaki, A. (1959) Variation unit cell parameters in wolframite series: Mineral. J., **5**, 375–390.

———, Yui, S. and Yamaguchi, M. (1975) New mineral, Kamiokite $Fe_2Mo_3O_8$: Ann. Meeting Mineral. Soc. Japan, Preprints, 9. (in Japanese)

Sawada, H. (1944) Crystallographic study on luzonite: J. Geol. Soc. Japan, **51**, 27. (in Japanese)

Sawada, M., Ozima, M. and Fujiki, Y. (1962) Magnetic properties of cubanite ($CuFe_2S_3$). J. Geomag. Geoelect., **14**, 107–112.

Sawamura, T. and Yoshinaga, M. (1953) Iron manganese deposits of the Kunimiyama mine, Kochi Pref.: Mining Geol., **3**, 207–219. (in Japanese with English abstract)

Schilling, J.G. (1973) Iceland mantle plume. Geochemical study of Reykjanes Ridge: Nature, **242**, 565–571.

Schmalz, R. F. (1959) A note on the system Fe_2O_3-H_2O: J. Geophys. Res., **64**, 575–579.

Schneiderhöhn, H. (1928) Die jungeruptive Lagerstätten Provinz in Servien, Siebenbürgen und Banat: Zbl. f. Min. 1928 A, 404–406.

——— (1941) Lehrbuch der Erzlagerstättenkunde: Gustav Fischer Verlag, Jena. 858p.

——— (1955) Erzlagerstätten, 3. Aufl.: Gustav Fischer Verlag, Stuttgart, 375p.

——— und Ramdohr, P. (1931) Lehrbuch der Erzmikroskopie, Bd. 2, Verlag Gebrüder Bointraeger, Berlin, 714p.

Schnetzler, C. C. and Philpotts, J. A. (1970) Partition coefficients of rare earth elements between igneous matrix material and rock-forming mineral phenocrysts II: Geochim. Cosmochim. Acta, **34**, 331–340.

Schreiber, E. (1963) The equilibrium boundary for the reaction $2MnCO_3 + SiO_2 = Mn_2SiO_4 + 2CO_2$: Ph. D. Thesis, N. Y. State College of Ceramics, Alfred, N. Y., 101p.

Sclater, J. G. and Francheteau, J. (1970) The implications of terrestrial heat flow observations on current tectonic and geochemical models of the crust and upper mantle of the earth: Geophys. J. Roy. Astr. Soc., **20**, 509–542.

Scott, H. S. (1948) The decrepitation method applied to minerals with fluid inclusions: Econ. Geol., **43**, 637–654.

Scott, S. D. and Barnes, H. L. (1972) Sphalerite-wurtzite equilibria and stoichiometry: Geochim. Cosmochim. Acta, **36**, 1275–1295.

Seki, T. (1972) Rb-Sr geochronological study of porphyries in the Kamioka mining district, central Japan: J. Japan. Assoc. Mineral. Petrol. Econ. Geol., **67**, 410–417.

Seki, Y. (1958) Chemical characteristics of glaucophanitic regional metamorphism: J. Japan. Assoc. Mineral. Petrol. Econ. Geol., **42**, 296–301. (in Japanese with English abstract)

———, Aiba, M. and Kato, C. (1960) Jadeite and associated minerals of meta-gabbroic rocks in the Shibukawa district, central Japan: Am. Mineral., **45**, 668–679.

Sékine, Y. (1959) Üder das Vorkommen von Magnetiten in den subvulkanisch-hydrothermalen Cu-Pb-Zn-Sn-W-Erzgängen der Grube Akenobe, Japan: N. Jb. Mineral., Abh., **93**, 220–239.

Shazly, E. M., Webb, J. S. and Williams, D. (1957) Trace elements in sphalerite, galena and associated minerals from the British Isles: IMM, Bull., **66**, 241–271.

Sheppard, S. M. F., Nielsen, R. and Taylor, H. P., Jr. (1971) Hydrogen and oxygen isotope ratios in minerals from porphyry copper deposits: Econ. Geol., **66**, 515–542.

——— and Taylor, H. P., Jr. (1974) Hydrogen and oxygen isotope evidence for the origin of water in the Boulder Batholith and the Butte ore deposits, Montana: Econ. Geol., **69**, 926–946.

Shibata, K. (1968) K-Ar age determinations on granitic and metamorphic rocks in Japan: Geol. Surv. Japan, Rept. *227*, 71p.

——— (1971) K-Ar age of the Ibaragi granitic complex: Chikyu-Kagaku [Geol. Sci.], *25*, 268–269. (in Japanese)

Shibata, K. (1973) K-Ar ages of volcanic rocks from the Hokuriku group: Geol. Soc. Japan, Mem., *8*, 143–149. (in Japanese with English abstract)

―― (1975) K-Ar age of the quartz porphyry from Yofuke, Nago City, Okinawa Island: Communication Bull. No. 1 for a Research Group on "Japanese Neogene Biostratigraphy and Radiometric Age", 67.

―― and Igi, S. (1966) Potassium-Argon age of the Maizuru metamorphic rocks (Komori metamorphic rocks): J. Geol. Soc. Japan, **72**, 358–360. (in Japanese)

―― and ―― (1969) K-Ar ages of muscovite from the muscovite-quartz schist of the Sangun metamorphic terrain in the Tari district, Tottori Pref., Japan: Bull. Geol. Surv. Japan, **20**, 707–709.

―― and Ishihara, S. (1974) K-Ar ages of the major tungsten and molybdenum deposits in Japan: Econ. Geol., **69**, 1207–1214.

―― and Nozawa, T. (1966) K-Ar ages of granites from Amami-Oshima, Ryukyu islands, Japan: Bull. Geol. Surv. Japan, **17**, 430–435.

―― and ―― (1968a) K-Ar ages of Osuzuyama acid rocks, Kyushu, Japan: Bull. Geol. Surv. Japan, **19**, 17–20.

―― and ―― (1968b) K-Ar age of Ominesan acid rocks, Kishu, Japan: Bull. Geol. Surv. Japan, **19**, 219–222.

―― and ―― (1968c) K-Ar ages of granitic rocks of Ashizuri-misaki, Takatsukiyama and Omogo, Shikoku, Japan: Bull. Geol. Surv. Japan, **19**, 223–228.

―― and ―― (1968d) K-Ar ages of acid rocks of Noma-misaki and Hioki Mountains, Kyushu, Japan: Bull. Geol. Surv. Japan, **19**, 232–236.

―― and ―― (1968e) K-Ar ages of Yakujima granite, Kyushu, Japan: Bull. Geol. Surv. Japan, **19**, 237–241.

―― and ―― (1968f) K-Ar age of Omi schist, Hida Mountains, Japan: Bull. Geol. Surv. Japan, **19**, 243–246.

―― and Ono, K. (1974) K-Ar ages of the Ono volcanic rocks, central Kyushu: Bull. Geol. Surv. Japan, **25**, 663–666. (in Japanese with English abstract)

―― and Togashi, Y. (1975) K-Ar ages of granodiorite stock (biotite) and altered rhyolite dyke (whole rock) from the western part of Amakusa-shimoshima island: Bull. Geol. Surv. Japan, **26**, 187–191. (in Japanese with English abstract)

――, Yamaguchi, S. and Sato, H. (1975) K-Ar ages of the Miocene to Pleistocene sereis in the Tokachi region, Hokkaido: Bull. Geol. Surv. Japan, **26**, 491–496. (in Japanese with English abstract)

Shikazono, N. (1973) Sphalerite-carbonate-pyrite assemblage in hydrothermal veins and its bearing on limiting the environment of their deposition: Geochem. J., *7*, 97–114.

―― (1974) Physicochemical properties of ore-forming solution responsible for the formation of Toyoha Pb-Zn deposit, Hokkaido, Japan: Geochem. J., **8**, 37–46.

―― (1975) Mineralization and chemical environment of the Toyoha lead-zinc vein-type deposits, Hokkaido, Japan: Econ. Geol., **70**, 694–705.

Shimada, M. (1955) On the folding structure in the eastern area of Hitachi mine. Study on the geology and ore deposits of the Hitachi mine: Mining Geol., **5**, 102–116. (in Japanese with English abstract)

Shimada, S. and Tsunori, T. (1962) Discovery of stannite in a fault zone in the Besshi mine, Ehime Pref., Japan: Mining Geol., **12**, 223–224. (in Japanese with English abstract)

Shimazaki, H. (1967) On zinc involved in skarn minerals: J. Geol. Soc. Japan, **73**, 87. (in Japanese)

―― (1968a) Tin content in the skarn mineral from the Tsumo mine: J. Geol. Soc. Japan, **74**, 126. (in Japanese)

―― (1968b) Genesis of pyrometasomatic ore deposits of the Tsumo mine, Shimane Pref.: Japan. J. Geol. Geogr., **39**, 73–87.

―― (1969) Pyrometasomatic copper and iron ore deposits of the Yaguki mine, Fukushima Pref., Japan: J. Fac. Sci. Univ. Tokyo, Sec. 2, **17**, Pt. 2, 317–350.

Shimazaki, H. (1974) Characteristics of tungsten mineralization in Japanese skarn deposits: *in* Stemprok, M. ed., Metallization Associated with Acid Magmatism, Geol. Surv., Praha. **1**, 312–315.

——— (1975) The ratios of Cu/Zn-Pb of pyrometasomatic deposits in Japan and their genetical implications: Econ. Geol., **70**, 717–724.

——— (1976) Granitic magmas and ore deposits (2). Oxidation state of magmas and ore deposits: Mining Geol., Spec. Issue **7**, 25–35. (in Japanese with English abstract)

——— (1977) Grossular-spessartine-almandine garnets from some Japanese scheelite skarns: Can. Mineral., **15**, 74–80.

——— and Bunno, M. (1976) Scheelite-bearing skarn containing "ferrobustamite" from the Kasugayama mine, Nagano Prefecture: Joint Meetings Soc. Mining Geol. Japan, Mineral. Soc. Japan, and Japan. Assoc. Mineral. Petrol. Econ. Geol., Preprints 44. (in Japanese)

——— and Yamanaka, T. (1973) Iron-wollastonite from skarns and its stability relation in the $CaSiO_3$-$CaFeSi_2O_6$ join: Geochem. J., **7**, 67–79.

Shimazu, M. (1973) On the Tsugawa-Aizu province in green tuff region of northeastern Japan: Geol. Soc. Japan, Mem., **9**, 25–38. (in Japanese with English abstract)

——— (1976) Development of igneous activity in the Japanese Islands from late Miocene to early Pleistocene: Chikyu-Kagaku [Earth Science], **30**, 61–66. (in Japanese)

——— Takizawa, M. and Takano, M. (1976) Some information on the Cenozoic volcanic activity in the Niigata district and its environs: Contr. Dept. Geol. Mineral., Niigata Univ., No. 4, 225–233.

Shiobara, K. (1961) Decrepitation temperatures and chemical characteristics of the mineral species from the Kamioka mine: Mining Geol., **11**, 344–349. (in Japanese with English abstract)

Shoji, H. (1933) Röntogenographische Untersuchungen über die Orientierungsanderung des Kristallgitters bei Modifikationsänderung einer Substanz: Z. Kristallogr., **84**, 74–84.

Shoji, T. (1969a) A zonal arrangement of magnetite, hematite and pyrite found in the Chichibu mine: Mining Geol., **19**, 66. (in Japanese)

——— (1969b) The anomalous crystal forms of magnetite from the Chichibu mine, Saitama Prefecture. 1. Foliate magnetite: J. Mineral. Soc. Japan, **9**, 301–310. (in Japanese)

——— (1970) The decrepitation pattern of the garnet from the Kamaishi mine, Iwate Prefecture: Mining Geol., **20**, 305–308. (in Japanese with English abstract)

——— (1971) Vesuvianite: synthesis and occurrence in skarn: Mining Geol., **21**, 457–460. (in Japanese with English abstract)

——— (1972) On feldspars in pyrometasomatic ore deposits: Mining Geol., **22**, 315–327. (in Japanese with English abstract)

——— (1974) $Ca_3Al_2(SiO_4)_3$-$Ca_3Al_2(O_4H_4)_3$ series garnet: composition and stability: J. Mineral. Soc. Japan, **11**, 359–372. (in Japanese)

——— (1975) Role of temperature and CO_2 pressure in the formation of skarn and its bearing on mineralization: Econ. Geol., **70**, 739–749.

——— (1976a) Role of H_2O-CO_2(-C) mixtures in the formation of zonal arrangement of skarn minerals: Joint Meetings, Soc. Mining Geol. Japan. Soc. Mineral. Japan and Japan. Assoc. Mineral. Petrol. Econ. Geol., Preprints, 43. (in Japanese)

——— (1976b) The stability of the assemblage calcite-quartz in H_2O-CO_2 mixtures: J. Japan. Assoc. Mineral. Petrol. Econ. Geol., **71**, 379–388.

——— (1977a) The stability of grossular in H_2O-CO_2 mixtures: J. Japan. Assoc. Mineral. Petrol. Econ. Geol., **72**, 30–41.

——— (1977b) Zoning found in the skarn-type ore deposits: *in* Prof. T. Miyazawa Memorial Volume on the Occasion of his Retirement, Pt. 2, B13–B28. (in Japanese)

——— (1977c) The stability of andradite in H_2O-CO_2 mixtures: J. Japan. Assoc. Mineral. Petrol. Econ. Geol., **72**, 399–411.

——— and Kihara, T. (1970) Correlation between decrepitation and fluid inclusion data of quartz: Ann. Meetings MMIJ, Preprints 238–239. (in Japanese)

———, Ootsuka, M. and Imai, H. (1969) Consideration on the formation of mineralized faults and non-mineralized fractures in the Chichibu mine, Saitama Prefecture: J. Mining Metallur.

Inst. Japan., **85**, 765–770. (in Japanese with English abstract)

Short, M. N. (1940) Microscopic Determination of Ore Minerals, U. S. Geol. Surv., Bull. *914*, 314p.

Shugurova, N. A. (1968) Chemical basis of method of gas analysis of individual inclusions in minerals: *in* Yermakov, N. P., ed., Mineralogical Thermometry and Barometry, v. **2**. Nauka, Moscow, 18–23. (in Russian)

Sillitoe, R. H. (1972) A plate tectonic model for the origin of porphyry copper deposits: Econ. Geol., **67**, 184–197.

—— and Sawkins, F. J. (1971) Geologic, mineralogic and fluid inclusion studies relating to the origin of copper-bearing tourmaline breccia pipes, Chile: Econ Geol., **66**, 1028–1041.

Simonen, A. (1948) On the petrology of the Aulanko area in southwestern Finland: Comm. Geol. Fin. Bull., No. *143*, 5–66.

Skinner, B. J. (1960) Assemblage enargite-famatinite, a possible geologic thermometer: Geol. Soc. Am. Bull., **171**, 1975.

—— (1966) The system Cu-Ag-S: Econ. Geol., **61**, 1–26.

——, Luce, F. D. and Makovicky, E. (1972) Studies of the sulfosalts of copper III. Phases and phase relations in the system Cu-Sb-S: Econ. Geol., **67**, 924–938.

Skippen, G. (1974) An experimental model for low pressure metamorphism of siliceous dolomitic marble: Am. J. Sci., **274**, 487–509.

Smirnov, V. L. (1960) Types of hypogene zonality of hydrothermal ore bodies: 21st. Geol. Congr. Report of the 21 session, Pt. 11, Norden., 181–191.

Smith, F. G. (1953) Historical Development of Inclusion Thermometry, University of Toronto Press, Toronto, 149p.

—— and Little, W. M. (1959) Filling temperatures of H_2O-CO_2 fluid inclusions and their significance in geothermometry: Can. Mineral., **6**, 380–388.

Soeda, A. (1960) Valleriite in the cubanite-bearing ores from the Chugoku district: Mining Geol., **10**, 346–358. (in Japanese with English abstract)

Sourirajan, S. and Kennedy, G. C. (1962) The system H_2O-NaCl at elevated temperatures and pressures: Am. J. Sci., **260**, 115–141.

Springer, G. (1968) Electronprobe analyses of stannite and related tin minerals: Mineral. Mag., **36**, 1045–1051.

—— (1969) Compositional variations in enargite and luzonite: Mineral. Deposita, **4**, 72–74.

—— (1972) The pseudobinary system Cu_2FeSnS_4-Cu_2ZnSnS_4 and its mineralogical significance: Can. Mineral., **11**, 535–541.

Steven, T. A., Luedke, R. G. and Lipman, P. W. (1974) Relation of mineralization to calderas in the San Juan volcanic field, southwestern Colorado: J. Res. U.S. Geol. Surv., **2**, 405–409.

Stollery, C., Botcsik, M. and Holland, H. D. (1971) Chlorine in intrusives: A possible prospecting tool: Econ. Geol., **66**, 361–367.

Strunz, H. (1966, 1970) Mineralogische Tabellen 4, 5 Aufl., Akad. Verlag., Leipzig, 560p., 621p

Stull, D. R. and Prophet, H. (1971) JANAF Thermochemical Tables, 2nd ed., Nat. Stand. Ref. Data Ser. *37*, Nat. Bur. Stand. (U.S.), 1141p.

Suga, K. (1952) The Osarizawa mine: Ore deposits as related to geologic structures: Mining Inst. Japan, Tech. Pub., No. *8*, 25–29. (in Japanese)

Sugaki, A. (1972) Thermal study on the inversion of cubanite. Mem. Fac. Eng., Yamaguchi Univ., **22**, 267–272. (in Japanese with English abstract)

——, Shima, H. and Kitakaze, A. (1976) Study on the chemical composition of enargite and minerals of luzonite-famatinite series from the Kasuga and Akeshi mines, Kagoshima Pref.: J. Mineral. Soc. Japan, **12**, 206–213. (in Japanese)

——, ——, ——, and Harada, H. (1975) Isothermal phase relations in the system Cu-Fe-S under hydrothermal conditions at 350°C and 300°C: Econ. Geol., **70**, 806–823.

Sugimoto, R. (1952) On the geology of the Teine mine, with some remarks on the mineral composition and paragenesis of the Mitsuyama, and Koganesawa ore deposits: J. Japan. Assoc. Mineral. Petrol. Econ. Geol., **36**, 72–84. (in Japanese with English abstract)

Sugisaki, R., Tanaka, T. and Hattori, H. (1970) Rubidium and potassium contents of geosynclinal basalts in the Japanese Island: Nature, 227, 1338–1339.

Sumita, M. (1969) Geology and ore deposit of the Ohmidani mine: Mining Geol., 19, 131–146. (in Japanese with English abstract)

Sun, S. S. and Hanson, G. N. (1975) Evolution of the mantle: Geochemical evidence from alkali basalt: Geol. 3, 297–302.

Sundius, N. (1935) On the origin of late magmatic solutions containing magnesia, iron and silica: Sver. Geol. Unders., Ser. C, No. 392, 1–24.

Sushchevskaya, T. M. and Ivanova, G. F. (1967) On the composition of mineral-forming solutions of some wolframite deposits in the eastern Transbaikal region: Geokhimiya, 1967, 1099–1105. (in Russian with English abstract)

Suwa, Y., Tamai, Y. and Naka, S. (1976) Stability of synthetic andradite at atmospheric pressure: Am. Mineral., 61, 26–28.

Suzuki, J. (1930) Petrological study of the crystalline schist system of Shikoku, Japan: J. Fac. Sci. Hokkaido Imp. Univ., Ser. IV, 1, 27–107.

────── (1932) The contact metamorphic ore deposit in the environs of the Ofuku mine, Province of Nagato, Japan: J. Fac. Sci. Hokkaido Imp. Univ., Ser. IV, 2, 69–131.

────── (1936) On the selective metamorphism: Kagaku [Sci.], 6, 147–151, 196–198. (in Japanese)

────── (1939) On the age of the Sambagawa system: Proc. Imp. Acad. 15, 56–59.

────── (1956) On the glaucophane schist facies: J. Geol. Soc. Japan, 62, 394–395. (in Japanese)

────── and Minato, M. (1952) On the schalstein in Japan: J. Geol. Soc. Japan, 58, 272. (in Japanese)

────── and Ohmachi, H. (1956) Manganiferous iron ore deposits in the Tokoro district of northeastern Hokkaido, Japan: 20th Intern. Geol., Congr. Manganeso Symp., t. 4, Mexico City, 199–204.

Suzuki, T., Kashima, N., Hara, S. and Umemura, H. (1972) Geosyncline volcanism of the Mikabu green rocks in the Okuki area, western Shikoku, Japan: J. Japan. Assoc. Mineral. Petrol. Econ. Geol., 67, 177–192.

Swanson, H. E. and Tatge, B. (1957) Standard X-ray diffraction powder patterns: N. B. S. Circular 539, 1, 32.

Taguchi, K. (1973) Volcanostratigraphy and radiometric age of the lower part of the Neogene Tertiary in Northeast Japan: Geol. Soc. Japan, Mem., 8, 183–193.

Taguchi, Y., Ohba, M. and Kizawa, Y. (1974) The jalpaite and mckinstryite: Mineral. J., 11, 345–359. (in Japanese with English abstract)

Takabatake, A. (1956) Genesis of manganiferous iron deposits in Japan: 20th Intern. Geol. Congr., Manganeso Symp., t. 4, Mexico City, 205–220.

Takahashi, K. (1963) Geochemical study on minor elements in sulfide minerals: Geol. Surv. Japan, Rept. No. 199, 67p. (in Japanese with English abstract)

Takahashi, T. (1960) Supergene alteration of zinc and lead deposits in limestone: Econ. Geol, 55, 1083–1115.

────── and Schreiber, E. (1965) Fugacity of CO_2 in the gaseous solutions of H_2O-CO_2: A discussion, in Symposium "Problems of Postmagmatic Ore Deposition," v. II, Geol. Surv. Czechoslovakia, Praha, 519–523.

──────, Takenouchi, S., Nishio, S. and Imai, H. (1955) Temperatures of mineral formation in some types of deposits in Japan, as measured by the decrepitation method (II): Mining Geol., 5, 9–17. (in Japanese with English abstract)

Takai, F. (1944) *Desmostylus* in the phosphate beds in the Noto Peninsula: Bull. Res. Inst. Natural Resources, 5, 59–62. (in Japanese)

Takeno, S. (1971) Magnetic transformations of cubanite: Carnegie Inst. Wash. Yearbook. 70, Geophys. Lab., 302–303.

Takenouchi, S. (1962a) Geology and vein formation of the Yaso mine in northeastern Japan: Mining Geol., 12, 94–104. (in Japanese with English abstract)

────── (1962b) Study of temperatures of mineral formation in the hydrothermal ore deposits by the liquid inclusion method: Mining Geol., **12**, 282–293. (in Japanese with English abstract)

Takenouchi, S. (1962c) Polyphase inclusions in the quartz from the Taishu mine, Nagasaki Pref., Japan: Mining Geol. **12**, 294–297. (in Japanese with English abstract)

────── (1968) Compositions of mineralized hydrothermal solutions inferred from the studies of liquid inclusions in minerals: Notes on Ore Deposits, **6**, 13–20. (in Japanese)

────── (1970) Fluid inclusion study by means of heating-stage and freezing-stage microscope: Mining Geol. **20**, 345–354. (in Japanese with English abstract)

────── (1971a) Study of CO_2-bearing fluid inclusions by means of the freezing-stage microscope: Mining Geol., **21**, 286–300. (in Japanese with English abstract)

────── (1971b) Hydrothermal synthesis and consideration of the genesis of malayaite: Mineral. Deposita, **6**, 335–347.

────── (1973) Carbonatite deposits: Mining Geol., **23**, 367–382, 437–451. (in Japanese)

────── (1975a) Fluid inclusion study of the Jishakuyama and Koganetsubo ore deposits and acidic igneous rocks at the Akagane mine, Iwate Prefecture: Mining Geol., **25**, 247–259. (in Japanese with English abstract)

────── (1975b, 1976a) Basic knowledge on studies of fluid inclusions in minerals: J. Gemmol. Soc. Japan, **2**, 25–33, 66–73, 110–121, 165–172; **3**, 25–31. (in Japanese with English abstracts)

────── (1976b) Porphyry copper deposits on the island arc system from Japan to Bougainville through Philippines: Proc. Joint Meeting MMIJ-AIME, Denver, v. 1, 47–64.

────── and Fujiki, Y. (1968a) Experimental studies on Cu-Fe-S minerals (1st report): J. Mining. Metallur. Inst. Japan., **84**, 1–6. (in Japanese with English abstract)

────── and ────── (1968b) Synthetic studies of the chalcopyrite-cubanite solid solution: Mem. Fac. Eng. Univ. Tokyo, Ser. A, No. 6, 50–51. (in Japanese with English abstract)

────── and Imai, H. (1971) Fluid inclusion study of some tungsten-quartz veins in Japan: Proc. IMA-IAGOD Meeting '70, IAGOD Vol. (Mining Geol., Spec. Issue *3*), 345–350.

────── and ────── (1974) Fluid inclusion study at the Fujigatani and Kiwada area of southwestern Japan: IAGOD 4 th Symposium 1975, Varna, Abstracts of paper, 278–279.

────── and ────── (1975) Glass and fluid inclusion in acidic igneous rocks from some mining areas in Japan: Econ. Geol., **70**, 750–769.

────── and Kennedy, G. C. (1964) The binary system H_2O-CO_2 at high temperatures and pressures: Am. J. Sci., **262**, 1055–1074.

────── and Shoji, T. (1969) The occurrence of malayaite in pyrometasomatic deposits of southwestern Japan: Mining Geol., **19**, 243–255. (in Japanese with English abstract)

Takeuchi, T. (1967) The characteristics of the pyrometasomatic deposits in Japan: J. Mining Metallur. Inst. Japan, **83**, 172–174. (in Japanese)

────── and Nambu, M. (1953) Cubanite from Kamaishi, Nodatamagawa and Yakuki mines in northern Japan—Studies on the cubanite-bearing ores in Japan. III: Bull. Res. Inst. Mineral. Dressing Metallur. Tohoku Univ., **9**, 31–36. (in Japanese with English abstract)

────── and ────── (1958) On cubanite in Japan—Studies on the minerals of Cu-Fe-S series in Japan, second report: Sci. Rept. Tohoku Univ., **6**, 1–10.

────── and Yamaoka, K. (1965) Genesis of the ore deposits of Omine Mine, Iwate Prefecture, Japan. Sci. Rept. Tohoku Univ. Ser. III, **9**, 277–312.

Takimoto, K. (1944) Studies on the tin deposits of Japan: Japan. J. Geol. Geogr., **19**, 195–241.

Taliaferro, N. L. (1943) Franciscan-Knoxville problem: Bull. Am. Assoc. Petroleum Geol., **27**, 109–219.

Tamanyu, S. (1975) Fission-track age determination of accessory zircon from the Neogene Tertiary tuff samples, around Sendai City, Japan: J. Geol. Soc. Japan, **81**, 233–246.

Tanaka, M. (1957) On the geological relation between the Sambagawa, Mikabu and Chichibu systems: Bull. Fac. Liberal Arts, Yamanashi Univ., **2**, 135–174. (in Japanese with English abstract)

Tanaka, T., Mori., H. and Sasaki K. (1971) Geology and ore deposits of the Ikuno mine, with special reference to the gold-silver ores: Mining Geol., **21**, 162–173. (in Japanese with English abstract)

Tatsumi, T. (1942) On the ore deposit Honzan, Seizyosi mine, Manchukuo, with the special reference to the mode of occurrence of silver: J. Geol. Soc. Japan, 49, 416–425. (in Japanese)
——— (1951) Liquid inclusion geothermometry: Kagaku [Science], 21, 655–659. (in Japanese)
——— (1953) Geology and genesis of the cupriferous iron sulphide deposits of the Makimine mine, Miyazaki Pref.: Sci. Paper. Collage. General. Educ. Univ. Tokyo, 3, 81–113, 201–247.
———, Sekine, Y. and Kanehira, K. (1970) Mineral deposits of volcanic affinity in Japan. Metallogeny: in Tatsumi, T., ed., Volcanism and Ore Genesis, Univ. of Tokyo Press, Tokyo, 3–47.
Tatsumoto, M. (1966) Genetic relations of oceanic basalts as indicated by lead isotopes: Sci., 153, 1094–1101.
Thompson, G. A. (1959) Gravity measurements between Hazen and Austin, Nevada: A study of Basin-Range structure: J. Geophys. Res., 64, 217–229.
Titley, S. R. (1975) Geological characteristics and environment of some porphyry copper occurrences in the southwestern Pacific: Econ. Geol., 70, 499–514.
Togo, F., Takase, H. and Mononobe, N. (1954) On the manganese-bearing hematite deposit at the Ohdake, Shizuoka Pref.: Mining Geol., 4, 45. (in Japanese)
Tokunaga, M. (1954) Geology and ore deposits of the Kasuga and Akeshi mines in the Makurazaki district, Kagoshima Pref.: Mining Geol., 4, 205–212. (in Japanese with English abstract)
——— (1965) On the zoned skarn including bustamite, ferroan johannsenite and manganoan hedenbergite from Nakatatsu mine, Fukui Pref. Japan: Sci. Rept., Tokyo Univ. of Eudcation, Ser. C, 9, 67–87.
——— (1970) Lead-zinc veins of the Toyoha mine: in Tatsumi, T. ed., Volcanism and Ore Genesis, Univ of Tokyo Press, Tokyo, 247–257.
Tolman, C. F. and Ambrose, J. W. (1934) The rich ore of Goldfield, Nevada: Econ. Geol., 29, 255–279.
Toriumi, M. (1975) Petrological study of the Sambagawa metamorphic rocks. The Kanto Mountains, central Japan: Univ. Museum. Univ. Tokyo, Mem. 9, 99p.
Tsuboya, K. (1928) Albitization of basic plagioclases in some igneous rocks in northeastern Japan: J. Geol. Soc. Japan, 35, 1–14. (in Japanese)
Tsusue, A. (1961) Contact metasomatic iron and copper ore deposits of the Kamaishi mining district, northeastern Japan: Fac. Sci. Univ. of Tokyo, Ser. II, 13, 133–179.
Tu Kwang-Chi and Liu Y-Mao (1965) Some problems pertaining to the genesis of wolframite genesis of wolframite deposits of southern Kiangsi, China: in Symposium "Problems of Postmagmatic Ore Deposition," Geol. Surv. Czechoslovakia, Praha, v. II, 263–267.
Tugarinov, A. I., and Naumov, V. B. (1972) Physicochemical parameter of hydrothermal mineral formation: Geochem. Intern., 9, 161–167.
Tuominen, H. V. (1951) Metamorphic concentration of magnesium and iron in the Orijärvi region: Bull. Comm. Géol. Fin., No. 154, 233–238.
——— and Mikkola, T. (1950) Metamorphic Mg-Fe enrichment in the Orijärvi region as related to folding: Bull. Comm. Géol. Fin., No. 150, 67–92.
Tupas, M. H. (1960) A preliminary study of the geology of Philippine copper deposits: Natural and Applied Sci. Bull. (Philippines), 17, 283–294.
Turneaure, F. S. (1960) A comparative study of the major ore deposits of central Bolivia: Econ. Geol., 55, 217–254, 574–606.
——— (1971) The Bolivian tin-silver province: Econ. Geol., 66, 215–225.
——— and Gibson, R. (1945) Tin deposits of Carguaicollo, Bolivia: Am. J. Sci., 243–A, 523–541.
——— and Welker, K. K. (1947) The ore deposits of the eastern Andes of Bolivia. The Cordillera Real: Econ. Geol., 42, 595–625.
Tuttle, O. F., and Bowen, N. L. (1958) Origin of granite in the light of experimental studies in the system $NaAlSi_3O_8$-$KAlSi_3O_8$-SiO_2-H_2O: Geol. Soc. Am. Mem. 74, 153p.
Uchida, N. (1967) Chemical composition of tuffaceous rocks in the Mikabu and Mamba formations: Seikei-ronso, No. 6, 206–220. (in Japanese)
Ueda, N. (1968) Unpublished master's thesis, Dept. of Geophysics, Univ. of Tokyo. Reconnaissance Survey of the Geochronology of the Korean Peninsula: 1–43.

Ueda, Y. and Suzuki, T. (1973) K-Ar ages of the glauconite and celadonite from northeast Japan: Geol. Soc. Japan, Mem., 8, 151–159. (in Japanese with English abstract)

———, Yamaoka, K., Onuki, H. and Tagiri, M. (1969) K-Ar dating on the metamorphic rocks in Japan (2), the Hitachi metamorphic rocks in southern Abukuma plateau: J. Japan. Assoc. Mineral. Petrol. Econ. Geol. 61,. 92–99. (in Japanese with English abstract)

———, Zimbo, N. and Tamiya, R. (1973) K-Ar ages of the Lowest Neogene welded tuffs in Yamagata Prefecture: J. Japan. Assoc. Mineral. Petrol. Econ. Geol., 68, 91. (in Japanese)

Ueno, H., Tonouchi, S. and Shibata, K. (in prep.) Paleomagnetical evidence for the genesis of the Chichibu pyrometasomatic deposits, Japan.

Uetani, K., Sakurai, K. and Kato, A. (1966) The occurrence of selenian sitbioluzonite from the Iriki mine, Kagoshima Pref., Japan: Bull. National Sci. Museum, 9, 609–613.

Uffen, R. J. (1959) On the origin of rock magma: J. Geophys. Res., 64, 117–122.

Uytenbogaardt, W. (1971) Tables for the Microscopic Determination of Ore Minerals, Elsevier Publ. Co., Amsterdam, 430p.

Vaasjoki, O. (1971) On the striation developed in cubanite by heating, a criterion for thermometamorphism: Mineral. Deposita, 6, 103–110.

Van der Veen, R. W. (1925) Mineragraphy and Ore Deposition: I. G. Naeff, Hague. 168p.

Van Wazer, R. (1961) Phosphorus and its Compounds, 2, Intersci. Pub., New York, 2046p.

Waldo, A. W. (1935) Identification of the copper minerals by means of X-ray powder diffraction patterns: Am. Mineral., 20, 575–597.

Walker, G. P. L. (1974) Eruptive mechanism in Iceland: in Kristjansson, L., ed., Geodynamics of Iceland and the North Atlantic Area, D. Reidel Pub., Dordrecht, Holland, 189–201.

Walker, R. T. and Walker, W. B. (1956) The Origin and Nature of Ore Deposit: Walker Associates, Colorado, 354p.

Wallace, S. R., Jonson, D. C., Navias, R. A. and Skapinsky, S. A. (1957) Ring fracture intrusion and mineralization at Climax, Colorado: Program Ann. Meeting 1957, Geol. Soc. Am., 139–140.

———, Muncaster, N. K., Jonson, D. C., Mackenzie, W. B., Bookstrom, A. A. and Surface, V. E. (1968) Multiple intrusion and mineralization at Climax, Colorado: in Ridge, J. D., ed., Ore Deposits of the United States, 1933–1967, AIME, N. Y., 605–640.

Wang, Y. (1955) Fracture patterns in Chinkuashih area, Taipei-Hsien, Taiwan: Acta Geol. Taiwanica, No. 7, 21–34.

——— (1973) Wallrock alteration of late Cenozoic mineral deposit in Taiwan: Proc. Geol. Soc. Taiwan, No. 16 (Apr.), 145–160, No. 16 (May), 1–29.

Warren, H. V. (1935) Distribution of silver in base metal ores: AIME, Trans., 115, 81–89.

Watanabe, M. (1932) Contact-Metasomatic Ore Depoists, Iwanami, Tokyo, 71p. (in Japanese)

——— (1941) Fibrous zinc sulphide mineral from the Ohmori gold mine, Fukushima Prefecture: J. Japan. Assoc. Mineral. Petrol. Econ. Geol., 25, 213–229. (in Japanese)

——— (1943) Modes of occurrences of luzonite at the Hokuetsu and the Kinkaseki mines: J. Japan. Ascoc. Mineral. Petrol. Econ. Geol., 30, 52–73. (in Japanese)

——— (1951) Luzonite from Japan and Formosa: Sci. Rept. Tohoku Univ., Ser., III, 4, 33–44.

Watanabe, T. (1940) On the occurrence of gold and silver in the ore from the Onzin mine, North Korea: J. Japan. Assoc. Mineral. Petrol. Econ. Geol., 23, 104–114. (in Japanese)

——— (1943a) Mode of occurrence of minerals of enargite group from the Teine mine, Hokkaido: J. Japan. Assoc. Mineral. Petrol. Econ. Geol., 30, 80–90. (in Japanese)

——— (1943b) Geology and mineralization of the Suian mining district: J. Fac. Sci. Hokkaido Imp. Univ., Ser. 4, 6, 205–303.

——— (1957) Genesis of bedded manganese deposits and cupriferous pyrite deposits in Japan: Mining Geol., 7, 87–97. (in Japanese with English abstract)

——— (1958) Characteristic features of ore deposits found in contact-metamorphic aureoles in Japan: Jubilee Pub. Commem. Prof. J. Suzuki, 169–191. (in Japanese) [Transl. in Internat. Geol. Rev., 2, 946–966]

Watanabe, T. (1959) The minerals of Noda-Tamagawa mine, Iwate Prefecture, Japan I. Notes on geology and paragenesis of minerals: Mineral. J., **2**, 408–421.

——, Iwao, S., Tatsumi, T. and Kanehira, K. (1970) Folded ore bodies of the Okuki mine: in Tatsumi, T., ed., Volcanism and Ore Genesis, Univ. of Tokyo Press, Tokyo, 105–117,

—— and Kato, A. (1966) Ore microscopy and electron probe microanalysis of some manganese minerals with vrendenburgite type intergrowth: Mineral. Soc. India, IMA Vol., 197–202.

——, Yui, S. and Kato, A. (1970a) Bedded manganese deposits in Japan: in Tatsumi, T., ed., Volcanism and Ore Genesis, Univ. of Tokyo Press, Tokyo, 119–142.

——, —— and —— (1970b) Metamorphosed bedded manganese deposits of the Noda-Tamagawa mine: in Tatsumi, T., ed. Volcanism and Ore Genesis, Univ. of Tokyo Press, Tokyo, 143–152.

Webber, B. N. (1929) Marcasite in the contact metamorphic ore deposit of the Twin Butte districts, Prima County, Arizona: Econ. Geol., **24**, 304–310.

—— (1948) Manganese deposits of Costa Rica, Central America: AIME, Trans., **178**, 339–345.

Weis, P. L. (1953) Fluid inclusions in minerals from zoned pegmatites of the Black Hills, South Dakota: Am. Mineral., **38**, 671–697.

Weisbach, A. (1874) Tscher. Mineral., Mitt., **3**, 257–258.

Wenk, E. (1949) Die Assoziation von Radiolarienhornsteinen mit ophiolithischen Erstarrungsgesteinen als petrographisches Problem: Experimentia, **5**, 226–232.

White, D. E. (1968) Environment of generation of some base-metal ore deposits: Econ. Geol., **63**, 301–335.

Whitney, J. A. (1975) Vapor generation in a quartz monzonite magma: A synthetic model with application to porphyry copper deposits: Econ. Geol., **70**, 346–358.

Wilson, L. K. (1943) Tungsten deposits of the Darwin Hills, Inyo County, Calif.: Econ. Geol., **38**, 543–560.

Wisser, E. (1936) Formation of the north-south fractures of the Real del Monte area, Pachuca silver district, Mexico: AIME, Tech. Pub., No. 753, 1–47.

—— (1960) Relation of Ore Deposition to Doming in the North American Cordillera: Geol. Soc. Am. Mem. 77, 117p.

Wolfe, J. A. (1972) Setting of porphyry copper deposits in Philippines: Joint Meeting MMIJ-AIME, Tokyo, TIb2, 12p.

Yajima, J. (1969) Fundamental problems in the research of fluid inclusions in minerals and rocks: Mining Geol., **19**, 376–388. (in Japanese)

—— (1977) New occurrence of the tin minerals from the Toyoha mine, Hokkaido, Japan. Studies on the ore minerals from the Toyoha mine, 1: Mining Geol., **27**, 23–30. (in Japanese with English abstract)

—— and Okabe, K. (1971) On the iron-rich banded ore from the Toyoha mine, Hokkaido, Japan: Mining Geol. **21**, 221–228. (in Japanese with English abstract)

Yamaguchi, K. (1939) Ore deposits of the Ikuno mine and their zonal arrangement: J. Japan. Assoc. Mineral. Petrol. Econ. Geol., **21**, 259–275. (in Japanese)

Yamaguchi, M. and Yanagi, T. (1970) Chronology of some metamorphic rocks in Japan: Eclogae Geol. Helvetiae, **63**, 371–388.

Yamamoto, M. (1974a) Distribution of sulfur isotope in the Iwami Kuroko deposits, Shimane Pref., Japan: Geochem. J., **8**, 27–36.

—— (1974b) Distribution of sulfur isotopes in the Ryusei vein of the Akenobe mine, Hyogo Pref., Japan: Geochem. J., **8**, 75–86.

Yamanaka, T. and Kato, A. (1976) Mössbauer effect study of ^{57}Fe and ^{119}Sn in stannite, stannoidite, and mawsonite: Am. Mineral., **61**, 260–265.

Yamaoka, K. (1958) Spectrographic investigation on the trace elements in pyrite: Fac. Sci. Kumamoto Univ., Ser. B, Sec. 1, **3**, 31–37.

—— (1962) Studies on the bedded cupriferous iron sulfide deposits occurring in the Sambagawa metamorphic zone: Sci. Rept. Tohoku Univ., Ser. III, **8**, 1–68.

Yamaoka, K. and Ueda, Y. (1974) K-Ar ages of some ore deposits in Japan: Mining Geol., **24**, 291–296. (in Japanese with English abstract)

Yamasaki, M. and Miyajima Y. (1970) Eruption age of moonstone rhyolite in Toyama Prefecture, central Honshu: J. Japan. Assoc. Mineral. Petrol. Econ. Geol., **63**, 22–27. (in Japanese with English abstract)

Yanai, K. (1972) The Late Mesozoic acidic igneous rocks of the northern Ashio mountainland pt. 1: J. Japan. Assoc. Mineral. Petrol. Econ. Geol., **67**, 193–202. (in Japanese with English abstract)

Yen, T. P. (1959) The stratigraphic distribution of the bedded cupriferous pyrite deposits and manganese deposits in Taiwan: Mining Geol., **9**, 19–24.

——— (1974) Structural controls of the metallic deposits in Taiwan: Proc. Geol. Soc. China, No. *17*, 111–122.

Yermakov, N. P. (1965) Studies of mineral-forming solutions, *in* Yermakov, N.P. and Roedder, E., eds., Researches on the Nature of Mineral-Forming Solutions, Intern. Series of Monograph in Earth Sciences, *22*, Pergamon Press, N.Y., 9–348.

Yoder, H. S., Jr. (1950) Stability relations of grossularite: J. Geol., **58**, 221–253.

——— (1952) Change of melting point of diopside with pressure: J. Geol., **60**, 364–374.

——— (1965) Diopside-anorthite-water at five and ten kilobars and its bearing on explosive volcanism: Carnegie Inst. Wash., Yearbook **64**, Geophys. Lab., 82–89.

——— (1973) Contemporaneous basaltic and rhyolitic magmas: Am. Mineral., **58**, 153–171.

Yokoyama, I. (1974) Calderas and their formation, in fracturing and igneous activity: Chidanken Senpo [Assoc. Geol. Collab. Japan Monogr.], No. *18*, 40–53. (in Japanese with English abstract)

Yoshii, T. (1973) Upper mantle structure beneath the north Pacific and the marginal seas: J. Phys. Earth, **21**, 313–328.

Yoshimura, T. (1952) Manganese Ore Deposits of Japan, Mangan-Kenkyukai, 567p. (in Japanese)

——— (1967, 1969) Supplements to manganese ore deposits of Japan: Sci. Rept. Fac. of Sci., Kyushu Univ., Geol., **9**, Spec. Issue *1*, 1–485, Spec. Issue *2*, 487–1004. (in Japanese)

Yoshitani, A., Yamaguchi, S., Kosaka, T. and Onishi, I. (1976) On the Neogene and Quaternary volcanic activities in Shimane Prefecture, Southwest Japan: Chikyu Kagaku [Earth Science], **30**, 95–101.

Young, P. A. (1967) The stability of the copper-iron sulphides: Mineral. Deposita, **2**, 250.

Yui, S. (1966) Stability relations among iron oxide, sulphide, and carbonate minerals during magmatic ore deposition with special reference to the role of graphite: Mining Geol., **17**, 16–27. (in Japanese with English abstract)

Yun, S. K. (1966) Relations of structural pattern and tungsten deposition in Sangdong Mine and its vicinity: J. Geol. Soc. Korea, **2**, 1–16. (in Korean with English abstract)

Yund, R. A. and Kullerud, G. (1961) The system Cu-Fe-S: Carnegie Inst. Wash., Yearbook 60, Geophys. Lab., 180–181.

——— and ——— (1966) Thermal stability of assemblages in the Cu-Fe-S system: J. Petrol., 7, 454–488.

Zatsikha, B. V. (1968) Study of postmagmatic formations of the Kamennye Mogily granite massif: Abstract of Reports of Third All Union Conference on mineralogical thermometry and geochemistry of deep-seated mineral-forming solutions, 140–142. (From Proc. Fluid Inclusion Research COFFI (1968), p. 55).

Zen, E-an (1969) The stability relations of the polymorphs of aluminum silicate: A survey and some comments: Am. J. Sci., **267**, 297–309.

INDEX OF SUBJECTS

(Italic page numbers indicate tables and figures)

A

advance of magma 95, 109, 121
Ag-pentlandite 187, 188
Akaishi Tectonic Line 240
Akiyoshi Orogenic Cycle 4
alabandite 346
albitization 67
alkali-rich basalt 293, *295*, 297, 299
alkaline basalt 11
Alpine Orogenesis 147
alunite 105, 150
andradite 178, 187, 207
——, stability of 206
angle of internal friction 71
anticlinorium 43, 184, 258
antithetic fault 15
apatite 289
argento-plumbojarosite 174–175
Arima Formation 87, 99
Ashio Rhyolite Mass 112
Ashio-Nikko Metallogenetic Province 109
assemblage of magnetite-pyrrhotite-pyrite 207
asthenosphere 293, *295*, 301
axial plane *236*, 237, 261
axinite 195

B

Back-scattered electron image 219
Basin and Range Province 89
bedded cupriferous iron sulfide deposit 233
—— cupriferous pyrite deposit 233, 252, 255
bedding thrust 200
—— reverse fault 16, 47, 48
—— schistosity *237*, 238, 255
beryl 33
bimodal coexistence of rhyolite and andesite 18
—— volcanism 9, 10

biotite 197
biotitization 200
bismalith 14
bixbyite 339
blanket vein 26
blue schist 242
boiling 19, 79, 273, 277, 313, 327
bornite 86, 93, 101, 108, *113*, 149, 167
branch fault 70
—— fissure 70
—— vein 75
braunite 246
breccia dike 191
—— pipe 15, 62, 124, 185, 324
brown stannite 86
brushite 289
buoyancy effect 305
bustamite 346
Butsuzo Line 240
bysmalith 149

C

calc-alkali rocks 293
calcium phosphate mineral 246
canbyite 174–175
canfieldite 76, *113*
carbon dioxide 205
—— mixing into H_2O-CO_2 fluids 207
carbonate apatite 289
cassiterite 24, 29, 31, 34, 76, 93, 108, *113*, 123, 197
cauldron subsidence 13, 15, 63, 115, 309
cerussite 174–175
chalcocite 99, 149, 150, 285
chalcophanite 174–175
chalcopyrrhotite 133, 217, 219, 221, 224, 226, 227
chertification 249
Chichibu Formation 19, 24, 123, 171
—— Geosyncline 241
—— Paleozoic sedimentary rocks 123
—— System 1, 2, 3, 234, 237, 240

383

Chichibu terrain   247, 255
Chichibu-type ore deposit   209
Chichibu-Shimanto terrain   233
chonolith   147
co-precipitated gel   218
collapse   124
collapsible gold tube   218
colloform   252
—— structure   253
complex explosion vent   156
composite dike   44
compressional gap   307
cooling-stage microscope   34
coral reef   285
corundum   114, 150
covelline   285
CO_2 fugacity for manganese minerals   343, 345
—— hydrate   319, 320, 321
—— liquid   34, 50
CO_2-rich fluid inclusion   317, 320, 322
crushing-stage method   325
Cu-As-S minerals   145
Cu-Fe-As-S minerals   145
Cu-Fe-S mineral syntheses   216–227
—— minerals   130–135
—— system   216, 217, 225, 226
Cu-Fe-S-O system   189
Cu-Fe-Sn-S minerals   161–169
Cu-Sb-S minerals   145
cubanite   7, 29, 40, 42, 87, 123, 130–135, 187, 190, 217, 219, 220, 224, 226, 227, 271
$CuFeS_2$-FeS system   218, 220, 221, 224, 227
$CuFeS_2$-(Cu_5FeS_4)-FeS-FeS_2 system   223, 225
cupola   20
curved shingles   71

D

Daijima Stage   9
decay system   298, *298*, 301
decomposition of cubanite   217
decrepitation   34
—— method   60, 123, 173, 312, 313
—— pattern   211
—— temperature   18, 22, 39, 96, 174, 182, 203, 204, 315
decrepitoscope   313
Desmostylus   287, 291
diabase schist   235
diapir   14
—— of magma   305, 308
diapiric intrusion   309
—— limestone   177

diapiric structure   121
diapirism   171, 178
diaspore   105, 114, 150
digenite   99, 150
dike-like intrusion of wedgeing effect   311
dimorphic   229
dimorphism   155
diopside   206
disorder in zonal distribution   50
distribution coefficient   297
domal uplift   308
dome structure   27, 31
doming   180
doming-up   58, 61, 148
drag fold   47, 173, 239, 252
dynamic mantle   293
$\delta^{34}S$ value   95

E

effect of mixing of H_2O and CO_2 in Mn-minerals   344
effective pressure   305
electron microprobe analysis   103–104, 131, 152–153, 218, 220
electrum tarnish method   167
emulsion texture   142
en echélon   20, 28, 305, *306*
—— shear fracture   307
—— tension fracture   307, 308
enargite   18, 62, 143–161
end-member isotope ratio   299
epitaxy   141, 151, 154
epitaxial relation   142, 152, 156, 160
epithermal deposit   160
—— type   13
—— vein   54
eugeosyncline   241
Eutektischer Verwachsung   126
evacuated rigid silica tube   218
exhalative sedimentary hypothesis   253
—— origin   247
expansion of penny-shaped inclusion   304
exsolution   30, 105, 131–135, 142

F

Faltung durch Horizontalverschiebung   48
famatinite   143–161
feather joint   71
felsite   93
ferberite   108, *113*
ferruginous chert   245, 247, 249
—— (manganiferous) chert   256
filling degree   19, 79
—— temperature   8, 39, 81, 119, 198, 202, 204

fissure pattern 171
flatly-lying vein 26, 31
flow cleavage *236*, 238, 252, 255
fluid inclusion 195, 202, 211, 265, 305
———inclusion study 34, 50, 76–85, 117, 312–334
——— pressure 303
fluid-filled fracture system 305
folding axis 233, 253
formation of fissures 13
——— of green rock 243
Fossa Magna 1, 5, 7
fractional melting 18
fracture systems caused by intrusion 192–193
——— zone 305
Franciscan Group 242
freezing-stage 313
——— method 315
——— microscope 34, 80, 119
f_{s_2}-f_{O_2} condition 76
f_{s_2}-T 122
f_{s_2}-T curve in Cu-Fe-Sn-S minerals 101–102
f_{s_2}-T in sulfide minerals 159
Funakawa Stage 11
funnel-shaped rhyolite body 309
——— ring normal fault 309

G

galaxite 346
Ganidake granitic rocks 184
garnet 173, 178, 186, 189, 194, 197
———, anisotropic 206
———, birefringence of 206
gas-rich inclusion 77, 117
gaseous inclusion 77, 79, 265, 271, 313
gash-like fracture 71
germanium 123, 157–158
glass inclusion 77, 117
glauconite 10, 287
glaucophane 241
——— schist 1, 3, 243, 245, 255
graben subsidence 311
graben-dome structure 311
grandite, stability of 206
granophyre 93
green rock 234–245, 252, 255, 260, 264
greenschist 91, 234–245, 255, 256, 260, 264
green tuff 8, 9, 65
Green Tuff Movement 8
greisen 34
greisenization 29
Griffith crack 304

Griffith inclusion 305
——— theory 303
grossular 178, 187
———, stability of 205
guano 291
——— phosphorite 285
gudmundite 124, 126–129

H

H. V. Faltung 48
halite 78
hausmanite 339, 346, 347
haycockite 226, 227
heated cubanite 219, 226
heating-stage microscope 34
heating-stage method 312, 313, 315
hedenbergite 173, 175, 178, 181, 193, 207
hematite 59, 67, 78, 173, 175, 227, 229, 230, 239, 243, 246, 257
hemimorphite 174–175
hessite 56
hexastannite 86
Hida Gneiss 172
——— Injection 4
——— Metamorphic Rocks 1
Hidaka Metamorphics 7
high-cubanite 219, 220
highly saline inclusion 265, 270, 315, 325, 334
hinge fault 193
hisingerite 174–175
homogenization temperature 18, 202, 273, 313, 315
hood 14
hornblendization 200
horse-tail 75, 146
horseshoe-shape fault 115
huebnerite 108
hydrothermal circulation 311
——— condition 224
hydrozincite 174–175
hypothermal type 13, 124
——— vein 17, 20
H_2O content 296
H_2O-CO_2 system 320

I

idaite 167
Ikuno Formation 87, 105
ilvaite 193
indium 96, 157–158
Inner Zone 1, 5, 233
intermediate principal stress 31, 121
intrusive breccia 191
iron-manganese ore 246

iron-wollastonite   208

J

jadeite   243
jalpaite   99
joint   13, 15, 193

K

Kajika-type deposit   114–117
Kamaishi type deposit   209
Kamioka-Nakatatsu type deposit   209
kamiokite   173
Kamuikotan belt   7
kaolinite   105, 150
keratophyre   258
Kieslager   233
klaprothite   149
kobellite   26
Kurihashi granodiorite   184
Kuroko   9, 11, 62, 150, 233, 253

L

lamellar magnetite   93
large ion lithophile element   293, 297, 301
lead isotope   299
left-handed   72
——— wrench fault   48, 91, 146
Lepidocyclina-Miogypsina fauna   9
linear structure   44
lineation   31, 44, 179, 233, *236*, 238, 239, 253, 255, 256, 261, 262, 263
liquid CO_2 79
——— immiscibility of magma   18
——— inclusion   77, 271, 279, 313
liquid-rich inclusion   77, 117
lithosphere   293
lithosphere-asthenosphere boundary   *294*
lithostatic pressure   303
lodestone   231
low gravity   92
low-velocity zone   295
ludwigite   186
luzonite   18, 62, 143–161

M

mackinawite   29, 33, 42, 123, 130–135, 271
maghemite   231
magmatic intrusion   308
——— upheaval   13, 57, 62
magnetite   42, 197, 227, 229, 230, 231, 239, 245, 259, 339
———, foliate   195
———, lamellar   93
———, short-columnar   195
——— colloid adherence   188

Maizuru Folded zone   40, 233, 258
——— zone   3, 4
malayaite   212–216
manganese mineral deposit   335
manganese-bearing ferrugenous chert   235, 245, 252
manganosite   337, 339, 346
mantle convection   294, 301
marcasite   113, 123, 136, 148, 149, 154, 181, 227, 229, 230, 263
martite   230
martitization   231
martitized   173
master fault   71, 72, 75, 146
matildite   173
mawsonite   86, 93, 99, 108, *113*, 161–169
maximum principal stress   31, 121
mckinstryite   56, 99
Median Tectonic Line   1, 3, 255
mélange   242
mesh-like fracture   71
——— vein   74
mesosphere   293
mesothermal deposit   160
——— type   13
——— vein   20, 41
metabrushite   289
Mg enrichment   243
Mg-Al metasomatism   243
Mg-Fe metasomatism   91, 243, 245, 256, 264
micro-folding axis   233
microcrack   305
mid-ocean ridge   293
——— ridge basalt   293, 295, 296, 299, 301
Mikabu Line   240
Mikabu System   3, 4, 234, 235, 237
minimum principal stress   31
mixed rock   193
mixing model   299
mixing of carbon for oxygen fugacity   210
Miyazu granite   41
molybdenite   31, 42, 93, 123, 173, 179, 197, 198, 199
monetite   289
monoascendant   95
monophase inclusion   79, 117
mooihoekite   226, 227
multiple dike   10, 44

N

NaCl equivalent concentration   34, 52, 79, 120, 316

nappe structure 115
native silver 94–97, 108
—— sulfur 150
Nd isotope ratio 301
negative anomaly in magnetic survey 92
—— pressure 80
Nishikurosawa Stage 9
Nishioga Stage 8
normal fault 54, 99, 106, 192
normal step-like pattern 307

O

oceanic crust, entrenched 242
ocean island basalt 293
Onnagawa Stage 10
Operculina-Miogypsina fauna 10
ophiolite 91, 241
ore-shell 271
oriented overgrowth 141, 151
orthomagmatic deposit 252
Outer Zone 1, 233, 258
overshoot 39
overshooting of decrepitation temperature 204
overturned fold 236, 238, 255
oxidized ore of zinc and lead 174
oxygen fugacity for manganese minerals 342
—— fugacity for rhodochrosite and hausmanite 348
—— fugacity in skarn 207–209
—— fugacity in copper mineralization 95

P

paired metamorphic belts 5
parallel intergrowth 142
paramorph 229
partial melting 293, 296, 297, 305
pegmatitic type 33
—— vein 20
pentlandite 42, 130–135, 188, 190
phosphate deposit 280, 284, 281–291
—— nodule (concretion) 288
phosphorite 283
phosphorus cycle 290–291
piedmontite schist 251
piedmontite-quartz schist 3, 255, 256
pillow lava 245
plate tectonics 293
—— thickness *294*
plunge 251
—— of anticlinal axis 258
—— of folding axis 238, 260
—— of ore body 233

plunging fold *236*, 255
pneumatolysis 129, 245
polyascendant 95
polymorph 155
polymorphism 217, 226
polyphase fluid inclusion 265, 270, 271, 276, 313
—— inclusion 78, 198, 268, 331, 332
porphyry copper deposits 265–280
post-magmatic brine 84
primary convection 294, 295
principal stresses 39
propylite 16, 17, 27, 56, 59, 63, 65, 67–68, 75, 150
propylitization 8, 67, 243
pulsation 95
—— of magma 109
pushing up by intrusion 91
—— of gabbro 239
—— of intrusive *238*
—— of magma 62, 93
—— of propylite 149
—— of quartz porphyry 76
pyrochroite 337, 339, 346
—— ore 347
pyrometasomatic deposit 160, 171–231, 245
—— origin 252
pyrophyllite 105, 114
pyroxenite 235
pyrrhotite, hexagonal 188, 189, 190, 221, 224–227
——, monoclinic 188, 189, 190, 225–227

R

radial fracture 308
radiated pattern of the vein 148
radiogenic isotope 299
radiolarian remain 245, 247, 250
rare earth element 297
reaction rim 149
recumbent fold 238
—— synclinal structure 261
red chert 239
rejuvenation 95
retreat of magma 95, 114, 121
reverse step-like pattern 307
reverse zoning 114
rhodochrosite 247, 335, 339, 346, 347
rhodonite 339, 346
rift system 311
right-handed 72
—— shearing stress 20
—— wrench fault 54

ring dike 63
—— fracture 15
roof 14
roquesite 93
rutile 150
Ryoke Injection 4
—— Metamorphic Rocks 1, 3
—— terrain 264
Ryoseki Formation 255

S

saddle reef 239, 248, 252, 255
Sakawa Orogenesis 254
—— Orogenic Cycle 4
salinity 79, 273, 313, 315
—— gradient 83
Sambagawa Metamorphism 241, 242
—— System 3, 4, 239, 256, 258
Sambagawa-Mikabu Complex 241, 250, 252, 254, 255
—— -Mikabu Metamorphism 4
—— -Mikabu terrain 233, 242, 255, 264
Sangun Metamorphic Rocks 1, 2, 3
—— Metamorphism 3, 4
—— terrain 233
sauconite 174–175
scapolite 20, 257
schalstein 91, 191, 241–243, 247
scheelite 29, 31, 34, 93, *113*, 124, 173, 179, 182, 195, 197–199, 323
—— quartz vein 29, 179, 181
scheelite-bearing skarn 207
selenium content of sulfide minerals 96, 157–158
shear fracture 303
—— of second order 15, 70
Shimanto System 4, 240
—— terrain 240, 255
siderite 339
similar fold 173, 181, 239
—— folding 171, 173, 179, 181
sister rock 18
skarn 7, 58, 173, 181, 182, 184, 186, 214
——, feldspar-bearing 201, 204, 210
——, ore-bearing 206
—— formation 201–212
skarnization 20, 211
slickenside 72, 148
small-scale convection 294–296, 299
solid crystal inclusion 117
solvus relation 154
spessartite 346
spilitization 249
spinel twinning 141
splay fault 70

$^{87}Sr/^{86}Sr$ ratio 299
stability calculation of manganese minerals 337
—— relations for manganese minerals 339, 342
stannite 26, 29, 76, 93, 99, 108, *113*, 123, 161, 167, 257
stannoenargite 150
stannoidite 86, 93, 99, 108, *113*, 161–169
stibioenargite 145
stibioluzonite 144
stock-like intrusion 309
strain ellipsoid 48
strata-bound deposit 233–264
striation 28, 54
strike-slip fault 15
stromeyerite 56, 99, 108
strontium isotope 299
structure etching 140
submarine exhalation 250
subvolcanic type 13
—— vein 17
sulfur deficient condition 155
—— fugacity for sulfide minerals 167
supergene minerals 176
supratenuous fold 242
surface barrier 308
sylvanite 149
sylvite 78
synchronous intrusion 39, 43
synclinal structure 262
synthetic normal fault 14, 57

T

talnakhite 226, 227
Tamba Formation 27, 87, 99
tectogene 242
tellurobismuthite 22
tennantite 93, 147, 148
tension crack 14, 60, 70, 71, 76, 93, 95, 112, 180–182
tension fracture 303
tephroite 335–337, 339, 346
tephroite ore 348
test-tube type pressure vessel 218
Tetori Group 1
—— Jurassic Formation 178
thermal model 294
thrust 307
—— fault 42
topaz 34, 93, 114
tourmaline 31, 123, 186
tourmaline-copper vein 130
trace element abundance 297, 301
transverse dike 311

tremolite-actinolite zone   271
troilite   188–190, 225, 226
trough reef   239, 252
Tsushima Basin   17, 124
two stage model   299

U

unmixing intergrowth   188
upheaval due to intrusion   99
—— of magma   59, 60, 95, 121
—— pressure   31, 39, 180
upward pressure   309
uraninite   7

V

vacant position   195
vein fissure   13
—— fracture   302
—— pattern   75
—— system   68–72
vein-fissure system   93
vent of andesite eruption   67
vesuvianite   193, 211
vivianite   285
volcanic neck   65
—— vent   156
volcanic-subaquatic   247

W

water convection   308
Western Kinki Metallogenetic Province   27, 40, 87
withdrawal of magma   95, 109, 114, 121

—— of ore-forming source   114
wolframite   24, 31, 34, 93, 123, 124, 197
wollastonite   181, 205, 211
wrench fault   15, 28, 42, 47, 70, 75, 105, 112
wrinking in the pelitic rock   261
wurtzite   67, 135–143, 148, 154

X

X-ray powder diffraction   104–105, 135, 151–152, 218–221, 225, 226, 245–246, 288–289
xanthophyllite   193
xenothermal type   13
—— vein   17, 86–123, 150
xonotlite   205

Y

Yakuno basic intrusives   41

Z

zigzag microfault   307
zinnwaldite   26
Zn-Pb-Cu-SiO$_2$-H$_2$O amorphous mineral   174–175
zonal distribution   18, 59, 64, 93, 104, 108, 188, 208
zone refining   297
zoned skarn   193, 205, 211
zoning   209
—— of ore body   209
—— of ore minerals   191
—— of ore deposits   199
zunyite   67, 114

INDEX OF MINES AND LOCALITIES

(Italic page numbers indicate tables and figures)

A

Akagane	*6*, 209, 226, 323, 330
Akatani	*203*
Akenobe	*6*, 15, 16, 19, 86, 89–97, 123, 161, *203*
Akeshi	*6*, 150
Angaur	289
Ani	*6*, 11, 13, 15, 59–61
Asahi (Hyogo Prefecture)	89
Asahi (Niigata Prefecture)	*314*
Asakawa	*6*, 254
Ashio	*6*, 15, 19, 86, 110–120, 123, 135, 161, *314*, 327

B

Bandojima	209
Bayerland	126
Besshi	*6*, 233, 240, 242, 252, 256–258
Broken Hill	253
Butte	143, 156

C

Cerro de Pasco	143, 156
Chichibu	*6*, 190–196, 209
Chinkuashih	*6*, 143, 147–148
Chitose	*6*, 17, *314*, 327
Choja	*6*, 254
Climax	15, 309
Cornwall	245

D

Darwin Hill	72

E

Ergani-Maden	253

F

Fujigatani	*6*, 179, 182–183, *203*, 327
Fujinokawa	245, *246*
Fukoku	86

H

Hanan	*6*, 125
Hirase	*314*
Hitachi	*6*, 7, 233, 260–264
Hiuchidani	*6*, 281
Hoei	*6*, 213, 214
Hokuetsu	*6*, 144, 150
Hol Kol	171
Hosogoe	184
Hosokura	*6*, 15, 18, 65–74, 135–143
Huanzala	159

I

Igashima	*203*
Iimori	*6*, 233
Ikuno	*6*, 17, 19, 86, 105–109, 123, 161, 311
Ilimpeia	245
Ilkwang	*6*, 123, 124–130
Inakuraishi	*6*, 15–16
Iriki	*6*, 158
Isakozawa	*6*, 246
Iwai	226
Iwami	11
Iwato	226

J

Jakobsbacken	128
Jiro	*6*, 240
Jishakuyama	331

K

Kagata	208
Kaize	*6*, 150
Kamaishi	*6*, 7, 183–190, *203*, 208, 209, 220, 226
Kameyama	226
Kamioka	1, 2, *6*, 171–176, 179, *203*, 209, 332
Kanan	125, See Hanan
Kaneuchi	*6*, 15, 16, 19, 27–40, 86, 123, *314*, *319*

Kasuga	150	Ohtani	6, 16, 19, 27–40, 86, 123, *314*, *319*, 322, 325, 326
Kasugayama	208		
Kaveltorp	227	Ohya	6, 14, 17, 20–23
Kawayama	6, 226–233	Okuki	6, 233–256, 258, 264
Kiangsi	114	Okura	226
Kinkaseki	147–148, See Chinkuashih	Omine	184, 185
Kiwada	6, 179, 181–183, *203*, *314*, 327	Oppu	*314*
Koganetsubo	323	Osarizawa	6, 15, 18, 58–59, *314*, 326
Kohnomai	6, 17	Outokumpu	253
Komori	6, 40–43, 86, 130–135, 190, 226		
Konjo	86		**P**
Koya	6, 240	Pirin	227
Kudo	197	Princess	179
Kuga	6, 16, 31, 179–182, *203*, 213, 215, *314*		**Q**
Kune	6, 233	Quisma Cruz	16
Kunimiyama	6, 247		**R**
Kushikino	6, 16, 17	Rammelsberg	253
		Rio Tinto	253

L

S

Lifudzin	114	Sado	6, 17, 54–56, 314
Lunchburg Tunnel	179	Sahinai	184
		Sampo	6, 208, 213, 215

M

		Sangdong	6, 196–200, *203*
Makimine	6, 226, 233, 254	Santo Tomas II	274–277
Mamut	265–274	Sasagatani	6, 227–232
Mankayan	143, 145–147	Sata	6, 233
Marcopper	277–280	Sazare	6, 233
Masutomi	6, 150	Seikoshi	6, 11, 17
Mikawa	6, 15, 56–58	Shibuki-Seki	6, 253
Minenosawa	233, 242	Shimokawa	6, 7, 226, 233
Minenosawa-Nago	6, 242	Shimonomoto	173
Mitate	6, 211, 213,	Shinyama	184, 186–190
Mochikura	210	Suan (Suian)	6, 171
Mount Morrison Pendant	179	Sulitelma	253
Mukuromi	184, 226	Suttu	6, 150
		Suzuyama	6, *314*, 328

N

T

Nakatatsu	6, 176–179, 209	Tada	6, 19, 86, 97–105, 161
Nanogawa	6, 254	Taishu	6, 15, 34, 40, 43–53, 75–85, *307*, *314*, 323, 326
Nikko	124–130, See Ilkwang		
Nippo	184, 185	Takamae	184
Noda-Tamagawa	6, 171, 226, 346–348	Takatama	11
Notojima	286	Takatori	6, 15, 16, 19, 24–27, 31, 34, 123, *314*, 316, *318*, *319*, 321, 322, 325, 326

O

Obira	6, 123, *203*	Taro	6, 233
Odake	6, 246, 248–249	Tawara	285
Odake-Chibayama	246	Teine	6, 15, 148–150
Oe	6, 15–16	Tenryu	226
Ofuku	210	Tokoro	6, 246
Ogoya	6, 71	Toroku	6, 213, 214
Ohmidani	89	Toyoha	6, 17, 18, 75–85, *314*
Ohmori	135		
Ohta	208		

Tri-State 249
Tsuchihata 6, 18, 62
Tsumo 6, 171, *203*, 208, 213
Twin Butte 229
Tyrny-Auz *203*

U

Udo 11

W

Wanibuchi 11

Y

Yaguki 6, 179, *203*, 209, 226
Yakuoji 6, 125
Yamaguchi 6, 7
Yamato 208
Yanahara 6, 233, 258–260
Yaso 6, 15, 63–65, *314*
Yellowknife 72
Yoshihara 226

COLUMBIA UNIVERSITY LIBRARIES

This book is due on the date indicated below, or at the expiration of a definite period after the date of borrowing, as provided by the library rules or by special arrangement with the Librarian in charge.

DATE BORROWED	DATE DUE	DATE BORROWED	DATE DUE
	NEW BOOK DOES NOT CIRCULATE UNTIL 7-22-80		
	— 83		

) 50M